Residential Hydronic Heating

Installation & Design

I=B=R
Guide RHH

Foreword . . .

The Hydronics Institute, a division of the Air Conditioning, Heating and Refrigeration Institute (AHRI), is a non-profit organization whose members are leading manufacturers of boilers, radiation and other equipment used with hydronic (steam and hot water) heating and allied cooling systems.

Rigid codes have been established by the Institute for the determination of ratings for boilers, baseboards, and finned tube units. These codes are based on research, sound engineering practices, and practical application in the field, and provide a uniform, scientific method of determining output and establishing dependable ratings. Approval of ratings is granted in writing on a unit, or series of units, and is controlled by a License executed between the manufacturer and the Institute. This rating approval program was established in 1940 and, since its inception, has been available to any manufacturer, whether a member of the Institute, or not.

Since 1957 tests on baseboards and finned tube (commercial) radiation have been conducted in the I=B=R Laboratory to establish and confirm the accuracy of the I=B=R Ratings on those products.

The technical material published by the Institute, founded on research, represents the consensus of the outstanding engineers in the industry who are employed by member companies. The information is made available to the entire industry by means of I=B=R Guides and the I=B=R School conducted throughout the United States and Canada.

In addition to I=B=R Guide, the Hydronics Institute Division of AHRI has also produced the following resources to be used in conjunction with the I=B=R Guide:
- H-22 Heat Loss Calculation Guide
- Heat Loss Worksheets for Guide H-22
- Set of ¼ - Scale House Plans

The I=B=R Guide, as well as the above teaching resources, are published and distributed by The Air Conditioning Contractors of America (ACCA).

ACCA also publishes the following materials:
- Residential and commercial HVAC load calculation, duct design and equipment selection manuals.
- Power points, videos, CDs, curriculums, lesson plans and other HVAC educator resources.
- HVAC systems design software
- NATE and 608 training materials

For more detailed descriptions and ordering information contact ACCA, 2800 Shirlington Road, Suite 300, Arlington, VA 22206. ACCA customer service: 888-290-2220 or visit www.acca.org/store.

Every effort has been made to assure the accuracy of this guide, Guide RHH. However, neither AHRI nor those responsible for the preparation of AHRI publications make any representations or guarantee, or assume or accept any responsibility or liability, with respect thereto.

I=B=R Guide

RHH

Residential Hydronic Heating . . .

Installation & Design

Hydronics Institute
Section of **AHRI**

Hydronics Institute
Section of **AHRI**

Guide RHH . . .

I=B=R Guide RHH, "Residential Hydronic Heating Installation & Design Guide," provides general information on the application and installation of hydronic water and steam systems for residential heating. Guide RHH is published in loose-leaf format to allow ease of periodic updating.

The chapters of the Guide are treated as separate modules, allowing the material to be adapted well to the needs of heating system installers throughout the continent. The Guide user can focus predominantly on the material which applies to his region.

This edition of Guide RHH updates the material to include current installation methods and newer systems and controls (such as radiant heating systems, electronic controls and condensing boilers). It also includes added emphasis on system troubleshooting, particularly when quoting boiler replacements.

Troubleshooting the system before quoting allows the installer to provide a comprehensive quotation to the owner — including not just a new boiler, but also quoting system modifications or repairs needed to improve system performance and assure long life of the new equipment. Many existing hydronic systems have aged significantly. And newer boilers cycle very differently from the original (probably coal-fired) boilers installed when the system was new. Achieving the best performance from the heating system requires some updating and repair of the system — not just a boiler replacement.

A successful installation means the customer's expectations must be met. Guide RHH provides information on hydronic system application, installation and troubleshooting that will assist installers in accomplishing this goal. See the customer expectations discussion on the following pages.

Included in Guide RHH is a **System survey** form for use by installers to present the owner with a complete analysis of the heating system, explaining where and why repairs or replacements may be needed. The **System Survey**, **Form 1530-W or 1530-S**, is available in quantity as a separate order item.

In addition to a good installation, installers also need to assure that the boiler matches the heat loss of the home (or the radiator count for steam systems). Apply **HI Guide H-22** or other method to determine the heat loss for hydronic water system-heated homes. The closer the boiler is matched to the heat loss, the higher the seasonal efficiency and the lower the fuel bills.

Meet customer expectations

Customer expects:
Fuel bills will be lower

When you have finished installing a new, high efficiency boiler, your customer expects his fuel bills will be lower. If you size the boiler to the heat loss, you will probably accomplish this. You need to do a heat loss calculation (water) or radiator heating surface count (steam) because your customer deserves it.

Every time a boiler or furnace cycles on and off some heat is lost. The more often the appliance cycles, the greater the amount of heat wasted. Appliances cycle because the heat demand is smaller than the appliance output. If the heat demand were higher than or equal to the appliance output, the appliance would stay on indefinitely.

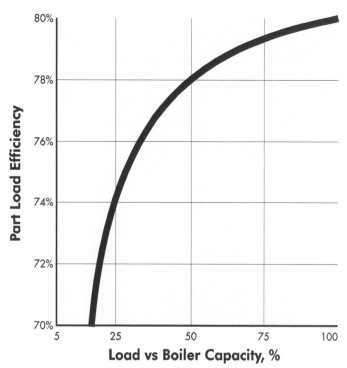

Based on typical boiler with 80% steady-state efficiency and 2% jacket loss.

The effective efficiency of the boiler as the demand is reduced is referred to as part load efficiency. The graph on this page shows what happens as the demand becomes smaller than the boiler output. Fuel economy (part load efficiency) drops quickly as demand falls.

To lower heating costs for your customers, install heating equipment sized as close as possible to the heating load. Installing a new boiler the same size as the old one, or using "rules of thumb" in selecting replacement boilers will cost your customer money in excess fuel bills. Oversized boilers cycle too often, so they waste heat.

Size the boiler to meet the building's heat demand —

Select a boiler just large enough to handle the demand.

Water systems — You can determine the heat loss for a building based on the building construction and geographic location and the customer's desired indoor temperature during the winter. HI Guide H-22 provides the information and method to calculate heat loss for most residential applications. You can also use any of the computer sizing programs available from manufacturers or the ASHRAE Handbook. Select a boiler with an I=B=R water net load rating just larger than the building heat loss (plus allowance for domestic water heating or any other additional demand).

Steam systems — Steam boilers cannot be sized to building heat loss. They must supply the steam demanded by the connected radiation (or heat exchangers). Size a steam boiler by totalling the heating surface of all radiators connected to the system. Then select a boiler with an I=B=R steam net load rating (in square feet) just larger than the total heating surface (plus any additional demand).

Customer expects:

More . . . .

Customers call a heating specialist when their hydronic system fails to work properly. They may think that the problem is with the boiler, since it is the most obvious candidate. In fact, the problem most likely originates with system or component failures. A successful installation means the customer is satisfied and the installer doesn't have to return for frequent callbacks. Meet the customer's expectations and the job should be profitable as well as successful.

So, what else does the customer expect in addition to lower fuel bills?

Customer expects:

All the problems will be gone

The customer probably has the impression that the problems with the system relate to the "broken boiler."

- He or she expects that all the problems will be gone when a new boiler is installed.
- He or she expects that the upstairs bedroom will be warm again, and that the knocking in the pipes in the dining room will disappear.
- The water will quit dripping from the relief valve, or the condensate tank will quit flooding.

Don't leave the customer with this expectation unless you inspect, troubleshoot and repair the system and its components as well as replacing the boiler.

The system and its components have aged. And many different technicians may have worked on or changed the system through the years. Inspect the system and talk to the owner to diagnose all the system ailments. Then quote a complete repair job. The customer may decline doing everything you recommend, but you've let the customer know so there won't be any nasty surprises later.

Customer expects:

The new boiler will last as long as the old one

Remember — The most likely reason the old boiler has failed is that something went wrong with the system, and this caused the failure.

Analyze the system and determine why the old boiler failed. If you don't, the new boiler won't last long. The same mechanism that caused the old boiler to fail is still there unless the installer makes the needed repairs or changes.

System failures are often due to leaking pipes or inoperative (or undersized) expansion tanks. These cause excessive water to enter the system. The water causes oxygen corrosion and lime deposits — either one can quickly destroy a new boiler.

Study Guide RHH for system application and installation, and the likely causes of boiler failure and system problems. Fill out a System survey when you quote a boiler replacement or service on a hydronic system. Chances are you'll have a satisfied customer and a heating system that is efficient and reliable.

Guide RHH contents

Part **1** **Harnessing heat**

This is an introductory section, including basic boiler design and operation and how heat is transferred. You will also find: • a brief history of hydronic heating • boiler construction differences (cast iron, copper, water tube, firetube, electric) • how combustion is applied and measured in boilers.

Part **2** **Selecting & placing a boiler**

Part 2 explains how boilers are rated and how to determine the boiler capacity required (based on heat loss or radiator heating surface count), plus: • flue gas venting (gas and oil) • sizing and placing air openings for combustion and ventilation • general information on applicable codes and standards.

Part **3** **Components of hydronic hot water heating systems**

Part 3 explains the construction and operation of the components used in hydronic water systems, including: • circulators (pumps) • heating units (baseboard and radiators, radiant panels, unit heaters and fan coil units) • expansion tanks • air vent devices • common controls.

Part **4** **Piping hydronic hot water heating systems**

Part 4 explains the importance of component placement and piping design. You will find: • examples of the most common piping systems — series-loop, one-pipe, two-pipe, and primary/secondary systems • radiant heating considerations • system comparisons • drawing piping schematics • pipe routing and support • freeze protection (glycol mixtures).

. . . continued on next page

Part **5**

Sizing hydronic hot water heating systems

Here you will find methods of sizing the piping and main components in your system — circulators and expansion tanks. Detailed and quick-selection methods are included. Even if you use a computer program to size and select components, this section will give you an understanding of how the program results are obtained.

Part **6**

Auxiliary heating loads

Part 6 discusses domestic water heating and other applications of the versatile hydronic heating system: • tankless heaters • indirect water heaters • sizing and piping water heaters • multi-purpose systems, including combined space heating and snow melting applications.

Part **7**

Condensing boilers

Part 7 discusses condensing boilers — how they operate and how they compare to conventional boilers, plus: • advantage of condensing boilers • application considerations • making sure the boiler condenses as much as possible • finned-tube baseboard applications • reset controls • application examples.

Part **8**

Radiant heating basics

Part 8 is an introduction to radiant heating systems: • radiant heating advantages • how radiant panels work • radiant system components (tubing, manifolds, pumps, controls) • installation considerations (for slab systems, above-floor, and below-floor systems) • timing installation with the other trades • controlling and piping basic radiant heating systems.

Guide RHH contents

Part 12 Piping hydronic steam heating systems

Part 12 discusses operation of gravity-return and pumped-return steam systems, including: • one-pipe counterflow and parallel upfeed and downfeed systems • critical piping (Hartford loop and dimension A) • two-pipe upfeed and downfeed systems, gravity- and pumped-return • steam system pipe sizing • feed pumps and feed systems • applying water level controls.

Part 13 System survey

Part 13 provides a summary of the system analysis and troubleshooting needed to complete Form 1530, the System survey. The information in the previous sections of Guide RHH provide the background needed to make this section clearer and more meaningful to the heating system installer. Forms 1530-W and 1530-S are included at the end of this section.

Hydronics Institute Section of AHRI

35 Russo Place

Berkeley Heights, NJ 07922-0218

I=B=R Guide RHH — Residential Hydronic Heating: Installation & Design

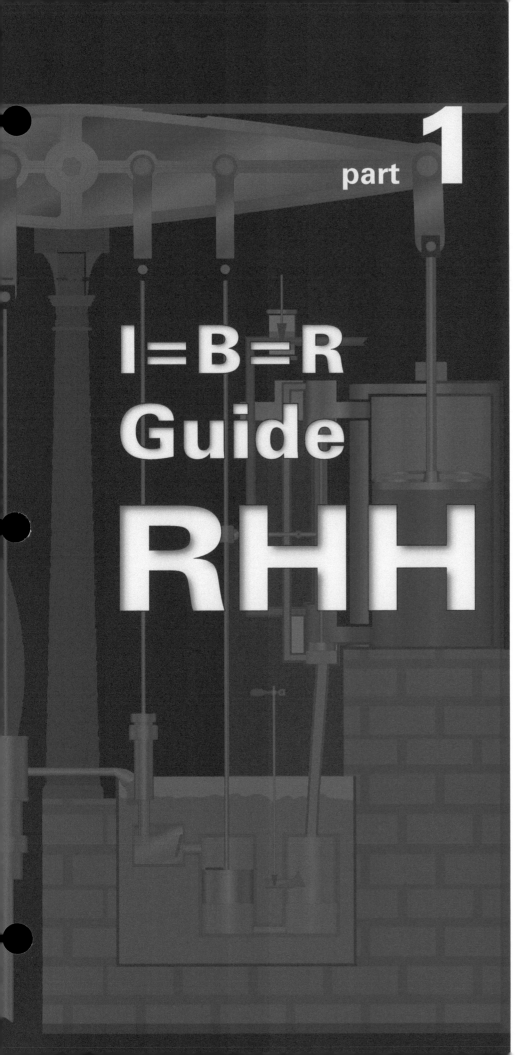

part **1**

I=B=R
Guide
RHH

Harnessing heat

Residential Hydronic Heating . . .

Installation & Design

Hydronics Institute Section of AHRI

I=B=R Guide RHH
Residential Hydronic Heating

Hydronics Institute Section of **AHRI**
35 Russo Place
Berkeley Heights, NJ   07922-0218

Contents – Part 1

Contents *– continued*

● Combustion & burners

● Heat exchangers

● Vent systems

Contents – continued

Contents *– continued*

Illustrations

Contents *– continued*

Tables

Powering the

Introduction

Today's hydronic heating systems exist because of the demand for steam engines during the Industrial Revolution. The Industrial Revolution (first in Britain, later in America and mainland Europe) drove the demand for means of operating industrial equipment with a more practical means than waterpower. Waterpower required locating mills along rivers, but the workforce was in the growing cities. The rapid development of practical and safe steam engines in the 18th century presented the solution. Steam engines needed steam boilers so boiler development paralleled the progress of steam engines. Mills also needed space heating, so steam available from industrial boilers heated mills as well by the beginning of the 18th century.

Trace the origins of steam engines and you'll encounter James Watt, an 18th century Scottish "inventor/engineer/contractor/scientist." James Watt pioneered the steam engine. He wasn't the first to build a practical engine. Thomas Newcomen did that. Newcomen's steam engines helped drain mines throughout the British Isles for nearly half a century before James Watt's engines came into use. James Watt's work is notable because he used applied science (much of it developed from his own testing) to solve the problem of the low efficiency and performance of the existing Newcomen engine. His innovations rendered the low-pressure steam engine practical, safe and cost-effective, making it ideal for the growing industries in Europe and America. Before talking about Watt and his engine, we'll talk about the century of slow development that preceded him.

1643 — Evangelista Torricelli

During the 17th century, the mining industry in Britain grew substantially. Mines were getting deeper. As the depth increased, flooding became a problem. The available pumps and methods couldn't move water high enough or fast enough to solve the problem. Evangelista Torricelli, a student of Galileo, noticed that lift pumps couldn't lift water more than about 32 feet. [Barometric pressure at sea level is normally 29.92 mm Hg, or 34.1 feet water column.] He speculated that this might be the result of atmospheric pressure. To test his theory, he designed a new instrument — the mercury barometer, in 1643. Torricelli's work showed the presence of atmospheric pressure and proved that a vacuum was possible.

Two years after Torricelli's invention of the barometer, Otto von Guericke developed the first air vacuum pump. He once demonstrated the power of a vacuum by placing two hollow bronze hemispheres together, then evacuating the air from inside. Two teams of eight horses each were unable to pull the hemispheres apart. About 1680, Denis Papin, a French physicist, developed the pressure cooker. He noted that the steam inside tended to lift the lid as pressure increased. He concluded that steam could be used to drive a cylinder and piston, and proposed a steam-driven pump to operate a water wheel in 1707. Papin also developed a counterweight-limited pressure relief valve, which seems to be the prototype for those used on most boilers throughout the 18th century.

Industrial Revolution

1698 — Thomas Savery

Thomas Savery, an English inventor, used Papin's work to develop his application of steam power to a pumping operation. These early engines were dubbed "fire engines" because they used fire to pump water. Savery filled a cylinder with steam, then closed a valve and poured water over the cylinder. The steam inside then condensed, creating a vacuum. Savery turned this concept into a functional steam-driven pumping engine. See Figure 1 for a schematic of Savery's engine. Savery obtained a patent covering all applications of the "fire engine" in 1698.

Savery's engine developed a vacuum in the vessel (Figure 1, item 1 or 2) by using manual valve 4 to spray cold water over the outside of a vessel filled with steam. With a vacuum in the vessel, atmospheric pressure pushed water up the suction line pipe into the vessel, through the check valve (item 5 or 6). While one vessel developed a vacuum, the operator used valve 3 to supply high-pressure steam to the other vessel. The high-pressure steam pushed water up and out the water discharge line. The operator manually changed valves 3 and 4 in this way to cycle the engine. As drawn in Figure 1 vessel 1 is condensing steam and pulling a vacuum. Vessel 2 is pushing water up and out due to the high-pressure steam supplied through valve 3.

Savery proposed his steam engine pump for use in mines, but had little success because of the risk of boiler explosion. In order to force the water up and out of the mine, the steam pressure had to exceed the required lift. For the water to be pushed up 100 feet, for example, the steam pressure would have to be greater than 100 feet water column, or 43 psig. The boiler construction used wasn't suitable for high pressures, and Savery did not equip his boilers with relief valves. Explosions occurred as a result. In his promotional pamphlet, "The Miner's Friend; or, an Engine to Raise Water by Fire," (London, 1702) Savery said: "The labor of turning these two parts of the engine, the regulator and water cock, and tending the fire, being no more than what a boy's strength can perform for a day together, and as easily learned as their driving a horse in a tub-gin; Yet, after all I would have men, and those too

Figure 1 Schematic of Thomas Savery's "fire engine," using high pressure steam to pump water

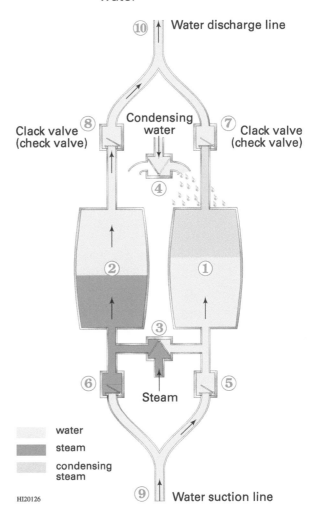

most apprehensive, employed in working of the engine, supposing them more careful than boys." Savery had a gift for understatement.

Savery's engine was applied to lift water for municipal and private water reservoir applications, but was seldom applied in mining because of the boiler explosion hazard. Jean Desaguliers refined the Savery engine in 1718, and subsequently applied several of them, including a water reservoir installation for the Czar of Russia.

1712 — Thomas Newcomen

Thomas Newcomen, an English blacksmith/inventor, improved the Savery concept by injecting water directly into the steam instead of cooling the vessel to cause condensation. His engine was the first "atmospheric steam engine," so named because the piston was driven by atmospheric pressure. He accomplished this by using a separate pump, driven by the steam cylinder and beam, to pump the water. The boiler only had to supply low-pressure steam (typically 10 psig) to fill the cylinder. Atmospheric pressure did the rest. When steam condensation caused a vacuum in the cylinder, atmospheric pressure pushed down on the piston, causing the pump piston to be pulled up by the engine beam. See Figure 2. Despite the huge improvements in engine design, Newcomen couldn't patent his engine because Savery's patent on "fire engines" was too broad. So Newcomen was forced to build his engines in partnership with Savery. They built their first engine-driven steam pump in 1712.

Newcomen collaborated with John Calley in the early years of development. Later, John Smeaton made significant improvements in the engine, allowing much larger cylinders and improved performance. Early steam cylinders were typically 22 to 23 inches in diameter. Smeaton's improvements allowed steam cylinders up to 5 feet in diameter (an increase to about 6.7 times the driving force).

See Figure 2, a schematic of Newcomen's engine, approximately 1712. Notice the drive beam (7) added to apply the steam piston movement (2) to the water pump (3). The Newcomen boiler (1) featured a spiral-shaped flue passage, with the coal fire under the steel plate construction. It was protected by a Papin-style counter-weighted relief valve. Instead of cooling the entire cylinder to cause condensation, Newcomen injected water into the cylinder, using a valve at 8. The engine is shown in mid-position of the condensing stroke (as on the hand-operated pumps of the day). The water reservoir (5) was mounted high enough to get the right water spray effect inside the cylinder (2). When the cylinder reached the bottom of the stroke, linkages closed valve 8 and opened valves 9 and 10. The weight of the pump rod and piston (3) plus slight steam pressure from the boiler caused the pump piston to drop and the steam piston to rise to the top, ready for another stroke. Air escaped through the reservoir at 6. The water in the reservoir (6) reduced likelihood of excessive air entering the cylinder.

An early Newcomen pumping engine lifted water 162 feet in a pipe approximately 4 inches diameter (total water column weight of 3,535 pounds). The pump piston was 8-inch diameter and the steam piston 24-inch diameter. Piston speed was about 75 feet per minute and the engine cycled 15 times per minute. [Raising 3,535 pounds of water 75 feet in a minute requires 75 x 3,535 or 265,125 foot-pounds per minute. One horsepower is 33,000 foot-pounds per minute, so this early engine was about 8 horsepower. One of Smeaton's Newcomen engines, erected at the Chasewater mine in Cornwall, 1775, used a 52-inch diameter steam piston and achieved 40 horsepower.]

Figure 2 Schematic of Thomas Newcomen's atmospheric engine — Newcomen's engine was reportedly so successful, it was applied in almost every large mine in England.

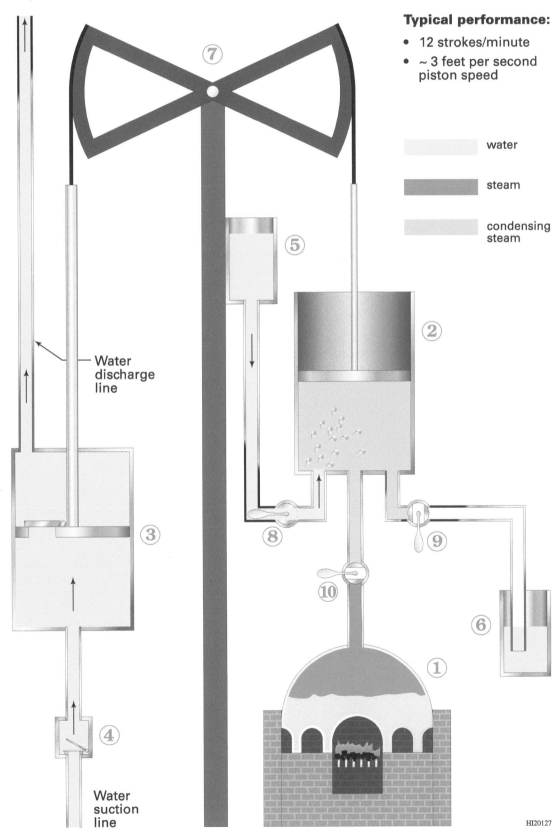

Typical performance:

- 12 strokes/minute
- ~ 3 feet per second piston speed

water

steam

condensing steam

Water discharge line

Water suction line

HI20127

1774 — James Watt

As a young man, James Watt trained in London to be a mathematical instrument maker. When he returned to Glasgow, Scotland, to pursue his practice, he was prevented from setting up shop by the local guilds because he hadn't apprenticed locally. A friend of Watt's, Dr. Black, a professor at the University of Glasgow solved the problem by having him work with the University on its premises. Watt studied chemistry at the university while there and worked with Dr. Black, whose research at the time led to the discovery of latent heat. Watt's own experiments led him to the conclusion of latent heat in steam because a small amount of steam could heat a large quantity of water.

Watt's work in trying to make the university's model of the Newcomen engine work led him to the development of his own engine. Because he was aware of the huge energy content of steam, he realized he needed to conserve it where possible. His breakthrough arose from finding the large energy required to heat up the steam cylinder. The Newcomen engine cooled the steam cylinder off in every cycle. Watt believed this to be an unacceptable waste of energy. His conclusion: use an external means to cause the steam to condense and pull a vacuum in the steam cylinder. He added a separate condenser (Figure 3, item 2). By incorporating this innovation and others, he increased the engine's efficiency nearly four times. [Watt's engine efficiency was about 2%, compared to steam engines by 1900 with efficiencies of about 17%. Today's steam turbines achieve a conversion efficiency as high as 47%. But it has taken over 200 years to get there.]

Figure 3 is a schematic of Watt's engine, about 1774. Compare it to Newcomen's engine in the previous figure to see how far Watt had come. This schematic doesn't do justice to the complete engine Watt (with others) developed. Not shown, for example is the parallelogram linkage between the beam and cylinder to cause parallel motion of the steam piston. Figure 3 does, though, show the key features:

- Separate condenser (2)
- Air pump (3)
- Flywheel (11)
- Inertial governor (5) — (an automatic speed control)
- Double-acting steam cylinder (1)
- Steam valve linkage (7)

The schematic in Figure 3 shows the engine approximately midway through the upward stroke of the steam cylinder (inlet valve routing steam below the cylinder and the upper passage routed to the condenser). This is a double-acting cylinder design because steam pressurizes one side at the same time the other side is condensing.

Note the inertial governor (5) incorporated in this design. As the governor's rotational speed increased, the balls swung out further, causing the linkage above to move. This movement was applied to adjust the steam flow at valve 4.

Pump 6 provided cooling water as needed for the condenser. The pump attached to linkage arm 14 removed excess water from the air pump/condenser assembly.

This engine could be used to drive a water pump or rotating shaft industrial application. The flywheel and speed regulator made the engine ideal for textile mill applications as well.

The Encyclopedia Britannica estimates that Watt and his partner, Matthew Boulton, built 500 engines, over 60% of them used for rotative applications, the remainder for pumping. Among the many applications of Watt's pumping engines was land

Figure 3 Schematic of James Watt's atmospheric engine, with external condenser, flywheel and inertial speed controller

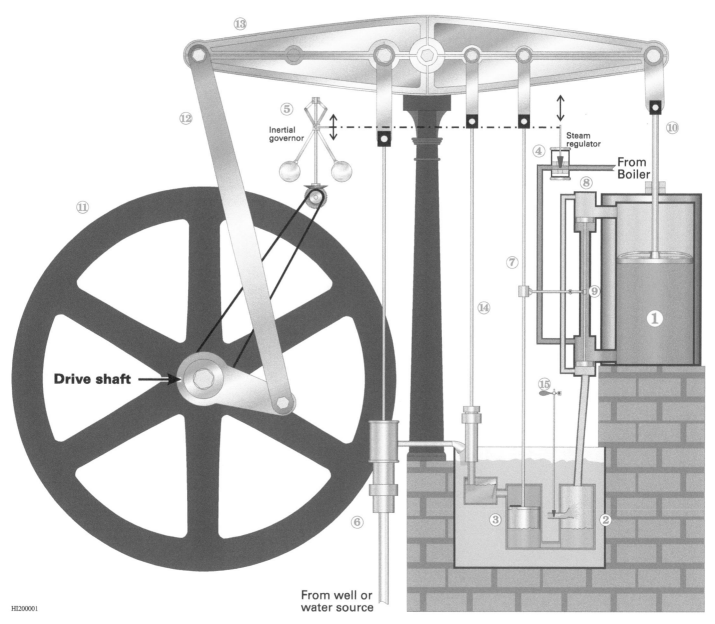

HI200001

drainage in Holland. The steam engine provided higher and more consistent power than the best of windmills.

James Watt didn't accomplish all of this alone. He struggled in the early years. (He had to give up his steam engine work for two years to earn some money. He spent everything he had on his work and prototypes.) Matthew Boulton, a wealthy industrialist, became Watt's partner in 1768. Boulton provided money, business knowledge and mechanical aptitude of his own. The two extended the value of their engines in 1800 when they applied steam

from a Watt engine boiler to heat a mill as well. They supplied low-pressure steam to flat-plate radiators (similar to the "mattress" radiator introduced by Samuel Gold in 1854) and hollow structural iron beams throughout the plant. This was the first major use of indirect steam space heating.

Watt and Boulton brought boilers into general industrial use to supply steam for their engines. Their well-designed and constructed mechanisms help speed industrial development through the late 18th and early 19th centuries.

Year	Name	Description
1632	Galileo Galilei	*"Dialogue Concerning the Two Chief World Systems — Ptolemaic and Copernican" — He nearly lost his life over this publication because it conflicted with Catholic church doctrine of a geocentric universe.*

| 1643 | Evangelista Torricelli | *Invented the mercury barometer.* |

| 1645 | Otto von Guericke | *Invented the first vacuum air pump.* |

| 1698 | Thomas Savery | *Obtained patent covering "fire engines" — engines operated by using fire to make steam.* |

| 1712 | Thomas Newcomen | *Prototyped first steam-engine-driven water pump.* |

| 1716 | Sir Martin Triewald | *Introduced greenhouse heating with direct-injection of steam.* |

| 1736 | Johnathan Hulls | *Patent for steam-engine-driven paddle-wheel boat.* |

| 1740 | Benjamin Franklin | *Developed the Franklin stove. He also did significant work in improvement of flue gas vent systems (chimney design).* |

| 1745 | Colonel William Cook | *Direct-injection steam heating of his home.* |

| 1769 | James Watt | *First prototype of an improved atmospheric steam engine — incorporating a separate condenser with a double-acting steam cylinder.* |

| 1773 | The Colonies | *Boston Tea Party.* |

Then 'til now — Progress of steam engines and boilers

Boilers have been in use for over 300 years — beginning with the first practical application in Thomas Savery's steam boiler, used to power his vacuum lift pump for mines.

Early boiler development benefited from the demand for steam to:

- Operate industrial equipment and pumps, allowing mills to be located near the labor source rather than being built along rivers.
- Provide propulsion for railroads, dredging equipment and farm machinery.
- Provide heat for greenhouses (because of the high demand for flowers and fresh plants in Europe).

Railroads

Richard Trevithick applied a high-pressure firetube steam boiler for rail and equipment propulsion from 1805 to 1812 (Figure 4).

Steamboats

Robert Fulton's "Clermont" steam-engine-driven paddlewheel steamboat went into service between New York City and Albany in 1807. (See Figure 5, a typical paddle-wheel steam boat.)

Ventilation fans

Before the introduction of electric motors, steam engines were used to drive ventilation fans.

Power generation

Babcock and Wilcox patented the first watertube boiler in 1867. This design evolved into the large boilers used for power plant generation and probably influenced cast iron boiler design as well (the Mills boiler, for instance).

Figure 4 Schematic of Richard Trevithick's high pressure steam locomotive

HI200002

1775 to 1883	*The Colonies*	*American Revolutionary War.*
1777	*Bonnemain*	*Gravity hot water heating applied to poultry incubators.*
1784	*James Watt*	*Designed and installed the first radiator (mattress-type) in his home. Steam system was one-pipe design with a manual air venting valve.*
1789 to 1799	*France*	*French Revolution and Napoleonic years.*
1793	*England*	*Assembly hall heated with perimeter indirect steam system. Terminal units consisted of a steam pipe inside an air pipe. Air discharge openings were spaced along the pipe around the room perimeter.*

Figure 5 Steam-driven paddle-wheeler, typical

1800	James Watt & Matthew Boulton	Used steam from cotton mill steam engine boiler to heat mill, with mattress-style radiators. They also routed steam through hollow cast iron structural members in the mill.
	Oliver Evans	Obtained patent for a high-pressure firetube steam boiler.

1801	Robert Fulton	First prototype (though unsuccessful because the engine broke through the hull) of a steam-engine-driven paddle-wheel boat, in France.

1805	Richard Trevithick	Used a double-acting cylinder high pressure steam engine to drive a prototype railroad engine. Trevithick's used a Cornish-style single-tube firetube boiler.

1807	Robert Fulton	The "Clermont", a steam-engine-driven paddle-wheel boat proven in a trip from New York City to Albany. The Clermont made the round trip in 32 hours. Before this, the best time was 4 days using sailing vessels.

1812-1815	United States	War of 1812 — The U. S. and Britain at it again.

1820	Marquis de Chabannes	Applied steam heating to heat the House of Lords, using tubular-style radiators.

1824	Thomas Tredgold	Published "Principles of Warming and Ventilating Public Buildings, Dwelling Houses, Manufactories, Hospitals, Hot Houses, Conservatories," a book on heating system design.

1825	A. M. Perkins	Invented pipe threads.

1831	A. M. Perkins	Developed a high-temperature hot water (350 °F) gravity heating system, based on printing press heating system developed by his father.

1835	United States	Gravity warm-air central heating introduced.

Space heating

Evolution of space heating

By the beginning of the 1800s boilers were in use for space heating:

- Sir William Cook used direct steam injection to heat his home in 1745 (though his wife must have eventually complained about soggy furniture and drapes).
- James Watt and Matthew Boulton distributed low pressure steam from a textile mill steam engine to radiators to heat the building. (Watt applied his radiator design in his home in 1784.) This was also the first large-scale use of radiators.
- Bonnemain applied gravity-circulated hot water heating to poultry incubators.
- The Marquis de Chabannes applied steam heating with tubular radiators to the House of Lords in London in 1820.

Residential heating

Central furnace-type heating systems were used in homes and public buildings in the early 19th century. But pioneers of steam and water heating systems searched for means to improve heat distribution, comfort, and indoor air quality over what these methods provided. Hot water and steam heat provided the answer, but the boiler needed to be made self-regulating to assure safe and practical application to residential heating.

A breakthrough came from an American inventor — Stephen Gold. In 1854, he introduced a self-regulating boiler and a practical radiator design. His radiator was similar to the flat-plate radiator developed by Watt and Boulton, and resembled a mattress because of the stay rods used to secure the plates together. So it inherited the nickname of the "mattress" radiator. It wasn't perfect, because it was noisy and tended to leak at the edge seams. But both the radiator and the boiler were practical and cost-effective enough to make them suitable for residential central heating.

Stephen Gold's son, Samuel, improved on the concept in 1859. With H. B. Smith Company, he introduced a cast iron sectional boiler, followed in 1862 by a new radiator design — the pin-style radiator.

Stephen Gold's boiler design also provided ventilation for the residence by heating fresh air across the upper boiler heating surfaces, ducting this air to the spaces above. This air heated the lower floors. They installed one-pipe steam radiators on the upper floors. (Other indirect system designs were already in use.)

Introduction of pumps

In the 20th century, the introduction of the pump (circulator) caused a switch from steam to water heating. Controls, pumps and piping designs have progressed significantly.

Today's boilers

Advances in materials engineering, pressure vessel design, combustion systems and electronics have brought significant changes in boiler performance and efficiency.

A heating contractor has many choices today, from conventional (non-condensing) boilers to fully condensing boilers. Many boilers are now available with modulating firing, providing greater fuel efficiency and precise control.

Today's contractor as pioneer

James Watt could be a role model for today's heating contractor. To be successful, a heating contractor has to "pioneer" his way through many of the systems he encounters, applying his knowledge of system operation.

A heating contractor is expected to deal with a wide range of systems — from simple single-loop water systems to large two-pipe steam systems in apartment houses and commercial buildings. Often, he's also expected to handle air conditioning equipment and furnaces as well.

Many of the systems he encounters may not be installed correctly, or could have been tinkered with many times over the years.

One solution is for the contractor to arm himself with knowledge of what makes systems work (and not work). He can then, like James Watt did, apply this knowledge to solving the system's problems . . . assuring a successful, profitable job and a satisfied customer.

Further reading

This section is a very brief overview of the contributions made to engines, boilers, terminal units and controls. For additional reading and research into heating history, try the books listed below. If you search the internet under "steam engine," you will receive many listings of sites with information on engines and the people who developed them. Some of the websites provide complete books on the subject. One website has animated graphics of the Savery, Newcomen and Watt engines.

- "The Beginning of a Century of Steam and Water Heating," Stifler, 1960, The H. B. Smith Co., Inc.
- "Energy Conversion: Developments of the Industrial Revolution: Steam Engines," Encyclopedia Britannica 1999 or Britannica Online.
- "Heat and Cold," Donaldson & Nagengast, 1994, ASHRAE, Atlanta, GA.
- "A History of the Growth of the Steam-Engine," Thurston, 1878, D. Appleton and Company, New York, NY [This is a rich history of the century and a half of development in all applications of the steam engine, including transportation as well as industrial applications.]
- "The Lost Art of Steam Heating," Holohan, 1992, Dan Holohan Associates, Inc., Bethpage, NY.
- "The Lost Art of Steam Heating Companion," Holohan, 1997, Dan Holohan Associates, Inc., Bethpage, NY.

1845	Sir William Fairbairn	Patented multi-tubular firetube boiler.

1848	George Brayton	Developed cast iron sectional boiler.

1854	Stephen Gold	Developed low pressure steam heating with "mattress"-style radiator, similar to Watt's 18th century device.

1859	Samuel Gold & H. B. Smith Co.	Developed cast iron section steam heating boiler.

1861 to 1865	United States	American Civil War

1862	Gold & Foskett	Introduced pin-style steam radiator.

1867	Babcock & Wilcox	Patented a watertube boiler design.

1880	United States	American Society of Mechanical Engineers founded.

1914	United States	First boiler code published by ASME.

1843 to 1930	United States	During this period, 750 patents were issued for hot water or steam radiator designs.

1957	United States	Term "hydronics" coined by the Institute of Boiler and Radiation Manufacturers (I=B=R), now the Hydronics Institute.

Boilers today

Fuel makes a difference

From early use until well into the 20th century, coal was the fuel of choice for residential boilers. The heating system designs of the time took the performance of these boilers into account:

- Coal-fired boilers couldn't be shut down instantly (as we can today with gas or oil-fired equipment).

- Starting the fire wasn't an easy job, so the fire was kept going almost continuously.

- Many functions that are automatic now were manual then — water make-up, for example.

Things are different now — Boilers may cycle frequently (often 6 times an hour or more). This is due to automatic controls and to the lower stored energy in current (smaller) boilers. Consider this when troubleshooting steam systems in particular. The biggest difference in performance due to frequent cycling is in how much and how often air must be moved out of a steam system. When the boiler cycles off, vacuums form in the system and air is pulled in. Slight modifications need to be made, particularly with a steam system, to make the system friendlier to modern boilers (adding main vents at the tops of risers, for instance).

Residential boiler evolution

It is likely that most residential boiler designs evolved from commercial/industrial units, scaled down for residential heating loads. Early 20th century design evolution included the switch from coal to gas or oil, and a gradual reduction in boiler physical size to ease residential installations.

Early in the 20th century, the American Society of Mechanical Engineers (ASME) developed design standards for boilers, resulting in safer, more reliable equipment.

Significant technological innovations prior to the 1980's included flame retention oil burners and associated burner head designs, advancements in atmospheric gas burners (from cast iron to tubular steel) and substantial boiler size reductions, facilitated by improved combustion systems.

Beginning in the 1980's, boiler design evolution accelerated, driven by the demand for improved fuel efficiency and implementation of energy standards. Significant advancements included:

Burner technology

Oil burner designs refined the retention head, yielding a wide range of operation with clean burning. The addition of advanced ignition and improved fuel atomization also contributed to improved performance.

Gas burners made the greatest advances, both in the performance of tubular atmospheric burners and in the most significant change of all — the introduction of premix gas combustion for residential boilers. Another significant technology was pulse combustion, in which the energy from fuel ignition was used to force flue products through the heat exchanger.

Premix gas burners

Premix gas burners are the heart of nearly all gas-fired condensing residential boilers. Premix combustion became possible in residential use because of the availability of small zero-governor gas valves and electronic controls to supervise burner operation.

Premix combustion had proven its importance in industrial applications, but was first introduced in residential applications in the 1990's. Premix combustion is effective because it allows a wide range of operation, a substantial reduction in excess air required, and major reduction in NOx emissions.

Electronic controls

Electronic device applications began to provide reliable automatic ignition, integrated fuel regulation and sophisticated temperature regulation. Microprocessor technology yielded cost-effective, yet versatile integrated control capabilities.

Combined with premix gas technology and new motor speed control electronics, boiler integrated controls were now capable of full modulation while still being cost effective.

Metallurgy

New material applications, such as stainless steel and cast aluminum, opened the door for high-efficiency condensing boiler designs.

Stainless steel, particularly AL29-4C, provided a versatile and reliable material for vent systems likely to condense. (Some condensing boilers are rated for use with PVC or CPVC vent pipe as well.)

Radiant floor/radiant panel heating

Radiant heating provided an efficient, ultra comfortable new application for hydronic systems. The special needs in radiant heating accelerated advances in controls, valves and piping systems.

Comfort control

New piping components, piping designs and electronic controls provided load-responsive heat input regulation. In addition to outdoor reset, new controls monitored water temperature response to input/load changes to fine tune boiler firing rate. Multiple-boiler systems and modulating boilers provided close input vs demand regulation.

Productive learning curve

Hydronic manufacturers and installers quickly learned what worked best and drove boiler/system design improvements.

Efficiency standards

The U. S. Department of Energy and ASHRAE introduced standards and guidelines for boiler efficiency.

State of the art

Non-condensing boiler designs have evolved to be very cost effective, with efficiency maximized, while ensuring appropriate vent system design. The ANSI standards defined testing methods to determine whether boilers can operate with a conventional, non-condensing vent. If the efficiency is too high, the boiler must be installed with a vent suitable for condensing operation, as specified in the boiler instructions.

Condensing gas boilers are available with full modulation, providing annual efficiencies up to the high 90's. Integrated boiler controls match boiler firing rate to demand. Some controls are even programmable, allowing customization of boiler operation to specific system needs.

Boiler application

Each boiler may have special requirements for control and piping, and will be specified by the boiler manufacturers, usually in their instruction manual. Pay careful attention to this information. Make sure your installation agrees with the boiler manufacturer's recommendations.

Overview

The following discussion in this section outlines **RESIDENTIAL BOILER** subsystems and how they operate. Note that the example boilers shown are only a sampling of what is available, and do not represent an endorsement of any particular design. They are chosen only to facilitate the explanation of basic residential boiler operation.

The discussion begins with basic boiler design, and follows with detailed discussion of the operation of each portion of the boiler.

Boiler anatomy

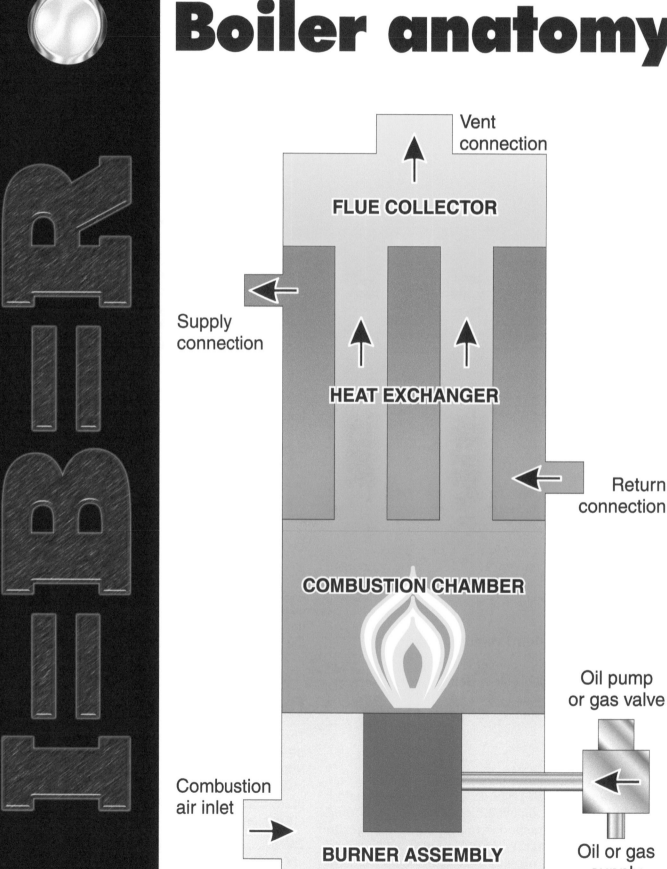

Vent connection

FLUE COLLECTOR

Supply connection

HEAT EXCHANGER

Return connection

COMBUSTION CHAMBER

Oil pump or gas valve

Combustion air inlet

BURNER ASSEMBLY

Oil or gas supply

Overview

The illustration on the opposite page is a schematic of a residential boiler, showing the primary components. All residential boilers, regardless of their specific designs, use the same basic components. The discussion below is an overview of each component. You will find detailed discussion related to each of the components on the following pages in this section, plus additional information in other sections of Guide RHH. [NOTE: Some boilers use two-stage heat exchangers — one for conventional operation, the other for condensing operation.]

Combustion chamber

The combustion chamber contains the flame, allowing the volume needed for the fuel and air to burn completely. It can be surrounded by water-backed surfaces (wet base design) or lined with refractory on all or just some sides (dry base design).

Burner

The burner mixes fuel and air, provides an ignition source, and controls combustion. Gas burners can be atmospheric type, premix, pulse combustion, or forced draft. Oil burners are most often forced draft, retention-head type.

Heat exchanger

The boiler heat exchanger is the pressure vessel that contains water (and steam in steam boilers) and provides heat exchange surfaces where the hot flue gases can transfer their heat to the water.

Fuel system

The fuel system includes the components required to supply fuel to the burner, such as gas or oil valves and fuel pumps. It includes pressure regulation devices (pressure regulator, for example) and may include fuel pressure sensors.

Combustion air system

All burners require air for combustion. The air system provides required air openings into the boiler and/or burner. The boiler/burner and their components establish how much air flows — atmospheric gas boilers rely on natural draft to pull the air into the burner assembly, while induced draft or forced draft boilers use blowers to pull or push combustion air into the boiler/burner.

Vent

The boiler includes a flue gas exhaust system, and may include part of the vent, such as a draft hood, vent damper or termination assembly. The boiler manual specifies the type of venting required. You must provide only the vent components/systems specified by the boiler manufacturer.

Controls

(This component is not shown in the boiler schematic.) Boiler controls regulate combustion and boiler water temperature (or steam pressure). Some boilers include electronic controls that provide additional system control functions. See Guide RHH Sections 10 and 11 for operation of typical boiler controls.

Water/steam piping

The boiler includes connections for return and supply, relief valve, limit controls, level controls, etc. The boiler manual will specify recommended or special system connection and piping/control design requirements. Read the boiler manufacturer's instructions carefully. Typical requirements you will find in the manual include:

- special pumps and/or piping — the circulator supplied with the boiler may be intended only for boiler loop circulation, for example.

- temperature mixing piping to prevent condensation caused by low-temperature return water.

- minimum flow to force water through the heat exchanger.

- special piping, such as primary/secondary connections.

Condensate system

(This component is not shown in the boiler schematic.) Condensing boilers require a condensate disposal system. It can consist of a tube connected to a floor drain, or may require a condensate pump if condensate cannot flow to drain by gravity. It may also include a condensate neutralizer.

Combustion & burners

Combustion

Combustion occurs when oxygen is combined with carbon and hydrogen (and trace amounts of sulfur) and an ignition source is present. When these reactions take place, heat and light are given off. Combustion self-sustains if fuel and oxygen continue to be supplied. Successful boiler operation requires that combustion be controlled — the flame needs enough oxygen to burn completely (called complete combustion), and the burner must shape the flame to fit in the combustion chamber.

In residential boilers, the oxygen for combustion is supplied by providing air to the flame. Air is 20.9% oxygen. The remaining 79.1% is mostly nitrogen.

See Figure 6, illustrating typical combustion with an atmospheric gas tubular burner. Two effects cause the air to be added to the flame:

- Primary air enters the burner throat with the gas because the jet action of the gas coming out of the orifice causes a slightly negative pressure. Air moves in to fill the vacuum, and mixes with the gas when it gets there. For typical atmospheric gas burners, only about 30% of the air needed enters as primary air.

- Secondary air enters the combustion chamber because the height of the boiler and its draft hood act like a chimney, causing a negative pressure in the chamber. Air comes in to break the vacuum. [Altering the boiler or shortening the draft hood height will lower the amount of air entering the chamber. The result will usually be incomplete combustion — a potentially dangerous situation be-

cause of carbon monoxide and soot production.] Secondary air flows up around the burner tubes and feeds the flame from the outside in.

Combustion products

H_2O — water vapor

When oxygen combines with hydrogen, the result is H_2O, or water vapor. Unless the boiler is designed for condensing operation, the controls and piping must be installed properly to prevent condensation on the boiler surfaces. Condensation can damage the boiler and may also affect combustion.

CO_2 — carbon dioxide

Complete combustion of carbon results in carbon dioxide, CO2. If combustion is interrupted, or if inadequate oxygen (air) is supplied, combustion will be incomplete. This will cause formation of carbon monoxide, CO, a very toxic, odorless gas.

NOx — oxides of nitrogen

Nitrogen has little effect on the flame, but it can combine with oxygen at high temperatures, as in the flame. This causes oxides of nitrogen (often called NOx) to exhaust with the flue products. Some areas regulate maximum NOx emissions, often requiring special burners and/or boiler designs.

SOx — oxides of sulfur

Sulfur in fuels results in oxides of sulfur in the flue products. When these sulfur oxides combine with water, they form sulfurous or sulfuric acid. Boiler heating surface temperatures need to be kept warm enough to prevent acid condensation on the boiler.

Figure 6   Combustion with a tubular atmospheric gas burner — mixing air and burning the fuel

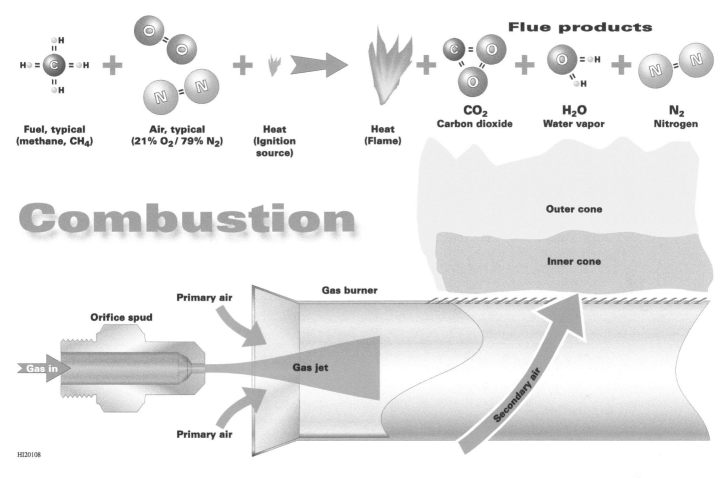

Fuel, typical
(methane, CH_4)

Air, typical
(21% O_2 / 79% N_2)

Heat
(Ignition
source)

Heat
(Flame)

Flue products

CO_2
Carbon dioxide

H_2O
Water vapor

N_2
Nitrogen

Combustion

Outer cone

Inner cone

Gas burner

Primary air

Orifice spud

Gas in

Gas jet

Primary air

Secondary air

HI20108

Figure 7   Typical atmospheric gas burner assembly (burners shown include integral venturi tubes
and precision-perforated, geometric-patterned ports)

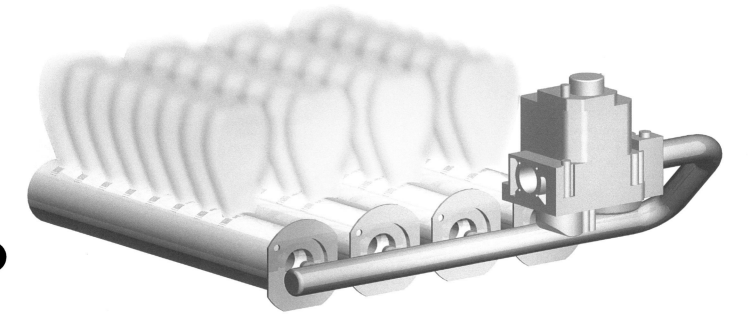

Combustion air flow

Atmospheric gas boilers control combustion air using either a draft hood (vertical bell-shaped device, as in Figure 8) or draft diverter (integral device with air opening from the bottom, either mounted on top or on side of appliance). Draft hoods and draft diverters operate in the same way, as described below.

See Figure 8, showing operation of a draft hood. With the chimney or vent pulling flue gases out as it should, the draft hood allows room air to enter the vent as draft (pull) increases. If the chimney or vent is exposed to a wind gust or the boiler room pressure goes negative, flue products flow downward in the vent. But this downward flow doesn't affect the proper movement of flue products from the boiler (provided the draft hood is installed per the boiler manufacturer's instructions, with no alterations).

WARNING

Figure 9, shows the importance of installing the draft hood as supplied, **with no modifications**. The height from the bottom of the boiler to the draft hood bell sets a minimum "chimney height" for the boiler. You must not change this height, or the boiler will not function properly. Incomplete combustion will result, with the potential of dangerous emission of carbon monoxide.

Figure 8 Draft diverter operation

Normal operation — negative pressure in chimney

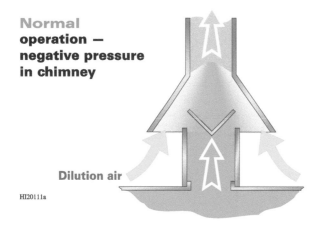

Dilution air

HI20111a

Downdraft operation — positive pressure in chimney

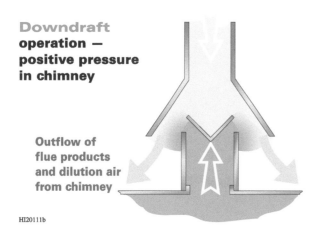

Outflow of flue products and dilution air from chimney

HI20111b

Figure 9 **NEVER** alter draft diverter

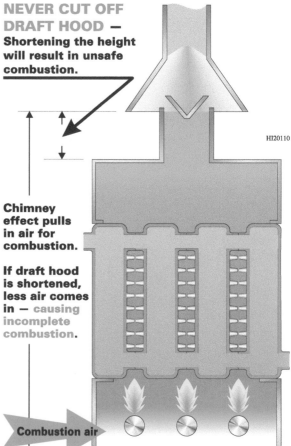

NEVER CUT OFF DRAFT HOOD — **Shortening the height will result in unsafe combustion.**

HI20110

Chimney effect pulls in air for combustion.

If draft hood is shortened, less air comes in — causing incomplete combustion.

Combustion air

Induced draft, **premix** or **power-fired** gas boilers control combustion air with the blower and venting system. Follow the boiler manual for installation of the boiler and vent to be sure of proper operation.

Direct vent (sealed combustion) boilers require that air be piped to the boiler, following the boiler manual instructions.

Primary and secondary air

Primary air is air that enters the combustion zone with the fuel. In atmospheric gas burners, this is the air pulled in by the venturi action at the throat of the burner (see Figure 6, Page 1–23).

Secondary air is air that mixes into the flame in the combustion zone.

In premix combustion, ALL air is primary air, mixed with the fuel before entering the combustion zone.

Fuel flow

The right amount of fuel is controlled by two things:

1. Orifice (or nozzle) size

    - The area of the orifice opening and the orifice geometry control the amount of fuel.

    - Carefully follow manufacturer's instructions for orifice sizing.

    - Never drill out orifices or change nozzle size beyond manufacturer's directions.

    - Don't alter orifices or nozzles. The result could be dangerous due to incomplete combustion or fire hazard.

2. Pressure at the orifice or nozzle

    - The higher the pressure at the orifice (or nozzle), the greater the fuel flow.

    - Make sure the pressure agrees with the boiler/burner manufacturer's instructions.

Figure 10   Premix burner fuel flow regulation

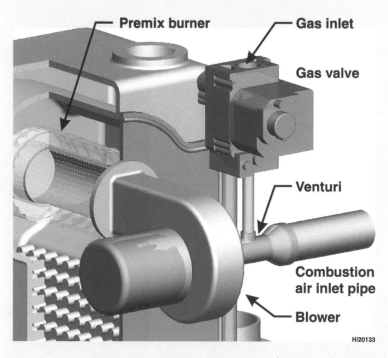

HI20133

Premix combustion gas flow

For gas premix combustion applications, fuel flow responds to the suction developed on the intake side of the premix blower. The gas valve is designed to regulate the valve outlet pressure at zero inches water column (referred to as a "zero governor" gas valve). [Fuel won't flow unless the blower causes a negative pressure.] When the gas valve senses a negative pressure, it allows enough gas to flow so the pressure at the valve outlet comes to zero. (This contrasts to atmospheric gas applications, where the valve outlet pressure is usually around 3.5 inches water column for natural gas.)

See Figure 10, showing a portion of a premix gas boiler. When the blower runs, air pulls in through the combustion air pipe. Air flowing through the throat of the **venturi** causes a negative pressure. The gas valve (if activated) senses the drop in pressure, and allows gas to flow until the valve outlet pressure rises to zero. An orifice (typically mounted in the valve outlet) limits gas flow to the correct input rate for the boiler. The gas mixes with the air and is pushed into the burner by the premix blower.

Oil burners

Clean burning of fuel oil requires first that the fuel be finely atomized — that is, the fuel must be split into small enough particles to be sure the carbon and hydrogen in the fuel will break out and find oxygen once in the flame. Light oil burners do this by spraying oil at high pressure through carefully-designed oil nozzles. The nozzle not only causes the oil to atomize, but shapes the oil pattern to properly work with the burner airflow and combustion chamber shape. See Figure 12 for a typical oil nozzle. [Use the nozzle(s) specified by the burner manufacturer. The nozzle specification includes nozzle flow rating (in gallons per hour), the spray angle, and the spray pattern (solid, hollow or semisolid). Each of these specifications will affect nozzle behavior and burner performance, so always use the recommended nozzle(s).]

See Figure 11, a cutaway of the main components in a residential oil burner. The burner housing (1) is shaped to cause the right airflow down the burner tube (7). The burner motor (6) rotates the blower (2) and the oil pump (not shown, but connected to the motor shaft with coupling, 4). Oil enters the burner from the oil valve and oil pump (not shown) through the oil tube to the nozzle (9). The oil is ignited when the ignition transformer or spark control (5) applies high voltage to the electrodes (8). Air enters the burner through the blower opening (3) and is directed down the burner tube (7). Air flows through the slots in the retention plate (10). Some air flows through the center of the plate and often around the plate.

The retention plate radial blades cause the air to rotate as it leaves the plate. The spin of the air helps mix the air and fuel together. As the air leaves the slots, it causes a slight negative pressure on the back side of the blades. This negative pressure tries to pull the flame back, holding the flame to the retention plate. With the flame retained at the plate, combustion is consistent and easier to set up and maintain.

NOTE: Always make sure the spacing from the oil nozzle to the back of the retention plate is set to the burner manufacturer's instructions. Incorrect positioning of the nozzle can cause dirty combustion (making soot), rough starts or rapid deterioration in burner performance due to sooting of the retention plate. Replace the oil nozzle annually or at any time the nozzle tip shows deposits that can affect the spray pattern.

Figure 11 Primary components of a residential oil burner (see discussion for legend to numbers)

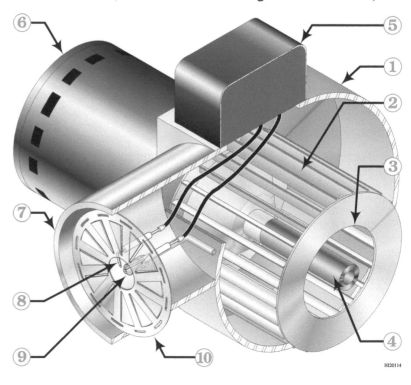

HI20114

Figure 12 Typical oil nozzle construction

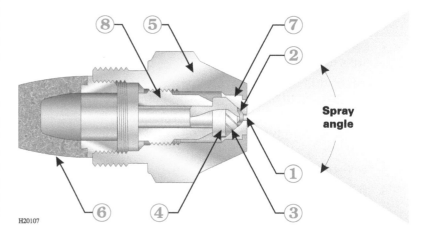

H20107

Spray angle

① Orifice ⑤ Nozzle body

② Swirl chamber ⑥ Filter

③ Tangential slots ⑦ Orifice disc

④ Distributor ⑧ Screw pin

Proving the flame

Residential and commercial boiler/burner units supervise flame (prove there is a flame\) in the chamber using either direct insertion methods (thermocouple or electronic flame rod) or visual sensor methods — a flame sighting device responsive to visual light (cad cell sensor or rectifying photocell), infrared light (IR sensor) or ultraviolet light (UV sensor). There is still some limited use of the "stack switch," a temperature sensing device mounted in the appliance vent pipe.

Thermocouples

Thermocouples "sense" flame because they are made using two specially selected metal wires (two different metals) that are fused together at the thermocouple tip. When this junction between the two metals is heated, the electrons of one metal tend to move to the other metal, causing one metal wire to be charged differently from the other. The higher the temperature at the junction, the greater the voltage difference. The amount of this difference is the same for any given temperature and the same two metals. The voltage difference caused is in millivolts. The main or pilot gas valve is designed to read this voltage difference and will allow pilot gas to flow if the millivolt value is high enough.

Thermocouples require time to cool off when flame is removed. This can cause the response of a thermocouple to be up to 90 seconds after loss of flame. For this reason, thermocouple supervision is limited to boilers 400 MBH and smaller. Larger boilers require an electronic supervision method. Electronic controls have almost instantaneous response to loss of flame.

Flame rods

A flame rod is a high-temperature sensing rod mounted so the flame wipes across it. The flame rod works with an electronic control to prove the flame, using a process called "flame rectification." Rectification means making a current flow one-directional (dc current) when the voltage supplied is two-directional (AC voltage). The line voltage adapters for calculators and battery-operated appliances do this, for example.

The flame rod causes a dc current when an AC voltage is applied if the grounded metal surface near the flame is large compared to the flame rod. The surface area of the grounded parts has to be at least 4 times the surface area of the flame rod. See Figure 14, Page 1–28. The grounded metal surfaces of an actual application would be

Figure 13 Flame rectification using a flame rod

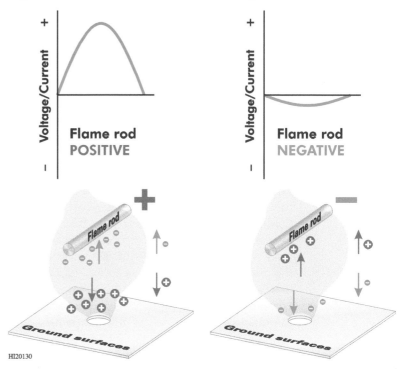

HI20130

the surfaces of the pilot burner or main burner. These parts are grounded because they are bolted to the boiler housing, which is grounded through the electrical supply and probably through the piping connections to the boiler as well.

The flame is so hot that some of the molecules in the flame exchange electrical charge when they collide — that is, there is so much energy available that some of the electrons break free from molecules and attach to other molecules. This is called flame ionization — there are charged particles floating around in the flame. When the electronic control applies a voltage to the flame rod, the charged particles in the flame move toward or away from the voltage as shown in Figure 13. When the voltage on the rod is positive, positive particles move away from the rod, toward ground. Negative particles move toward the flame rod. The opposite occurs when the voltage on the rod is negative.

When the ground surface area is larger than the flame rod area, current flows easier when the flame rod is positive than when it is negative. Notice the graphs of current vs. voltage in Figure 13. Because there is more current on the top of the line (positive flame rod) than below the line (negative flame rod), this looks like dc current. To cause enough of this effect, the grounded surface area near the flame has to be at least 4 times the flame rod surface area. [The typical dc current level needed for electronic flame rectification controls is about 2 microamperes.]

If there were a short from flame rod to ground, or if the flame rod insulator were dirty enough to become conductive, the current could flow as easily during the negative cycle as during the positive cycle — no rectification. The control would shut down because the current would be AC current, not dc. This is why flame rectification is effective — it can tell the difference between a short and a flame.

Rectifying photocell

The rectifying photocell is a visible light-sensor that rectifies the current in a vacuum tube. See Figure 14. Current flow is rectified because only one of the electrodes in the tube is coated with a material that responds to light. When light strikes the material, the light energy causes electrons to break free. Electrons are negatively-charged. So, when the uncoated (upper) electrode is positive, the electrons are attracted to it, causing a current. When the uncoated electrode is negative, the electrons are pushed away, so no current flows.

Cad cell

Visual flame sensors are usually cylinders with a lens on the sighting end and a sensing surface or sensing tube inside the cylinder. The cad cell, for example, has a plate inside that is coated with cadmium sulfide. When visible light hits the surface of the cadmium sulfide, the light energy boosts electrons free to move about in the material. This makes the cadmium sulfide conduct current more easily. The flame supervisory control senses this change. If the change is large enough, the cad cell control is satisfied and allows fuel to continue flowing. Cad cell controls/sensors are used only on oil burners. The light given off by gas flames doesn't cause enough response by the cadmium sulfide.

Figure 14 Rectifying photocell

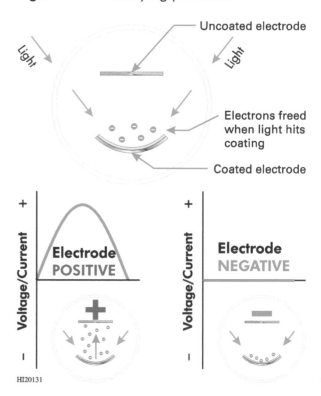

HI20131

Infrared and ultraviolet sensors

Infrared and ultraviolet sensors are not often used on residential-sized boilers. You will find the ultraviolet sensor frequently on power gas or gas/oil burners. The infrared sensor responds only to infrared light. The ultraviolet sensor responds only to ultraviolet light. Both of these sensors require an electronic control, ranging from pilot-flame-proving controls to programming controls that prove pilot flame, main flame and many other burner/boiler conditions, depending on the attached controls.

Igniting the burner — atmospheric gas

Most gas boilers will use gas pilot or hot surface ignition of the main flame, though standards allow direct spark of the main flame on units under 400 MBH input.

Figure 15 shows a standing pilot burner, with thermocouple supervision. This pilot is lighted manually, with pilot gas controlled by the main automatic gas valve or a separate thermocouple-operated pilot valve. The pilot (or main) gas valve has to be manually held open until the thermocouple heats enough to prove flame is there. When the main gas valve uses the thermocouple to prove flame, main gas will flow once the thermocouple is heated enough and the main gas valve knob or lever is moved to the main flame position.

Figure 16 shows another standing pilot, similar to Figure 17, except this pilot uses a thermocouple only to prove pilot. The pilot gas is controlled by a separate thermocouple-operated pilot valve. An electronic control wired to the flame rod controls the main gas valve. This provides quick shutoff of main gas if pilot flame should be lost.

Figure 17 shows a spark-ignited pilot. An electronic flame supervisory control operates a pilot gas valve and main gas valve (which often will be in the same valve body). On call for heat, the control opens the pilot valve and applies a high voltage (typically 6,000 volts) to the spark electrode. This causes a spark to jump from the electrode to the pilot hood. The heat of the spark ignites the pilot gas. The control applies an AC voltage to the flame rod. If flame is present (and there are no shorts to ground from the flame rod), the current through the flame will be rectified to a dc current. If the dc current level is high enough, the control will be satisfied that flame is present and will then open the main gas valve. (Some pilot burners are also ignited by hot surface igniters.)

Figure 15

Standing pilot burner with thermocouple supervision

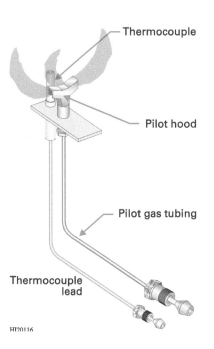

Figure 16

Standing pilot burner with thermocouple pilot supervision and flame rod for main gas control

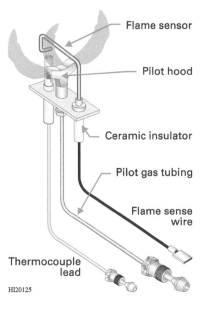

Figure 17

Spark-ignited pilot with flame rod supervision

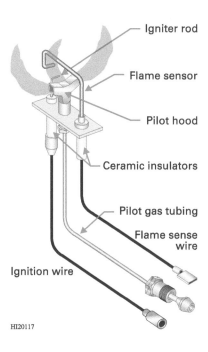

Figure 18, Page 1–30 shows a hot surface igniter. This is typically a piece of silicon carbide or silicon nitride connected to a voltage source. Voltage is applied to the silicon carbide (or silicon nitride) for a controlled amount of time (or the current flow is monitored until it reaches the right value) to ensure the igniter surface is hot enough to ignite gas. Then the gas valve is opened to allow ignition. Flame supervision is usually done using flame rectification, with the igniter as the flame rod or using a separate rod in the flame.

Some gas boilers are equipped with spark electrodes used to directly ignite the main flame, without the use of a pilot burner. This is called direct spark ignition, and is limited by current standards to boilers 400 MBH and smaller.

Figure 18   Hot surface igniter, typical

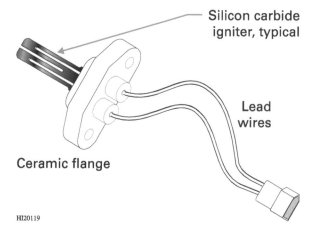

Silicon carbide igniter, typical

Lead wires

Ceramic flange

HI20119

Igniting the burner — oil

Most residential oil burners are ignited using direct spark of the oil flame. Gas pilot ignition of oil is usually reserved for large commercial or industrial burners. Flame sensing is usually done with a cad cell visible light sensor, with the cad cell mounted inside the burner housing, pointed at the flame.

See Figure 19, Page 1–31 for a typical arrangement of spark electrodes and oil nozzle. Carefully follow the burner manufacturer's instructions for setting and verifying the three critical dimensions:

1 Electrode tip-to-tip spacing. This is critical. If too close, the spark will be too focused to light the oil. If too far apart, the spark will be too weak. Electrodes wear with use and must be adjusted or replaced periodically.

2 Electrode tip-to-nozzle vertical spacing.

3 Electrode tip-to-nozzle horizontal spacing. Carefully set dimensions 2 and 3 to be sure the spark is placed into the right location of the oil spray pattern.

Check these electrode dimensions on every inspection of the burner, at least during the annual inspection/cleaning service of the boiler/burner.

Figure 19 Typical oil burner nozzle and electrode layout

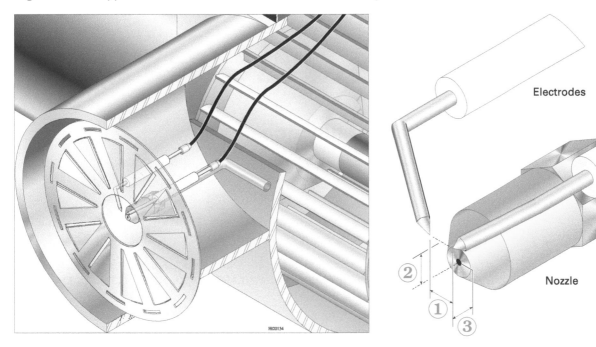

Electrodes

Oil tube

Nozzle

② ① ③

Igniting the burner — power gas or gas/oil

Power gas and gas/oil burners can be ignited using:

- Direct spark ignition of both fuels (smaller burners only)
- Direct spark on oil, gas pilot on gas – or – gas pilot ignition of both fuels, depending on codes or job requirements.

Flame supervision is usually done with flame rectification (flame rod) or ultra-violet sensor. (Note that cad cell sensing cannot be used on gas flames, because the sensor cannot see the radiation emitted by the gas flame.)

Pilot assemblies in power burners operate similarly to those on atmospheric gas applications, but they are shaped differently to deal with the air currents at the exit of a power burner.

Heat exchangers

Heat exchanger functions

1. Boiler heat exchangers provide surfaces with hot flue gases on one side and boiler water on the other. The hot flue gases give up their heat to the boiler water.

2. Boiler heat exchangers contain boiler water and distribute water and flue gas flow. And the boiler heat exchanger is a pressure vessel —

   • The boiler heat exchanger is designed and rated as a pressure vessel, intended to contain boiler water/steam and withstand water/steam pressures up to its design limit.

   • Residential boilers must comply with the ASME (American Society of Mechanical Engineers) Code, Section IV, Heating Boilers. This code covers direct-fired pressure vessels up to 160 psig water, 250 °F, and up to 15 psig steam. The code specifies design requirements and verification/testing procedures. These boilers are stamped with the ASME "H" stamp.

   • For higher pressure or temperature boilers, ASME Section I, Power Boilers, applies. Power boilers are stamped with the ASME "S" stamp. Residential boilers are not likely to be designed to Section I requirements.

Distributing flue gases

Boilers are designed to distribute the flow of flue gases for the most effective transfer of heat. You will find this done by the geometry of the flue passageways, with baffles, and other methods.

Many condensing boilers also pass the exiting flue gases over the return water section of the heat exchanger instead of the outlet water section. Because the return water is cooler, this increases the temperature difference for greater heat transfer. Heat transfer is proportional to temperature difference.

Distributing water flow

The water side of the boiler heat exchanger must distribute the flow of water to keep the boiler heating surfaces cooled and ensure the best possible temperature difference between water and flue gases throughout the heat exchanger.

Some boilers have natural internal circulation. As water is heated, it rises and is replaced with cooler water from adjacent regions in the boiler. These boilers may be tolerant of low flow rates. Other boilers, such as straight horizontal tube types, may require forced flow at a minimum rate to ensure proper cooling. Always read the boiler manual to ensure the installation meets boiler requirements.

Heat transfer

Heat always moves from a hot object to a cold object. It gets there in up to three ways:

Conduction (heat transfer due to direct contact).

Convection (heat transfer due to movement of a fluid).

Radiation (heat transferred by infrared light waves sent out by the hotter object).

See the illustrations on this page for a visual explanation of these methods.

Heat transfer in a boiler occurs through all of the above methods.

- Flue gases give off heat to the boiler surfaces by direct contact, flue gas movement and radiation.

- The water in the boiler absorbs the heat, thus cooling the boiler surfaces. The water absorbs this heat mostly through conduction (direct contact) and convection (water movement).

- There is very little radiation heat transfer on the water side of the boiler.

Enhancing heat transfer

Extended heating surface

Because water is dense and very conductive compared to flue gases, heat transfers much more easily on the water side than on the flue gas side. In fact, the difference in effectiveness can be about 100 times better on the water side.

To compensate for this, many boilers incorporate extended surfaces on the flue side. You will see this as pins, fins, ribs, etc. These extended surfaces provide more contact area, thus increasing heat transfer.

Turbulence and mixing flue gases

Mixing and disturbing the flue gases as they pass along the heat exchanger keeps them from forming stagnant layers that reduce heat transfer. Pins and other extended surfaces help on some boilers. Other boilers use baffles or turbulators (bent lengths of steel in boiler tubes) to force mixing and turbulence.

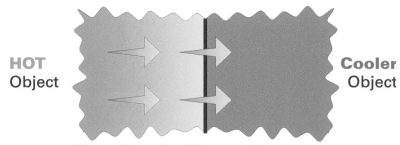

Conduction

HOT Object   **Cooler Object**

Heat flow inside a substance
or
Between objects directly in contact

- -

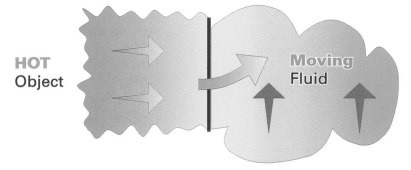

Convection

HOT Object   **Moving Fluid**

Heat given off or absorbed
by a moving fluid (such as air or water)

Natural convection —
fluid moves only because density changes as it heats or cools.

Forced convection —
fluid moved by external force, such as a pump or blower.

- -

Radiation

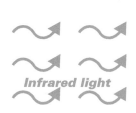

HOT Object   Cooler Object

Infrared light

HT20112

Heat carried by infrared light

Vent systems

Functions of the vent system

1. The vent system removes flue products from the boiler and the building.

2. The vent system may also contribute to combustion air flow, as with atmospheric gas boilers. The draft created by the boiler's draft hood or draft diverter works like a chimney to pull air and flue products through the boiler.

3. Direct vent (sealed combustion) boilers also require air piping. The air piping is the only source of combustion air for these boilers.

The vent system is determined by the boiler

Boiler design determines the required vent system. Atmospheric gas boilers require a natural draft (gravity) vent, typically using type B venting. Forced draft and induced draft boilers usually require positive-pressure-rated venting. Condensing boilers and other high efficiency boiler designs require venting suitable for condensation.

You will find detailed discussion of vent systems and ANSI vent categories in Guide RHH Section 2.

Vent materials and acceptable types are specified in the boiler manual. You must use only the options listed by the boiler manufacturer. For examples:

- Using PVC or CPVC vent on a conventional boiler could result in vent failure from flue gases that are too hot.

- Using type B vent on an application requiring pressurized venting will cause flue gases to leak into the building.

Pay attention to the boiler manual

Vent systems are specific to the boiler. You must carefully read the instruction manual and follow all guidelines to ensure a successful installation.

WARNING

You must install the vent system (and combustion air piping, when required) using the materials and methods specified in the boiler manual. Follow all applicable codes. The vent system both removes flue products and affects boiler operation. Failure to properly install the vent system can result in severe personal injury or death.

Draft regulation — gas boilers

You will find draft regulation on residential gas boilers done with:

• Draft diverters for atmospheric gas boilers. See Page 1–24 for discussion.

• Induced draft (on vent outlet) or power blower (blowing into combustion chamber) on fan-assisted gas boilers. No additional draft regulation is needed if the boiler and vent are installed per the boiler and vent manufacturer's instructions.

• Forced draft burners (including premix gas burners), in which the boiler's blower pushes flue products through the boiler and through the vent system.

Draft regulation — oil boilers

Residential oil boilers often require installation of a barometric damper (Figure 20) on the vent. The barometric damper can be set to regulate the available draft at the boiler outlet within a couple of hundredths of an inch water column. This assures the flue gas flow in the boiler will be stable, helping to maintain good combustion.

Figure 20 Barometric damper, typical

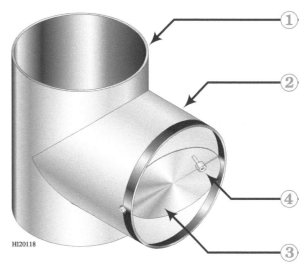

HI20118

| 1 | Vent pipe | 3 | Damper blade |
| 2 | Damper collar | 4 | Counterweight |

The barometric damper operates similarly to a draft hood in that it allows room air to flow into the vent to control draft levels. Oil boilers are usually equipped with single-acting barometric dampers — they can swing in, but not out. In order to allow adjustment of the damper for many different vent conditions, it is equipped with an adjustable counterweight. As the weight is screwed out (away from the damper blade), the damper opens more, reducing draft. As the weight is screwed in, the damper closes more, increasing draft. Adjust the damper to obtain the boiler manufacturer's recommended draft.

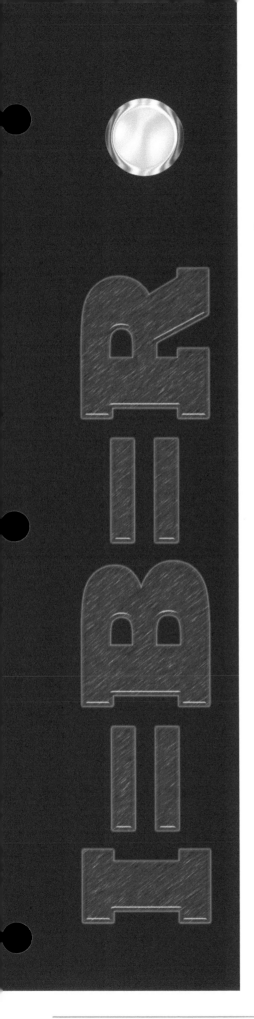

Condensate systems

Flue gas condensate installations

1. Flue gas condensate traps are used on condensing vent applications to collect condensate that accumulates in the vent piping. Follow the boiler and vent manufacturers' instructions for installation.

2. Flue gas condensate piping/tubing is required for ALL condensing boilers. The boilers are equipped with integral condensate traps, and provide a connection for condensate tubing or piping. Follow the boiler manufacturer's instructions to install the piping/tubing and any related devices.

3. Condensate pumps — When the condensate cannot flow by gravity to a suitable drain, you will need to install a condensate pump, discharged to an appropriate drain.

4. Condensate is acidic — Local codes may require use of a condensate neutralizer. The neutralizer contains a chemical filter that makes the condensate more neutral (less acidic).

> **⚠ WARNING**
>
> You must install the condensate system using the materials and methods specified in the boiler manual. Follow all applicable codes. The condensate must be discharged to a suitable drain, and must be neutralized if necessary to prevent corrosion of the drain piping. Failure to properly install the condensate system can result in severe personal injury or death.

Residential boiler examples

Overview

The following information provides simplified representations of typical residential boilers, selected as a cross section of boilers available. The boilers selected lend themselves to a generic discussion of how boilers operate, and are not intended as an endorsement of any particular design.

Residential boilers can be generally categorized as tubular (firetube, watertube, copper tube, electric immersion element, etc.) or plate style — designs where the flue gases pass between exchanger surfaces (cast iron sectional, spiral geometry, etc.). Some boiler heat exchanger designs may incorporate both tubular and plate-type configurations.

The examples discussed in this Section are: firebox, horizontal firetube, vertical firetube, cast iron sectional, copper tube, and a cast aluminum condensing boiler design. Other examples not shown include watertube boilers, stainless exchanger boilers (firetube style or stainless spiral designs, for examples) and steel spiral plate boilers.

Firetube and firebox boilers

Firetube and firebox boilers route flue products inside tubes that are surrounded by water, as in the simplified boiler illustrations of Figure 21, Page 1–39, Figure 22, Page 1–39 and Figure 23, Page 1–39. [In contrast, watertube boilers route water through the tubes, with flue gases passing around them.]

Watertube boilers

The example shown in Figure 24, Page 1–40 is a simplified model of a finned copper tube boiler, commonly used in residential boiler applications. Other watertube boiler designs may incorporate steel tubes, and tubes may be bent as well as straight.

Cast iron boilers

Figure 25, Page 1–40, Figure 26, Page 1–41 and Figure 27, Page 1–41 are simplified illustrations of typical cast iron boilers. They show dry base and wet base versions.

Condensing premix gas boiler

Figure 28, Page 1–42 shows a typical condensing boiler, an aluminum cast heat exchanger version in this example.

Figure 21   Vertical firetube boiler, simplified

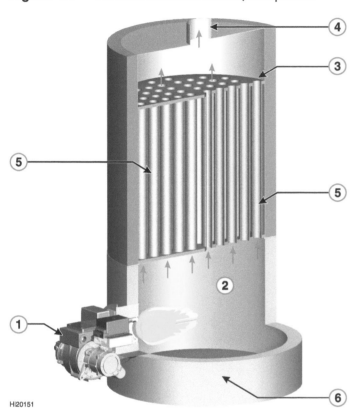

HI20151

1	Burner, oil or gas	**4**	Vent connection
2	Combustion chamber (dry-base design in this example)	**5**	Boiler tubes
3	Heat exchanger	**6**	Refractory floor and chamber

Figure 22   Horizontal firetube boiler, simplified

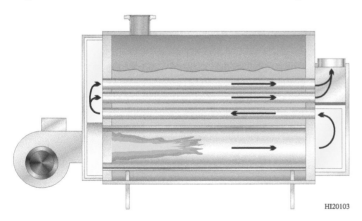

HI20103

This is a simplified 3-pass firetube boiler, dry-back turnaround (area at rear of combustion chamber where flue products turn to enter the tubes). Other horizontal firetube boilers may incorporate a water-backed rear turnaround and up to 4 passes.

Figure 23   Firebox boiler, simplified

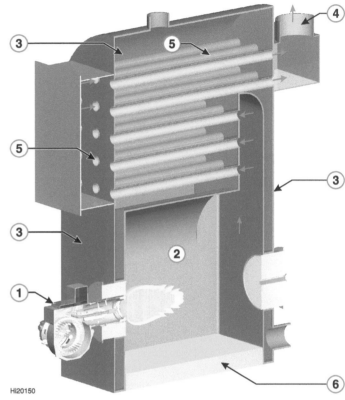

HI20150

1	Burner, oil or gas	**4**	Vent connection
2	Combustion chamber (water-backed on sides, front and rear in this example)	**5**	Boiler tubes
3	Heat exchanger (includes all water-backed surfaces)	**6**	Refractory floor

Figure 24 Copper finned-tube boiler, atmospheric gas type, simplified

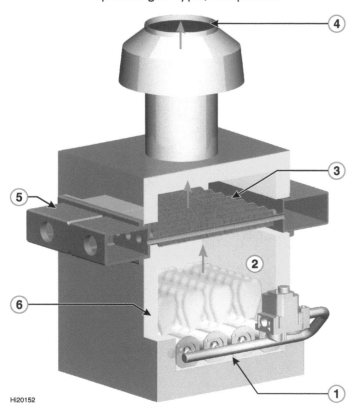

HI20152

1   Burner assembly —
    atmospheric gas with tubular
    burners in this example

2   Combustion chamber (dry-
    base design)

3   Heat exchanger, comprised
    of finned copper tubes and
    headers

4   Vent connection, with bell-
    type draft diverter for gravity
    venting

5   Tube headers

6   Refractory-lined combustion
    chamber/base

Integral-fin copper tubing, as used in this example, can have up to 7 or 8 times as much surface on the finned side (flue gas side) as on the water side (inside of tube) because of the large surface are of the fins. Other copper tube boiler designs may incorporate induced draft or forced draft combustion.

Figure 25 Cast iron sectional boiler, atmospheric gas type, dry base, simplified

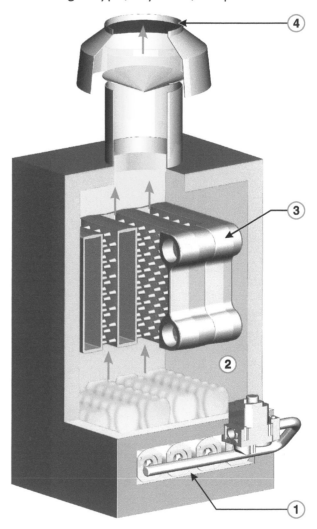

HI20153

1   Burner assembly —
    atmospheric gas with tubular
    burners in this example

2   Combustion chamber (dry-
    base design)

3   Heat exchanger, comprised
    of cast iron sections, typically
    with pins or fins on the flue
    side

4   Vent connection, with bell-
    type draft diverter for gravity
    venting

Figure 26 Cast iron sectional boiler, induced draft gas type, dry base, simplified

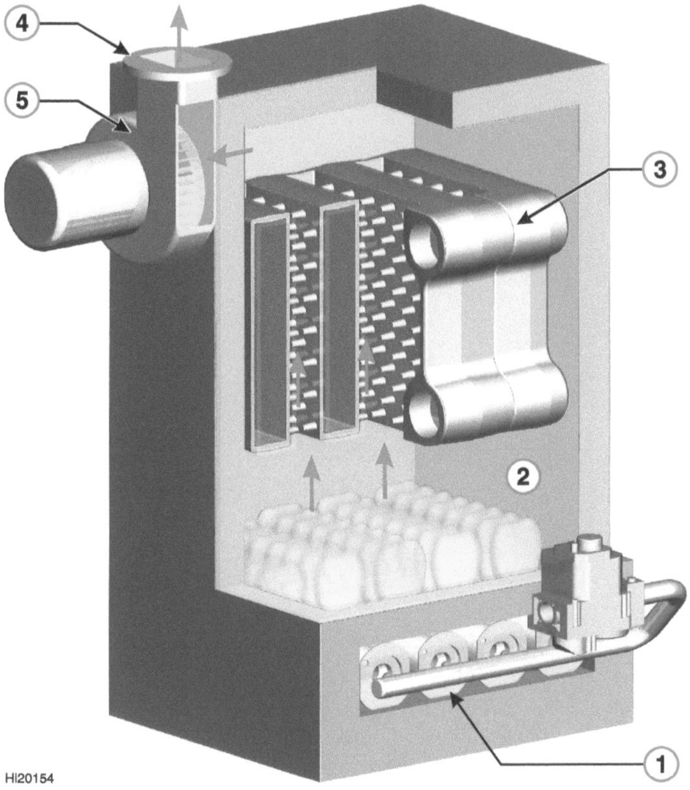

HI20154

Figure 27 Cast iron sectional boiler, forced draft gas or oil type, wet base, simplified

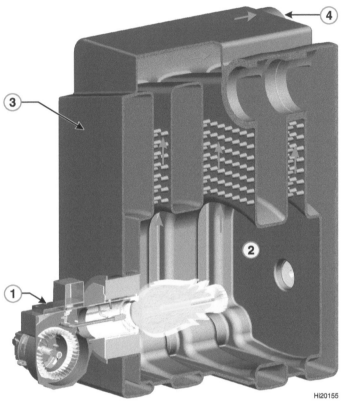

HI20155

1 Burner assembly — atmospheric gas with tubular burners in this example

2 Combustion chamber (dry-base design)

3 Heat exchanger, comprised of cast iron sections, typically with pins or fins on the flue side

4 Vent connection, off of induced draft fan — may be pressurized or gravity vented, depending on boiler manual instructions

5 Induced draft fan

1 Forced draft burner, oil or gas

2 Combustion chamber (water-backed on all surfaces in this design)

3 Heat exchanger, comprised of cast iron sections, typically with pins or fins on the flue side

4 Vent connection

Figure 28 Condensing premix gas boiler, cast aluminum heat exchanger, simplified

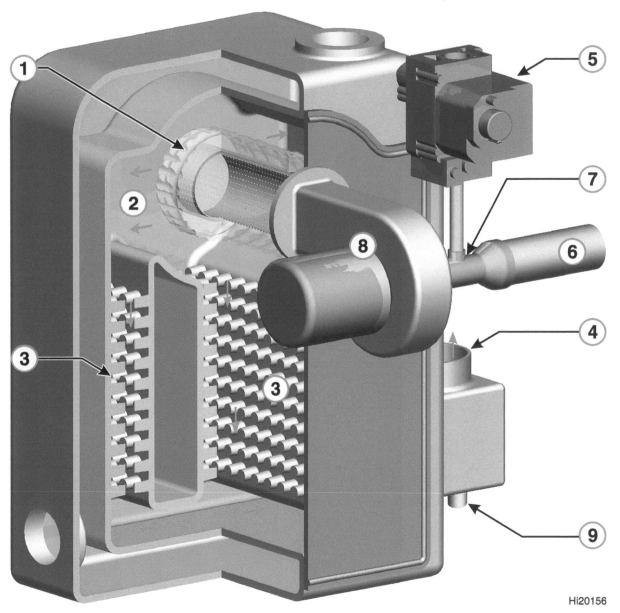

Hi20156

1 Premix gas burner — Gas and air are premixed in the blower and injected into the burner; ignition is typically done with hot surface ignitor or spark electrodes

2 Combustion chamber, all water-backed in this design

3 Heat exchanger, aluminum casting in this design; other condensing boilers may use stainless steel or other materials, and the geometry of the exchanger will vary

4 Vent connection, forced draft; requires vent piping suitable for condensing operation, as per boiler manual

5 Gas valve, zero-governor type; valve controls outlet pressure at zero inches water column — amount of gas flow depends on how much air flows through the venturi, item 7

6 Air inlet pipe — many condensing boilers will be direct vent, requiring air piping from outside to boiler, per boiler manual

7 Venturi — Air flowing through the neck of the venturi causes a negative pressure — Gas valve senses drop in pressure and allows gas to flow in proportion to the negative pressure — Gas and air combine in the venturi and pass into the premix blower, item 8

8 Premix blower — Blower injects gas/air mixture into the burner and provides force to push flue products through the boiler and the vent system

9 Condensate connection — Connect inside the boiler jacket to the boiler's internal condensate trap — Outlet of trap is fitted with a connection for attachment of condensate line by installer

Instruments

Checking combustion

Atmospheric gas boiler combustion depends on the height of the boiler draft hood and the boiler construction. On these boilers, make sure the orifice size and gas pressure are as specified by the boiler manufacturer and the draft hood and venting systems comply with the instructions. (There are usually no other adjustments available.) This is generally adequate to be sure the boiler will perform properly. To verify acceptable operation of a gas appliance, you can use a carbon monoxide tube or electronic carbon monoxide tester to sample the flue products leaving the boiler.

On oil-fired boilers, always use combustion test instruments. Because oil burners must be adjustable to allow for the wide variations in installation conditions, you have to adjust the burner on the jobsite. To be sure it is set correctly, check the smoke reading and CO_2, making sure they agree with the boiler/burner manufacturer's instructions.

Smoke spot testing

The smoke spot gun is a pneumatic pump used to pull flue samples through a special filter paper. Be sure to check the smoke spot under good lighting. Most manufacturers recommend a zero or trace (less than #1 on the Bacharach scale).

CO_2 or O_2 testing

The reason for checking the percentage of CO_2 or O_2 in the vent is to be sure the burner is getting enough, but not too much, air.

- Not enough air can mean incomplete combustion and/or sooting.
- Too much air will rob efficiency, increasing fuel usage.

When checking CO_2, make sure the reading indicates there is enough air. CO_2 can read 11% to 13% both when there is enough air and when there is not enough air. To make sure the value is good, take a reading — then open the burner air damper slightly and take another reading. If the CO_2 reading was good, the CO_2 will drop when you open the air. Otherwise, the reading will increase when you open the air. If this happens, open the air gradually until the CO_2 value is good.

You usually check O_2 levels with an electronic tester. These testers use special cells which must be replaced periodically for the instrument to work correctly. Carefully follow the tester manufacturer's instructions for maintaining and checking the tester and its sensing cells.

Appendix

Heating terminology

Energy units

In the United States, the most common energy term is the **Btu**, or British Thermal Unit. This term was originally defined as the heat needed to raise the temperature of 1 pound of water 1 degree Fahrenheit. This amount of heat depends on the starting water temperature, though. It can vary about 1%. So the term Btu has been redefined more accurately in terms of its metric equivalent, or 1054.8 joules.

Figure 29 shows the common metric energy terms. The **calorie** was originally defined like the Btu, equal to the energy required to raise the temperature of 1 gram of water 1 degree Centigrade. The calorie also has been redefined for precision. It is still the energy required to raise 1 gram of water 1 °C, but this is done at a specified average water temperature. The most common value, the 15-degree calorie, is the energy required to raise the temperature of water from 14.5 °C to 15.5 °C, equal to 4.1858 joules.

Dietary ratings use the term **Calorie**, but this is actually a kilogram-calorie — the energy required to raise the temperature of 1 kilogram (not 1 gram) of water 1 °C.

Metric energy ratings usually are given in joules. The **joule** is equal to .0009480 BTU.

Figure 29   Energy terms

Energy terms

Btu (British Thermal Unit)
Energy required to raise 1 Pound of Water 1 °F
Metric equivalent = 252 calories, or 1,055.1 joules.

calorie (metric system)
Energy required to raise 1 Gram of Water 1 °C
NOTE: The Calorie, or "large" calorie, used for dietary ratings, is actually 1,000 calories, or a "kilogram-calorie."
British equivalent = .003968 BTU

Joule (metric system)
Metric unit of work or energy, equal to 0.7377 foot-pounds or 0.00094782 Btu.

These definitions have changed, because there is about a 1% variation in these energy values, depending on the water temperature. So, the calorie is now defined at a specific temperature of water (usually the 15° calorie, to raise 1 gram of water from 14.5 °C to 15.5 °C, equal to 4.1858 joules). The BTU is now defined as about 1054.8 joules.
HI20124

Btuh (British Thermal Unit per hour)
Metric equivalent = 0.00029307 kilowatts, or 0.29307 watts.

MBH (1,000 Btuh)
Metric equivalent = 0.29307 kilowatts.

Bhp (Boiler horsepower)
English unit equivalent = 33,446 Btuh, 33.446 MBH
Metric equivalent = 9.8095 kilowatts.

hp (Brake horsepower)
English unit equivalent = 2542.5 Btu, 2.5425 MBH, 1,980,000 foot-pounds/hr
Metric equivalent = 0.7457 kilowatts.

kW (kilowatt, or 1,000 watts)
English unit equivalent = 3412.1 Btuh, 3.4121 MBH
Metric equivalent = 1,000 joules/second, 859.86 kilogram-calories/hour.

Boiler heating capacity

In the United States, boiler ratings are usually shown in **Btuh**, or Btus per hour. You will often see the term **MBH**, meaning 1,000s of Btu per hour.

Metric ratings will usually be given in kilowatts (kW). A **kilowatt** equals 1,000 joules per second, and is equivalent to 3,412.1 Btuh (Btus per hour) [One MBH = 0.29307 kW].

Commercial and industrial boilers are also rated in boiler horsepower. One boiler horsepower is equivalent to 33,446 Btuh. Don't confuse this with horsepower (or brake horsepower) used to rate motors. One horsepower equals 2542.5 Btuh, or 0.7457 kW.

Heat content and heat from combustion

Water can exist in three states — **solid** (ice), **liquid** (water) or **steam** (vapor). Water can be changed to any of these states with the right combination of temperature (and pressure).

When water is heated or cooled, causing a temperature change, the heat added or removed is called **sensible heat**. For water, the sensible heat required to change the temperature 1 °F is approximately 1 Btu per pound of water. See Figure 30.

When water changes from ice to liquid (melting), or liquid to ice (fusion), the change takes place without any change in temperature. The same is true when water changes from liquid to vapor (evaporation), or vapor to liquid (condensation). The heat that is given up or taken away during these changes is called **latent heat**. The **latent heat of fusion** for water/ice is 143.5 Btu per pound of water/ice. The **latent heat of vaporization** for water at 212 °F and 14.696 psia is 970.3 Btu per pound of water/vapor. [Note that psia refers to absolute pressure. The absolute pressure of the air around us is called atmospheric pressure. The term psig refers to what a pressure gauge would read, such as on a boiler. Gauge pressure, in psig, is the difference between absolute pressure in the boiler and the room atmospheric pressure around the gauge. Barometers read absolute pressure, psia. Pressure gauges read gauge pressure, psig.]

Figure 31 gives the heat available from combustion of common fuels. This is the amount of heat given off as the fuel's carbon and hydrogen (and trace amounts of other combustible elements) combine with oxygen in the flame.

Figure 30   Energy needed to heat water

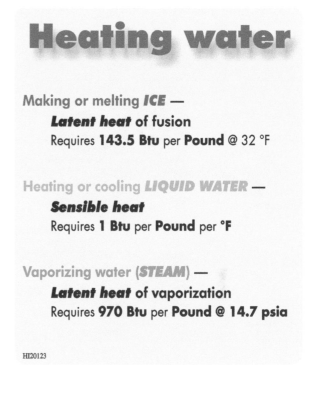

Figure 31   Energy content of fuels

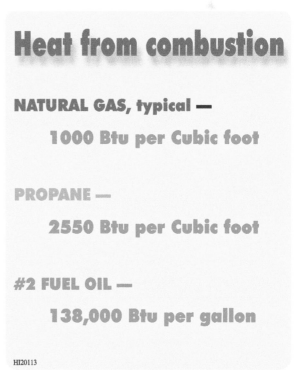

Table 1   Conversion multipliers for common English and metric units

Unit to convert —	Abbrev.	Metric units	Abbrev.	Multiply by:	English units	Abbrev.	Multiply by:
Area							
square centimeter	cm^2	square meter	m^2	0.0001	square inch	in^2	0.155
square inch	in^2	square centimeter	cm^2	6.4516	square foot	ft^2	0.0069444
square foot	ft^2	square centimeter	cm^2	929.03	square inch	in^2	144
		square meter	m^2	0.092903			
square meter	m^2	square centimeter	cm^2	10,000	square foot	ft^2	10.7639
Energy and heating capacity							
boiler horsepower	bhp	kilowatt	kW	9.802	British Thermal Unit per hour	Btuh	33,446
					1,000's of Btuh	MBH	33.446
British Thermal Unit	Btu	calorie	cal	251.9957	foot-pound	ft-lb	778.17
		joule	j	1055.056	horsepower-hour	hp-h	0.0003929
		kilowatt-hour	kW-h	0.00029307			
		watt-second	W-s	1055.056			
British Thermal Unit per hour	Btuh	calorie per second	cal	0.069999	foot-pound per hour	ft-lb/h	778.17
		watt (joule per second)	j	0.29307	horsepower (brake)	hp	0.0003929
		kilowatt	kW-h	0.00029307	boiler horsepower	bhp	.000029899
1,000's of Btuh	MBH	calorie per second	cal	69.999	foot-pound per hour	ft-lb/h	778.17
		watt (joule per second)	j	293.07	horsepower (brake)	hp	0.3929
		kilowatt	kW-h	0.29307	boiler horsepower	bhp	.029899
calorie (gram-calorie)	cal	joule	j	4.1868	British Thermal Unit	Btu	0.0039683
		watt-second	W-s	4.1868			
horsepower	hp	kilowatt	kW	0.7457	BTU per hour	Btuh	2542.5
					1,000's of Btuh	MBH	2.5425
joule	j	calorie	cal	0.23885	British Thermal Unit	Btu	0.00094782
kilowatt	kW	joule per second	j	1,000	BTU per hour	Btuh	3412.1
		calorie per second	cal/s	238.85	1,000's of Btuh	MBH	3.4121
					horsepower (brake)	hp-h	1.341
					boiler horsepower	bhp	0.10202
kilowatt-hour	kW-h	joule	j	3,600,000	British Thermal Unit	Btu	3412.1
		calorie	cal	859,860	horsepower-hour	hp-h	1.341
Length							
centimeter	cm	meter	m	0.01	inch	in	0.393701
inch	in	millimeter	mm	54.4	foot	ft	0.08333
		centimeter	cm	2.54			
		meter	m	0.0254			
foot	ft	centimeter	cm	30.48	inch	in	12
		meter	m	0.3048			
meter	m	centimeter	cm	100	inch	in	39.3701
					foot	ft	3.28084

Table 1   Conversion multipliers for common English and metric units *(continued)*

Unit to convert —	Abbrev.	Metric units	Abbrev.	Multiply by:	English units	Abbrev.	Multiply by:
Mass							
grain	*gr*	kilogram	*kg*	0.0000648	pound	*lb*	0.00014286
kilogram	*kg*				pound	*lb*	2.20462
ounce	*oz*	kilogram	*kg*	0.02835	pound	*lb*	0.0625
pound	*lb*	kilogram	*kg*	0.45359	grain	*gr*	7000
					ounce	*oz*	16
Pressure							
atmosphere	*atm*	kilogram force per square centimeter	*kgf/cm²*	1.03323	pounds per square inch	*psi*	14.696
		kilopascal	*kPa*	101.325	inches water column	*in w.c.*	407.19
					feet water column	*ft w.c.*	33.933
feet water column	*ft w.c.*	millimeters mercury (32 °F)	*mm Hg*	22.398	pounds per square inch	*psi*	0.43309
					inches mercury	*in Hg*	0.8818
inches mercury	*in Hg*	millimeters mercury (32 °F)	*mm Hg*	13.609	pounds per square inch	*psi*	0.491154
inches water column	*in w.c.*	millimeters mercury (32 °F)	*mm Hg*	1.8665	pounds per square inch	*psi*	0.036091
					inches mercury	*in Hg*	0.073483
millimeters mercury (at 32 °F)	*mm Hg*	kilogram force per square centimeter	*kgf/cm²*	0.0013595	pounds per square inch	*psi*	0.0193368
		kilopascal	*kPa*	0.13332	inches water column	*in w.c.*	0.53578
					feet water column	*ft w.c.*	0.044648
pounds per square inch	*psi*	millimeters mercury (32 °F)	*mm Hg*	51.715	inches water column	*in w.c.*	27.708
		atmosphere	*atm*	0.068046	feet water column	*ft w.c.*	2.309
		bar	*bar*	0.068948	inches mercury	*in Hg*	2.036
		kilogram force per square centimeter	*kgf/cm²*	0.07030696			
		kilopascal	*kPa*	6.8948			
Speed							
feet per minute	*fpm, f/m*	meters per second	*m/s*	18.288	miles per hour	*mph*	0.011364
					feet per second	*fps, f/s*	0.016667
feet per second	*fps, f/s*	meters per second	*m/s*	0.3048	miles per hour	*mph*	0.6818
					feet per minute	*fpm, f/m*	60
miles per hour	*mph*	kilometers per hour	*km/h*	1.6093	feet per second	*fps, f/s*	1.46667
					feet per minute	*fpm, f/m*	88
kilometers per hour	*km/h*	meters per second	*m/s*	0.27778	feet per second	*fps, f/s*	0.9113
					feet per minute	*fpm, f/m*	54.68
					miles per hour	*mph*	0.62137
Volume							
cubic foot	*ft³*	litre (= 1,000 cc)	*l*	28.3169	cubic inch	*in³*	1728
		cubic meter	*m³*	0.028317	gallon	*gal*	7.481
cubic inch	*in³*	cubic centimeter	*cc, cm³*	16.387	cubic foot	*ft³*	0.0005787
					gallon	*gal*	0.004329
gallon	*gal*	litre (= 1,000 cc)	*l*	3.7854	cubic inch	*in³*	231
		cubic meter	*m³*	0.0037854	cubic foot	*ft³*	0.13368

Hydronics Institute Section of AHRI

35 Russo Place

Berkeley Heights, NJ 07922-0218

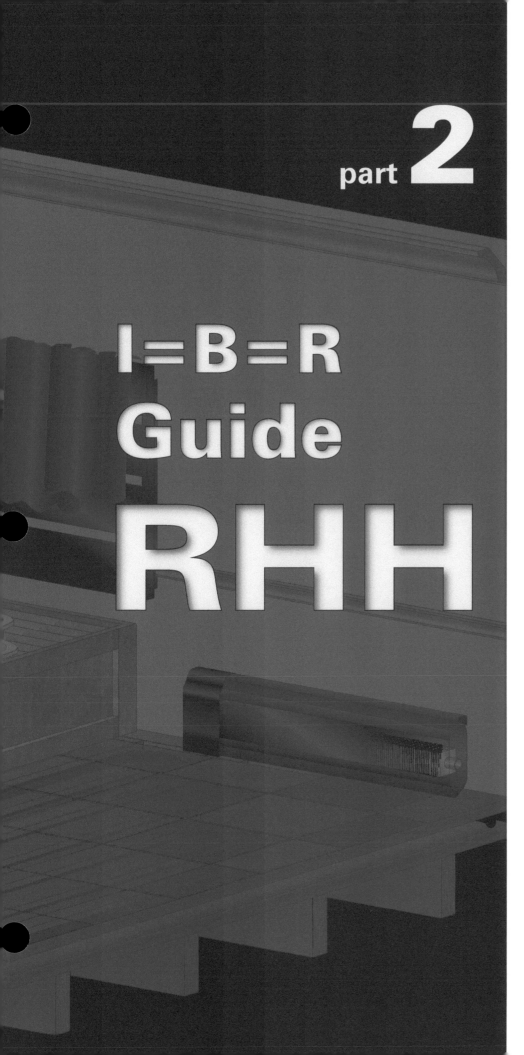

part **2**

Selecting & placing a boiler

Residential Hydronic Heating . . .

Installation & Design

Hydronics Institute
Section of **AHRI**

I=B=R Guide RHH
Residential Hydronic Heating

Hydronics Institute Section of **AHRI**
35 Russo Place
Berkeley Heights, NJ 07922-0218

Contents – Part 2

Contents – *continued*

Illustrations

Contents – continued

Tables

Hydronics

What is it?

Hydronic heating means using water or steam to carry heat to terminal units. The terminal units can be baseboard heaters, radiators, convectors, fan coil units, metal or plastic tubing or radiant panels, indirect water heaters or heat exchangers.

Water and steam are effective, versatile and efficient for carrying heat — **a single 1-inch diameter pipe can carry as much heat as a 10" x 18" rectangular duct carrying hot air** (at 130 °F). This means multiple zones and applications are easy and cost-effective to handle.

Figure 1   The versatility of hydronic heating — some possibilities

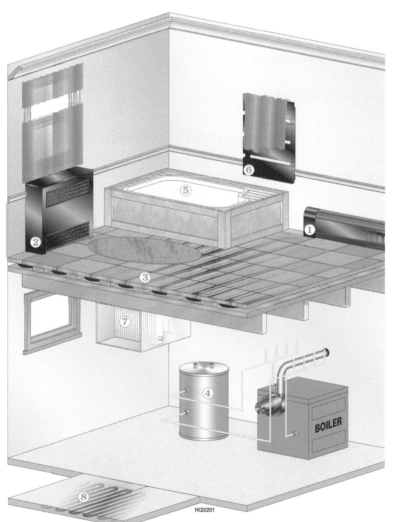

1
Baseboard heating

2
Convectors or radiators

3
Radiant floor heating

4
Domestic water heating

5
Spa and pool heating

6
Radiant panels (towel warmers)

7
Fan coil and in-duct units

8
Snow melting

HI20201

What you can do with it

A hydronic heating system can range from a simple single-zone series loop heating system to a multi-use, even multi-boiler, application. Look at Figure 1, Page 2–6, for instance, showing just a few of the many possibilities for hydronic heating:

1 Space heating with finned-tube or radiant baseboard

2 Space heating with convectors or radiators

3 Space heating with a radiant floor system (above-floor "sleeper" system shown, usable for retrofit as well as new construction)

4 Indirect water heating using a companion water heater (or tankless coil) — connected to a 100,000 Btuh input boiler, a 40-gallon water heater has a first hour rating over twice that of a typical gas-fired or electric water heater

5 Spa or pool heating (using the high recovery available from an indirect water heater)

6 Radiant panels (here shown using a towel warmer radiant panel heater)

7 Fan coil units or heating coils for air distribution systems

8 Tubing embedded in concrete or below asphalt for snow melting and/or comfort heating in garages, for example

Look at the advantages . . .

Comfortable, clean and quiet:

- *Independent zoning* — Each zone (room or group of rooms) can have its heating controlled separately, allowing fine-tuning of the heat to the space.

- *Minimum temperature fluctuation* — Heat can be matched to the space and even matched to varying heat loss.

- *Less moisture loss from the space than ducted-heat systems due to air leakage from ducts and space* — Easier humidity control.

- *Ductless* — No ducts to accumulate and distribute dust, odor and bacteria through the house. The "whole-house efficiency" of hydronic systems surpasses warm-air ducted systems — the distribution losses and uncontrolled infiltration caused by leaky/ uninsulated ducts are eliminated.

- *Quiet* — No air noise or large ducts to conduct human, burner, and vibratory noises as with ducted systems.

Economical:

- Individual room temperature control

- High efficiency plus control options for even more savings (outdoor reset, for example)

- Low transmission losses compared to ducted heating systems — small pipes lose very little heat

- Lower air infiltration and no air loss from duct

- No ducts to clean or filters to change

Easily-controlled

Piping and controls can be matched to the system needs, allowing independent control of each space.

Reliable

Hydronic heating systems have a long history of reliability, and have proven dependable in millions of homes. When designed and installed properly, service problems are rare. The life expectancy of a boiler is typically 25 years, at least 10 years longer than the typical furnace.

Flexible and expandable

- As shown in Figure 1, Page 2–6, a single boiler (or multiple-boiler system) can provide heat for a wide range of applications.

- Terminal units can be selected for the needs of the space and the client.

- Piping and control options allow matching the needs of virtually any building or user.

- Systems can easily be extended to cover new applications and new living spaces. The small pipes used to connect heating units can be run inside walls or through closets.

Space-saving installation:

- Fewer limitations in furniture arrangements — With radiant heating, nearly all of the floor space is usable. Even with finned-tube baseboard heaters, furniture can be placed next to it as long as there is enough room for airflow (though 6 inches clearance will enhance heat flow). Unlike electric baseboard, clearance from the baseboard is not a concern.

- In residential applications, the majority of these uses can be handled using small-diameter copper tubing (mostly 3/4" diameter) or PEX (polymer) tubing for distribution. No space is needed for large ductwork.

- The boiler requires minimal space — many can even be located in a crawl space.

What about air conditioning?

Comfort in a home depends mostly on the heating system for the majority of North America. Considerably more time is spent heating than cooling. But cooling is often one of the comfort needs of the homeowner.

You don't give up air conditioning when you heat hydronically. On the contrary, you can design the cooling system only for cooling. You don't have to compromise the design or layout in order to heat with the same system. Select the cooling system needed for the space, using one or a combination of:

- Through-the-wall units
- Ductless split-systems
- "High-velocity" (small duct) systems
- Valance systems (finned-tube units mounted at the ceiling)
- Separate cooling system with traditional ductwork (but registers can be located where best for cooling, with no concern for heating).

Hydronics —

No other system can match the performance of hydronics, particularly when paired with a well-designed cooling system to complete the job, providing year-round comfort.

Selecting a boiler

Summary of steps

Use the following sequence to be sure the boiler you select will meet your customer's needs and yield a profitable job for you.

Step 1   **Determine required heating capacity**

Do a heat loss calculation and determine the required boiler output or net load rating.

Step 2   **Check location/talk with homeowner**

Check the boiler installation site, customer needs and code requirements to determine which type of boiler to use.

Step 3   **Replacement boilers — Do a System Survey**

Do a complete Hydronic System Survey if this is a replacement boiler. Identify why the old boiler failed and locate any and all system problems. See Part 13 of Guide RHH for details.

Step 4   **Select piping, controls, & trim**

Select system piping method, controls, and trim.

Step 5   **Quote a complete job**

Quote the installation, including any contingencies (such as possibility of additional system repairs or having to re-vent existing appliances left on an old vent, for examples).

Step 1 — Determine heating capacity (Water only)

Preferred method

Do a room-by-room heat loss calculation on the building, whether it is for a replacement boiler or a new installation. You can be sure the equipment will be as efficient as possible, ensuring minimum fuel bills and maximum equipment life (by reducing cycling). Knowing the heat loss of each space, you can select the most effective space-heating units. Use Hydronics Institute Guide H-22 or apply one of the computer-based methods available.

Alternate method 1

Whole-house heat loss. This method will usually provide a good basis for boiler size. But it won't help in sizing space-heating units. You will find a whole-house method in Part 2 Appendix, Page 2–24 through Page 2–31.

Step 1 —
Heating capacity (Water only)

(continued)

Alternate method 2

(Not recommended) — Size the boiler to the installed heating unit capacity or existing boiler size.

- For copper baseboard, assume 600 Btuh for standard baseboard or 800 Btuh for high-capacity baseboard.

- For cast iron radiation at 180 °F average water temperature, multiply the total square feet of radiation (determined from radiator area chart) times 150 Btuh and add 15 to 25% additional (for the extra piping and pick-up allowance needed).

- You can also size the new boiler to match the output of the old boiler, but the old boiler may have been oversized (particularly if the house has had energy-saving improvements such as insulation, weather-stripping or new windows or doors).

- None of these alternate approaches is accurate, because there is no assurance the boiler, baseboard or radiation were sized correctly when installed. To be sure, avoid shortcuts. Do a heat loss calculation instead.

Select a boiler to handle the load

See the following pages for selecting the correct boiler capacity once you have a heat loss or required load.

Step 1 —
Heating capacity (Steam only)

Steam boilers have to provide the steam demanded by the connected radiation. If they can't supply enough steam, some areas will not be heated adequately. Size steam boilers based on the total square feet of EDR (Equivalent Direct Radiation) of all connected radiation.

Equivalent Direct Radiation is the method used to equate the performance of a radiator or convector to the performance of a "standard square foot of radiation," capable of delivering 240 Btuh per square foot when supplied with low pressure steam.

See Table 1, Page 2–20 through Table 5, Page 2–21 in Part 2 Appendix for the EDR of typical radiators, cast iron baseboard and convectors. See the following pages to select the required boiler capacity.

For steam systems, you can size the radiation to the heat loss of each space. But the boiler must be sized based on the total connected radiation — not on the building heat loss.

Step 2 —
Check location/talk with homeowner

Check the planned boiler location and talk to the homeowner. Select a boiler that can be installed in the available space and will meet the efficiency and cost requirements set by the owner.

Compliances

Make sure the boiler installation complies with all applicable local codes and with the installation instructions on the boiler and all system components.

Vent system and combustion air:

Codes and standards

Consider all applicable local codes and the National Fuel Gas Code. See code listings in Part 2 Appendix, Table 6, Page 2–22.

Air openings to the boiler room

Will the installation provide the minimum combustion air openings needed for a gravity-vented unit? If not, you will have to use a direct vent (sealed combustion) boiler.

Combustion air contamination

If the boiler is intended for installation in a laundry room, pool area, workshop or craft area, there may be serious risk of damage to the boiler and vent system due to chlorine and fluorine in the air. You may have to use a direct vent boiler if you cannot isolate the contaminants from the boiler room or locate the boiler in an uncontaminated space.

Some codes require isolation of combustion air

Some local codes require the boiler to be located in an isolated space unless the unit is direct vented (sealed combustion).

Masonry chimneys

If venting into an exterior masonry chimney, make sure the chimney is vented in accordance with the National Fuel Gas Code (or applicable local codes).

Existing vent systems

- Is the existing vent system in usable condition? If not, you will have to install a new vent system or purchase a boiler that can be sidewall vented.

- If another appliance is to be left on a common vent and the new boiler will not be using the common vent, you may have to revise the common vent or

Step 2 —
Check location/talk with
homeowner (continued)

install a new vent for the other appliance. See Figure 3, Page 2–15 for procedure.

- If the existing vent system is masonry or type B vent, make sure the appliance is rated for Category I venting, approved for use with type B vents and masonry chimneys.

- For boilers equipped with draft hoods or diverters — will there be enough height to install an elbow on the draft hood or diverter outlet and still provide the required rise (usually 1/4 inch per foot) connecting to the chimney entrance? If not, select another boiler with a lower height or an induced-draft or fan-assisted boiler. NEVER alter the draft hood or diverter.

Also consider:

Boiler location

- Consult homeowner and builder or architect to be sure the boiler location, heating units, piping, and venting will not interfere with plans for future space use.

Foundation/flooring

- Must the boiler be installed on a combustible floor? If so, make sure the boiler you purchase is rated for this use. Follow local codes and the boiler instruction manual carefully when installing on combustible flooring. Never install any boiler on carpeting.

- If the area or basement is prone to flooding, locate the boiler to ensure minimum risk of flooding the boiler, controls or components.

- Provide a solid, level foundation for the boiler. If necessary, provide a concrete slab, making sure to let the slab cure completely before starting the boiler.

Will you need to use multiple or modular boilers?

- If the heating load will vary greatly, consider using two boilers with a staging control. A system with a large snowmelt load (more than 25% of the boiler output) is an example. The snowmelt demand occurs for only a small time during the year.

Step 3 —
Replacement boilers —
Do a System Survey

Use the method of Part 12 of Guide RHH to completely inspect and troubleshoot the existing system. This will avoid unpleasant surprises later for both you and your customer and will virtually guarantee a profitable and successful installation.

Step 4 —
Select piping, controls and trim

Do a rough piping layout based on the heating units you plan to use. Use a piping design that will serve the needs of each connected unit.

Select the optional boiler accessories and controls needed to meet the job requirements. This will include:

- Vent system components (especially Category II, III and IV boilers)

- Circulators (for circulator zoning, primary/secondary circuits, etc.)

- Zone valves (for zone valve zoning)

- Zoning controls or relays to sequence the circulators or zone valves

- Thermostats

- Pipes, valves, and fittings

Step 5 —
Quote a complete job

Quote a complete installation. Itemize recommended system repair work if necessary to allow the customer to choose. But be sure to let the customer know everything that should be done. If the customer elects not to spend the money, at least you have informed him and shouldn't be blamed later for not advising. This will improve customer relations and will probably increase the revenue from each hydronic installation and reduce callbacks.

Boiler ratings

You may find the number of different ratings for boilers confusing. Look at any boiler brochure and you'll find data on steady-state efficiency (or combustion efficiency), AFUE, input, gross output, net load ratings, and more. So what are you supposed to use?

Once you have determined the required heating load, use the following procedures.

Net Load Rating

Determine the required boiler size from the Net Load Rating. Net Load Rating includes allowance for heat losses from piping and additional time needed to warm the piping.

Water

For water boilers, select a boiler with a Water Btuh Net Load Rating just higher than the required heating load (building heat loss plus any additional capacity needed for domestic water heating, snowmelt, etc.). The Net Load Rating includes a 15% allowance to provide for piping losses and pick-up (extra capacity to reduce time to bring system to temperature from a cold start).

Steam

For steam boilers, select a boiler with a Steam Square Foot Net Load Rating just higher than the total connected square feet of radiation. If additional load is needed for domestic water, snowmelt, etc., use the Steam Btuh Net Load Rating. Multiply total radiation square feet times 240. Then add additional Btuh needed for the extra loads.

The other Btuh rating data include Input (total Btuh energy input) and DOE Output or Heating Capacity (equals Steady-State Efficiency during DOE testing times Input) .

Efficiency

AFUE

Use efficiency data to compare cost of operation. Residential boilers with higher AFUE ratings will generally cost less to operate. AFUE is a measure of the seasonal average heat output compared to the fuel required. It accounts for off-cycle losses out the vent system as well as the losses during operation.

Steady-State Efficiency

Steady-State Efficiency is determined during DOE testing. It is equal to 100% minus the heat contained in the flue products leaving the boiler vent outlet.

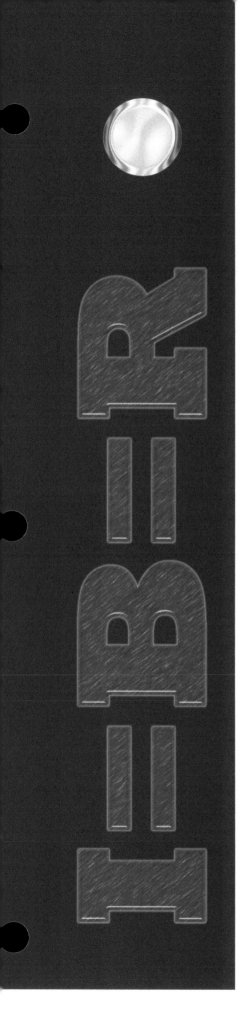

Venting & air

Boiler venting

Boilers are vented either using gravity systems (conventional chimneys or B vent) or are fan-assisted (induced draft or forced draft). Many gas boilers today require special vent systems, depending on their ANSI vent category rating.

There is no vent category system in place for oil-fired boilers. But standard NFPA 31 includes a non-mandatory appendix for use in sizing chimney liners for vent systems which experience condensing or poor draft problems.

Existing vent systems in many homes often use large cross-section chimneys or vents originally intended to meet the needs of coal-fired equipment. In many other homes, the existing vent systems have deteriorated from long use without repair or maintenance. To aggravate the problem further, today's high-efficiency appliances vent lower-temperature flue gases that may not provide enough heat to make some older vent systems operate correctly and/or prevent damaging condensation.

PRECAUTIONS:

1. **Always inspect the vent system thoroughly**. If there is any sign of deterioration, be sure the homeowner has the vent repaired. If the vent system shows ongoing condensation or poor draft, the homeowner may need to have a corrosion-resistant metal liner installed. Refer to NFPA 54 for gas boilers or NFPA 31 for oil boilers for guidance in sizing the vent or liner.

2. **When removing a gas boiler from an existing combined vent system**, perform the test prescribed in the boiler manual, as required by ANSI standard Z21.13. Figure 3, Page 2–15.

3. **Read the instruction manual carefully**. Pay close attention to the vent (and air piping, if used) installation requirements. High-efficiency appliances vary in their specific installation needs.

   • Vent materials must only be of the type and manufacturer listed in the boiler manual.

   • DO NOT mix vent piping components from different vent manufacturers without specific approval from the boiler and vent manufacturers. Each vent may have a different sealing mechanism.

   • Vent terminations must use only the specified components; terminations and air intakes must be located with the clearances and arrangements specified.

 WARNING — High-efficiency appliances require special venting, and may require air piping as well. The boiler installer must ensure that the complete installation complies with the instructions from the boiler manufacturer and other vendors used. Read and follow all instructions carefully and completely.

ANSI vent categories

The ANSI standards responded to the need for better guidance in appliance venting. The National Fuel Gas Code (ANSI Z223.1/NFPA 54) now includes an extensive section on vent sizing. And the standards for gas-fired appliances rate boilers, furnaces, etc., based on efficiency and required venting pressure. Appliances are assigned to one of four vent categories (I, II, III or IV) as explained below.

The vent category alerts the installer to any special venting requirements. Type B vent or masonry chimneys can only be used for Category I appliances, provided the boiler manufacturer specifies this in the installation instructions. Pay close attention to a boiler's vent category and venting instructions in the boiler manual.

Special vent systems are often considerably more expensive than B vent and may require special techniques for installation. Be sure to include the vent materials and installation in your quotation and to purchase all of the vent piping and components needed. Don't attempt to use any vent method other than that shown in the manufacturer's instructions. The results of an incorrect vent system could be serious.

Even if the boiler allows type B vent or masonry chimney, you should test the vent system to ensure it is functional and in good condition. If not, the vent system must be repaired or replaced, or you must install a boiler which can use a different venting method.

Boilers are classified as "Condensing" not because flue condensation occurs in the boiler, but because the flue products are likely to condense in the vent system. See Figure 2.

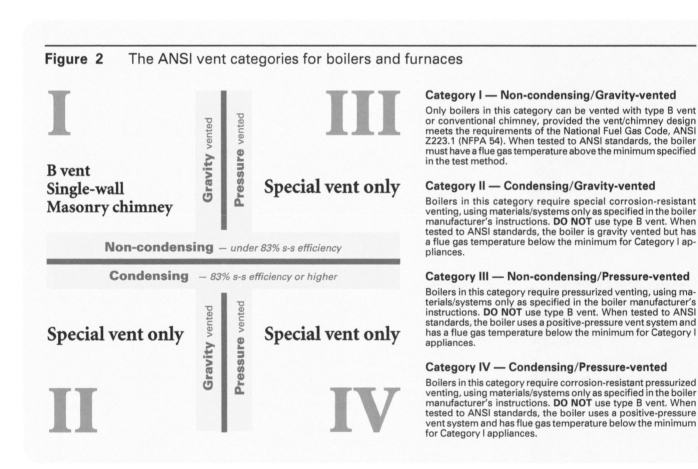

Figure 2   The ANSI vent categories for boilers and furnaces

I

B vent
Single-wall
Masonry chimney

Gravity vented | Pressure vented

III

Special vent only

Non-condensing — *under 83% s-s efficiency*

Condensing — *83% s-s efficiency or higher*

Special vent only

Gravity vented | Pressure vented

II

Special vent only

IV

Category I — Non-condensing/Gravity-vented
Only boilers in this category can be vented with type B vent or conventional chimney, provided the vent/chimney design meets the requirements of the National Fuel Gas Code, ANSI Z223.1 (NFPA 54). When tested to ANSI standards, the boiler must have a flue gas temperature above the minimum specified in the test method.

Category II — Condensing/Gravity-vented
Boilers in this category require special corrosion-resistant venting, using materials/systems only as specified in the boiler manufacturer's instructions. **DO NOT** use type B vent. When tested to ANSI standards, the boiler is gravity vented but has a flue gas temperature below the minimum for Category I appliances.

Category III — Non-condensing/Pressure-vented
Boilers in this category require pressurized venting, using materials/systems only as specified in the boiler manufacturer's instructions. **DO NOT** use type B vent. When tested to ANSI standards, the boiler uses a positive-pressure vent system and has a flue gas temperature below the minimum for Category I appliances.

Category IV — Condensing/Pressure-vented
Boilers in this category require corrosion-resistant pressurized venting, using materials/systems only as specified in the boiler manufacturer's instructions. **DO NOT** use type B vent. When tested to ANSI standards, the boiler uses a positive-pressure vent system and has flue gas temperature below the minimum for Category I appliances.

Figure 3   Procedure for ensuring proper performance of a common vent system when boiler is removed

---

When removing boiler from existing common vent system:

(from ANSI Z21.13 — required information in boiler or furnace manuals)

At the time of removal of an existing boiler, the following steps shall be followed with each appliance remaining connected to the common venting system placed in operation, while the other appliances remaining connected to the common venting system are not in operation.

a. Seal any unused openings in the common venting system.

b. Visually inspect the venting system for proper size and horizontal pitch and determine there is no blockage or restriction, leakage, corrosion or other deficiencies which could cause an unsafe condition.

c. Test vent system — Insofar as is practical, close all building doors and windows and all doors between the space in which the appliances remaining connected to the common venting system are located and other spaces of the building. Turn on clothes dryers and any appliance not connected to the common venting system. Turn on any exhaust fans, such as range hoods and bathroom exhausts, so they will operate at maximum speed. Do not operate a summer exhaust fan. Close fireplace dampers.

d. Place in operation the appliance being inspected. Follow the lighting/operating instructions. Adjust thermostat so appliance will operate continuously.

e. Test for spillage at draft hood relief opening after 5 minutes of main burner operation. Use the flame of a match or candle.

f. After it has been determined that each appliance remaining connected to the common venting system properly vents when tested as outlined above, return doors, windows, exhaust fans, fireplace dampers, and any other gas-burning appliance to their previous conditions of use.

Any improper operation of the common venting system should be corrected so the installation conforms with the National Fuel Gas Code, ANSI Z223.1 - latest edition. Correct by resizing to approach the minimum size as determined using the appropriate tables in Part 11 of that code. Canadian installations must comply with CAN/CGA B149.1 or B149.2 Installation Code.

Mechanical draft/direct vent

You should be aware of two special venting methods defined by the National Fuel Gas Code — Mechanical draft venting and Direct vent.

Mechanical draft venting

Mechanical draft venting means the vent products are forced out the vent termination under pressure. This can be by means of a forced-draft or induced-draft blower on the boiler or an induced-draft blower in the vent system. Mechanical draft venting is sometimes referred to as direct exhaust venting.

Sidewall venting of oil boilers

When sidewall venting oil boilers, using vent terminations and components from other than the boiler manufacturer, ensure that the installation complies with requirements from both the the boiler manufacturer as well as the vent component manufacturer.

Direct vent

Direct venting means all air for combustion is ducted directly to the boiler and flue gases are ducted to outside. This venting method is often called sealed combustion.

Boilers in ANSI vent category III or IV will be either mechanical draft vent or direct vent. Some boilers in category I may be designed for direct vent.

Pay close attention to boiler manual directions for installation and location of the vent termination of mechanical draft vent and direct vent boilers. The termination must be located to maintain minimum clearances from windows, doors, building air intakes and public walkways as required by the National Fuel Gas Code and as described in the boiler manual.

Combustion/ventilation air openings

For homes with loose construction (single-pane windows, no vapor-barrier and no weather-stripping, for example), air infiltration is easy and there is little concern for the air needs of gas or oil appliances. Homes are built much tighter now. New construction (and energy-conscious remodeling) uses double-pane windows, tight weather-stripping and vapor barriers. And many homes have been retrofitted to make them tighter and more energy-efficient. You have to pay very close attention to provision for combustion and ventilation air.

You can find required air opening sizes and locations in the National Fuel Gas Code (ANSI Z223.1/NFPA 54) for gas appliances or NFPA 31 for oil appliances. Also carefully read the boiler instruction manual for any special requirements. Even boilers installed as direct vent may require air openings into the boiler space for room cooling and ventilation.

Figure 5, Page 2–17 gives a summary of the ANSI standard required openings for draft-hood-equipped gas boilers and oil-fired boilers. **Read and follow the boiler instruction manual** regarding the requirements for air openings as specified by the boiler manufacturer.

The information in Figure 5 is taken from the ANSI/NFPA standards. These requirements apply to appliances located in confined spaces, applicable to most boilers. (A confined space is any space less than 50 cubic feet volume per 1,000 Btuh of all appliances in the space.)

Notice that the air opening minimum size depends on how fresh air gets to the boiler room. For homes with tight construction, with combustion/ventilation air taken from inside the home, be sure there are combustion air openings in the home to provide needed air. If this can't be provided, you must duct air to the boiler room or use a boiler which can be direct vented (uses outside air piped directly to the boiler). Some local codes require direct venting of appliances unless they are located in an isolated space.

Figure 5 shows only the arrangements for using two air openings. The National Fuel Gas Code also provides an alternative for the use of a single opening (located within 12 inches of the top of the boiler enclosure) when the installation provides certain minimum clearances from the appliance. Refer to the National Fuel Gas Code and the boiler manual for application details and opening sizing information.

Exhaust fans

For homes equipped with exhaust fans or whole-house fans, you must size the combustion air openings to the boiler space and/or home to **be sure the air openings will provide enough air for the fan air movement as well as the air required for boiler combustion, space ventilation, and draft system dilution**. If you don't, dangerous flue gas spillage could occur when the fan operates.

Using a direct-vented boiler is another, possibly surer, solution if there is a risk the boiler area could be under a negative pressure due to exhaust fan or whole-house fan operation.

Combustion air dampers

Some boiler rooms are equipped with motor-operated louvers or dampers on the outside air openings. Wire these devices so they operate on a call for heat. Wire an end switch on the damper/louver that only allows the boiler to fire after the damper/louver is fully open. If you don't do this, a dangerous lack of combustion air could occur. See Figure 4.

Figure 4   Air damper/louver interlocks

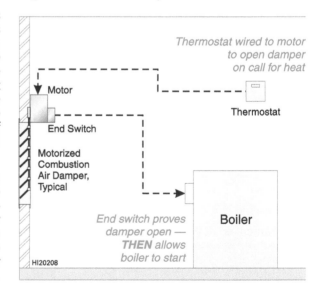

Figure 5 ANSI/NFPA standards minimum requirements for combustion and ventilation openings

Case 1 All air from inside house

(House must have adequate infiltration or be equipped with air openings to the outside sized same as those to boiler room)

Free area min. *each* opening:

1 sq. Inch *per* **1,000** Btuh

Combined input of all gas appliances

Oil appliances: 1 sq. Inch per 1,000 Btuh combined input of all appliances.

Case 2 All air directly from outdoors through ventilated attic

(Attic must be ventilated at both ends of house)

Free area min. *each* opening:

1 sq. Inch *per* **4,000** Btuh

Combined input of all gas appliances

Oil appliances: 1 sq. Inch per 4,000 Btuh combined input of all appliances.

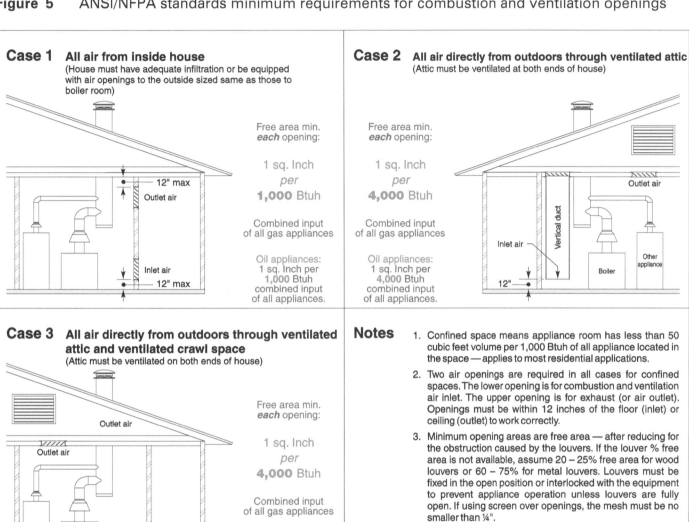

Case 3 All air directly from outdoors through ventilated attic and ventilated crawl space

(Attic must be ventilated on both ends of house)

Free area min. *each* opening:

1 sq. Inch *per* **4,000** Btuh

Combined input of all gas appliances

Oil appliances: 1 sq. Inch per 4,000 Btuh combined input of all appliances.

Notes

1. Confined space means appliance room has less than 50 cubic feet volume per 1,000 Btuh of all appliance located in the space — applies to most residential applications.

2. Two air openings are required in all cases for confined spaces. The lower opening is for combustion and ventilation air inlet. The upper opening is for exhaust (or air outlet). Openings must be within 12 inches of the floor (inlet) or ceiling (outlet) to work correctly.

3. Minimum opening areas are free area — after reducing for the obstruction caused by the louvers. If the louver % free area is not available, assume 20 – 25% free area for wood louvers or 60 – 75% for metal louvers. Louvers must be fixed in the open position or interlocked with the equipment to prevent appliance operation unless louvers are fully open. If using screen over openings, the mesh must be no smaller than ¼".

4. Obstruction — Make sure there is no blockage of the air openings by furniture, debris, or any other. Remind homeowner of this to prevent problems due to inadequate air.

Case 4 All air directly from outdoors through horizontal ducts

(Duct cross sectional area no smaller than free area of inlet)

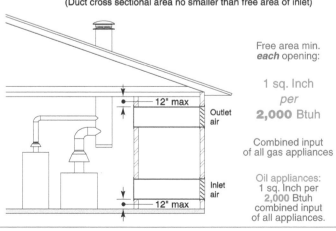

Free area min. *each* opening:

1 sq. Inch *per* **2,000** Btuh

Combined input of all gas appliances

Oil appliances: 1 sq. Inch per 2,000 Btuh combined input of all appliances.

Case 5 All air directly through outside wall

Free area min. *each* opening:

1 sq. Inch *per* **4,000** Btuh

Combined input of all gas appliances

Oil appliances: 1 sq. Inch per 4,000 Btuh combined input of all appliances.

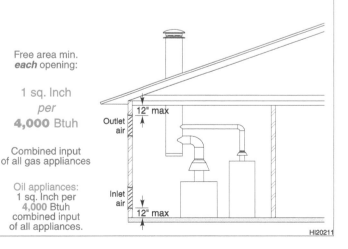

HI20211

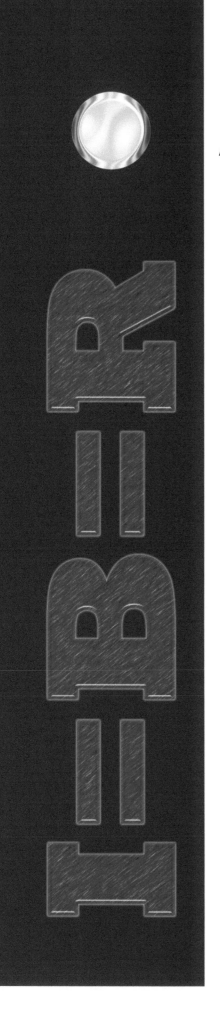

Appendix

Estimating radiator square feet EDR

Use Table 1 through Table 5 to determine the square feet EDR of radiators or convectors when manufacturer's data is not available.

Figure 6　Cast iron baseboard

HI20206

To determine EDR of radiator:

1. Measure length of baseboard
2. Read baseboard square feet (EDR) from Table 1

Table 1　Square feet EDR

Cast Iron Baseboard		
Width (inches)	**Height** (inches)	**Square feet per linear foot**
2 1/2	10	3.40

Figure 7　Radiant convector

HI20205

To determine EDR of radiator:

1. Measure depth and height of radiator
2. Read square feet (EDR) from Table 2
3. Multiply table value times the number of radiator sections

Table 2　Square feet EDR

Radiant Convector		
Depth (inches)	**Height** (inches)	**Square feet per section**
5	20	2.25
7½	20	3.40

Figure 8　Tubular radiators

HI20203

To determine EDR of radiator:

1. Count the number of tubes (illustration is 5-tube radiator)
2. Measure height of radiator, floor to top
3. Read square feet (EDR) from Table 3
4. Multiply table value times the number of radiator sections

Table 3　Square feet EDR

Tubular Radiation					
Height (inches)	3 Tube	4 Tube	5 Tube	6 Tube	7 Tube
14					2.67
17					3.25
20	1.75	2.25	2.67	3.00	3.67
23	2.00	2.50	3.00	3.50	
26	2.33	2.75	3.50	4.00	4.75
32	3.00	3.50	4.33	5.00	5.50
38	3.50	4.25	5.00	6.00	6.75

Figure 9　Columnar radiators

HI20204

To determine EDR of radiator:

1. Count the number of columns (illustration is a 3-column radiator)
2. Measure height of radiator, floor to top
3. Read square feet (EDR) from Table 4
4. Multiply table value times the number of radiator sections

Table 4　Square feet EDR

Columnar Radiation					
Height (inches)	1 column	2 column	3 column	4 column	5 column
14					4.00
17					4.00
18			2.25	3.00	5.00
20	1.50	2.00			5.00
22			3.00	4.00	6.00
23	1.67	2.33			
26	2.00	2.67	3.75	5.00	
32	2.50	3.33	4.50	6.50	
38	3.00	4.00	5.00	8.00	
44				10.00	
45		5.00	6.00		

Table 5   Convectors, copper or cast iron — estimated square feet EDR for typical units

Copper convectors
Cabinet dimensions (inches)
(Typical data — EDR varies slightly with manufacturer)

Approx. Cabinet Depth (inches)	Approx. Cabinet Length (inches)	Cabinet height (inches)					
		18	20	24	26	32	38
4	20	10.4	11.3	13.1	13.3	14.0	14.6
	24	12.8	13.9	16.1	16.4	17.2	17.9
	28	15.2	16.5	19.1	19.4	20.4	21.3
	32	17.6	19.1	22.1	22.5	23.6	24.6
	36	20.0	21.7	25.2	25.5	26.8	28.0
	40	22.4	24.3	28.2	28.6	30.0	31.3
	44	24.8	26.9	31.2	31.6	33.2	34.7
	48	27.2	29.5	34.2	34.7	36.4	38.0
	56	32.0	34.7	40.2	40.8	42.8	44.7
	64	36.8	39.9	46.3	46.9	49.2	51.4
6	20	15.3	16.3	18.4	18.8	19.7	20.6
	24	189.8	20.1	22.7	23.1	24.2	25.4
	28	22.3	23.8	26.9	27.4	28.7	30.1
	32	25.8	27.6	31.1	31.7	33.3	34.8
	36	29.3	31.3	35.4	36.0	37.8	39.6
	40	32.8	35.1	39.6	40.3	42.3	44.3
	44	36.3	38.8	43.8	44.6	46.8	49.0
	48	39.8	42.6	48.1	48.9	51.3	53.8
	56	46.8	50.1	56.5	57.5	60.4	63.2
	64	53.8	57.6	65.0	66.1	69.4	72.7
8	20	18.7	20.0	22.5	23.0	24.5	25.9
	24	22.9	24.5	27.6	28.2	30.1	31.8
	28	27.2	29.1	32.8	33.5	35.6	37.7
	32	31.4	33.6	37.9	38.7	41.2	43.6
	36	35.6	38.1	43.0	43.9	46.8	49.5
	40	39.9	42.7	48.2	49.2	52.3	55.4
	44	44.1	47.2	53.3	54.4	57.9	61.3
	48	48.4	51.8	58.4	59.6	63.5	67.2
	56	56.9	60.9	68.7	70.1	74.6	79.0
	64	65.4	70.0	78.9	80.6	85.8	90.8
10	20	20.4	22.0	25.2	25.7	27.5	29.3
	24	25.1	27.1	31.0	31.7	33.9	36.1
	28	29.8	32.2	36.8	37.7	40.3	42.9
	32	34.6	37.3	42.7	43.7	46.7	49.7
	36	39.3	42.4	48.6	49.6	53.1	56.5
	40	44.0	47.6	54.4	55.6	59.5	63.3
	44	48.8	52.7	60.2	61.6	65.9	70.2
	48	53.5	57.8	66.1	67.6	72.3	77.0
	56	63.0	68.0	77.8	79.6	85.1	90.6
	64	72.5	78.2	89.5	91.5	97.9	104.2

Cast iron convectors
Cabinet dimensions (inches)
(Typical data — EDR varies slightly with manufacturer)

Approx. Cabinet Depth (inches)	Approx. Convector Length (inches)	Cabinet height (inches)					
		18	20	24	26	32	38
4 (No. 3)	18	8.4	9.1	10.5	11.0	11.8	12.3
	23	10.9	11.8	13.5	14.2	15.2	15.9
	28	13.3	14.4	16.5	17.4	18.6	19.4
	33	15.8	17.1	19.7	20.6	22.1	23.0
	38	18.2	19.7	22.7	23.8	25.5	26.5
	43	20.6	22.3	25.7	26.9	28.9	30.1
	48	23.1	25.0	28.7	30.1	32.3	33.6
	53	25.5	27.6	31.8	33.3	35.7	37.2
	58	28.0	30.3	34.8	36.5	39.1	40.7
	63	30.5	33.0	37.9	39.7	42.5	44.3
6 (No. 5)	18	12.3	13.5	15.4	16.2	17.5	18.2
	23	15.9	17.4	19.9	20.9	22.6	23.5
	28	19.5	21.3	24.4	25.6	27.7	28.8
	33	23.1	25.2	28.9	30.4	32.9	34.1
	38	26.7	29.2	33.4	35.1	38.0	39.4
	43	30.3	33.1	37.9	39.8	43.1	44.7
	48	33.9	37.0	42.4	44.5	48.1	50.0
	53	37.5	40.9	46.8	49.2	53.3	55.3
	58	41.1	44.8	51.3	53.9	58.4	60.6
	63	44.7	48.7	55.8	58.7	63.5	65.9
8 (No. 7)	18	16.0	17.1	19.4	20.4	22.5	23.7
	23	20.8	22.2	25.0	26.4	29.1	30.6
	28	25.4	27.2	30.7	32.4	35.7	37.5
	33	30.0	32.2	36.4	38.4	42.3	44.5
	38	34.7	37.2	42.1	44.3	48.9	51.4
	43	39.6	42.3	47.8	50.3	55.5	58.4
	48	44.2	47.3	53.5	56.3	62.0	65.3
	53	48.8	52.3	59.2	62.3	68.6	72.3
	58	53.5	57.3	64.9	68.3	75.2	79.2
	63	58.1	62.3	70.6	74.3	81.8	86.1
10 (No. 9)	18	19.2	20.6	23.4	24.6	27.3	28.8
	23	24.9	26.7	30.3	31.8	35.3	37.2
	28	30.8	32.8	37.2	39.1	43.3	45.7
	33	36.2	38.9	44.2	46.3	51.4	54.2
	38	41.8	45.0	51.1	53.6	59.5	62.7
	43	47.8	51.1	58.0	60.8	67.5	71.2
	48	53.3	57.2	64.9	68.1	75.6	79.6
	53	58.0	63.3	71.8	75.4	83.6	88.1
	58	64.7	69.4	78.7	82.6	91.6	96.6
	63	70.4	75.5	85.6	89.8	99.7	105.1

Common codes and standards

Table 6   Common codes and standards applying to boiler and furnace installation

General	**Before beginning any project, find out:** • what codes apply? • who has local jurisdiction? • what are local code differences?	**Venting standards** **ANSI Z223.1/NFPA 54** • Part 7, Venting of Equipment • Part 11, Sizing of Category I Venting Systems • Appendix G, Sizing of Venting Systems Serving Appliances Equipped with Draft Hoods, Category I Appliances, and Appliances Listed for Use with Type B Vents
Oil installations	Use **NFPA 31** as the primary code order as *Standard for Installation of Oil-Burning Equipment, ANSI/NFPA 31*	
Gas installations	Use **NFPA 54** as the primary code order as *National Fuel Gas Code, ANSI Z223.1/NFPA 54*	**National Fuel Gas Code Handbook** • Various commentary on above code sections • Supplement 1, Development of Revised Venting Guidelines
Electrical installations	Use **NFPA 70** as the primary code order as *National Electrical Code, ANSI/NFPA 70*	

Ordering codes	CSA International, Inc.
	8501 Pleasant Valley Road Cleveland, Ohio 44131-5575
		NFPA 31
		• Chapter 1 General Provisions
		  1-6 Disposal of Flue Gases
		  1-7 Chimneys and Chimney Connectors
		  1-8 Special Venting Arrangements
		• Appendix F Relining Masonry Chimneys

	National Fire Protection Association One Batterymarch Park P. O. Box 9101 Quincy, MA 02269-9101	
	Underwriters Laboratories, Inc. Publication Stock 333 Pfingsten Road Northbrook, IL 60062	
BOCA	National Mechanical Code	**UL 103** • Standard for Chimneys, Factory-Built, Residential Type and Building Heating Appliances
	(International Mechanical Code)	
	International Plumbing Code	**UL 441** • Standard for Gas Vents
	Building Officials and Code Administrators International, Inc. 4501 West Flossmoor Road Country Club Hills, IL 60478-5795	
ICBO	Uniform Mechanical Code	**UL 641** • Standard for Low-Temperature Venting Systems, Type L
	(International Mechanical Code)	
	International Plumbing Code	**UL 1738** • Standard for Venting systems for Gas Burning appliances, Categories II, III and IV
	International Conference of Building Officials 5360 Workman Mill Road Whittier, CA 90601-2298	
SBCCI	Standard Mechanical Code	**UL 1777** • Chimney Liners
	(International Mechanical Code)	
	Standard Gas Code	**Combustion air standards**
	International Plumbing Code	**ANSI Z223.1/NFPA 54** • Part 5 Equipment Installation 5.3 Air for Combustion and Ventilation (Equipment Located in " Confined" and "Unconfined Spaces") ("Unusually Tight Construction")
	Southern Building Code Conference International, Inc. 900 Montclair Road Birmingham, AL 35213-1206	
IAPMO	Uniform Mechanical Code	
	Uniform Plumbing Code	**NFPA 31** • Chapter 1 General Provisions 1-5 Air for Combustion and Ventilation
	International Association of Plumbing and Mechanical Officials 20001 South Walnut Drive Walnut, CA 91789	

Figure 10 Model codes adopted by states as of December, 1999

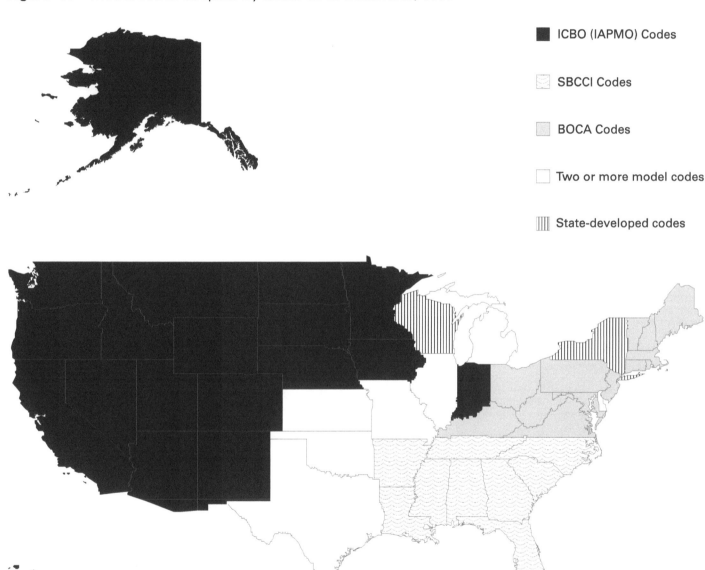

The National Fuel Gas Code (ANSI Z223.1) is used in many areas for guidance on the proper installation of gas-fired equipment.

Verify information with the state or local building department. Many areas use different codes for other construction areas.

HI20202

Estimating heat loss — whole house method

Note: This is provided for estimating purposes only. You should do a complete heat loss for the house if you plan to install or replace heating units. You will need room-by-room heat loss to size heating units or radiant heating panels.

Background

To determine how much heat is needed to maintain a residence comfortably during the heating season, find the Outdoor Design Temperature (**ODT**) for the location. The ODT is the standard design heating season outdoor temperature for the area. You will find ODT data listed by city on pages 2-26 to 2-29 and in the ASHRAE Handbook, state codes, and Hydronics Institute Guide H-22. (The ODT is usually taken from the 97½% confidence level data — a temperature low enough that it is unlikely the outdoor temperature would be colder for more than 2½% of the time during the coldest months.)

A house loses heat through walls, ceilings, floors, windows and doors that are heated on the inside and exposed to the outdoors on the other side. (Interior walls, floors and ceilings don't lose heat if both sides are heated.) These losses are called **"transmission losses"** because heat passes through the construction.

In addition to the transmission losses, you must account for cold air that enters the house by infiltration. These are called **"infiltration losses."** The amount of infiltration depends on tightness of the building envelope. Loose construction causes high infiltration. Tight construction reduces infiltration.

The total heat loss for the house is the sum:

Total heat loss = Transmission + Infiltration.

The heating equipment in the home must replace the lost heat and provide additional heat needed for piping losses and pick-up needs (extra heat needed to heat house and piping from a cold start).

The usual allowance is 15% for water boilers. The Net Load Rating for a water boiler provides this allowance. You may find you need additional allowance if the system has a large water volume (gravity system or converted steam system) or if the house will be used intermittently during the heating season. Under these circumstances, you may want to add an additional allowance of 20% to provide for quick pick-up. See Part 2 of Guide RHH for further discussion on boiler sizing.

You do not size **steam boilers** based on building heat loss. You have to size steam boilers to meet the requirements of the connected heating units. See Part 2 of Guide RHH for procedure.

If the system includes an **indirect-fired water heater**, you usually do not need to add additional capacity because the piping and pick-up allowance is normally adequate. If heavy domestic water usage is expected, see Part 6 of Guide RHH to determine the extra boiler capacity needed. (If the domestic water heating load is more than 25% of the boiler capacity, see Part 6 of Guide RHH.)

Heat loss estimate — Instructions

1. Dimensions — Measure dimensions of all heat loss surfaces (walls, windows, doors, floor, ceiling) and fill in Table 9, page 2-25.

2. Areas and volumes — Complete the calculations of Table 9 to find square feet, cubic feet, etc., of heat loss surfaces.

3. Copy areas and volumes to Table 8, below.

4. Heat loss factors — Find the factor for each heat loss surface from Table 9 and enter data in Table 8. Find the infiltration factor based on the house construction from Table 7 and enter in Table 8. (For types of construction not listed, see Hydronics Institute Guide H-22, "Heat Loss Calculation Guide".)

5. Temperature difference, TD, equals inside design temperature (usually 70°F) minus ODT. Enter this value in Table 8 and complete all calculations.

Table 8 Heat loss estimate calculation

	Multiply	x	Factor	x	TD	=	Btu per hour
Windows/doors	Sq ft	x		x	°F	=	
Net wall	Sq ft	x		x	°F	=	
Cold ceiling	Sq ft	x		x	°F	=	
Cold floor (If applicable)	Sq ft	x		x	°F	=	
Slab floor (If applicable)	Lin ft	x		x	°F	=	
Basement floor (If applicable)	Sq ft	x		x	°F	=	
Basement walls above grade (If applicable)	Sq ft	x		x	°F	=	
Basement walls below grade (If applicable)	Sq ft	x		x	°F	=	
Infiltration	Cub ft	x		x	°F	=	
TOTAL						=	

Table 7 Infiltration heat loss factors

Fully detached residence	Attached residence	Assumed air changes	Factor
New construction Very tight Extra heavy insulation	Front & back exposed Heavy insulation	½ per hour	0.009
Tight construction Heavy insulation	Front & back exposed Insulated	¾ per hour	0.011
Standard house Some insulation	Front, back, 1 side exposed Insulated	1 per hour	0.018
Older house Good condition	Front, back, 1 side exposed No insulation	1½ per hour	0.027
Loose construction Air leaks		2 per hour	0.036

Table 9   Calculations and heat loss factors for whole-house heat loss estimate

Gross cold wall (including windows and doors)

	Height	x	Length		=	Square feet
Front		x			=	
Left side		x			=	
Right side		x			=	
Rear		x			=	
				Gross wall:		

Windows & outside doors

Height	x	Width	x	Quantity	=	Square feet
	x		x		=	
	x		x		=	
	x		x		=	
	x		x		=	
		Total windows & outside doors:				

Gross Wall:	
Subtract Total windows & outside doors:	
Net Wall:	

Cold ceiling (attic or roof above)

Length	x	Width	=	Square feet
	x		=	
	x		=	
	x		=	
	x		=	
	Total cold ceiling:			

Cold floor (over cold basement or crawl space)

Length	x	Width	=	Square feet
	x		=	
	x		=	
	Total cold floor:			

— or —

Floor on concrete slab (cold edge only)

Length	+	Width	+	Length	+	Width	=	Linear Feet
	+		+		+		=	
	+		+		+		=	
			Linear ft. Total cold floor:					

Heated (finished) basement

	Width	x	Length	=	Square feet
Floor		x		=	

	Height	x	Length	=	Square feet
Wall above grade		x		=	
Wall below grade		x		=	

Construction	Factor
Glass & outside doors:	
Single	1.14
Double glass	0.65
Storm windows	0.56
Walls not insulated:	
Wood frame	0.24
Brick	0.31
Walls, insulated:	
2" Batts (R-7)	0.10
3 5/8" Batts (R-11)	0.08
6" Batts (R-19)	0.06

Construction	Factor
Uninsulated	0.30
2" Batts (R-7)	0.11
3 5/8" Batts (R-11)	0.07
6" Batts (R-19)	0.05
8" Batts (R-28)	0.03
12" Batts (R-40)	0.02

Construction	Factor
Floor over unheated basement:	
Uninsulated	0.15
2" Batts	0.05
3 5/8" Batts	0.04
— or —	
Floor over cold vented crawl space:	
Uninsulated	0.36
2" Batts	0.09
3 5/8" Batts	0.06
— or —	
Slab floor on grade (x linear ft. perimeter):	
No edge insulation	0.86
1" Edge insulation	0.69
2" Edge insulation	0.53

Construction	Factor
Floor (square feet) 0.04	
Foundation wall, uninsulated:	
Above grade	0.53
Below grade	0.06
Foundation wall, 2" insulation:	
Above grade	0.13
Below grade	0.04

Table 10   Outdoor design temperatures (AP=airport; AFB=air force base; CO=city office)

Location	ODT	Location	ODT	Location	ODT	Location	ODT
ALABAMA	ODT	Fairfield-Travis AFB	32	Leadville	-14	Moultrie	30
Alexander City	22	Fresno AP	30	Pueblo AP	0	Rome AP	22
Anniston AP	22	Hamilton AFB	32	Sterling	-2	Savannah-Travis AP	27
Auburn	22	Laguna Beach	43	Trinidad AP	3	Valdosta-Moody AFB	31
Birmingham AP	21	Livermore	27	**CONNECTICUT**	ODT	Waycross	29
Decatur	16	Lompoc, Vandenburg AFB	38	Bridgeport AP	9	**HAWAII**	ODT
Dothan AP	27	Long Beach AP	43	Hartford, Brainard Field	7	Hilo AP	62
Florence AP	21	Los Angeles AP	43	New Haven AP	7	Honolulu AP	63
Gadsden	20	Los Angeles CO	40	New London	9	Kaneohe Bay MCAS	66
Huntsville AP	16	Merced-Castle AFB	31	Norwalk	9	Wahalwa	59
Mobile AP	29	Modesto	30	Norwich	7	**IDAHO**	ODT
Mobile CO	29	Monterey	38	Waterbury	2	Boise AP	10
Selma, Craig AFB	25	Napa	32	Windsor Locks, Bradley Field	4	Burley	2
Montgomery AP	26	Needles AP	33	**DELAWARE**	ODT	Coeur D'Alene AP	-1
Talladega	22	Oakland AP	36	Dover AFB	15	Idaho Falls AP	-6
Tuscaloosa AP	23	Oceanside	43	Wilmington AP	14	Lewiston AP	6
ALASKA	ODT	Ontario	33	**DISTRICT OF COLUMBIA**	ODT	Moscow	0
Anchorage AP	-18	Oxnard	36	Andrews AFB	14	Mountain Home AFB	12
Barrow	-41	Palmdale AP	22	Washington National AP	17	Pocatello AP	-1
Fairbanks AP	-47	Palm Springs	35	**FLORIDA**	ODT	Twin Falls AP	2
Juneau AP	1	Pasadena	35	Belle Glade	44	**ILLINOIS**	ODT
Kodiak	13	Petaluma	29	Cape Kennedy AP	38	Aurora	-1
Nome AP	-27	Pomona CO	30	Daytona Beach AP	35	Belleville, Scott AFB	6
ARIZONA	ODT	Redding AP	31	Fort Lauderdale	46	Bloomington	-2
Douglas AP	31	Redlands	33	Fort Myers AP	44	Carbondale	7
Flagstaff AP	4	Richmond	36	Fort Pierce	42	Champaign/Urbana	2
Fort Huachuca AP	28	Riverside-March AFB	32	Gainesville AP	31	Chicago, Midway AP	0
Kingman AP	25	Sacramento AP	32	Jacksonville AP	32	Chicago, O'Hare AP	-4
Nogales	32	Salinas AP	32	Key West AP	57	Chicago CO	2
Phoenix AP	34	San Bernadino, Norton AFB	33	Lakeland CO	41	Danville	1
Prescott AP	9	San Diego AP	44	Miami AP	47	Decatur	2
Tucson AP	32	San Fernando	39	Miami Beach CO	48	Dixon	-2
Winslow AP	10	San Francisco AP	38	Ocala	34	Elgin	-2
Yuma AP	39	San Francisco CO	40	Orlando AP	38	Freeport	-4
ARKANSAS	ODT	San Jose AP	36	Panama City, Tyndall AFB	33	Galesburg	-2
Blytheville AFB	15	San Luis Obispo	35	Pensacola CO	29	Greenville	4
Camden	23	Santa Ana AP	39	St. Augustine	35	Joliet	0
El Dorado AP	23	Santa Barbara MAP	36	St. Petersburg	40	Kankakee	1
Fayetteville AP	12	Santa Cruz	38	Sanford	38	LaSalle/Peru	-2
Fort Smith AP	17	Santa Maria AP	33	Sarasota	42	Macomb	0
Hot Springs	23	Santa Monica CO	43	Tallahassee AP	30	Moline AP	-4
Jonesboro	15	Santa Paula	35	Tampa AP	40	Mt. Vernon	5
Little Rock AP	20	Santa Rosa	29	West Palm Beach AP	45	Peoria AP	-4
Pine Bluff AP	22	Stockton AP	30	**GEORGIA**	ODT	Quincy AP	3
Texarkana AP	23	Ukiah	29	Albany, Turner AFB	29	Rantoul, Chanute AFB	1
CALIFORNIA	ODT	Visalia	30	Americus	25	Rockford	-4
Bakersfield AP	32	Yreka	17	Athens	22	Springfield AP	2
Blythe AP	33	Yuba City	31	Atlanta AP	22	Waukegan	-3
Burbank AP	39	**COLORADO**	ODT	Augusta AP	23	**INDIANA**	ODT
Chico	30	Alamosa AP	-6	Brunswick	32	Anderson	6
Concord	27	Boulder	0	Columbus, Lawson AFB	24	Bedford	5
Covina	35	Colorado Springs AP	2	Dalton	22	Bloomington	5
Crescent City AP	33	Denver AP	1	Dublin	25	Columbus, Bakalar AFB	7
Downey	40	Durango	-1	Gainesville	21	Crawfordsville	3
El Cajon	44	Fort Collins	1	Griffin	22	Evansville AP	9
El Centro AP	38	Grand Junction AP	7	La Grange	23	Fort Wayne AP	1
Escondido	41	Greeley	4	Macon AP	25	Goshen AP	1
Eureaka/Arcata AP	33	LaJunta AP	3	Marietta, Dobbins AFB	21	Hobart	2

Table 10 Outdoor design temperatures (AP=airport; AFB=air force base; CO=city office) (continued)

City	ODT	City	ODT	City	ODT	City	ODT
Huntington	1	Madisonville	10	Marquette CO	-8	Sedalia, Whiteman AFB	4
Indianapolis AP	2	Owensboro	10	Mt. Pleasant	4	Sikeston	15
Jeffersonville	10	Paducah	12	Muskegon AP	6	Springfield AP	9
Kokomo	0	**LOUISIANA**	**ODT**	Pontiac	4	**MONTANA**	**ODT**
Layfayette	3	Alexandria AP	27	Port Huron	4	Billings AP	-10
LaPorte	3	Baton Rouge AP	29	Saginaw AP	4	Bozeman	-14
Marion	0	Bogalusa	28	Sault Ste. Marie AP	-8	Butte AP	-17
Muncie	2	Houma	35	Traverse City AP	1	Cut Bank AP	-20
Peru, Bunker Hill AFB	-1	Lafayette AP	30	Yipsilanti	5	Glasgow AP	-18
Richmond AP	2	Lake Charles AP	31	**MINNESOTA**	**ODT**	Glasgow AP	-18
Shelbyville	3	Minden	25	Albert Lea	-12	Glenclive	-13
South Bend AP	1	Monroe AP	25	Alexandria AP	-16	Great Falls AP	-15
Terre Haute AP	4	Natchitoches	26	Bemidji AP	-26	Havre	-11
Valparaise	3	New Orleans AP	33	Brainerd	-16	Helena AP	-16
Vincennes	6	Shreveport AP	25	Duluth AP	-16	Kalispell AP	-7
IOWA	**ODT**	**MAINE**	**ODT**	Fairbault	-12	Lewiston AP	-16
Ames	-6	Augusta AP	-3	Fergus Falls	-17	Livingston AP	-14
Burlington AP	-3	Bangor, Dow AFB	-6	International Falls AP	-25	**NEBRASKA**	**ODT**
Cedar Rapids AP	-5	Caribou AP	-13	Mankato	-12	Beatrice	-2
Clinton	-3	Lewiston	-2	Minneapolis/St. Paul AP	-12	Chadron AP	-3
Council Bluffs	-3	Millinocket AP	-9	Rochester AP	-12	Columbus	-2
Des Moines AP	-5	Portland	-1	St. Cloud AP	-11	Fremont	-2
Debuque	-7	Waterville	-4	Virginia	-21	Grand Island AP	-3
Fort Dodge	-7	**MARYLAND**	**ODT**	Wilmar	-11	Hastings	-3
Iowa City	-6	Baltimore AP	13	Winona	-10	Kearney	-4
Keokuk	0	Baltimore CO	17	**MISSISSIPPI**	**ODT**	Lincoln CO	-2
Marshalltown	-7	Cumberland	10	Biloxi, Keesler AFB	31	McCook	-2
Mason City AP	-11	Frederick AP	12	Clarksdale	19	Norfolk	-4
Newton	-5	Hagerstown	12	Columbus AFB	20	North Platte AP	-4
Ottumwa AP	-4	Salisbury	16	Greenville AFB	20	Omaha AP	-3
Sioux City AP	-7	**MASSACHUSETTS**	**ODT**	Greenwood	20	Scottsbluff AP	-3
Waterloo	-10	Boston AP	9	Hattiesburg	27	Sidney AP	-3
KANSAS	**ODT**	Clinton	2	Jackson AP	25	**NEVADA**	**ODT**
Atchison	2	Fall River	9	Laurel	27	Carson City	9
Chanute AP	7	Framingham	7	McComb AP	26	Elko AP	-2
Dodge City AP	5	Gloucester	5	Meridian AP	23	Ely AP	-4
El Dorado	7	Greenfield	7	Natchez	27	Las Vegas AP	28
Emporia	5	Lawrence	0	Tupelo	19	Lovelock AP	12
Garden City AP	4	Lowell	1	Vicksburg CO	26	Reno AP	10
Goodland AP	0	New Bedford	9			Reno CO	11
Great Bend	4	Pittsfield AP	-3			Tonapah AP	10
Hutchinson AP	8	Springfield, Westover AFB	0	**MISSOURI**	**ODT**	Winnemucca AP	3
Liberal	7	Tauton	9	Cape Girardeau	13		
Manhattan, Fort Riley	3	Worcester AP	4	Columbia AP	4		
Parsons	9	**MICHIGAN**	**ODT**	Farmington AP	8		
Russel AP	4	Adrian	3	Hannibal	3		
Salina	5	Alpena AP	-6	Jefferson City	7		
Topeka AP	4	Battle Creek AP	5	Joplin AP	10	**NEW HAMPSHIRE**	**ODT**
Parsons	9	Benton Harbor AP	5	Kansas City AP	6	Berlin	-9
KENTUCKY	**ODT**	Detroit	6	Kirksville AP	0	Claremont	-4
Ashland	10	Escanaba	-7	Mexico	4	Concord AP	-3
Bowling Green AP	10	Flint AP	1	Moberly	3	Keene	-7
Corbin AP	9	Grand Rapids AP	5	Poplar Bluff	16	Laconia	-5
Covington AP	6	Holland	6	Rolla	9	Manchester, Grenier AFB	-3
Hopkinsville, Campbell AFB	10	Jackson AP	5	St. Joseph AP	2	Portsmouth, Pease AFM	2
Lexington AP	8	Kalamazoo	5	St. Louis AP	6	**NEW JERSEY**	**ODT**
Louisville AP	10	Lansing AP	1	St. Louis CO	8	Atlantic City CO	13

Table 10 Outdoor design temperatures (AP=airport; AFB=air force base; CO=city office) (continued)

Location	°F	Location	°F	Location	°F	Location	°F
Long Branch	13	Syracuse AP	2	Warren	5	Williamsport AP	7
Newark AP	14	Utica	-6	Wooster	6	York	12
New Brunswick	10	Watertown	-6	Youngstown AP	4	**RHODE ISLAND**	**ODT**
Paterson	10	**NORTH CAROLINA**	**ODT**	Zanesville AP	7	Newport	9
Phillipsburg	6	Ashville AP	14	**OKLAHOMA**	**ODT**	Providence AP	9
Trenton CO	14	Charlotte AP	22	Ada	14	**SOUTH CAROLINA**	**ODT**
Vineland	11	Durham	20	Altus AFB	16	Anderson	23
NEW MEXICO	**ODT**	Elizabeth City AP	19	Ardmore	17	Charleston AFB	27
Alamagordo, Holloman AFB	19	Fayetteville, Pope AFB	20	Bartlesville	10	Charleston CO	28
Albuquerque AP	16	Goldsboro, Seymour-Johnson AFM	21	Chickasha	14	Columbia AP	24
Artesia	19	Greensboro AP	18	Enid-Vance AFB	13	Florence AP	25
Carlsbad AP	19	Greenville	21	Lawton AP	16	GeOrgetown	26
Clovis AP	13	Henderson	15	McAlester	19	Greenville AP	22
Farmington AP	6	Hickory	18	Muskogee AP	15	Greenwood	22
Gallup	5	Jacksonville	24	Norman	13	Orangeburg	24
Grants	4	Lumberton	21	Oklahoma City AP	13	Rock Hill	23
Hobbs AP	18	New Bern AP	24	Ponca City	9	Spartanburg AP	22
Las Cruces	20	Raleigh/Durham AP	20	Seminole	15	Sumter-Shaw AFB	25
Los Alamos	9	Rocky Mount	21	Stillwater	13	**SOUTH DAKOTA**	**ODT**
Raton AP	1	Wilmington AP	26	Tulsa AP	13	Aberdeen AP	-15
Roswell, Walker AFB	18	Winston-Salem AP	20	Woodward	10	Brookings	-13
Santa Fe CO	10	**NORTH DAKOTA**	**ODT**	**OREGON**	**ODT**	Huron AP	-14
Silver City AP	10	Bismark AP	-19	Albany	22	Mitchel	-10
Socorro AP	17	Devil's Lake	-21	Astoria AP	29	Pierre AP	-10
Tucumcari AP	13	Dickinson AP	-17	Baker AP	6	Rapid City AP	-7
NEW YORK	**ODT**	Fargo AP	-18	Bend	4	Sioux Falls AP	-11
Albany AP	-1	Grands Forks AP	-22	Corvallis	22	Watertown AP	-15
Albany CO	1	Jamestown AP	-18	Eugene AP	22	Yankton	-7
Auburn	2	Minot AP	-20	Grants Pass	24	**TENNESSEE**	**ODT**
Batavia	5	Williston	-21	Klamath Falls AP	9	Athens	18
Binghamton AP	1	**OHIO**	**ODT**	Medford AP	23	Bristol-Tri City AP	14
Buffalo AP	6	Akron-Canton AP	6	Pendleton AP	5	Chattanooga AP	18
Cortland	0	Ashtabula	9	Portland AP	23	Clarksville	12
Dunkirk	9	Athens	6	Portland CO	24	Columbia	15
Elmira AP	1	Bowling Green	2	Roseburg AP	23	Dyersburg	15
Geneva	2	Cambridge	7	Salem AP	23	Greenville	16
Glen Falls	-5	Chilicothe	6	The Dallas	19	Jackson AP	16
Gloversville	-2	Cincinnati CO	6	**PENNSYLVANIA**	**ODT**	Knoxville AP	19
Hornell	0	Cleveland AP	5	Allentown AP	9	Memphis AP	18
Ithaca	0	Columbus AP	5	Altoona CO	5	Murfreesboro	14
Jamestown	3	Dayton AP	4	Butler	6	Nashville AP	14
Kingston	2	Def lance	4	Chambersburg	8	Tullahoma	13
Lockport	7	Findlay AP	3	Erie AP	9	**TEXAS**	**ODT**
Massena AP	-8	Fremont	1	Harrisburg AP	11	Abilene AP	20
Newburg-Stewart AFB	4	Hamilton	5	Johnstown	2	Alice AP	34
NYC-Central Park	15	Lancaster	5	Lancaster	8	Amarillo AP	11
NYC-Kennedy AP	15	Lima	4	Meadville	4	Austin AP	28
NYC-La Guardia AP	15	Mansfield AP	5	New Castle	7	Bay City	33
Niagra Falls ANYC-LaP	7	Marion	5	Philadelphia AP	14	Beaumont	31
Olean	2	Middletown	5	Pittsburgh AP	5	Beeville	33
Oneonta	-4	Newark	5	Pittsburgh GO	7	Big Springs AP	20
Oswego CO	7	Norwalk	1	Reading CO	13	Brownsville AP	39
Plattsburg AFB	-8	Portsmouth	10	Scranton/Wilkes-Barre	5	Brownwood	22
Poughkeepsie	6	Sandusky GO	6	State College	7	Bryan AP	29
Rochester AP	5	Springfield	3	Sunbury	7	Corpus Christ! AP	35
Rome-Griffiss AFB	-5	Stubenville	5	Uniontown	9	Corsicana	25
Schenectady	1	Toledo AP	1	Warren	4	Dallas AP	22
Suffolk County AFB	10			West Chester	13	Del Rio, Laughlin AFB	31

Table 10 Outdoor design temperatures (AP=airport; AFB=air force base; CO=city office) (continued)

Location	°F	Location	°F	Location	°F	Location	°F
Denton	22	Staunton	16	Rock Springs AP	-3	Truro CO	-5
Eagle Pass	32	Winchester	10	Sheridan AP	-8	Yarmouth AP	9
El Paso AP	24	**WASHINGTON**	**ODT**	Torrington	-8	**ONTARIO**	**ODT**
Fort Worth AP	22	Aberdeen	28	**CANADA**		Belleville	-7
Galveston AP	36	Bellingham AP	15	**ALBERTA**	**ODT**	Chatham	3
Greenville	22	Bremerton	25	Calgary AP	-23	Cornwall	-9
Harlingen	39	Ellensburg AP	6	Edmonton AP	-25	Hamilton	1
Houston AP	32	Everett-Paine AFB	25	Grande Prairie AP	-33	Kaupuskasing AP	-28
Houston CO	33	Kennewick	11	Jasper	-26	Kenora AP	-28
Huntsville	27	Longview	24	Lethbridge AP	-22	Kingston	-7
Killeen-Gray AFB	25	Moses Lake, Larson AFB	7	McMurray AP	-38	Kitchener	-2
Lamesa	17	Olympia AP	22	Medicine Hat AP	-24	London AP	0
Laredo AFB	36	Port Angeles	27	Red Deer AP	-26	North Bay AP	-18
Longview	24	Seattle-Boeing Fld	26	**BRITISH COLUMBIA**	**ODT**	Oshawa	-3
Lubbock AP	15	Seattle CO	27	Dawson Creek	-33	Ottawa AP	-13
Lufkin AP	29	Seattle-Tacoma AP	26	Fort Nelson AP	-40	Owen Sound	-2
McAllen	39	Spokane AP	2	Kamloops CO	-15	Peterborough	-9
Midland AP	21	Tacoma-McChord AFB	24	Nanaimo	20	St. Catharines	3
Palestine CO	27	Walla Walla AP	7	New Westminister	18	Sarnia	3
Pampa	12	Wenatchee	11	Penticton AP	4	Sault Ste. Marie AP	-13
Pecos	21	Yakima AP	5	Prince George AP	-28	Sudbury AP	-19
Plainview	13	**WEST VIRGINIA**	**ODT**	Prince Rupert CO	2	Thunder Bay AP	-24
Port Arthur AP	31	Beckley	4	Trail	0	Timmins AP	-29
San Angelo, Goodfellow AFB	22	Bluefield AP	4	Vancouver AP	19	Toronto AP	-1
San Antonio AP	30	Charleston AP	11	Victoria CO	23	Windsor AP	4
Sherman-Perrin AFB	20	Clarksburg	10	**MANITOBA**	**ODT**	**PRINCE EDWARD ISLAND**	**ODT**
Snyder	18	Elkins AP	6	Brandon	-27	Charlottetown AP	-4
Temple	27	Huntington CO	10	Churchill AP	-39	Summerside AP	-4
Tyler AP	24	Martinsburg AP	10	Dauphin AP	-28	**QUEBEC**	**ODT**
Vernon	17	Morgantown AP	8	Flin Flon	-37	Bagotville AP	-23
Victoria AP	32	Parkersburg CO,	11	Portage [a Prairie AP	-24	Chicoutimi	-22
Waco AP	26	Wheeling	5	The Pas AP	-33	Drummondville	-14
Wichita Falls AP	18	**WISCONSIN**	**ODT**	Winnipeg AP	-27	Granby	-14
UTAH	**ODT**	Appleton	-9	**NEW BRUNSWICK**	**ODT**	Hull	-14
Cedar City AP	5	Ashland	-16	Campbellton CO	-14	Megantic AP	-16
Logan	2	Beloit	-3	Chatham AP	-10	Montreal AP	-10
Moab	11	Eau Claire AP	-11	Edmundston CO	-16	Quebec AP	-14
Ogden AP	5	Fond du Lac	-8	Fredericton AP	-11	Rimouski	-12
Price	5	Green Bay AP	-9	Moncton AP	-8	St. Jean	-11
Provo	6	LaCrosse AP	-9	Saint John AP	-8	St. Jeirome	-13
Richfield	5	Madison AP	-7	**NEWFOUNDLAND**	**ODT**	Sept. Iles AP	-21
St. George CO	21	Manitowoc	-7	Corner Brook	0	Shawinigan	-14
Salt Lake City AP	8	Marinette	-11	Gander AP	-1	Sherbrooke CO	-21
Vernal AP	0	Milwaukee AP	-4	Goose Bay AP	-24	Thetford Mines	-14
VERMONT	**ODT**	Racine	-2	St. John's AP	7	Trois Rivieres	-13
Barre	-11	Sheboygan	-6	Stephenville AP	4	Val d'Or AP	-27
Burlington AP	-7	Stevens Point	-11	**NORTHWEST TERR.**	**ODT**	Valleyfield	-10
Rutland	-8	Waukesha	-5	Fort Smith AP	-45	**SASKATCHEWAN**	**ODT**
VIRGINIA	**ODT**	Wausau AP	-12	Frobisher AP	-41	Estevan AP	-25
Charlottesville	18	**WYOMING**	**ODT**	Inuvik	-53	Moose Jaw AP	-25
Danville AP	16	Casper AP	-5	Resolute AP	-47	North Battleford AP	-30
Fredericksburg	14	Cheyene AP	-1	Yellowknife AP	-46	Prince Albert AP	-35
Harrisonburg	16	Cody AP	-13	**NOVA SCOTIA**	**ODT**	Regina AP	-29
Lynchburg AP	16	Evanston	-3	Amherst	-6	Saskatoon AP	-31
Norfolk AP	22	Lander AP	-11	Halifax AP	5	Swift Current AP	-25
Petersburg	17	Laramie AP	-6	Kentville	1	Yorkton AP	-30
Richmond AP	17	Newcastle	-12	New Glasgow	-5	**YUKON TERRITORY**	**ODT**
Roanoke AP	16	Rawlins	-4	Sydney AP	3	Whitehorse AP	-43

Definitions

Table 11   Definitions

Annual fuel utilization efficiency (AFUE)	determined for boilers using U. S. Department of Energy testing procedures, compliant with the National Appliance Energy Conservation Act (NAECA); allows comparison of whole-season performance of a boiler; equals 100% minus the losses from the boiler vent both during ON and OFF cycles and minus the losses due to air infiltration needed to replace air used for combustion and draft regulation
Classification of gas-fired boilers	Category I — a boiler that operates with a negative vent pressure and a steady-state efficiency less than 83% — (usually approved for use with Type B or masonry vent)
	Category II — a boiler that operates with a negative vent pressure and a steady-state efficiency of 83% or greater — (special venting required; cannot use Type B vent)
	Category III — a boiler that operates with a positive vent pressure and a steady-state efficiency of less than 83% — (special venting required; cannot use Type B vent)
	Category IV — a boiler that operates with a positive vent pressure and a steady-state efficiency of 83% or greater — (special venting required; cannot use Type B vent)
Commercial boiler	per U. S. Department of Energy, compliant with the Energy Policy Act of 1992 — any boiler with an input 300,000 Btuh or greater
Conduction	heat flow through solid materials
Confined space	from National Fuel Gas Code (ANSI Z223.1/ NFPA 54) — any space whose volume is less than 50 cubic feet per 1,000 Btuh input of all appliances in the space; an unconfined space is one with 50 cubic feet or more per 1,000 Btu
Convection	heat transferred by the movement of a fluid (water, air, etc.); natural convection is due to movement of fluids caused by density difference (warmer air rises or cooler air falls, for example); forced convection occurs when fluid is moved by another means, such as a pump or blower
Damper	an electromechanical device installed after the boiler draft hood or diverter that closes off the vent after the operating cycle
Degree day	used for estimating energy consumption; a day is a 24-hour period and the number of degrees equals 65 °F minus the average temperature for the 24-hour period; (example: for average outdoor temperature of 25 °F, the degree days for that day would be 65 − 25 = 40 degree days); the total degree days for a given period is the sum of all days during the period
Design temperature	the temperature required in a heated or cooled space (Indoor design temperature, or IDT) or outdoor conditions to operate against (Outdoor design temperature, or ODT) for the most extreme conditions considered necessary for the region and installation
Design temperature difference	the difference between the Indoor and Outdoor design temperatures = IDT − ODT
Direct vent	venting system in which the air for an appliance is ducted from outside directly to the appliance in a sealed pipe; the vent is sealed as well; no inside air is required for combustion or operation of the appliance
Equivalent direct radiation (EDR)	the equivalent heating surface of a radiator, convector or baseboard unit; each square foot EDR is rated 240 Btuh for steam or 150 Btuh for water (at 170 °F average)
Flow rate	volume of fluid moved per period of time; gpm (gallons per minute) or cfm (cubic feet per minute), for example
Flow velocity	speed of a fluid inside a pipe; fps (feet per second), for example
Flue damper	an electromechanical device installed between the boiler and its draft hood or diverter that closes off the vent after the operating cycle
Forced draft	operating condition in which air is blown into the combustion chamber and the chamber is maintained at a positive pressure
Gross output	commercial boilers — equals the total Btuh delivered to the water or steam at the boiler flow outlet
Head	mechanical energy a pump adds to a fluid
Head loss	mechanical energy lost by fluid due to friction in pipes and components
Heat	thermal energy, detected by temperature
Heat loss factor (HLF)	the quantity of heat (in Btu per hour) flowing through 1 square foot of wall, ceiling, floor, etc., for each 1 degree Fahrenheit difference between the outside air temperature and the inside air temperature; also known as the "U" factor, the coefficient of transmission
Induced draft	operating condition in which air is pulled into the combustion chamber by a blower located at the vent outlet of the boiler
Infiltration	any uncontrolled air leakage into a building
Latent heat	heat required to cause a material to change from vapor to liquid, liquid to vapor, solid to liquid, or liquid to solid

Table 11   Definitions (continued)

Term	Definition
Net load rating	the recommended maximum heating load (or radiation load) for a boiler; determined by dividing the boiler gross output by one plus the piping and pickup factor; i.e., divide by 1.15 for water boilers, or by 1.288 to 1.33 for steam boilers, depending on size; given in boiler ratings as Btuh water, Btuh steam, and Square feet steam
Overall efficiency	commercial boilers — equals the measure Gross output divided by the boiler input, at steady-state conditions
Oxygen permeation	the ability of oxygen to pass through a material when the oxygen concentration on one side of the material is higher than on the other
Pressure drop	a decrease in pressure as head is removed from a fluid
Perm	unit of measurement of permeability of a material; equal to 1 grain per square foot per hour per inch mercury pressure difference
Piping and pickup factor	allowance added to the load of a building to account for the heat loss from the piping and the extra energy needed to heat up the piping when starting a heating cycle; I=B=R factor for water boilers is 15%; for steam boilers, factor ranges from 28.8% to 33% depending on boiler size; see Net load rating
Radiation	heat transmitted by light (usually infrared)
Radiant barrier	in building construction, a surface which reflects long wavelength radiant energy; typically a composite of aluminum laminated to a plastic film, available attached to the face of fiberglass insulation or as separate rolls
Residential boiler	per U. S. Department of Energy, compliant with the National Appliance Energy Conservation Act — any boiler with an input less than 300,000 Btuh
Sealed combustion	see Direct vent
Sensible heat	heat added or removed that causes a change in temperature
Sidewall venting	vent system in which the vent ejects directly out the side wall, with not vertical vent system required
Specific heat (c)	a measure of the ability of a material to hold heat; commonly given in Btu per pound per degree F; water specific heat is approximately 1
Static pressure	pressure at a point in a fluid due to the weight of the fluid column above the point plus any additional pressure applied to the fluid; the pressure you would read on a pressure gauge located at the point
Steady state efficiency	equals 100% minus the energy contained in the flue products leaving the appliance
Temperature	a means of measuring heat content
Thermal conductance (C)	a number that indicates the ability of a material to allow heat by conduction
Thermal conductivity (k)	the number of Btuh that will pass through 1 square foot of material 1 inch thick for each 1 °F difference in temperature from one surface of the material to the other
Thermal mass	the heat stored in a material due to its mass and specific heat
Thermal resistance (R)	the inverse of the "U" value (R = 1/U); measures the resistance of a material to heat transfer; the greater the R-factor of a material or construction, the lower the heat loss through it
Thermal resistivity (r)	the inverse of thermal conductivity, k (r = 1/k); measure of the ability of a material to resist heat transfer through it
Transmission	in heat load calculations, refers to general heat movement due to conduction, convection or radiation or any combination
Transmission heat loss	amount of heat moving from inside to outside due to the temperature difference; commonly given in Btuh
U-factor	or **Overall coefficient of heat transfer**; a measure of the tendency of a material or construction (wall, ceiling, floor, etc.) to allow heat to move through; U-factor equals the number of Btuh that will pass through a material on construction for each square foot of area and each degree Fahrenheit difference in temperature
Vapor barrier	a material that retards the transmission of water vapor; permeability must be no more than 1 perm
Vent damper	an electromechanical device installed after the boiler draft hood or diverter that closes off the vent after the operating cycle

Hydronics Institute Section of AHRI

35 Russo Place

Berkeley Heights, NJ 07922-0218

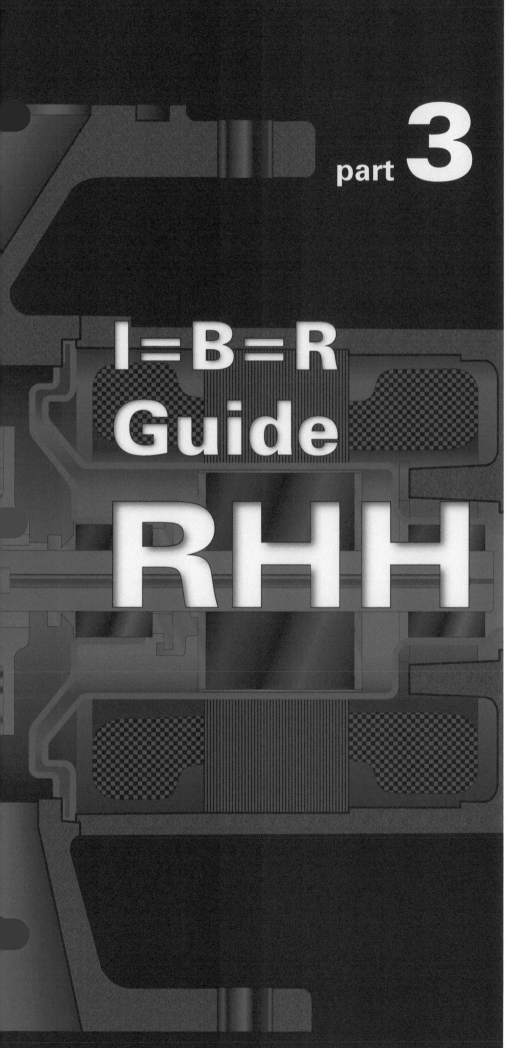

part 3

Components
of hydronic hot water heating systems

Residential
Hydronic
Heating . . .

Installation
& Design

Hydronics Institute
Section of **AHRI**

I=B=R Guide RHH
Residential Hydronic Heating

Hydronics Institute Section of **AHRI**
35 Russo Place
Berkeley Heights, NJ   07922-0218

Contents – Part 3

Contents *– continued*

Contents – *continued*

Illustrations *(continued)*

Tables

Heating units

In Part 3 you will find information on the components of hydronic water systems — how they are used and guides to selection. In the following part, Piping hydronic water systems, we explain how to organize these components into a heating system. Part 5, "Sizing hydronic hot water heating systems", provides information for sizing circulators and piping.

Space heating components

The following devices or methods can be used singly or in many combinations to adapt to the needs of virtually any space. Hydronic heating devices receive heat from the hot water delivered to them and give off this heat to the surroundings to heat the space. Use your room-by-room heat loss calculation for the building and the heating unit manufacturer's ratings to select and size the required units to meet the heat loss of each space.

Baseboard heaters

See Figure 1, Page 3–7. In the upper illustration, the room is heated by finned-tube baseboard, placed around the outside perimeter of the room. Finned-tube baseboard manufacturers provide enclosure components for the most common wall angles (90° standard or 60° to handle corners at bay windows, for example).

Sizing baseboard heaters

You will find that catalogued baseboard output is given in Btuh per linear foot of finned element. The table in Figure 1, Page 3–7, shows the typical range of ratings for each type of baseboard heating device. See Table 1, Page 3–8, reproduced from a typical baseboard rating chart. Notice the output depends heavily on the supply water temperature. Figure 2, Page 3–8, shows how quickly output of finned-tube baseboard drops as supply or average water temperature is reduced. Be sure to consider the temperature available to the heater when sizing.

Increasing flow rate will slightly increase output. So ratings are given both for 1 gpm (one gallon per minute) and 4 gpm through the heater. Temperature will depend on the controls, piping design and flow rate through the circuit. Flow rate will depend on the circulator, pipe sizing and piping design. See Parts 4 and 5 for more detail.

Determine the required length of the finned-tube elements based on the manufacturer's catalogue ratings. Do this by dividing the heat loss for the space by the output per linear foot of baseboard. The result is the required minimum number of feet of finned element.

Length required = Heat loss ÷ Output per linear foot

Example: Use the data from Table 2, Page 3–10 for 1 gpm with 200 °F average water temperature, showing an output of 680 Btuh per linear foot of typical ¾-inch finned-tube baseboard. If the space heat loss is 15,000 Btuh, the required length of finned-tube baseboard is 15,000/680 = 22 feet.

Figure 1   Typical baseboard heaters and application

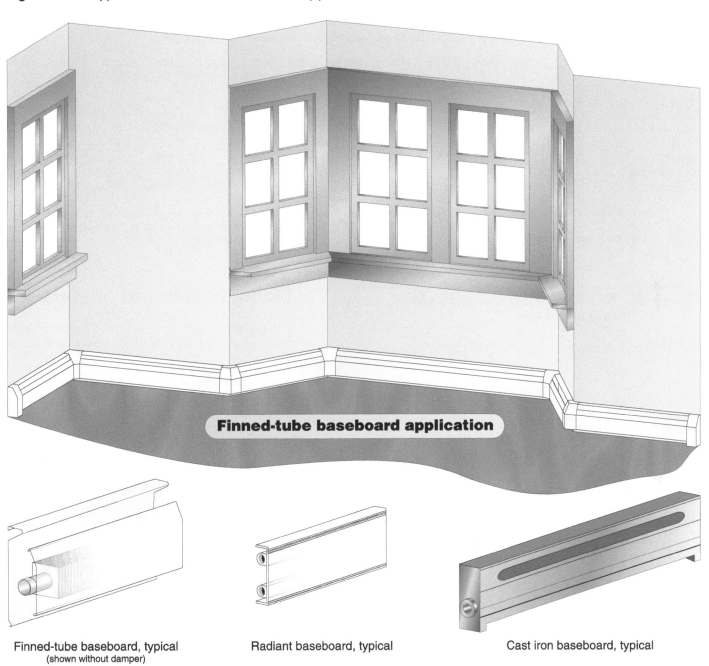

Finned-tube baseboard, typical
(shown without damper)

Radiant baseboard, typical

Cast iron baseboard, typical

Kickspace heater, typical

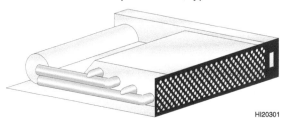

Baseboard heaters	Typical heating capacity @ 200 °F	
	Btuh per linear foot of baseboard	
	1 GPM Flow	4 GPM Flow
Copper finned-tube, standard	670 – 780	710 – 820
Copper finned-tube, high capacity	810 – 1090	860 – 1110
Radiant baseboard	270	290
Cast iron baseboard	700	
Kickspace heater	5900 – 8700 *total* output, typical	

HI20301

Table 1   Typical finned-tube baseboard heater rating data

			(Hot water ratings, Btuh per linear foot, with 65°F entering air)									
Element	Water flow	Pressure drop (Millinches per foot)	140°F	150°F	160°F	170°F	180°F	190°F	200°F	210°F	215°F	220°F
Type A Baseboard with ¾" element	1 gpm	47	290	350	420	480	550	620	680	750	780	820
	4 gpm	525	310	370	440	510	580	660	720	790	820	870
Type B Baseboard with ½" element	1 gpm	260	310	370	430	490	550	610	680	740	770	800
	4 gpm	2880	330	390	450	520	580	640	720	780	810	850

NOTE: Ratings are for element installed with damper open, with expansion cradles. Ratings are based on active finned length (5" to 6" less than overall length) and include 15% heating effect factor. Use 4 gpm ratings only when flow is known to be equal to or greater than 4 gpm; otherwise, 1 gpm ratings must be used.

Figure 2   Finned-tube baseboard output vs. temperature and flow

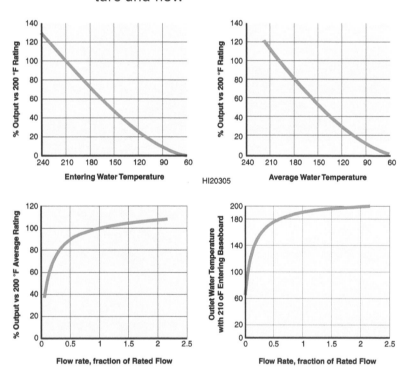

HI20305

(Based on typical baseboard heater sized for 210 °F Inlet and 20 °F Drop)

Sizing baseboard heaters (cont.)

It is generally not good practice to run finned-tube elements around the entire outer wall just to fill the space. This will usually cause over-sizing of the heating units. Use only the finned-tube element length needed for the heat loss of the space. Place the finned-tube elements evenly around the outside perimeter, though you may want to concentrate finned elements under large windows.

To achieve the uniform look of Figure 1, Page 3–7, purchase the enclosure components separate from the elements. Install baseboard enclosure continuously around the outer perimeter. Space the elements evenly within the enclosure. Then connect between elements with soldered copper tubing.

Finned-tube heating units are also available as kickspace heaters (for use under cabinets) as shown in Figure 1, Page 3–7, and as floor box heaters (useful in front of sliding doors, for example).

Finned-tube baseboard is the most common type sold. Other types available are cast iron and radiant. Cast iron baseboard delivers similar or higher output compared to standard finned-tube baseboard. But radiant baseboard output is less than half of typical finned-tube baseboard.

• Use radiant baseboard where heat loss is lower or when needed for supplemental heat — lower-output radiant floor or ceiling installations, or as supplemental heat in hydro-air (heating coil in duct) systems, for example.

• Size cast iron or radiant baseboard using the same formula as for finned-tube baseboard.

Figure 3   Hydronic heating units (heat emitters)

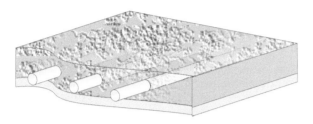

**Radiant floor heating
Tubing in concrete slab**
(typically 15 to 50 Btuh per sq. ft.)

**Radiant floor heating
Below-floor staple-up or with plates**
(typically 15 to 50 Btuh per sq. ft.)

**Radiant floor heating
Above-subfloor sleeper system**
(typically 15 to 50 Btuh per sq. ft.)

**Radiant floor heating
Above-subfloor embedding**
(typically 15 to 50 Btuh per sq. ft.)

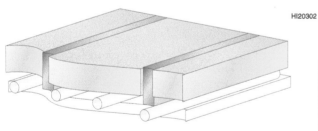

HI20302

Radiant ceiling or wall heating
(typically 15 to 50 Btuh per sq. ft.)

Radiators and convectors
(150 Btuh per sq. ft. EDR @ 170 °F average)

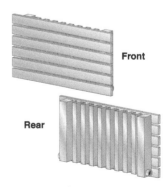

Front

Rear

**Indirect
water heaters**
(30 to 119 gallons, typical)

**Tankless
coils**
(3 to 6 GPM, typical ratings)

**Heating coils
& fan coil units**
(wide range of ratings available)

**Radiant panel
heaters**
(wide range of ratings available)

Table 2 Radiator output vs. temperature

Average temperature (°F)	Btuh per Sq Ft EDR
170	150
175	160
180	170
185	180
190	190
195	200
200	210
205	220
210	230
215	240

Radiant heating

Radiant heating can be done using tubing embedded or attached to the floor, wall, or ceiling. See Figure 3, Page 3–9 for typical applications of tubing to floors and ceilings. Figure 3, Page 3–9, also shows a typical configuration of radiant panel heaters, designed for mounting on the wall. Radiant panel heaters are also made for specialty applications, such as towel warmers. See Part 8 for more information on radiant heating.

The heat given off by a radiant surface depends on the surface temperature. The amount of heat per square foot of heated floor surface is approximately 2 Btuh per degree Fahrenheit difference between the radiant surface and the room temperature.

> Example: A 15-foot x 20-foot room with radiant tubing embedded has a floor temperature of 85 °F and room temperature of 70 °F. The heat given off per square foot of floor surface is:
>
> 2 x (85 - 70) = 30 Btuh per square foot
>
> For a room 15 feet by 20 feet, the floor area is 300 square feet and the heat given off by the floor would be 300 x 30 = 9000 Btuh.

The water temperature in the tubing, the amount of tubing used, and the construction of the floor and floor coverings will determine the heating capacity of a radiant system. Pay close attention to the tubing manufacturer's guidelines for design and installation.

When applying radiant heat to floors, walls or ceilings, be sure to use piping and controls that will limit the water temperature as required by the tubing and the radiant design. Excessive water temperature will cause tubing damage and uncomfortable surface temperatures.

Size radiant panels based on the catalogued ratings, selected to provide heat output at least equal to the heat loss for the space unless used as supplemental heat.

Radiators and convectors

Radiator or convector output depends on the average water temperature in the heating unit. Use manufacturer's rating information when available. Otherwise, you can use Table 2 to estimate the heating capacity of existing radiation. Multiply the square feet EDR of the radiator or convector by the output per square foot from Table 2 to determine heating capacity of the unit. (See Part 2 Appendix, Table 1 through Table 5, for square feet data if not available.)

Heating coils and fan coil units

Select and size heating coils and fan coil units. Be sure to consider water temperature and flow rate available to the unit when determining its capacity from the manufacturer's ratings. The heat provided to the heated space with a duct-mounted coil will depend on the flow rate of air across the coil as well as water temperature and flow in the coil.

Indirect water heating

You can supply domestic hot water using:

• tankless coil inserted in the boiler

• tankless coil in boiler plus external tank

• indirect water heater (incorporating a coil or heat exchanger inside).

See Part 6 for sizing and application information. As with other hydronic heating units, the output of the heater will depend on water temperature and flow rate. Be sure to consider when sizing.

Circulators

Moving water

For most residential systems, the circulator is the critical component for moving hot water through the piping. If your system is designed for gravity flow, then no pump may be required if the boiler manufacturer's instructions indicate this is acceptable. With most residential boilers today, however, the manufacturer will recommend installing a circulator even on gravity systems. This is generally because of the size of the supply and return connections at the boiler.

Circulators

A **circulator** is a centrifugal pump designed for residential and small commercial flow applications. Centrifugal pumps operate as shown in Figure 4. When the pump motor rotates the impeller:

- The impeller blades push water outward into the pump volute.

- When the water pulls away from the center (eye) of the impeller, the pressure drops as the water leaves, trying to pull a vacuum.

- This low pressure causes water to be pulled into the eye of the impeller, continuing the flow.

- The more water that flows through the pump, the lower the pressure in the impeller has to be to make this happen.

- Note: If the pump suction side is piped near the system expansion tank, then the pressure difference the pump generates will be added to the system. This is the advantage of locating the pump with its inlet (suction) side near the expansion tank. If the pump were to pump toward the expansion tank, the pump pressure would subtract (lower) the system pressure. See the discussion on expansion tanks beginning on Page 3–14.

Figure 4   Centrifugal pump (circulator) operation

1 **Impeller eye** — fluid is pulled into the eye because of the low pressure caused by water leaving

2 **Impeller blades** — blades push the water outward

3 **Pump volute** — enclosure around impeller that directs water to the pump discharge

4 **Entering fluid** — fluid enters the inlet (suction) connection of the pump and flows to the impeller eye

5 **Pump outlet** or discharge — water leaves pump with the energy imparted by the impeller blades

Figure 5   Oil-lubricated circulator, typical

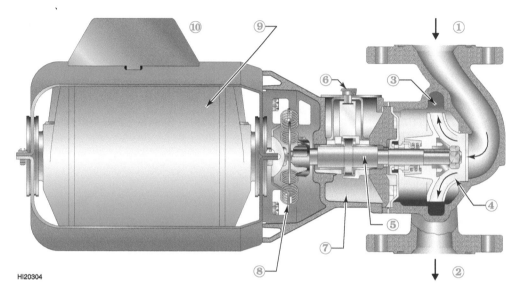

1 Inlet (suction)

2 Discharge (outlet)

3 Circulator volute

4 Impeller

5 Shaft and bearing

6 Oil cup

7 Oil reservoir

8 Coupling

9 Motor

10 Wiring compartment

HI20304

Circulator function

When water moves through piping and fittings, it loses energy to friction. The faster the water moves (velocity), the greater the energy loss. Water velocity increases as flow rate (gpm) increases or as pipe diameter is decreased.

Circulators inject the energy needed for the water to travel around the system and return to the pump. The amount of energy injected by the circulator has to equal the energy lost by the water's trip around the system.

The quantity, **pump head**, usually given in feet of water column, represents the energy added by the pump. (Note: You can think of the energy here as the amount of work the pump would do if it lifted a given number of gallons each minute a height equal to the number of feet head.)

How much flow?

The flow rate needed for a hydronic system depends on the piping design and the heat load handled by the system. For many systems, the flow rate should be about 1 gpm for each 10,000 Btuh heat load. The general formula for flow rate of water is approximately:

Flow rate, gpm = (Heat load, Btuh) ÷ (500 x DT),

where **DT** = temperature drop through the system or circuit, °F.

The most common design temperature drop, DT, is 20 °F. To achieve the design temperature drop, you must size the piping and select the circulator for the right flow rate. If you simply install a boiler with a factory-installed circulator on a system, the flow rate that actually occurs will depend on the system piping and the circulator capacity. The temperature drop probably will not be 20 °F, but some other value. You can apply the approximate formula for temperature drop of water if you know the flow rate:

DT, °F = (Heat load, Btuh) ÷ (500 x gpm)

The flow rate, gpm, is determined by circulator head capacity and system piping.

Figure 6   Water-lubricated circulator, typical

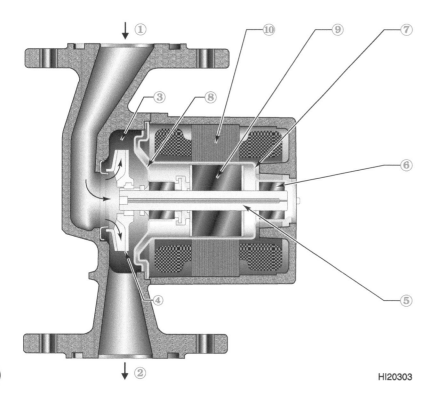

1 **Inlet (suction)**

2 **Discharge (outlet)**

3 **Circulator volute**

4 **Impeller**

5 **Hollow shaft** — allows system water to flow into rotor can to lubricate moving parts

6 **Bearings**

7 **Rotor can**

8 **Rotor can face plate**

9 **Rotor**

10 **Stator**

HI20303

Sizing circulators

You will find sizing information for circulators (and piping) in Part 5. For our discussion now, just consider:

- Circulator head indicates the ability of the pump to move the water.

- Circulator flow and head ratings are not determined by circulator connection sizes, but by a circulator motor, impeller and volute designs. **Do not line size circulators**.

- Residential circulators use fractional horsepower motors. Most are limited to from 8 to 10 feet maximum head, with flow rates of less than 20 gpm.

- The circulator mounted on (or supplied with) a residential boiler will often, but not always, be large enough to provide all the flow requirements for a system.

- See Parts 4 and 5 for system layout and circulator sizing.

Circulator types

You will find circulators available as oil-bearing type (some with permanent-lubed bearings) or as water lubricated.

Figure 5, Page 3–12 shows a typical 3-piece, oil-lubricated circulator. (Make sure the owner is aware of the need to oil this type circulator periodically.)

Always mount this circulator with the oil cup up and the shaft horizontal. The bolt-on design allows rotation of the pump volute to allow pumping upward, downward or horizontally.

Figure 6 shows a typical water-lubricated circulator. In this design, the lubrication between moving parts is done with water from the system. These circulators incorporate a hollow shaft to allow water to flow into the rotor can, and usually include a particulate filter in the shaft to help reduce solids. Always mount these circulators with the shaft horizontal. The bearings are not designed for thrust operation and the rotor can must be horizontal to ensure its water level is correct.

Integrated-function circulators — Circulators are available with integral relay (allowing the room thermostat to be connected directly to the circulator without a separate circulator relay), built-in flow/check valve, and other options to simplify installation.

Variable-speed circulators — Circulators are available with integral speed controllers, and can be operated based on sensor inputs or by a control signal from a remote controller.

Expansion

Overview

When water is heated, it expands. When it cools, it contracts. Water expands about **5%** when heated from 60 °F to 240 °F. **You have to provide somewhere in your system for this extra volume of water to go.** If the water content of your system is 50 gallons, for example, you need room for 5% or more extra water, or about 2.5 gallons.

You control expansion in hydronic water systems using an **expansion tank**. Expansion tanks are partially filled with air. When attached to the system piping, they allow water to flow into the tank as the system volume expands. By making the tank large enough, you can control how much the system pressure increases as the water expands into the tank, compressing the air. If you don't install a properly sized, functioning expansion tank, the system pressure will go too high. At the least this will cause the boiler relief valve to pop or weep. At the worst it can actually crack boiler sections or result in dangerous conditions.

There are three types of expansion tanks:

1. **Open tanks** (usually only on old systems)

2. **Air cushion tanks** (air control systems)

3. **Diaphragm tanks** (air removal systems)

1. Open tanks (usually only on old systems)

These tanks are installed at the top of the house. The tank is open to the atmosphere, allowing water to rise into the tank as system water expands. Such tanks are generally not recommended for new systems. Because the tank is open to atmosphere, a lot of air dissolves in the water. This causes corrosion of boilers and systems components, and often causes heating problems due to air pocketing.

TIPS when using open tanks:

- Install an efficient air separator.

- Install automatic air vents.

- Install a tank fitting (note in Figure 7, Page 3–15) in the tank to reduce gravity circulation of the cool tank water.

- Encourage your customer to purchase your installation of a new, diaphragm-type expansion tank.

- Use only manual fill for open tank systems to prevent risk of flooding

- Some boilers may not be approved for use with open tank systems..

2. Air cushion tanks

See Figure 7. Air cushion tanks are closed cylindrical tanks. They are not exposed to the atmosphere, but the water surface is in direct contact with the air. This allows the air to dissolve in the water as with open tanks. With air cushion tanks, however, the system must always return the air to the tank. If the air doesn't return, the tank will eventually become waterlogged, and will fail to control expansion.

TIPS when using air cushion tanks:

• **NEVER use automatic air vents**. They allow air to escape from the system, and will quickly cause the tank to waterlog.

• Install an efficient **air separator**, but pipe the air separator discharge to the expansion tank, not to an automatic air vent.

• Air cushion tanks should fill to about 1/2 full at initial fill pressure.

• **Always** use a tank fitting. It will reduce recirculation of cool, air-saturated water to the system.

• Connect from the air separator (or boiler air vent tapping) to the expansion tank with at least 3/4" pipe to allow room for air to float up the pipe to the tank. Pitch pipe at least ½" per foot up toward tank to aid air movement.

Sizing air cushion tanks

Use the manufacturer's sizing guides when possible. See Figure 16, Page 3–28 to calculate expansion tank size based on system volume and fill pressure required.

As a rule of thumb, use an air cushion tank with a volume at least 1 gallon for each 5,000 Btuh of total heat load.

Figure 7   Air cushion expansion tank

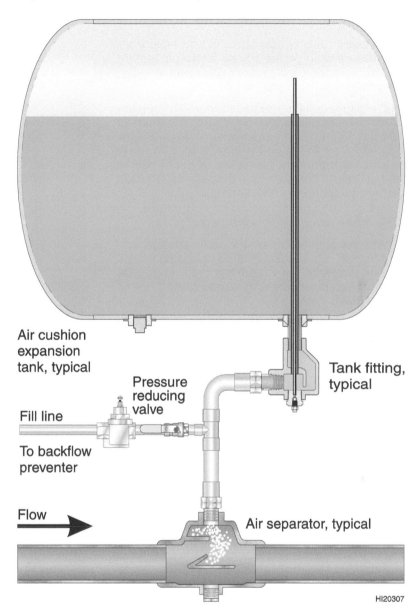

Air cushion expansion tank, typical

Pressure reducing valve

Fill line

To backflow preventer

Flow

Tank fitting, typical

Air separator, typical

HI20307

3. Diaphragm tanks

See Figure 8.

Diaphragm (or bladder) tanks use an elastomer membrane to separate the air from the system water. Air is not in contact with the system water, so it can't dissolve in the water.

You will find that residential-sized diaphragm expansion tanks are factory pre-charged (usually to 12 psig). This is the most common fill pressure for residential systems, and is suitable for homes up to 2 stories high. For other cases, you may need a higher charge pressure (and possibly higher than 30 psig boiler relief valve setting).

TIPS when using diaphragm tanks

- **Always** mount the tank with its system tapping UP — never on its side or upside down. This ensures the diaphragm will always be wet, avoiding cracking due to drying out of the elastomer.

- **Always** install an efficient air separator.

- **Always** install automatic air vents.

Sizing diaphragm or bladder tanks

Use the tank manufacturer's sizing guides when possible. (Select tank based on acceptance volume.)

See Figure 16, Page 3–28 to calculate tank size based on system volume and fill pressure required. To be sure the tank is adequate, assume 60 °F for cold temperature and 240 °F for hot temperature when calculating required tank acceptance volume. This allows for about 5% expansion of the system water volume. Assuming 240 °F for the hot temperature allows the the possibility of the boiler limit control being set to its highest setting.

As a rule of thumb, use a diaphragm or bladder tank with a volume at least 1 gallon for each 7,000 Btuh of total heat load.

Figure 8   Diaphragm expansion tank

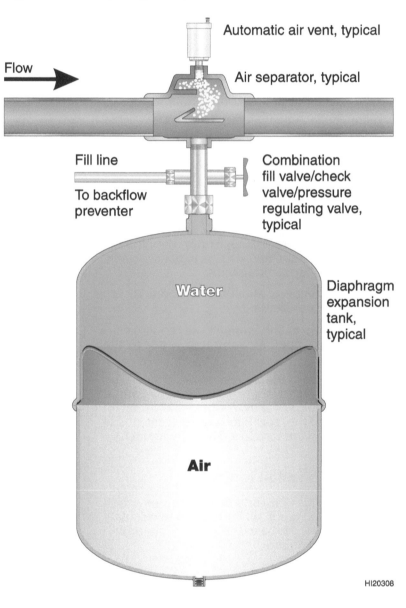

HI20308

Expansion tank location

The only way the pressure at the expansion tank can change is by adding or removing water volume from the system. This has to be, because the only way the air inside the tank expands or contracts is by changing the volume of water inside the tank. This means the point where the expansion tank connects to the system can be called the "**point of no pressure change.**"

The circulator can't change the pressure at the tank.

- If the circulator pumps toward the tank, the pressure (head) it creates can't raise the pressure at the tank. So this pressure has to lower the pressure at the pump inlet. In other words, it will lower the pressure in the system. Water will leave the expansion tank point at the expansion tank pressure and lose pressure as it works its way around the system and back to the circulator.

- **The best place to locate the circulator is with its inlet (suction side) connected near the expansion tank**. The pressure (head) the circulator creates will add to the expansion tank pressure, increasing the pressure in the system.

Why does it matter what the system pressure is?

- Pressure keeps the air dissolved in water. The higher the pressure, the more air can be held in. So you want the lowest pressure where the air is separated - at the air separator/expansion tank connection to the system. If the circulator pumps toward the expansion tank, it lowers the pressure at the top of the system, making the air want to come out of solution there. If the air comes out of solution, it will air bind the heating units. No matter what you do, you will never solve this problem unless you relocate the circulator. This problem isn't as likely with low-head (under 10 feet head) circulators, but it will always be a problem with high-head circulators.

- Many systems operate with water temperatures up to 220 °F or higher. Lower the pressure too much and the water can flash to vapor in the pipes or damage the circulator due to flashing in the impeller.

Air control

Air removal/control

Air enters the system with fresh water. Like a glass of water from the tap, air entering the boiler has dissolved air in it. Air comes out of solution when heated, and can cause corrosion and/or heating problems if not dealt with correctly.

Air separators

Every system should have an efficient air separator. This ensures the air will be kept out of heating units, preventing heating problems and flow noise. Locate the air separator on the hot water supply connection to the system. This is where the water is hottest. (The hotter the water, the lower the solubility of air, making it easier to separate.)

See Figure 9, Page 3-19. You will find air separators of mechanical design (air scoops) and high-surface design. Figure 9 illustrates two typical high-surface air separators.

Tank fittings

See Figure 7, Page 3-15 If the system uses an air cushion or open expansion tank, always install a tank fitting. The tank fitting will:

* Reduce air bubbling up through the tank water.

* Reduce gravity circulation of cool, air-saturated water back to the system.

* Allow filling the tank to the correct level.

Automatic air vents

Automatic air vents (see Figure 8, Page 3-16) use a float-operated valve mechanism to allow removal of air from the system. Locate these vents on boiler air vent tappings, on top of air separators, and at the top of the system for best control of air.

NEVER use an automatic air vent on a system equipped with an air cushion tank.

Manual air vents

Install manual air vents on heating units at the top of the system. They may be useful in speeding air elimination on initial startup and for trouble-shooting air problems. They are not effective in ongoing air control.

Boiler dip tubes

Where recommended by the boiler manufacturer, install a dip tube as in Figure 10. This allows gravity separation of air, using the boiler as an air accumulator. Make sure not to insert the nipple any further than suggested by the boiler manufacturer. Inserting too far could cause an intermittent low water condition in the boiler or drop the water line below the limit control sensing bulb, causing the limit to sense air, not water temperature. (Some boilers incorporate a built-in dip tube. Refer to the manufacturer's instructions for correct piping.)

Purge valves

Locate drain cocks correctly in the system to allow forcing the air out of each circuit as the system is filled. See Part 4 for typical applications and purging procedures.

Figure 9   Typical air separators

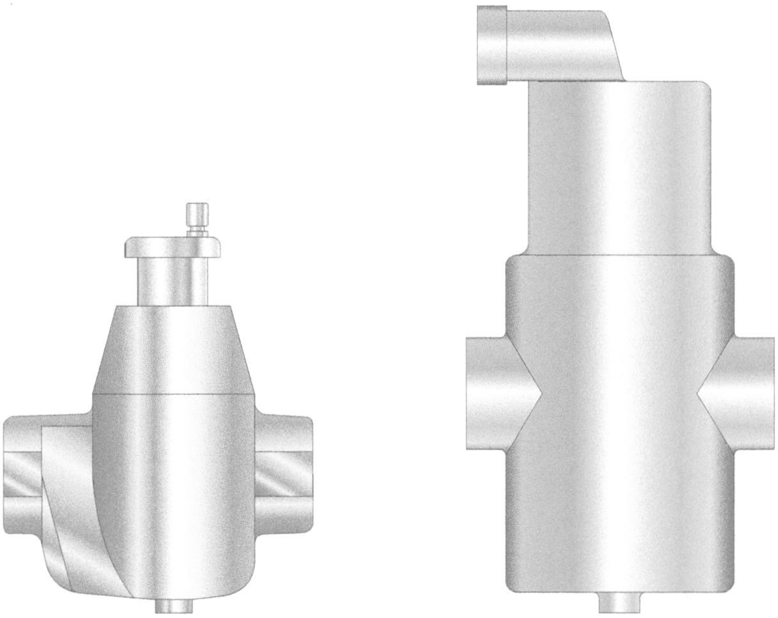

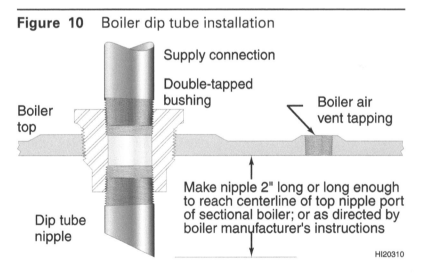

Typical high-surface air separators   HI20309

Figure 10   Boiler dip tube installation

Supply connection

Double-tapped bushing

Boiler top

Boiler air vent tapping

Dip tube nipple

Make nipple 2" long or long enough to reach centerline of top nipple port of sectional boiler; or as directed by boiler manufacturer's instructions

HI20310

Fill line

Automatic fill valve (or pressure reducing valve)

Except for systems with glycol/water mixture (see following), some installers provide an automatic fill valve (or pressure reducing valve) to replace water if system pressure drops. As air is removed from a system, make-up water is needed to fill the void left. This occurs mostly within the first few days of operation. See Figure 11, Page 3–21.

Some insurance companies prefer the fill line be manually closed off after the initial fill. If a leak occurs in the system, the system pressure will drop. This lets the owner know there is a problem. With automatic fill, the leak will be masked because the pressure will never drop. This increases the risk of water damage to the residence. You may find it better to close off the fill line after the system is running normally.

Water meter

Install a water meter on the make-up water line. Have the owner check the meter periodically to make sure it doesn't change much over time. A large change indicates there is a leak in the system that must be located and fixed.

Check valve

Install a check valve if not integral to the fill valve or pressure reducing valve used. The backflow preventer, if used, will eliminate the need for another check valve.

Backflow preventer

Many codes require a backflow preventer. Install and pipe the backflow preventer as required. Even where not required by code, install a back-flow preventer when needed to prevent possible contamination of the fresh water lines. Always install a backflow preventer when any additives are used in the boiler. See Figure 11.

Glycol/water systems

Glycol/water systems may require a different approach. Glycol leaks through an opening more easily than water. Consequently, if a system filled with a glycol/water mix leaks, what leaves through the leak isn't the glycol/water mix, but mostly glycol. If the leak continues, the percent of glycol will be reduced as fresh make-up water enters. A solution is to use manual make-up only. If a leak occurs, the system pressure will drop, no-tifying the owner of a problem. Another solution is the use of a water meter on the make-up line. Tell the owner to check the meter periodically to make sure the meter reading doesn't change much over time. Alternatively, you could use a pressure switch to sense system pressure and activate a glycol feed system when pressure drops.

Figure 11   Typical fill line installation

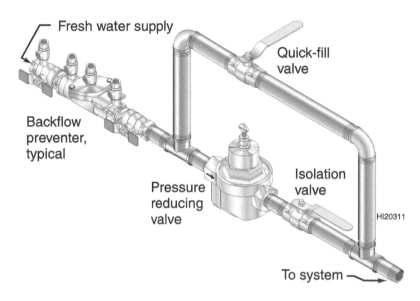

Fill line connection to system

ALWAYS install the fill valve connection at the same point where the expansion tank is connected. Any other location in the system will have a varying pressure, and will cause false readings for the fill valve.

CAUTION: When filling the boiler with the quick-fill by-pass, watch the boiler pressure gauge to avoid over-pressurizing the boiler.

Components

Valves & fittings

Isolation valves

Install isolation valves:

- On boiler supply and return connections
- On supply and return of each zone of a multi-zone system
- On both sides of each circulator.

Control valves

See Figure 12, Page 3–23. Depending on system design, you may need 2-way, 3-way, or 4-way control valves to regulate supply temperature to the system or return temperature to the boiler, or both.

Two-way valves regulate flow in a single path. They can be used for injecting heat into a circuit or changing temperature drop by changing flow rate.

Three-way valves are usually used for mixing hot water from one line with cooler water from another to yield the desired water temperature leaving the third valve port.

Three-way diverting valves redirect flow from one port to another.

Four-way valves use flow from two ports to regulate temperature out the other two ports. These valves can be used to both regulate supply temperature to the system and return temperature to the boiler, all in a single valve.

Tempering or mixing valves

Use a tempering or mixing valve on the outlet of indirect water heaters or tankless coils as shown in Part 6. This will help ensure correct temperature of domestic water supplied to fixtures.

Thermostatic valves

You can use thermostatic valves to individually zone spaces without having to wire room thermostats and zone valves.

Balancing valves

On multi-circuit systems, install balancing valves as suggested in Part 4. These allow adjustment for different pressure drops in the circuits, ensuring correct flow in each. Use square-head cocks or memory-stop ball valves.

Figure 12 Control valve port configurations

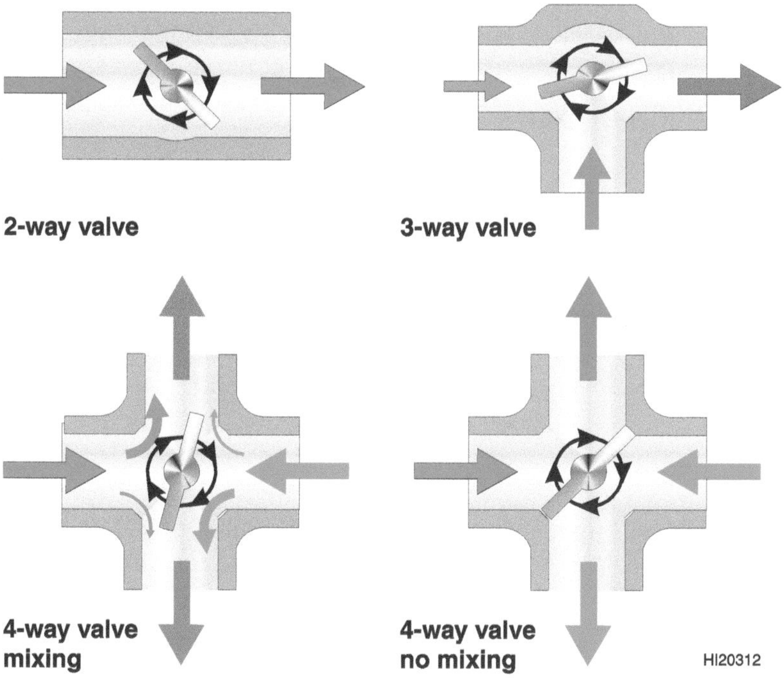

2-way valve

3-way valve

4-way valve mixing

4-way valve no mixing

HI20312

Figure 13 Relief valve, typical

HI20313

Flow control valves (flow/check valves)

Install flow control (flow/check) valves as shown in Part 4 on circulator-zoned systems and primary/secondary-piped systems. Flow control valves incorporate a check valve with a weighted disk. The weight on the disk prevents gravity circulation. The amount of weight is enough to offset the pressure difference caused by the difference in density between supply and return water in the pipes of a typical residence.

Bypass pressure regulators

On two-pipe multi-zone systems with zone valves, install a bypass pressure regulator to limit the pressure if the circulator operates with all zone valves closed. See Part 4 for details. (An alternate method is to use one three-way zone valve in the system. Connect the extra port of this valve to the return line using ½-inch pipe to provide a flow path when all valves are closed.)

One-pipe fitting (diverter tee)

As discussed in Part 4, a one-pipe fitting is used to add a slight pressure drop through the main line to cause the correct amount of water to flow through a heating unit, located in a branch line.

Relief valves

See Figure 13. Every boiler is equipped with a pressure relief valve, usually set at 30 psig. This valve will discharge if water pressure exceeds its setting. Many boilers are available with 45 psig relief valves when needed for special cases. Never use a relief valve with a higher pressure setting than the boiler Maximum Allowable Working Pressure (on boiler nameplate).

Radiator valves

Radiators and convectors are often fitted with valves used for isolation and flow regulation. These valves can often be used to help balance the system.

Controls

The following information is a brief introduction to hydronic controls. See Part 8 for more detail.

Thermostats

You will find both electromechanical (operated by a bi-metal devices) and electronic thermostats.

The simplest devices use a mercury switch connected to a bi-metal spring. As the spring temperature changes, the bi-metal tries to expand or contract, causing the spring to move the switch.

Electronic thermostats use electronic temperature sensors with an electronic control circuit. Some electronic thermostats require a constant voltage supply. Make sure to check this when applying to boiler and control circuits, since many of these circuits may interrupt the power to the thermostat.

Thermostat voltage and loading

Most thermostats used on hydronic systems are 24-volt, with low current contact ratings. Use caution when applying a thermostat to be sure the load is suitable for the thermostat's rating.

Thermostat anticipator setting

Set the anticipator to the current (milliamps) drawn by the connected boiler (see boiler manual) or control (such as relay or zoning control).

Temperature limit controls

Every water boiler is equipped with a temperature limit control. This device will shut off the burner or gas valve when the water temperature reaches its setting. You may use additional controls for special applications, such as domestic water heating, automatic circulator operation or other.

Flow switches

You may find a flow switch used on some boilers to ensure proper flow through the heat exchanger. This device usually uses a paddle in the flow stream attached to a switch.

Water level controls

Always install a low water cutoff if the boiler is located at the top of the system. As an additional precaution against a low water condition, you may want to install a low water cut-off on other applications, particularly if the boiler is likely to operate for long periods with no occupancy of the building.

You will find both float-type and probe-type water level controls. When using probe-type controls, take care when selecting and wiring the control to correctly wire into the boiler control circuit.

Circulator relays

Circulators are always line voltage (120 vac or higher). Most thermostats used on hydronic systems are low voltage only. In order to cycle a circulator, install a circulator relay. The thermostat operates the relay coil. The relay contacts switch the circulator on and off.

Zone valves

Use zone valves to zone systems with a single circulator. Pay close attention when wiring 3-wire zone valves to the boiler thermostat circuit. Incorrect wiring could result in damage to components on some boilers. (See the simple method of proving correct wiring in Part 8, page 8-25.) Use 4-wire zone valves when possible. They isolate zone valve power circuit from the thermostatic circuit.

Electronic controls

Electronic controls are available to provide any of the functions typical of electromechanical controls, plus the ability to provide proportional operating signals to boilers and system components, DHW priority operation and other functions. Controls are often programmable and adjustable to match system requirements.

Electronic zone controllers provide the ability to regulate operation of multiple system components and provide boiler return water protection as well as system temperature control.

Integrated boiler controls

Many boilers are equipped with multi-function control systems. This can include circulator operating terminals (for system, boiler and DHW circulators), multiple temperature sensor inputs (for outdoor sensor, room sensors, etc.) and other functions integrated into the boiler control. Many of these controls are programmable, allowing adjustment of control operating characteristics to fit the specific application.

Refer to the boiler instructions for details of operation and adjustment.

Appendix

Copper fittings and dimensions

Figure 14 Copper fittings and dimensions

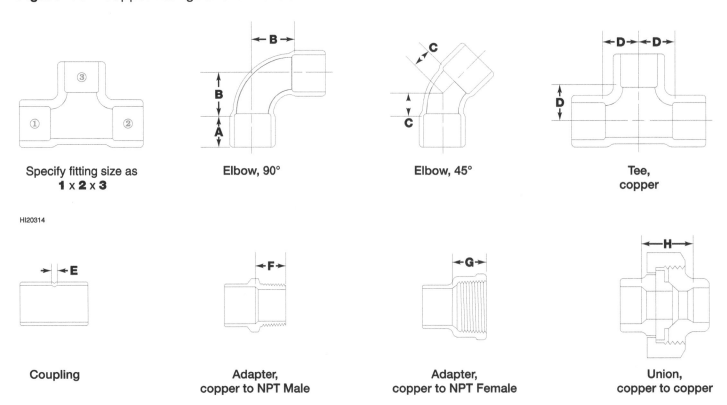

Specify fitting size as
1 x **2** x **3**

Elbow, 90°

Elbow, 45°

Tee,
copper

HI20314

Coupling

Adapter,
copper to NPT Male

Adapter,
copper to NPT Female

Union,
copper to copper

Copper tubing size (inches)	Solder Cup End short radius copper to copper	90° Elbow short radius copper to copper	90° Elbow long radius copper to copper	45° Elbow copper to copper	Tee copper to copper	Coupling copper to copper	Adapter copper to NPT male	Adapter copper to NPT female	Union copper to copper
	A (inches)	**B** (inches)	**B** (inches)	**C** (inches)	**D** (inches)	**E** (inches)	**F** (inches)	**G** (inches)	**H** (inches)
1/2	1/2	5/8	7/8	5/16	7/16	1/16	5/8	27/32	41/64
3/4	3/4	7/8	1-3/16	3/8	9/16	1/16	47/64	3/4	1
1	29/32	1-1/8	1-1/2	1/2	3/4	1/16	15/16	15/16	1-1/16
1 1/4	31/32	--	1-7/8	3/4	15/16	1/16	1-1/16	31/32	1-1/16
1 1/2	1-3/32	--	2-1/4	15/16	1-1/16	1/16	15/16	1	1-7/16
2	1-11/32	--	3	1-1/4	1-5/16	1/16	1-7/32	1-1/32	1-1/4

Malleable iron fittings and dimensions

Figure 15 Malleable iron fittings and dimensions

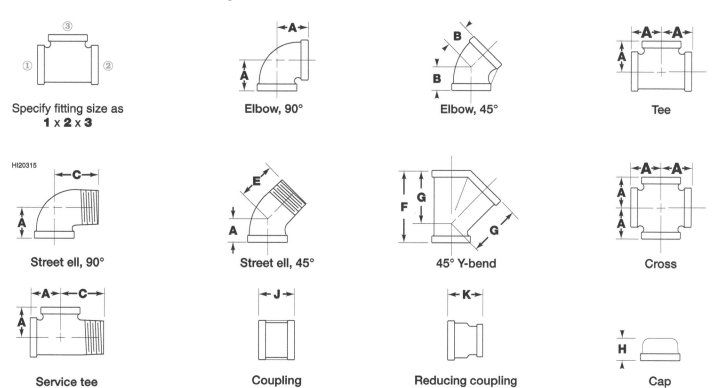

Pipe size	Thread make-up allowance (inches, appr.)	A (inches)	B (inches)	C (inches)	D (inches)	E (inches)	F (inches)	G (inches)	H (inches)	J (inches)	K (inches)
1/8	1/4	0.69	-	1.00	-	-	-	-	0.53	0.96	-
1/4	3/8	0.81	0.73	1.19	0.73	0.94	-	-	0.63	1.06	1.00
3/8	7/16	0.95	0.80	1.44	0.80	1.03	1.93	1.43	0.74	1.16	1.13
1/2	1/2	1.12	0.88	1.63	0.88	1.15	2.32	1.71	0.87	1.34	1.25
3/4	9/16	1.31	0.98	1.89	0.98	1.29	2.77	2.05	0.97	1.52	1.44
1	11/16	1.50	1.12	2.14	1.12	1.47	3.28	2.43	1.16	1.67	1.69
1 1/4	11/16	1.75	1.29	2.45	1.29	1.71	3.94	2.92	1.28	1.93	2.06
1 1/2	11/16	1.94	1.43	2.69	1.43	1.88	4.38	3.28	1.33	2.15	2.31
2	3/4	2.25	1.68	3.26	1.68	2.22	5.17	3.93	1.45	2.53	2.81
2 1/2	1 1/8	2.70	1.95	3.86	1.95	0.26	6.25	4.73	1.70	2.88	3.25
3	1 3/16	3.08	2.17	4.51	2.17	3.00	7.26	5.55	1.80	3.18	3.69
3 1/2	1 1/4	3.42	2.39	-	-	-	-	-	1.90	-	-
4	1 1/4	3.79	2.61	5.69	2.61	3.70	8.98	6.97	2.08	3.69	4.38
5	1 3/8	4.50	3.05	6.86	-	-	-	-	2.32	-	-
6	1 1/2	5.13	3.46	8.03	-	-	-	-	2.55	-	-

Equations for calculating minimum expansion tank volume

Figure 16 Equations for calculating minimum expansion tank volume

Air cushion expansion tank sizing

$$P_{fill} = 5 + \left[\left(H_{system} - H_{tank} \right) \times \left(\frac{D_{cold}}{144} \right) \right]$$

$$V = \frac{V_{system} \times \left(\frac{D_{cold}}{D_{hot}} - 1 \right)}{14.7 \times \left(\frac{1}{P_{fill} + 14.7} - \frac{1}{P_{reliefvalve} + 14.7} \right)}$$

Diaphragm expansion tank sizing

$$P_{charge} = 5 + \left[\left(H_{system} - H_{tank} \right) \times \left(\frac{D_{cold}}{144} \right) \right]$$

$$V = V_{system} \times \left(\frac{D_{cold}}{D_{hot}} - 1 \right) \times \left(\frac{P_{reliefvalve} + 9.7}{P_{reliefvalve} - P_{charge} - 5} \right)$$

Insert the values for the following in the above equations to calculate required tank volume.

V	= minimum required expansion tank volume, gallons	P_{fill}	= pressure at air cushion tank at initial cold water fill, psig
V_{system}	= total volume of system, gallons	P_{charge}	= diaphragm charge air pressure, psig
		$P_{reliefvalve}$	= boiler relief valve pressure setting, psig
D_{cold}	= density of system water at fill temperature, lbs/cu ft	H_{system}	= height from boiler room floor to highest system component, feet
D_{hot}	= density of system water at operating temperature, lbs/cu ft	H_{tank}	= height from boiler room floor to expansion tank water connection

Note: The equations above provide for a minimum cold fill pressure of 5 psig at the top of the system.

Water density values for D_{cold} and D_{hot} at water temperatures shown:
(generally use 60 °F and 240 °F for D_{cold} and D_{hot}, providing for about 5% expansion)

60 °F — 62.34	100 °F — 62.00	110 °F — 61.84	120 °F — 61.73	130 °F — 61.54	140 °F — 61.39	150 °F — 61.20
160 °F — 61.01	170 °F — 60.79	180 °F — 60.57	190 °F — 60.39	200 °F — 60.13	220 °F — 59.63	240 °F — 59.10

Calculating system water volume, V_{system}:

Copper piping (gallons per foot of pipe):	½" — 0.016	¾" — 0.027	1" — 0.046	1¼" — 0.068	1½" — 0.096	2" — 0.165
Steel piping (gallons per foot of pipe):	½" — 0.026	¾" — 0.028	1" — 0.045	1¼" — 0.078	1½" — 0.106	2" — 0.175

Estimates for heating unit volume:	Cast iron radiation — columnar 0.114 gallons per square foot EDR
	Cast iron radiation — tubular 0.056 gallons per square foot EDR
	Cast iron convector — 0.15 gallons per 1,000 Btuh output @ 200 °F
	Cast iron baseboard — 0.47 gallons per 1,000 Btuh output @ 200 °F
	Fan coil unit or unit heater — 0.02 gallons per 1,000 Btuh output @ 180 °F
	Copper convector — 0.064 gallons per 1,000 Btuh output @ 200 °F
	Copper baseboard, ¾" element — 0.037 gallons per 1,000 Btuh output @ 200 °F

Estimating boiler water content:	If not available in boiler literature, estimate volume as 0.186 gallons per 1,000 Btuh boiler *input*

Hydronics Institute Section of AHRI

35 Russo Place

Berkeley Heights, NJ 07922-0218

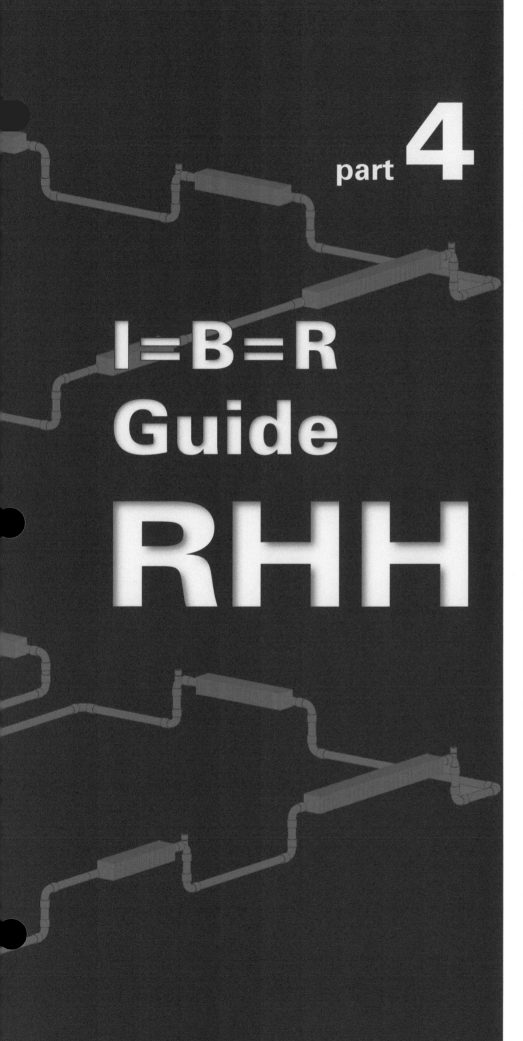

part **4**

Piping
hydronic
hot water
heating systems

I=B=R
Guide

RHH

Residential
Hydronic
Heating . . .

Installation
& Design

Hydronics Institute
Section of **AHRI**

I=B=R Guide RHH
Residential Hydronic Heating

Hydronics Institute Section of **AHRI**
35 Russo Place
Berkeley Heights, NJ   07922-0218

Contents – Part 4

● Selecting the right system

● Near-boiler piping

● Piping schematics

● System applications

Contents *– continued*

Illustrations

Contents *– continued*

Tables

Selecting the

Overview

In Part 2 (page 2-9), we suggested a procedure for selecting a boiler and system to meet your customer's needs. To review, the sequence is:

Step 1 — Determine required heating capacity

Step 2 — Check location/talk with homeowner

Step 3 — Replacement boilers — Do a System Survey

Step 4 — Select piping, controls, & trim

Step 5 — Quote a complete job

Guide Parts 4 through 6 will provide the information you need to complete Step 4. With the system design complete, you can prepare your quotation to complete Step 5.

Once you have selected the boiler and heating units (Guide Parts 2 and 3), place them on a scale plan of the building. Use this plan to

- layout your vent system, control locations and water piping
- determine circuit lengths and size piping and components.

Part 4 provides information to help you decide how to pipe from the boiler to the heating units.

Select the simplest piping design that will provide heat to the building and meet your customer's budget and performance needs.

If your customer wants domestic water heating as well as space heating, see Part 6 for suggestions. Be sure to consider the domestic water heater (or tankless coil) when deciding on and laying out your piping.

If you are using radiant heating, you will find some general information in Part 8. For design and layout, use the tubing manufacturer's guidelines.

Once you have decided on the piping plan, apply one of the methods in Part 5 for sizing the piping and circulator(s).

See Part 10 for suggestions on controls.

For complex jobs, you may want to call the boiler manufacturer's representative or wholesaler heating specialist as well — just to double-check your plan.

Choosing a piping system

Always install the simplest, most cost-effective system needed to meet your customer's needs. For a glance at some of the options, see Figure 1, Page 4–8 and Figure 2, Page 4–9. These drawings show three different piping methods with two variations of each method, all applied to the same single-story home. For simplicity, the drawings show only baseboard heaters, but you will apply the same methods with other space heating units, domestic water heating, snow melt, and any of the other versatile options with hydronic heating.

right system

Introducing common systems

The systems shown in Figure 1, Page 4–8 and Figure 2, Page 4–9 (are:

Series-loop piping — Figure 1, top

Series-loop means units are connected from one to the next, outlet to inlet, around the entire system.

One-pipe system — Figure 1, center

One-pipe systems use a single pipe that runs around the entire system. Heating units are connected to the main pipe with tees (one tee and a one-pipe fitting, called a diverter tee, or two diverter tees).

Two-pipe system — Figure 1, bottom

Two-pipe systems use two distribution pipes - one for supply and one for return. Heating units connect with their supply side to the supply pipe and return side to the return pipe.

Series-loop piping, split-loop — Figure 2, top

Split-loop systems break the system in multiple (two shown) branches as shown in the diagram. This helps typical residential boiler circulators (as supplied with typical packaged residential boilers) to pump more water by splitting the flow. It also helps provide more uniform supply water temperature to the heating units.

One-pipe split-loop system — Figure 2, center

This is a one-pipe system split into multiple (2 shown) branches to improve use of the pump and provide hotter water to terminal units.

Two-pipe zoned system — Figure 2, bottom

Two-pipe zoned piping breaks the system into multiple circuits. The heating needs of heating units on a circuit should be similar to allow more comfortable control. For simplicity, zones in this system use heating units which are close together in the home. In practice, you can combine heating units anywhere in the home into a zone by piping them together. Notice that each of the zones is a small series loop, connected on the series supply to the common system supply manifold and on the series return to the common system return manifold.

Primary/secondary systems

Another common piping system, not shown in Figure 1 or Figure 2 is primary/secondary piping. Primary/secondary piping consists of multiple circuits connected to a common distribution pipe (or pipes), each circuit having its own circulator. You will find more discussion on primary/secondary beginning on page 4-20.

Figure 1 Three piping methods applied to a typical single-story hydronic heating installation

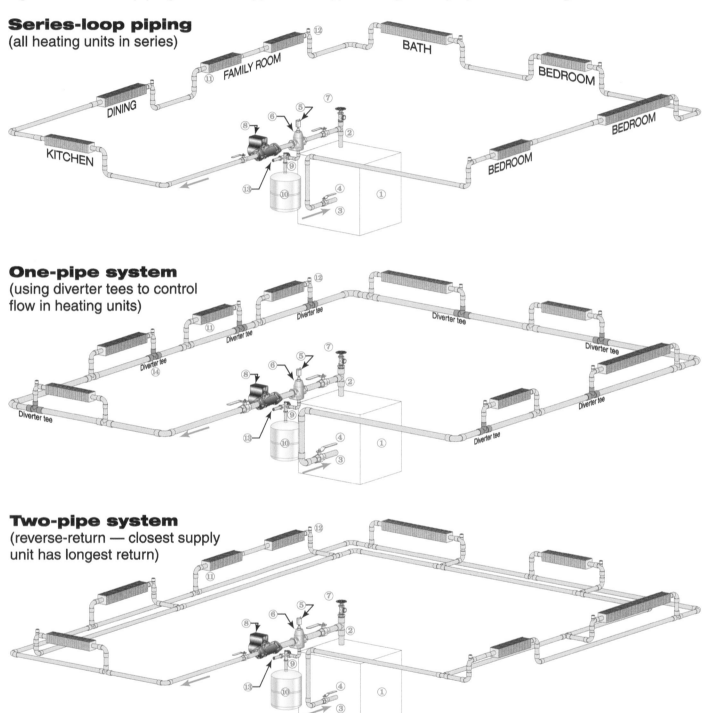

Series-loop piping
(all heating units in series)

One-pipe system
(using diverter tees to control flow in heating units)

Two-pipe system
(reverse-return — closest supply unit has longest return)

HI20401

① Hot water boiler	⑤ Automatic air vent, typical	⑨ Fill valve, typical	⑬ To fresh water make-up line
② Boiler outlet (hot water supply)	⑥ Air separator, typical	⑩ Diaphragm expansion tank, typical	⑭ Diverter tees (one-pipe fittings)
③ Boiler inlet (water return)	⑦ Hose bibb purge valve	⑪ Heating units, baseboard typical	
④ Isolation valves (handle shows flow)	⑧ Circulator, typical	⑫ Manual air vents	

Figure 2   Variations of Figure 1 piping systems

Series-loop piping
(split-loop — 2 circuits)

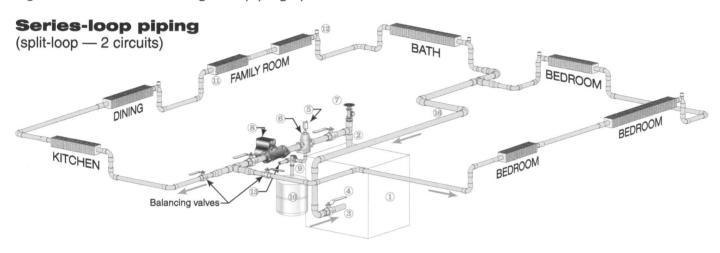

One-pipe system
(split-loop — 2 circuits)

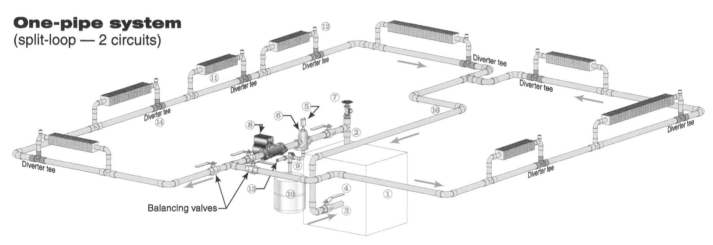

Two-pipe system
(zoned — 3 zones)

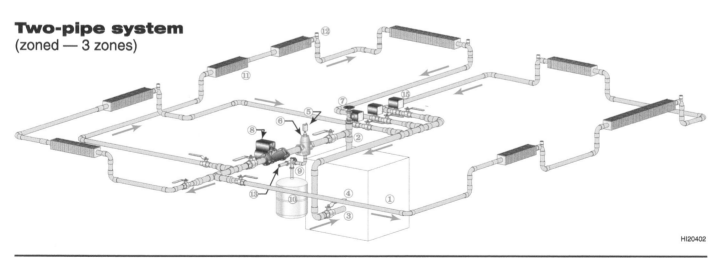

HI20402

① Hot water boiler	⑤ Automatic air vent, typical	⑨ Fill valve, typical	⑬ To fresh water make-up line
② Boiler outlet (hot water supply)	⑥ Air separator, typical	⑩ Diaphragm expansion tank, typical	⑭ Diverter tees (one-pipe fittings)
③ Boiler inlet (water return)	⑦ Hose bibb purge valve	⑪ Heating units, baseboard typical	⑮ Zone valves
④ Isolation valves (handle shows flow)	⑧ Circulator, typical	⑫ Manual air vents	⑯ Common return (trunk line)

About the systems

Series-loop piping

Series-loop piping requires the fewest number of fittings and least amount of pipe, and has a lower installed cost than any other system. Most small residences will work successfully using series-loop piping. Use series-loop piping when possible.

The main principle is to keep it simple. So, why not use a series loop in all cases? Series-loop piping may not always work. In a series-loop system, all the water flows through every heating unit, causing a higher pressure drop than other systems. The typical circulator supplied with most residential boilers may not be able to supply enough flow. Heating units near the end of the system may not provide enough heat — rooms at the beginning of the loop may overheat while those at the end are cold.

Split-loop series-loop piping

If a single-loop series system won't work for you, consider a split-loop system. You can use a split-loop system, with either a common return (as in Figure 2, Page 4–9, top) or common supply (not shown). The pressure drop per gpm flow can be reduced to as low as one-fourth of a single-loop system. The circulator can probably provide about twice the flow to each branch compared to a single loop. And the supply water to heating units will increase significantly.

If you encounter a single-loop system with heat distribution problems, you can usually convert it to a split-loop system with very little piping and probably cure the heat distribution problems. Before making this piping change, check the pipe sizing and circulator capacity as shown in Part 5 to be sure it will work for you.

You can control the loops in a split-loop system as zones. Install a thermostat in a strategic location for each loop. Then install a zone valve on each loop, with the valves operated by the loop thermostat. Turn the boiler and circulator on and off with the zone valve end switches.

One-pipe system

Fine-tune your heat distribution by moving up to a one-pipe system as the next best alternative to a series loop. The one-pipe system helps control pressure drop by using a larger-diameter distribution line, with heating units branched off. By selecting the right size and location of the diverter tees, you limit the flow to each heating unit to only the amount needed. But each of these diverter tees adds a pressure drop — you must select the tees correctly. If not, you will encounter heat distribution problems.

You can provide individual control of heating units on a one-pipe system by using thermostatic valves or zone valves on the heating units (see Figure 7, Page 4–17). Always select control valves with a low pressure drop. Otherwise you may be unable to cause enough flow through the units with the diverter tees. [Valves are often rated by their Cv value. Cv is the number of gpm flow through the valve that will cause a pressure drop of 1 psi (2.309 feet water column). When choosing between valves, select valves with the highest Cv value.]

When using heating units which have high flow or pressure drop requirements, be sure you use diverter tees with a high enough pressure drop and that the circulator can provide the flow and pressure drop needed for the entire system. Pay close attention to the diverter tee manufacturer's sizing procedures. Then double-check your circulator's capacity to be sure it will work.

Improve one-pipe system performance by installing multiple loops (split-loop system) as shown in Figure 2, Page 4–9, center. With split-loop piping, you can zone each loop with a zone valve or install control valves at each heating unit as desired.

Comparisons

See Table 2, Page 4–40, in Appendix for a listing of advantages and disadvantages of the system types discussed in Part 4. See further discussion of systems in "System applications", beginning on Page 4–16.

Two-pipe system

For single-circulator applications, probably no method is more forgiving than the two-pipe system. The piping splits water flow into many branches, making the best use of the limited head (pressure drop) of typical residential circulators. You pay for the forgiveness and flexibility of this method in using more pipe and fittings than the simpler series-loop or one-pipe systems, as you can see by comparing the systems in Figure 1, Page 4–8. But the effort and extra installation cost is often justified because of the "performance insurance" provided.

The heating units in this system are piped in parallel. That is, supply connections of all circuits pipe to a common supply; return connections pipe to a common return. (You can see this in the piping schematics of Figure 5, Page 4–15 and Figure 8, Page 4–19.) [Always pipe parallel circuits using reverse-return — first-supplied, last-returned piping. Connect the first unit supplied to the end (NOT the beginning) of the distribution piping. Why? — This makes the distance traveled by the water in each loop about the same length. With the length the same, the pressure drop would be about the same. So the flow in each loop will be similar. This is called a "balanced" system.] You could add ball valves to each loop to allow fine-tuning flow balance, but it is almost never needed in residential systems. Changing the flow in a heating unit makes much less difference in output than changing the temperature (see Part 3, Figure 2, Page 4–9 and discussion). Even reducing flow by half will only reduce output of a typical heating unit by about 10%! (Exception: When using water-to-water heat exchangers or indirect water heaters, flow affects temperature changes and may greatly affect performance. Ensure that the piping provides the correct flow as well as water temperature.) Don't take this flow tolerance for granted though, and pipe the system with direct return (first-supplied, first-returned). The difference in performance from one end of the system to the other could cause heat imbalance in the rooms.

To fine-tune heating with two-pipe systems, add thermostatic valves or zone valves to individual circuits. (See Figure 8, Page 4–19.)

Zoned two-pipe system

Zoned two-pipe systems split the piping into multiple, small series-loop branches, as shown in Figure 2, Page 4–9. You control flow to each branch using zone valves (zone-valve zoning) or individual circulators (circulator zoning) (see Figure 9, Page 4–19). Because the flow paths in such zoned systems may often be very different, use ball valves or square-head cocks where needed to balance pressure drops when using zone valves. When using ball valves for isolation and balancing, use memory-stop valves or mark operating position in some way to ensure it will always be returned to the correct position.

Select which heating units to pipe into combined zones by ensuring that thermostat location and temperature will represent all spaces of the zone and that the following considerations are similar for heating units and spaces:

- space usage and occupancy patterns (family room or kitchen occupancy and usage differs from bedrooms, for example)

- solar heat gain or internal heat gain from appliances (rooms with a lot of window area may gain heat depending on sun location; kitchens or laundry rooms gain heat from appliances)

- inside space temperature required (bathroom temperature is usually higher than bedrooms, so a common thermostat wouldn't be likely to work)

- supply water temperature needs of heating units (radiant floor systems may require 100 °F supply, while a water heater or heat exchanger might require at least 180 °F supply)

- radiant systems — flooring, floor covering, and construction

Near-boiler piping

Supply and return connections

Follow the boiler manufacturer's recommendations for sizing and location of supply and return connections to the boiler. If piping is not installed and sized correctly, the boiler and/or the system may not operate correctly and could possibly be damaged.

Pay special attention to the boiler manufacturer's recommendations for protecting the boiler against low return water temperature (discussed later in this section).

Circulator location

Many residential boilers include a pre-installed circulator, usually located on the return connection. Some residential boilers are available with the circulator shipped loose, allowing location of the circulator on either the supply connection or the return connection. As discussed in Part 3, the preferred location for any circulator is on the supply side because it allows the best location for the expansion tank and the best removal of air from the system water.

Figure 3, Page 4–13, shows the near-boiler piping for either a supply-mounted or return-mounted circulator. Be sure to follow two rules when piping the expansion tank and fresh water make-up line:

1. Always connect the make-up water line to the same point as the expansion tank connection to the system. This ensures that the fill valve will always sense the correct expansion tank pressure.

2. Always locate the expansion tank on the suction side, NOT the discharge side, of the circulator, regardless of the type of circulator used.

Expansion tank

Always install a diaphragm-type (or bladder-type) expansion tank if possible, on both new installations and replacement boiler installations. At least quote it as an extra so the owner can have the choice.

Diaphragm-type tanks allow the use of automatic air vents located anywhere in the system you find necessary. You cannot use automatic air vents on an air cushion tank. So air control is more difficult on systems equipped with air cushion tanks.

The drawings in Part 4 all show the use of diaphragm-type tanks. If using an air cushion tank instead, refer to the tank manufacturer's recommendation for tank location. If the circulator is on the boiler outlet as in Figure 3, Page 4–13, top, you can mount an air cushion tank off of the top of the air separator. If the circulator is on the boiler return and you plan to use an air-cushion tank, you should relocate the circulator to the boiler supply. If you don't, and locate the tank system connection off of the boiler or the supply line, you may encounter air removal problems at the top of the system because the system pressure will drop when the circulator starts. With low-head circulators as usually supplied with residential boilers, this problem may not occur, pumping away from the expansion tank in all cases gives greater assurance of a trouble-free system.

On some systems, the best location for the expansion tank may be in the system piping, as shown in some of the piping examples in Part 4.

Air separator

Always install an air separator in the system. Install an automatic air vent in the top of the separator only if the system uses a diaphragm-type (or bladder-type) expansion tank. Do not use automatic air vents if the system uses an air-cushion tank.

The piping drawings in Part 4 show preferred locations for the air separator. Consult the boiler manual for suggestions as well.

Make-up water line

Provide an automatic fill valve or pressure-reducing valve, a check valve and backflow preventer (if required by codes). Combination fill valves often handle all the functions of pressure regulation, check valve, and quick-fill operation. (If the system is filled with a glycol/water mixture, you may want to provide manual fill only by shutting off the fill line except for servicing. (If the system leaks, the boiler pressure gauge will show a decrease in pressure, notifying the owner of a problem. An alternative is to install a water meter on the fill line. Have the owner check the meter periodically to be sure no appreciable amount of make-up water is being added.)

Install (or at least quote) a water meter on the make-up water line of any system. The meter will let the owner know if his system develops a leak. This could save a lot of money by preventing damage to the boiler, system or home in the event of a leak.

Relief valve

Follow the boiler and relief valve manufacturers' recommendations on installing and piping the relief valve. Follow these guidelines:

1. Never leave an operating boiler installation unless you have connected piping from the relief valve to a safe drainage location. Someone could be seriously injured if the relief valve discharged on them.

2. Always pipe from the relief valve with metal pipe — NEVER use plastic pipe of any sort.

3. Piping must be full size of the relief valve outlet and must never have a shut-off valve installed.

4. Make sure there is no way in which the relief valve discharge piping could become plugged or blocked.

Purge valve(s)

Install a purge valve (or valves) in every system. This allows the fastest removal of air from the entire system. Note the suggested location of valves throughout Part 4. In all cases, locate the purge valve and an isolation valve so the isolation valve can be closed to separate the fill source from the purge valve, forcing the water to flow in the desired path through the system piping.

Figure 3   Near-boiler piping

Circulator on supply (preferred)

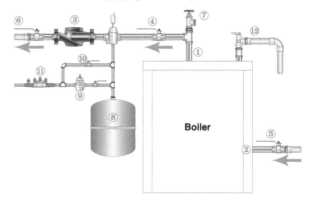

Circulator on return (alternate)

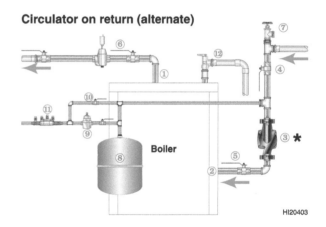

HI20403

★ **CAUTION:** If using a high-head circulator, ALWAYS mount circulator on supply line — NEVER on the return. A high-head circulator can cause the boiler relief valve to open if mounted on return with expansion tank.

① Boiler outlet (supply)

② Boiler inlet (return)

③ Circulator

④ Close valve when purging

⑤ Return isolation valve

⑥ Isolation valve

⑦ Purge cock (hose bibb)

⑧ Diaphragm expansion tank

⑨ Pressure reducing valve (if not using combination-type valve)

⑩ Quick-fill bypass valve

⑪ Backflow preventer (if required)

⑫ Pipe relief valve to drain, using only metal pipe full size of valve outlet connection.

Piping schematics

Piping schematics simplify piping layout and design. They can be helpful to you in troubleshooting heating system problems as well. If you sketch out the system layout, you will have a clearer understanding of how it is behaving.

When installing a new job, start with a schematic of the system. Decide how you want to pipe and control the system. Then you can quickly do a layout on the building floor plan to determine piping dimensions. Modify the schematic if you have to make any revisions during installation. Then keep this in your file for future use to simplify service and troubleshooting of the installation (and leave a copy with the owner). With the schematic in hand you'll have a better idea what to do before you get to the job, just from having the owner describe the problem over the phone.

Schematic symbols may vary slightly, depending on the source. The drawings in this guide use the symbols in Figure 4 below.

To compare schematic drawings with the three-dimensional layouts, see Figure 5, Page 4–15, — a piping schematic of each of the six systems shown in Figure 1, Page 4–8 and Figure 2, Page 4–9. The two-pipe schematics probably look the most different. Because the heating circuits in the two-pipe systems are in parallel, they are drawn as parallel lines in the schematic.

Figure 4   Piping schematic symbols used in Guide RHH

Symbol	Description	Symbol	Description
	Automatic air vent		Backflow preventer
	Air separator		Ball valve
	Circulator		Check valve, swing check
	Cross		Check valve, flow/check
	Diverter tee		Drain (purge) valve
	Elbow		Fill valve, combination type
	Expansion tank, air cushion type		Gate valve
	Expansion tank, diaphragm type		Globe valve
	Baseboard heater		Mixing valve, 3-way
	Immersion heating coil		Mixing valve, 4-way
	Heat exchanger		Pressure reducing valve
	Pressure or temperature gauge		Thermostatic radiator valves
	Tee		Square-head cock (for balancing)
	Union		Zone valve (or 2-way control valve)

HI20404

Figure 5 Piping schematics of systems in Figure 1, Page 4–8 and Figure 2, Page 4–9

Series-loop piping (all heating units in series)

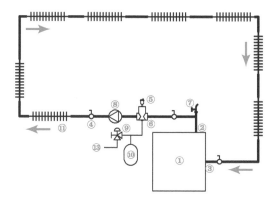

Series-loop piping (split-loop — 2 circuits)

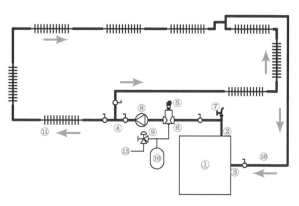

One-pipe system (using diverter tees)

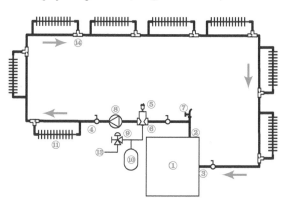

One-pipe system (split-loop — 2 circuits)

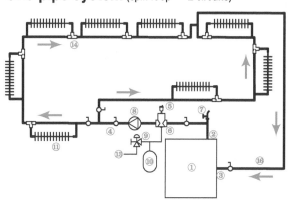

Two-pipe system (reverse-return)

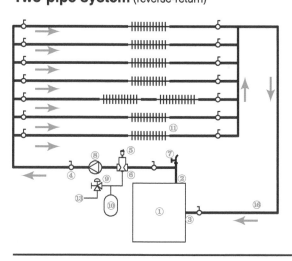

Two-pipe system (zoned — 3 zones)

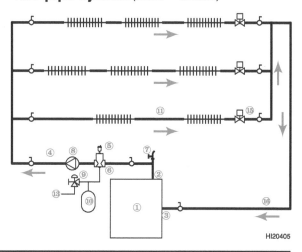

HI20405

① Hot water boiler	⑤ Automatic air vent, typical	⑨ Fill valve, typical	⑬ To fresh water make-up line
② Boiler outlet (hot water supply)	⑥ Air separator, typical	⑩ Diaphragm expansion tank, typical	⑭ Diverter tees (one-pipe fittings)
③ Boiler inlet (water return)	⑦ Hose bibb purge valve	⑪ Heating units, baseboard typical	⑮ Zone valves
④ Isolation valves (handle shows flow)	⑧ Circulator, typical	⑫ Manual air vents (not shown)	⑯ Common return (trunk line)

System applications

Series-loop systems

The series-loop system is the simplest possible choice. Use single-loop series piping only when zoning isn't required. Check the system length as explained in Part 5 to be sure the circulator will provide sufficient flow. You may have to use a higher-head circulator than the one supplied with the boiler in some cases.

If a single loop is too long, check the same circulator if used on a split loop. This will improve flow and could be zoned (two zones).

If the owner wants an indirect domestic water heater or radiant floor heating, for example, you cannot use series-loop piping. Use a system that can be zoned and can control the water temperature as required. Though you cannot apply an indirect water heater on a series-loop system, you can use a tankless coil in the boiler (with domestic water piped to a storage tank if desired).

(See the following pages for other system types.)

One-pipe systems

Figure 6, next page, shows the internal construction of a diverter tee and two typical applications. The tee incorporates an internal taper, as shown. The diverter tee application data will provide guidance on where to place the tees and the pressure drop with the tee in the forward or reverse positions shown in Figure 6. When the heating unit is located below the supply line, as in Figure 6, bottom, use a reverse-positioned diverter tee at the supply side of the heating unit as shown. This will provide enough pressure difference to overcome the buoyancy of the hot supply water. Also see application suggestions from the diverter tee manufacturer.

Obtain selection information for the diverter tees from your supplier. Carefully select the fittings and pipe size,

using the information provided by the diverter tee manufacturer and the guidelines for pipe sizing from Part 5 or another pipe sizing procedure.

One-pipe systems allow individual zoning of heating units, as shown in Figure 7. The system in Figure 7 includes baseboard units and a fan coil unit. Each one is independently controlled (using zone valves or thermostatic valves). You must include the pressure drop through the control valve when selecting and sizing diverter tees. Always select control valves with the lowest possible pressure drop (highest Cv). You will probably have to specially select a circulator for a system like that of Figure 7. The low-head circulator supplied with most boilers would probably not be able to handle the pressure drop at the required flow.

Figure 6 Diverter tee (one-pipe fitting)

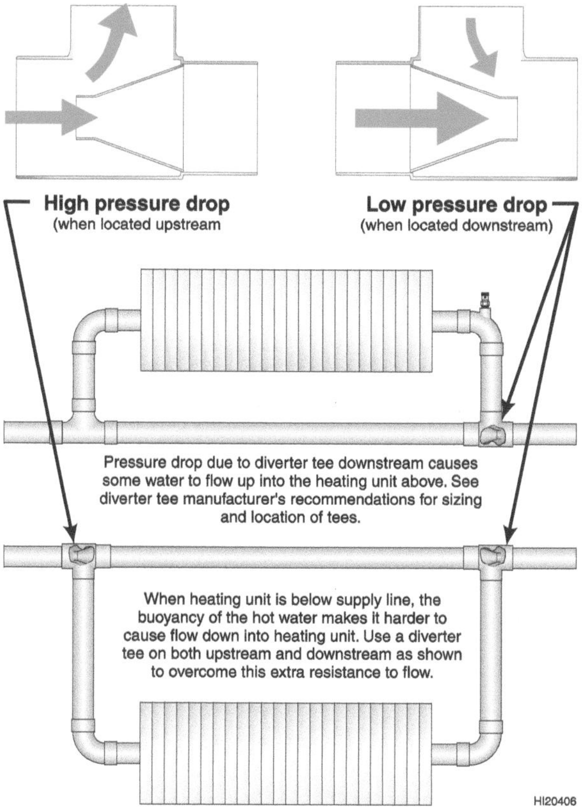

Figure 7 One-pipe heating system

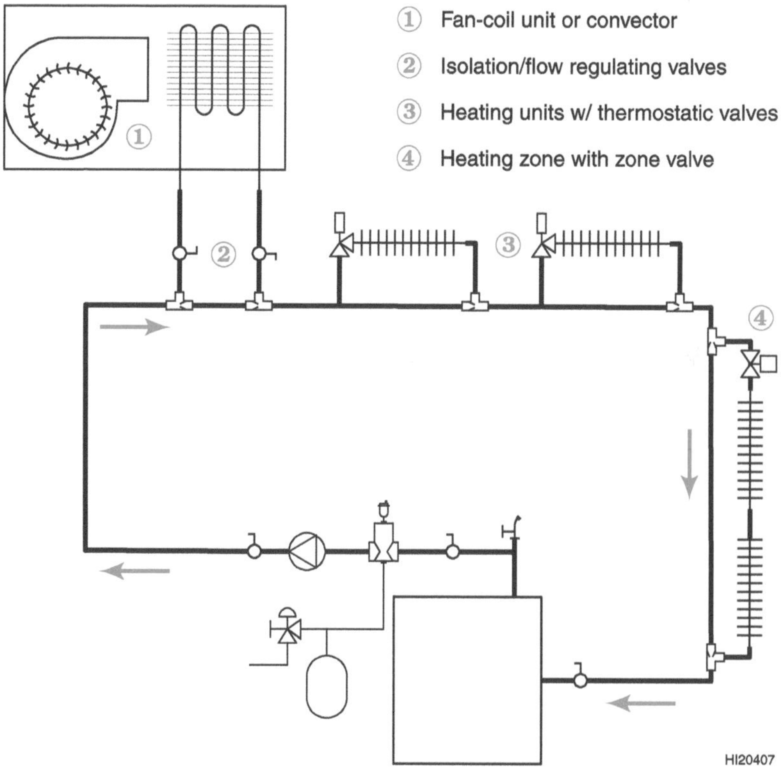

① Fan-coil unit or convector
② Isolation/flow regulating valves
③ Heating units w/ thermostatic valves
④ Heating zone with zone valve

Two-pipe systems

Figure 8, Page 4–19 is a schematic of a two-pipe system, with most of the heating units individually connected to the supply and return mains. Note the reverse-return piping. The first unit to receive supply water has the longest distance back to the boiler through the return line.

In the example of Figure 8, each heating unit is equipped with a thermostatic valve, which will sense room temperature and close off flow to the heating unit when the valve sensor is satisfied. One circuit, item 1, includes two heating units without a control valve. This system operates using a thermostat located in the circuit 1 space — a space in the home that is likely to call for heat most often. The thermostat would operate the boiler and circulator. Leaving one circuit open assures there is always a flow path for the water, even if all thermostatic valves close.

An alternative to the control method of Figure 8 is to install a thermostatic valve on every circuit and operate the circulator continuously. If you choose this alternative, install a bypass line from system supply to return, with a differential pressure bypass valve (bypass pressure regulator) in the line. Set the valve for a pressure below the maximum head for the pump and below the maximum close-off pressure of the zone valves. The bypass valve will open to limit circulator discharge pressure if all of the zones close. Also, you should install an outdoor temperature sensing control that will shut off the circulator when the outside temperature is above 65 °F (or an appropriate setting for the home and geographic area).

In Figure 8, the right side is a schematic of a two-pipe direct-return system. DO NOT pipe parallel circuits this way. You will encounter too many problems with unbalanced flow and heat distribution. It may be a minor problem on most residential jobs, but on large residential installations or commercial installations, it can be a serious problem.

Figure 8 Two-pipe system, typical (reverse-return and direct-return piping)

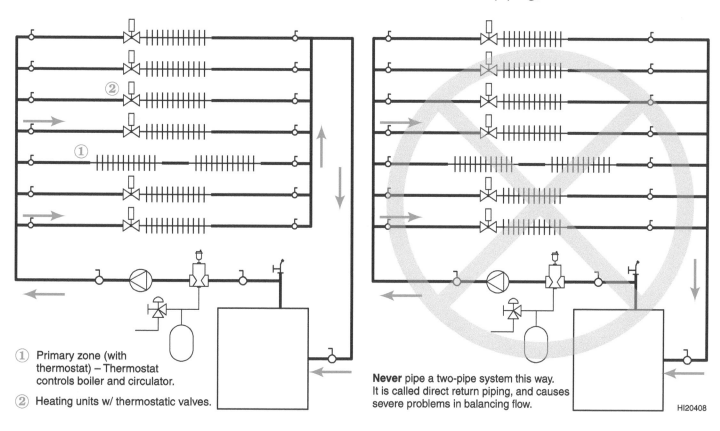

① Primary zone (with thermostat) – Thermostat controls boiler and circulator.

② Heating units w/ thermostatic valves.

Never pipe a two-pipe system this way. It is called direct return piping, and causes severe problems in balancing flow.

HI20408

Figure 9 Two-pipe zoned systems — zoning with zone valves & zoning with circulators

Zoning with zone valves

Zoning with circulators

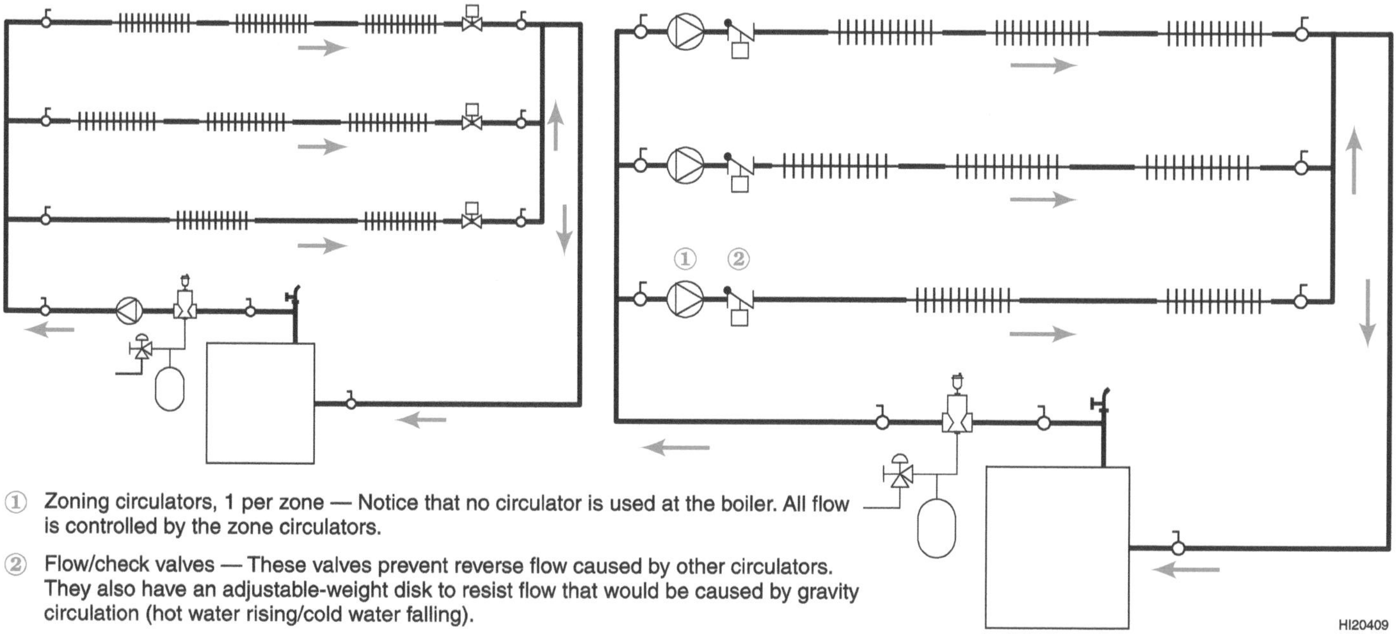

① Zoning circulators, 1 per zone — Notice that no circulator is used at the boiler. All flow is controlled by the zone circulators.

② Flow/check valves — These valves prevent reverse flow caused by other circulators. They also have an adjustable-weight disk to resist flow that would be caused by gravity circulation (hot water rising/cold water falling).

HI20409

Two-pipe zoned systems —
Zoning with zone valves

The system in Figure 2, Page 4–9, is a zone-valve-zoned system. You will see the same system repeated in Figure 9, Page 4–19 for comparison to circulator zoning of the same application.

When using the standard pump supplied with most residential boilers, you should use zone valves with the lowest available pressure drop. This will assure that you will provide sufficient flow to each of the heating circuits.

Generally, use zone valves with an end switch. Install 24-volt transformers as needed to supply the power required by the zone valves. Pay close attention to the va (volt-amp) requirement of each valve. Be sure the transformer va capacity exceeds the sum of all connected valves.

Service tip: When using 3-wire zone valves, pay careful attention to your wiring connections. If you mis-wire the connections from the valve to the boiler, you can damage boiler components with over-voltage. To check whether the valves are all wired to the boiler thermostat connection correctly, disconnect the wires from the boiler. Place a 24-volt voltmeter across the lead wires. Then operate the zones one at a time and together using the zone thermostats to activate the zone valves. Under no circumstances should the voltmeter show a voltage reading across the lead wires to the boiler. If you see a voltage reading, one or more of the wires is wrong. Backtrack which zone(s) caused the voltage to appear and correct the wiring.

If the home contains more than two or three zones, you may not want every zone to control the boiler. Consider only using primary zones (those most likely to have the longest duration of heating requirement) to operate the boiler. Let smaller zone thermostats only control the zone valve. Smaller zones will only be able to heat when the primary zones are calling for heat. (This method is called master/slave zoning.) If you select the primary zones correctly, you will find the boiler will cycle less. This will save fuel and provide smoother performance than if every zone, even the small ones, were to turn the boiler on and off on demand. You can use two-wire zone valves (no end switch) on zones which don't operate the boiler.

Figure 10, Page 4–21, bottom, shows a variation on zone-valve zoning. Here the boiler is piped on a secondary loop, using its own circulator. You would need to install an additional circulator, item 12, to provide flow for the system. Apply the method of Figure 10 when the system requires a larger circulator than supplied with the boiler.

Figure 10 Two-pipe zoning examples

Zoning with circulators (boiler on secondary loop)

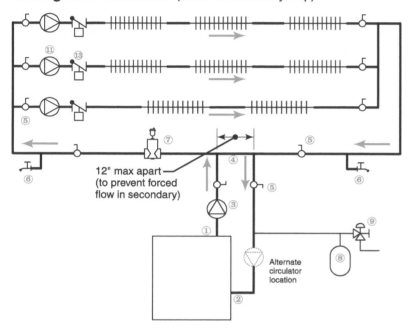

Zoning with zone valves (boiler on secondary loop)

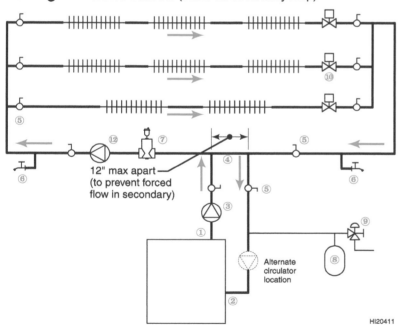

HI20411

① Boiler outlet (supply)
② Boiler inlet (return)
③ Boiler loop circulator
④ Secondary connection (boiler circuit)
⑤ Isolation valves
⑥ Purge cocks (hose bibb)
⑦ Air separator with automatic air vent
⑧ Diaphragm expansion tank
⑨ Combination fill valve
⑩ Zone valves
⑪ Zone circulators (circulator zoning)
⑫ System circulator (zone valve system)
⑬ Flow/check valves (on both supply and return to prevent gravity flow as well as reverse flow in circuit)

Two-pipe zoned systems — Zoning with circulators

See the right side of Figure 9, Page 4–19. Here you see the system of Figure 2, Page 4–9, piped using circulators instead of zone valves (as in Figure 9 left side).

Each circuit also requires a flow/check valve. The check function is needed to stop reverse flow in a circuit when its circulator is off, but other zone circulators are running. If the check mechanism weren't there, the pressure drop through the boiler and connected piping caused when another zone runs would push water through idle zones. The flow/check valve also has a disk with an adjustable weight. The weight on the disk holds it down enough to prevent gravity circulation. That is, with cold water in the circuit and hot water in the piping below, the hot water would try to rise into the circuit as the cold water tried to fall. You can set the weight just enough to overcome this buoyancy effect.

For residential applications, you can usually use low-head circulators. If using high pressure drop heating units, select a circulator that will provide the correct flow against the pressure drop of the heating unit.

Figure 10, top, shows a variation on circulator zoning. Here the boiler is piped on a secondary loop, using its own circulator. Install a circulator and flow/check valve on each circuit, sized for the heating unit flow and pressure drop required.

You can apply zone valve or circulator two-pipe methods to almost any application, including multiple applications such as space heating with domestic water heating, snow melt, pool heating or any other. Just select the pumps and piping needed to provide the required temperature and flow.

See Guide RHH Part 5 for methods to use in sizing piping and circulators.

Primary/secondary systems

See Figure 12, Page 4–23 and Figure 13, Page 4–24.

Primary/secondary piping has been used in commercial systems for many years, but only recently has been used in residential applications. The growth of primary/secondary systems is probably due to the wide range of applications in residential heating, multiple boiler installations and the popularity of radiant heating. Primary/secondary piping is the best choice for radiant heating because it provides the greatest flexibility in control of flow and temperature, both for system supply and boiler return.

You will find two variations in primary/secondary piping — one-pipe and two-pipe, as shown in Figure 12 and Figure 13.

Secondary connections

A critical aspect of primary/secondary piping is the secondary connection — the piping between the secondary circuit supply and return connections to the main. The supply and return connection must be no further aPart than 12 inches (or four pipe diameters, whichever is smaller). With the supply and return pipes close together, the flow in the primary piping won't cause much of a pressure drop flowing between the connections. This prevents forced flow in idle circuits which otherwise would be caused by flow in the primary loop.

It might appear that water would just bypass right back into the circuit, but it doesn't. When a secondary circulator operates, it pulls water from the primary loop. The same amount of water is put back into the primary loop. If the primary pump were not flowing, this water would just bypass right back. But, with the primary pump flowing, the water injected into the primary is moved away by the flow.

There are three possibilities for flow in the connection piping at a secondary connection:

1. Primary flow and secondary flow the same (Figure 11, top): All the water pulled by the secondary is taken from the primary flow and put back into the primary. There is no flow at all in between the connections while both pumps operate.

2. Primary flow higher than secondary flow (Figure 11, center): The water pulled into the secondary and put back into the primary subtracts from the primary flow. The flow in between the connections is the difference between the two.

3. Primary flow less than secondary flow (Figure 11, bottom): The water pulled into the secondary is more than the primary flow, so it pulls all of the primary water plus some water backwards through the connecting piping. The flow in the connecting piping will be the difference between the two, flowing in the opposite direction to the primary!

Size the connecting piping to whichever is larger — the primary or the secondary piping. You can then be sure the pipe size will always be correct for the flow.

Figure 11 Flow in secondary connections

Primary flow = secondary flow (P=S)

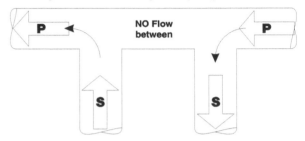

Primary flow more than secondary flow (P>S)

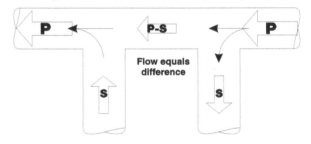

Primary flow less than secondary flow (P<S)

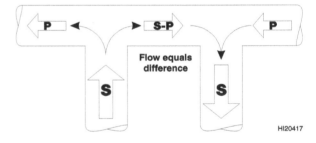

HI20417

Flow/check valves

Use flow/check valves where shown in the piping diagrams. Flow/check valves prevent flow in idle zones that might otherwise be caused by the pressure difference between the supply and return connections caused by water flow through the primary piping. Use flow/check valves on both sides of the secondary circuits. This will prevent gravity flow of hot water, which can occur even in a single pipe.

You do not need flow/check valves on secondary heating circuits located below the primary piping. For these circuits, no gravity circulation occurs because the heating unit water is colder than the supply water in the pipes above. The weight difference of the cold water in the heating units versus the hot water in the pipes above is also enough to prevent induced flow due to the pressure drop through the connecting piping.

One-pipe primary/secondary piping

One-pipe primary/secondary systems use a single-pipe primary loop (Figure 12, right). Heat sources (boiler, heat exchangers, etc.) and heating load circuits connect off the primary in a series fashion, similar to a one-pipe system (without diverter tees).

Size the primary piping and circulator for a flow rate high enough to ensure that all secondary circuits will receive water at a high enough temperature. This is because the supply loop temperature drops as secondary circuits remove heat from the primary. See Part 5 for the sizing procedure for one-pipe primary/secondary systems.

To minimize the impact of the temperature drop around the primary loop, pipe the circuits requiring the highest temperature at the beginning of the loop, with lower-temperature circuits (such as radiant heating) at the end.

Each secondary circuit operates independently of the others, and has no effect on the rest of the system except to remove heat from the primary circuit. Size the circulator in each secondary circuit to meet the needs of the heating units. If the boiler is equipped with a factory-supplied circulator, use it for the boiler secondary loop circulator.

This piping method uses about the same amount of piping as a circulator-zoned two-pipe system, with the exception of the additional pumps needed for the boiler loop and primary loop.

The controls for the system must operate the primary circulator any time any zone calls for heat. (The exception is a domestic water heater piped directly off of the boiler secondary circuit. See Part 6 for details.)

Figure 12, Page 4–23 and Figure 13, Page 4–24 show three methods of temperature control applied in the secondary circuits. Since each secondary circuit operates independently, any temperature control or piping method can be applied in the secondary circuit. For low-temperature heating systems, such as radiant floor heating, variable-speed injection pumping provides effective control and can also regulate the heat taken into the radiant heating circuit to control the return water temperature going back to the boiler.

Figure 12   One-pipe primary/secondary system

One-pipe primary/secondary piping

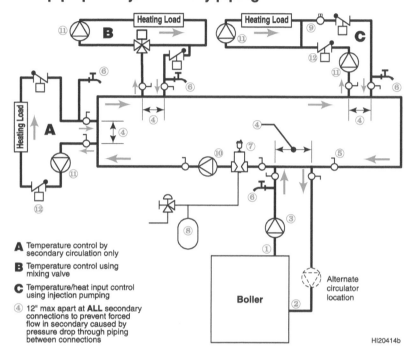

HI20414b

A Temperature control by secondary circulation only

B Temperature control using mixing valve

C Temperature/heat input control using injection pumping

④ 12" max apart at **ALL** secondary connections to prevent forced flow in secondary caused by pressure drop through piping between connections

① Boiler outlet (supply)
② Boiler inlet (return)
③ Boiler loop circulator
④ Secondary connections (see note above)
⑤ Isolation valves
⑥ Purge cocks
⑦ Air separator with automatic air vent

⑧ Diaphragm expansion tank w/ fill valve
⑨ Balancing valves (square-head cocks)
⑩ System circulator
⑪ Secondary circulators
Injection pump (fixed or variable speed)
⑫ Flow/check valves (to prevent gravity or pressure-drop-induced flow in secondary circuits)

Two-pipe primary/secondary piping

Two-pipe primary/secondary systems use a supply manifold and a return manifold, with a crossover piping connection for each secondary circuit, as shown in Figure 13, right.

Balance the flow to each of the crossover lines using square-head cocks or flow balancing valves where shown in Figure 13. Two-pipe primary/ secondary piping has a major advantage over the one-pipe equivalent — since every circuit has water available at about the same temperature, circulator and line sizing is easier. And, since return water from secondary and primary circuits has no impact on other circuits, the temperature drop in the primary circuit can be as large as desired. The higher the temperature drop, the lower the flow. So two-pipe primary/secondary systems can use much smaller primary circulators. See Part 5 for sizing procedure for two-pipe primary/secondary systems.

The operation of secondary circuits is the same as described for one-pipe primary/secondary circuits.

Figure 13   Two-pipe primary/secondary piping

Two-pipe primary/secondary piping

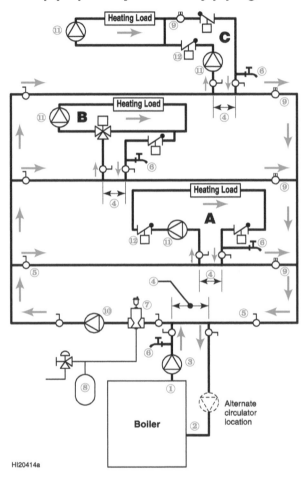

HI20414a

① Boiler outlet (supply)
② Boiler inlet (return)
③ Boiler loop circulator
④ Secondary connections
⑤ Isolation valves
⑥ Purge cocks
⑦ Air separator with automatic air vent
④ 12" max apart at **ALL** secondary connections to prevent forced flow in secondary caused by pressure drop through piping between connections

⑧ Diaphragm expansion tank with fill valve
⑨ Balancing valves (square-head cocks)
⑩ System circulator
⑪ Secondary circulators
   Injection pump (fixed or variable speed)
⑫ Flow/check valves (to prevent gravity or pressure-drop-induced flow in secondary circuits)

A Temperature control by secondary circulation only
B Temperature control using mixing valve
C Temperature/heat input control using injection pumping

Modular and multiple boilers

Using more than one boiler allows sequencing the heat input to match the load. These applications only fire all of the boiler units during maximum demand. As the system heat load reduces, boiler units are turned off. This can greatly improve the efficiency of a system, particularly if it is applied to widely varying loads (such as snow melting or spa heating, for examples).

A **modular boiler** is a group of boiler units connected to a common piping supply and return header, with no intervening valves. If the boiler units are isolated from the headers with valves, they are then called **multiple boilers**.

A modular boiler pipes to the system as if one unit, and can be substituted for any of the single boilers in the piping diagrams of Part 4.

Install multiple boilers in a secondary loop as shown in Figure 14. You can apply this piping to any system type by piping the secondary loop off of the primary as shown. Each boiler requires its own circulator (usually supplied with the boiler). This piping method ensures there is no flow through idle boilers.

Use multiple boilers instead of a single boiler when the heating load for a home can vary significantly. For example, if the hydronic system includes domestic water heating and a heat exchanger for snow melting, the large snow-melt load would only occur during a small portion of the year. A single boiler would be well over-sized for the remainder of the year. Another example would be space heating plus heating an outside pool or a large whirlpool heating load. The pool or whirlpool loads may be large and only used during limited times. For the remainder of the season, the heating system would be more efficient if using a smaller boiler. By using multiple boilers, with the boilers staged on based on load, heating capacity is matched closely to need. This reduces cyclic losses and increases seasonal efficiency, particularly if piped as in Figure 14, ensuring no flow through idle boilers.

Figure 14 Piping multiple boilers

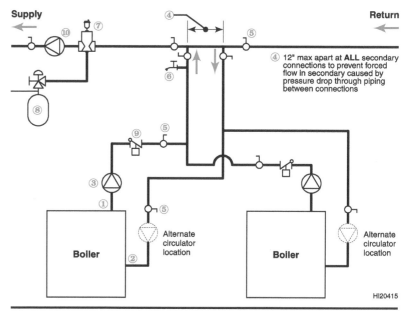

① Boiler outlet (supply)
② Boiler inlet (return)
③ Boiler loop circulators
④ Secondary connection
⑤ Isolation valves
⑥ Purge cock
⑦ Air separator with automatic air vent
⑧ Diaphragm expansion tank w/ fill valve
⑨ Flow/check valves
⑩ System circulator

Multi-use systems

See Figure 15, a schematic of a two-pipe circulator-zoned system. This schematic shows the versatility of even a simple hydronic system. This system provides heat for radiant floor heating, domestic water heating, snowmelt and baseboard. All you have to do is to size the circulators and the piping to handle the flow through each.

As noted in Figure 15, install protective piping at the boiler if the return water temperature could drop below the minimum recommended by the boiler manufacturer. See the following discussion on how to install bypass piping and temperature regulation for the return water.

Figure 15 Multi-use system

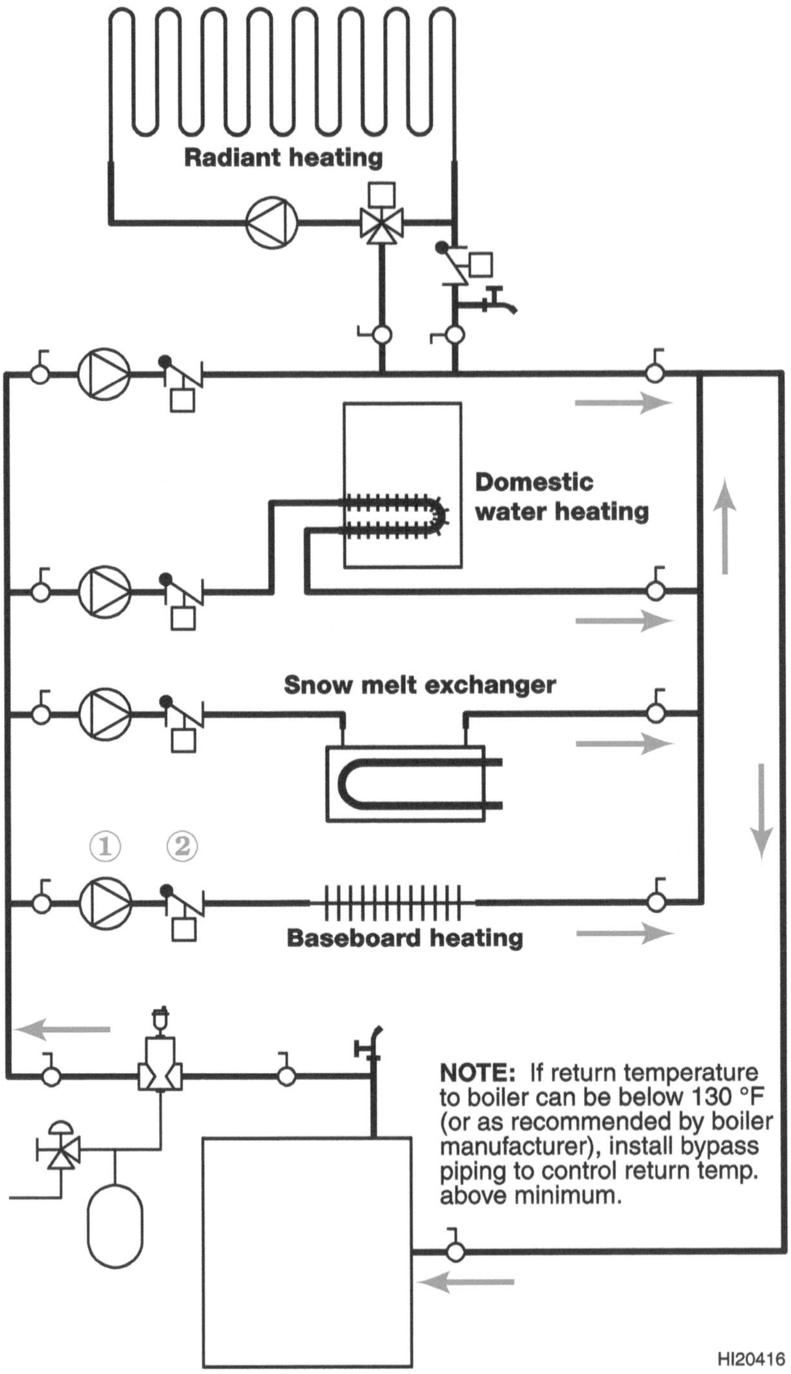

Radiant heating

Domestic
water heating

Snow melt exchanger

Baseboard heating

NOTE: If return temperature to boiler can be below 130 °F (or as recommended by boiler manufacturer), install bypass piping to control return temp. above minimum.

HI20416

Water temperature control methods

Water temperature to the system or to a heating loop can be controlled using mixing valves or injection mixing. This sections gives typical examples.

Notice the use of a 3-way valve to control the supply temperature to a radiant heating loop in Figure 15, Page 4–26. The 3-way valve is used here to reduce and control water supply temperature to the radiant loop because the other heating loops in this system require higher temperature water. Note that this valve could be used for outdoor reset as well as fixed-temperature.

Figure 16 is the same system, but heat input control to the radiant circuit uses an injection pump instead of a 3-way valve. The injection pump is operated by a temperature control in the radiant loop. The pump can be on/off or variable speed.

- When the radiant loop needs heat, the injection pump temperature controller control turns on the injection pump. Hot water from the main heating loop flows to the suction side of the radiant loop pump.

- Any water in the injection loop that doesn't flow to the radiant circuit returns through the balancing valve back to the main system loop.

- The balancing valve must be adjusted to limit the flow rate in the injection mixing loop. This limits the amount of heat that can enter the loop.

- The injection pump temperature control turns the pump on/off or regulates pump speed to meet heat demands in the radiant loop.

Other temperature control methods include using either automatic or manually-set control valves (2-way, 3-way or 4-way). Refer to recommendations from the boiler manufacturer and temperature controller manufacturer for alternatives.

Figure 16 Multi-use system (using injection mixing)

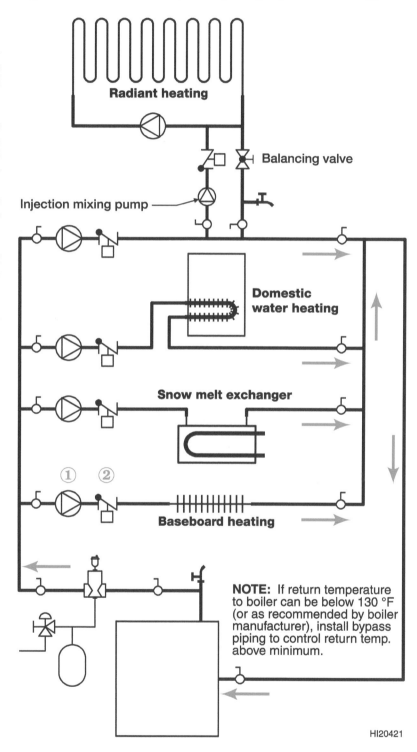

Radiant heating

Balancing valve

Injection mixing pump

Domestic water heating

Snow melt exchanger

① ②

Baseboard heating

NOTE: If return temperature to boiler can be below 130 °F (or as recommended by boiler manufacturer), install bypass piping to control return temp. above minimum.

HI20421

Bypass piping

<table>
<tr><td>

WARNING
Prevent boiler condensation

⚠️ Non-condensing boilers must be installed with piping and controls that prevent sustained operation with low return water temperature. Follow instructions in the boiler manual for piping and return water control. This section explains bypass piping for temperature regulation.

</td></tr>
</table>

Definitions

Bypass piping refers to piping near the boiler intended to either:

1. Divert flow around the boiler —

   BOILER bypass   — or —

2. Divert flow around the system —

   SYSTEM bypass.

Illustrations

Figure 17, Page 4–29 shows simplified flow diagrams for the two bypass methods.

Figure 18, Page 4–30 covers **Boiler bypass** piping. Figure 19, Page 4–31 covers **System bypass** piping.

Figure 20, Page 4–32 shows the correct method to connect an indirect water heater when using boiler bypass piping.

Boiler bypass:

- Mixes system return water with boiler supply water, so system supply temperature is lower than boiler outlet temperature. (The boiler runs at a hotter temperature than the system.)

- Increases flow through system and reduces flow through boiler. This causes a large temperature rise through the boiler, with a smaller temperature rise through the system.

- The adjusting valves will usually be set to divert about 2/3 of the flow around the boiler, making system flow 3 times boiler flow.

System bypass:

- Mixes boiler outlet water with boiler return water, to increase boiler return water temperature.

- Decreases flow through system and increases flow through boiler. This causes a large temperature drop through the system, with a smaller temperature rise through the boiler.

- System flow will be less than boiler flow. The amount of bypass is usually determined by how much flow is needed to raise boiler return temperature above the minimum temperature

that will prevent flue gas condensation in the boiler.

Boiler bypass applications

Large water content systems (gravity water systems or converted steam systems).

(Also see Page 4–33.) The boiler needs a large temperature rise (usually 60 °F) on large water content systems to prevent condensation because the return temperature is low through most of the heating season. (The system will be satisfied with water at 130 °F or less most of the time, and it takes a long time to heat the system up.) Boiler bypass allows the system to operate with, typically, a 20 °F temperature drop while the boiler operates with a 60 °F temperature rise. If the system flow rate were reduced to meet the boiler's needs, heat in the system would be uneven and uncomfortable.

Multi-zone baseboard systems

Boiler bypass piping will reduce radiation/piping expansion noises because hot water from boiler blends with system return water when a zone opens. The radiation heats up gradually, not all at once.

High-mass radiant heating systems

Radiant heating systems with tubing embedded in thick concrete slabs cause low return water temperature on start-up and usually require relatively low temperature supply water. Boiler by-pass piping allows the system to run cooler than the boiler, satisfying system needs while preventing boiler condensation.

System bypass applications

Systems using outdoor reset controls

Because the system operating temperature changes during the season, flow rates must be automatically controlled. Use an automatic 3-way or 4-way mixing valve to keep the boiler return temperature above 130 °F regardless of the system operating temperature.

A 3-way valve (as in Figure 19, Page 4–31) will supply water at boiler outlet temperature to the system.

A 4-way valve (as in), with separate circulator for boiler and system, allows boiler outlet water to mix with system return water, providing lower supply temperature to the system.

Figure 17 Boiler bypass piping

BOILER BYPASS

Simplified piping

Notice that circulator must be OUTSIDE the bypass loop.

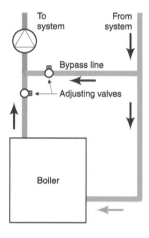

Circulator on supply

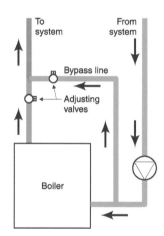

Circulator on return

Typical flow ratio

Sys Flow = 3 x Blr Flow
(typical)

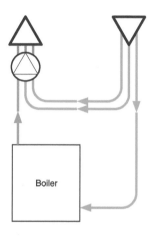

Flow diagram

HI20420

SYSTEM BYPASS

Simplified piping

Notice that circulator must be INSIDE the bypass loop.

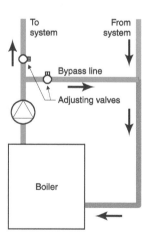

Circulator on supply

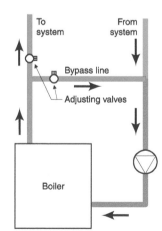

Circulator on return

Typical flow ratio

Sys Flow LESS THAN Blr Flow

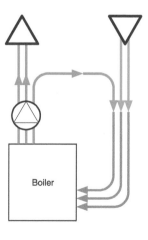

Flow diagram

Figure 18   Boiler bypass piping details

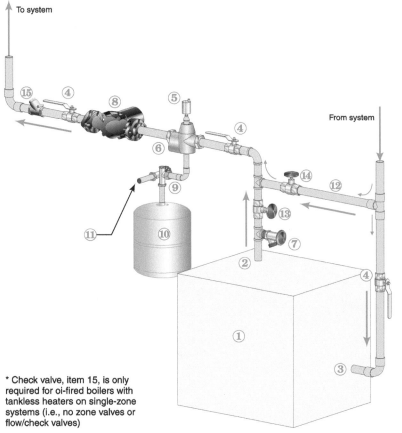

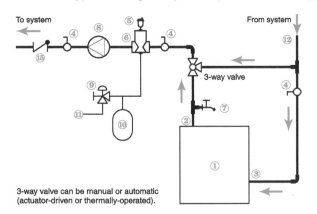

* Check valve, item 15, is only
required for oi-fired boilers with
tankless heaters on single-zone
systems (i.e., no zone valves or
flow/check valves)

A. Circulator on supply, as at left

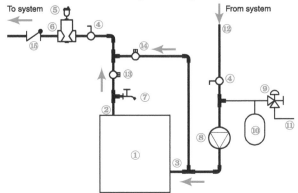

B. Circulator on return (alternate)

C. Boiler bypass using 3-way valve (circulator on supply)

3-way valve
To system
From system

3-way valve can be manual or automatic
(actuator-driven or thermally-operated).

Adjusting valves

Adjust valves (or 3-way manual valve) to obtain the
required flow through boiler and bypass. For high-water
content systems, such as gravity water or converted
steam systems, you will usually set the valves so boiler
flow is about 1/3 of system flow. This will cause boiler
temperature rise to be 3 times system temperature rise.

Install thermometers in the system supply and return
lines to simplify setting flows.

① Hot water boiler	⑤ Automatic air vent, typical	⑨ Fill valve, typical
② Boiler outlet (hot water supply)	⑥ Air separator, typical	⑩ Diaphragm expansion tank, typical
③ Boiler inlet (water return)	⑦ Hose bibb purge valve	⑪ To fresh water make-up line
④ Isolation valves (handle shows flow)	⑧ Circulator, typical	⑫ Bypass line

⑬ Globe or plug valve
⑭ Globe or plug valve
⑮ Check valve*

HI20417a

Figure 19   System bypass piping details

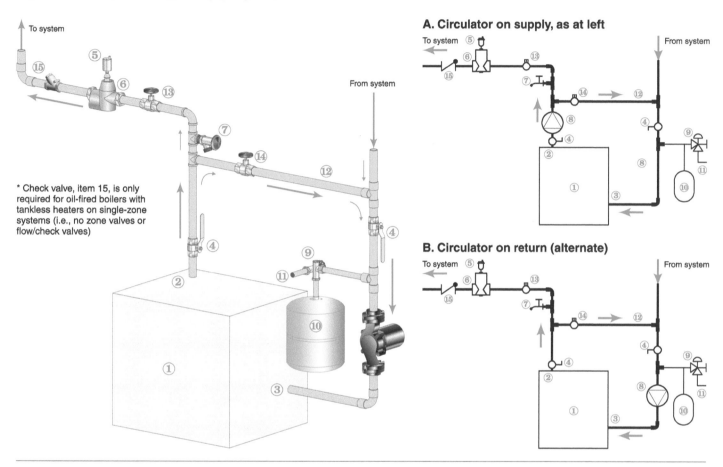

* Check valve, item 15, is only required for oil-fired boilers with tankless heaters on single-zone systems (i.e., no zone valves or flow/check valves)

A. Circulator on supply, as at left

B. Circulator on return (alternate)

Adjust valve(s) to obtain the required minimum return water temperature to the boiler. Install a thermometer in the boiler return line to simplify setting flows.

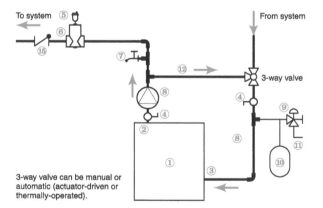

C. 3-way valve option (circulator on supply)

3-way valve can be manual or automatic (actuator-driven or thermally-operated).

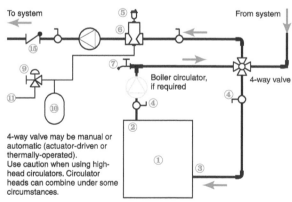

D. 4-way valve option (circulator on supply)

4-way valve may be manual or automatic (actuator-driven or thermally-operated).
Use caution when using high-head circulators. Circulator heads can combine under some circumstances.

① Hot water boiler	⑤ Automatic air vent, typical	⑨ Fill valve, typical	⑬ Globe or plug valve
② Boiler outlet (hot water supply)	⑥ Air separator, typical	⑩ Diaphragm expansion tank, typical	⑭ Globe or plug valve
③ Boiler inlet (water return)	⑦ Hose bibb purge valve	⑪ To fresh water make-up line	⑮ Check valve*
④ Isolation valves (handle shows flow)	⑧ Circulator, typical	⑫ Bypass line	

HI20418

Indirect water heater application

Indirect water heaters usually require hotter supply water than the space heating system. Figure 20 shows how to connect an indirect water heater to system bypass piping so it will receive hot water directly from the boiler.

The bypass adjusting valves (items 13 and 14) will usually be adjusted for system flow three times boiler flow. Install thermometers in the system supply and return lines to simplify setting flows.

The two flow/check valves (item 14) prevent gravity flow or induced flow. They also prevent the space heating circulator from pulling water through the water heater

piping and prevent the water heater circulator from pulling water through the system piping.

The water heater piping will usually have a lower pressure drop than the system piping. If both circulators should run at the same time, flow will favor the water heater. This gives a slight advantage to water heater demand. To ensure full boiler capacity will be dedicated to the water heater on DHW demand (domestic priority), use a control or relay to disable the space heating circulator during DHW call for heat.

Figure 20 Indirect water heater application using **Boiler bypass** piping for space heating

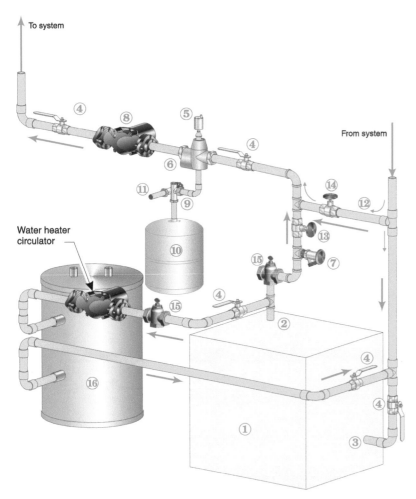

A. Schematic diagram, circulator on supply

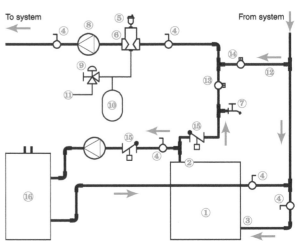

B. Simplified piping diagram to show flows

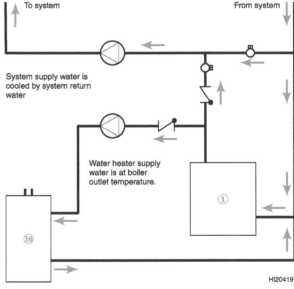

① Hot water boiler
② Boiler outlet (hot water supply)
③ Boiler inlet (water return)
④ Isolation valves (handle shows flow)
⑤ Automatic air vent, typical
⑥ Air separator, typical
⑦ Hose bibb purge valve
⑧ Circulator, typical
⑨ Fill valve, typical
⑩ Diaphragm expansion tank, typical
⑪ To fresh water make-up line
⑫ Bypass line
⑬ Globe or plug valve
⑭ Globe or plug valve
⑮ Flow/check valve
⑯ Indirect water heater

HI20419

Figure 21 Piping converted gravity systems

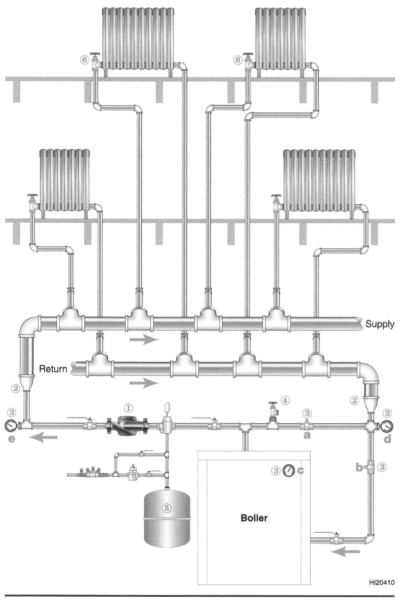

① Circulator

② Install reducing fittings so near-boiler piping will be sized as recommended by boiler manufacturer (typical ¾" to 1½" for residential boilers).

③ Install square-head cocks a and b to slow down the flow in the boiler by bypassing some of the water. Set valves so boiler outlet water temperature (gauge c) is about 60°F higher than the entering temperature (gauge d). Gauge e shows the water temperature going to the system. Check boiler manual for other information.

④ Purge bibb valve

⑤ Install a diaphragm-type expansion tank and air separator if possible to improve air control in the system.

⑥ Check upper floor radiator hand valves to see if flow restrictors are installed. Remove these restrictors to prevent heat distribution problems.

Converted gravity systems

On large water content systems, such as gravity flow systems retrofitted with a new boiler and forced circulation (Figure 21), the system takes a long time to heat up. Many boilers require a return temperature of at least 130 °F or higher. Starting with the water in the system at room temperature, the boiler will have to heat for an extended time before it can raise all of the system water above this minimum. During the long heating period, the return water temperature will be below recommended minimum. This will risk damage to the boiler from flue gas condensation.

The piping of Figure 21 protects the boiler by slowing the flow enough to cause a high water temperature leaving the boiler, even when return water is cool. Note also that the near-boiler piping is sized per the recommendations from the boiler manufacturer for a typical forced-flow application.

Make sure to check the boiler manual for the manufacturer's recommendations for low temperature applications.

Wire the circulator for these large water content systems to operate continuously to avoid long heat-up times. Never line-size the pump. For residential systems, a standard residential boiler packaged circulator should suffice because of the low pressure drop in these systems.

NOTICE: Many gravity systems were equipped with orifice plates in the upper-story radiators to reduce the flow there. You may have to remove these plates and balance flow in radiators throughout the system using the radiator supply valves.

NOTICE: Adjust balancing cocks as explained in the legend for Figure 21. Make sure when set that the system temperature drop (difference between gauges d and e) is no less than 20 °F. This ensures the circulator flow has been throttled enough to prevent motor overload.

Freeze protection

Freeze protection

Some boiler applications require the system to be filled with an antifreeze mixture to protect the system from freeze-up due to harsh conditions or in the event the boiler should shut off during a non-occupancy period.

See Table 1, Page 4–39, in the Appendix for tips on freeze protection and use of glycol.

Notice the warning about ethylene glycol (primary component of most automotive antifreeze). Foremost, ethylene glycol is very toxic and, unfortunately, appealing to animals. What is more, ethylene glycol will dissolve or soften most rubber (elastomer) materials, including the gaskets throughout your system. For these reasons, always use propylene glycol. Though propylene glycol is more viscous than ethylene, it is an excellent substitute and is totally nontoxic. Never use automotive antifreeze. Only use a product specifically designed for heating systems.

If you are mixing the glycol yourself, carefully calculate (or measure) the system volume to ensure the right proportion of glycol. Inform the owner that the system must be checked annually to ensure the concentration is correct and the inhibitor is still active. Without the inhibitor, any glycol is highly corrosive.

WARNING —

Maintain antifreeze integrity

Ensure that the system water is checked at least annually to ensure that the antifreeze inhibitors have not degraded and that the antifreeze concentration is maintained. Drain the system and replace the water/antifreeze if necessary due to inhibitor break-down or other major failure.

USE ONLY ANTIFREEZE INTENDED FOR USE IN HYDRONIC SYSTEMS.

Glycol inhibitors can deteriorate over time. Without the inhibitors required, glycol is corrosive and will damage system components and cause leaks. Test the system water regularly to ensure the inhibitor concentration is correct.

WARNING: UNHEATED SPACES

Crawl spaces, unheated attics, eaves, etc.

When possible, **DO NOT** install piping or heating units (fan coil units, for example) in unheated spaces. Water in the piping or heating units can freeze, causing piping/equipment failure and leaks. Resultant leaks will cause significant damage, including potential for structural failures.

If any portion of the hydronic system is located in an unheated space, the entire system must be filled with an antifreeze mixture suitable for the lowest potential exposure temperature. Make sure the piping/equipment is accessible for periodic inspections.

Installation

Equipment installation

1. Place major equipment first — boiler, circulator(s), mixing valves, for example. These items set measuring distances for piping. Be sure to consider venting requirements and routing when placing boiler and piping.

2. Provide access clearance around all components that have to be serviced. Pay close attention to boiler manufacturer's required clearances to combustible materials and clearances required for service and operation.

Supporting pipe and tubing

1. Support pipe and tubing from structural members — supports no further aPart than shown in Table 3, Page 4–42 and Table 4, Page 4–43, in Appendix.

2. Where heavy components (circulators, expansion tanks, etc.) are installed in runs, provide a piping support immediately on either side of the components if not independently supported.

3. Allow for expansion (see Table 3 and Table 4, in Appendix). **Install supports so pipe can move. If supports are too tight, noise and abrasion will result.**

4. Drill floor penetration holes at least ¼" larger diameter than tube diameter. For long runs of copper baseboard, drill the holes large enough to allow for the full expansion of the tubing listed in Table 3, in Appendix. If holes are too small, tubing will rub when heated, causing noise and possible damage to tubing or baseboard units.

5. Drill holes through joists at least ¼" larger than diameter of tube, but never larger than 1½" (see warning on Page 4–37).

6. Locate holes through joists or structural members no closer than 2" to the surface — to avoid possible penetration from nails or fasteners. Locate as near the center of members as possible.

Routing and alignment

1. Pre-build subassemblies (where possible) on a workbench, vise or flat surface. Work will proceed faster and results will be better.

2. Use longest lengths possible to reduce fittings (using 20-foot lengths where possible).

3. Pitch branch piping — pitch up to heating units above the circuit, pitch down to heating units below the circuit.

4. Keep direction changes to a minimum — each fitting adds more pressure drop.

5. Use soft copper for branch piping and direction changes where possible — to eliminate elbows. Use a bending tool to prevent kinks.

6. Ream the ends of all piping — burrs cause pressure drops and loose pieces can enter the flow stream and damage components and gaskets.

7. Measuring for cut length — allow for take-up at solder socket or thread joint (see Table 3, Page 4–42 and Table 4, Page 4–43, in Appendix).

8. Measure from centerline, not edge of piping when laying out. This makes sure even different pipe diameters will line up.

9. Allow space around piping as needed if piping will be insulated.

10. Align similar near-by components — zone valves, circulators, etc.

11. Use a level and/or plumb bob for accurate, neat-appearing pipe runs.

12. Route water piping below electrical wiring — leaks won't drip on wiring conduit or wires.

WARNING —

Drilling through structural members

When routing piping through joists or other structural members, **never drill holes larger than 1½" diameter**. Larger holes could reduce the strength of the structure. Never notch framing members or structural supports. The member will be weakened and could fail.

Appendix

Freeze protection tips

Table 1   Freeze protection tips

Glycol concentration limit	Never exceed 60% glycol by volume. Higher concentrations will result in sludge and equipment damage. Have glycol concentration checked annually. Verify that inhibitor level is correct. Deterioration of inhibitor will result in highly corrosive solution.
Constant circulation in cold areas	Use constant circulation in heating circuits which might be exposed to low temperatures for extended periods (heated garages, for instance). Circulating water is less likely to freeze because it doesn't stay in the cold area long enough to cool below the freezing point.
WARNING	Use only **inhibited propylene glycol**, designed specifically for heating systems. DO NOT use ethylene glycol or automotive antifreeze. Ethylene glycol is toxic and will damage most rubber materials, including gaskets.
No galvanized or aluminum pipe or fittings	Never use galvanized piping if glycol will be used. The coating reacts with glycol. Never use aluminum piping or components.
Clean the system	Thoroughly clean the system with trisodium phosphate or other chemical cleaner. Residue and sludge can compromise the glycol. Use caution when working with chemicals, following all instructions and warnings of the supplier.
Monitor water make-up	Install a water meter (with ½-gallon resolution) on the fill line, instructing the owner to check the meter periodically to ensure there are no system leaks. Alternatively, close off the automatic fill valve, allowing manual fill. This way, if the system leaks, the pressure will drop, making the owner aware of a problem. If the system is allowed to fill without being checked, the glycol concentration will reduce with time as fresh water enters the system. The problem is made worse because glycol leaks more easily than water, so the leak will be mostly glycol. Another alternative is to use a pressure switch and glycol feeder. If the system leaks, the pressure drops, and the switch activates the feeder.
No chromate water treatment	Chromate reacts with glycol. Never use chemicals containing chromate in the system.
Circulators should never packing gland seals	Packing glands can leak easily, allowing glycol to escape. Use only wet-rotor circulators or other circulators equipped with mechanical seals.
Increase expansion tank size	For most systems, increase the expansion tank size by at least 20% compared to water-only applications. Refer to tank manufacturer's sizing guidelines for more information.
Increase heating unit size or flow	Glycol/water does not transfer or carry heat as well as water alone. Rule-of-thumb: For most systems operating between 180 and 220 °F, increase design flow rate by about 14% when using 50/50 glycol/water to compensate for the lower heat transfer. Consult heating unit manufacturer's guidelines, particularly for heat exchangers.
Pressure drop for glycol/water is lower than for water only	Rule-of-thumb: For operating temperatures between 140 and 220 °F, the pressure drop for 50/50 glycol/water will be about 10% less than for water-only at the same conditions.

Advantages and disadvantages of piping systems

Table 2   Advantages and disadvantages of piping systems

System	Advantages	Disadvantages
Series-loop	• Simple. • Inexpensive. • Simple controls (single thermostat).	• Cannot be zoned. • High pressure drop due to long circuit. • Supply temperature to last heating units is lower than supply temperature to first units in loop. • May have low flow/high temperature drop, causing end-of-system heating problem's unless heating units are upsized at end of system to compensate. • If heating units are over-sized, will have heat distribution problems, overheating spaces heated by first heating units in flow.
Split-loop series-loop	• Simpler than most other systems. • Inexpensive. • Can be zoned (only one zone per loop). • An effective way to improve performance in an existing single-loop series-loop system experiencing heat distribution problems.	• If heating units are over-sized, will have heat distribution problems, even with split-loop piping. • Supply temperature to last heating units is lower than supply temperature to first units in loop. • Limited number of zones will make it difficult to locate thermostats for uniform comfort. • Heating units can only be grouped by proximity — cannot combine similar spaces for best zoning.
One-pipe	• Each heating unit can be separately controlled, using zone valves or thermostatic valves. • Usually requires less pipe than other multi-zoned methods. • Over-sized heating units not as likely to cause heating problems as in series-loop systems.	• More complex to design than series-loop or two-pipe systems. • Supply temperature to last heating units is lower than supply temperature to first units in loop. • Pressure drop through diverter tees increases as heating unit flow is closed off, causing reduced flow in system. • Heating units at the end of the system may need to be up-sized to account for lower available supply water temperature. • Circulator must run constantly if using thermostatic zone valves. • Often more difficult to purge air than other systems.
Split-loop one-pipe	• Each heating unit can be separately controlled, using zone valves or thermostatic valves. • Usually requires less pipe than other multi-zoned methods. • Over-sized heating units not as likely to cause heating problems as in series-loop systems. • Less likely to have heat reduction on end-of-system heating units because supply temperature to units is higher than single-loop one-pipe systems.	• More complex to design than series-loop or two-pipe systems. • Supply temperature to last heating units is lower than supply temperature to first units in loop. • Pressure drop through diverter tees increases as heating unit flow is closed off, causing reduced flow in system. • Circulator must run constantly if using thermostatic zone valves. • Often more difficult to purge air than other systems.
Two-pipe reverse-return	• Each heating unit (or series group of units) can be zoned, using thermostatic valves or zone valves. • More flow will occur in the branches than with series-loop or one-pipe when using standard-equipment residential boiler circulators. • Each heating unit (or series group of units) receives about the same supply water temperature. • Flow through each branch will be about the same for similar heating units without having to use balancing valves. No balancing may be needed, even when some branches have higher pressure drop than others. Good system for high-pressure-drop heating units. • Effective for applications with low supply temperature available.	• Uses more piping than series-loop or one-pipe systems. • If using high-head circulator, will require a bypass pressure regulator to protect the circulator as flow through branches closes off. • Large number of branches may make flow calculations difficult or impossible. Two-pipe zoned systems or primary/secondary systems will usually be easier to calculate. • Circulator must run when any zone calls for heat.

Table 2   Advantages and disadvantages of piping systems **(continued)**

System	Advantages	Disadvantages
Two-pipe zone valve zoning	• Each circuit (zone) can be individually controlled. • Piping is simpler than circulator-zoned or primary/secondary systems because no flow/check valves are required. • Heating units anywhere in the home can be combined into a zone to take advantage of similar space heating requirements.	• Does not provide individual zoning of heating units unless significant piping is added, i.e., fewer zones than two-pipe reverse-return or one-pipe systems. • Uses more piping than series-loop or one-pipe systems. • Circulator must run when any zone calls for heat. • Zone flow rates will depend on operation of other zones. Must be careful when selecting circulator not to cause excess flow in zones as others close.
Two-pipe circulator zoning	• Each circuit (zone) can be individually controlled. • Heating units anywhere in the home can be combined into a zone to take advantage of similar space heating requirements. • Can usually use low-head standard circulator rather than high-head circulator needed in some zone valve zoned systems. • Pumping power usage reduces as zones shut off. • If one circulator fails, the rest of the system can still operate. • Flow rate in each zone remains about the same, even as other zones turn on and off.	• Does not provide individual zoning of heating units unless significant piping is added, i.e., fewer zones than two-pipe reverse-return or one-pipe systems. • Uses more piping than series loop or one-pipe systems. • Requires more piping than zone valve systems because of need for flow/check valves and additional isolation valves (for circulators). • Requires more sophisticated controls than zone valve system to operate 120-VAC circulators. • May encounter problems if high-head and low-head circulators are used on the same system.
Primary/ secondary one-pipe	• Each circuit (zone) can be individually controlled. • Heating units anywhere in the home can be combined into a zone to take advantage of similar space heating requirements. • Good choice for radiant heating sytems, allowing simpler control of supply temperature to radiant loop and return temperature to boiler. • Allows flexible designing — each circuit can operate in a different manner and even use different piping methods. Some circuits could be series-loop while others are two-pipe. • Can usually use low-head standard circulators in secondary circuits. • Pumping power usage reduces as zones shut off. • If a zone circulator fails, the rest of the system can still operate. • Flow rate in each zone remains about the same, even as other zones turn on and off.	• Does not provide individual zoning of heating units unless significant piping is added, i.e., fewer zones than two-pipe reverse-return or one-pipe systems. • Uses more piping than series-loop or one-pipe systems. • Requires more piping than other zoned systems because of flow/check valves and extra circulators. • Requires more circulators than other zoned systems, i.e., one per circuit plus primary circulator. • Requires more sophisticated controls than zone valve system to operate 120-VAC circulators. • Primary circulator must run when any circuit calls for heat.
Primary/ secondary two-pipe	• Each circuit (zone) can be individually controlled. • Every circuit receives supply water at about the same temperature. • Primary circuit flow rate and circulator will usually be smaller than a one-pipe primary/secondary system. • Heating units anywhere in the home can be combined into a zone to take advantage of similar space heating requirements. • Good choice for radiant heating systems, allowing simpler control of supply temperature to radiant loop and return temperature to boiler. • Heating units anywhere in the home can be combined into a zone to take advantage of similar space heating requirements. • Can usually use low-head standard circulators in secondary circuits. • Pumping power usage reduces as zones shut off. • If a zone circulator fails, the rest of the system can still operate. • Flow rate in each zone remains about the same, even as other zones turn on and off.	• Does not provide individual zoning of heating units unless significant piping is added, i.e., fewer zones than two-pipe reverse-return or one-pipe systems. • Uses more piping than series loop or one-pipe systems. • Requires more piping than other zoned systems because of flow/check valves and extra circulators. • Requires more circulators than other zoned systems, i.e., one per circuit plus primary circulator. • Requires more sophisticated controls than zone valve system to operate 120-VAC circulators. • Primary circulator must run when any circuit calls for heat.

Type M copper tubing

Table 3   Type M copper tubing installation information

Type M tubing size	Diameter OD/ID Inches	Weight Pounds per foot empty	Weight Pounds per foot full	Volume Gallons per foot	Fitting take-up allowance Inches		Max distance between supports Inches (Note 1)	
					Solder	Thread	Horizontal	Vertical
½"	.625/.569	0.204	0.312	0.013	$^1/_2$"	$^1/_2$"	5 feet	10 feet
¾"	.875/.811	0.328	0.553	0.027	$^3/_4$"	$^9/_{16}$"	5 feet	10 feet
1"	1.125/1.055	0.465	0.840	0.045	$^{15}/_{16}$"	$^9/_{16}$"	6 feet	10 feet
1¼"	1.375/1.291	0.682	1.248	0.068	1"	$^5/_8$"	6 feet	10 feet
1½"	1.625/1.527	0.940	1.731	0.095	1 $^1/_8$"	$^5/_8$"	8 feet	10 feet
2"	2.125/2.009	1.460	2.834	0.165	1 $^3/_8$"	$^{11}/_{16}$"	8 feet	10 feet

Type M tubing size	Approximate expansion allowance (inches) for each 10 feet of pipe to allow for water temperature of: (Note 2)				Offset distance and length for U-shaped offset expansion joint to absorb total expansion of:			
	150 °F	175 °F	200 °F	225 °F	½"	1"	1½"	2"
½"	0.11	0.14	0.17	0.20	17	24	29	33
¾"	0.11	0.14	0.17	0.20	20	28	34	40
1"	0.11	0.14	0.17	0.20	22	32	39	45
1¼"	0.11	0.14	0.17	0.20	25	35	43	50
1½"	0.11	0.14	0.17	0.20	27	38	47	54
2"	0.11	0.14	0.17	0.20	31	44	53	62

Note 1: Support piping next to heavy components — immediately on either side of component.
Note 2: Assumes copper tubing installed at 50 °F or warmer. Use 225 °F column unless special controls will ensure temperature can go no higher than column selected.
Note 3: Expansion data based on 2000 ASHRAE Handbook, "HVAC Systems and Equipment."

Schedule 40 steel pipe

Table 4   Schedule 40 steel pipe installation information

Sch. 40 steel pipe size	Diameter OD/ID Inches	Weight Pounds per foot **empty**	Weight Pounds per foot **full**	Volume Gallons per foot	Fitting take-up allowance Inches **Threaded**	Max distance between supports Inches (Note 1) **Horizontal**	**Vertical**
½"	.840/.622	0.850	0.983	0.016	1/2	"7 feet	9 feet
¾"	1.050/.824	1.130	1.363	0.028	9/16	"7 feet	9 feet
1"	1.315/1.049	1.680	2.055	0.045	11/16	"7 feet	9 feet
1¼"	1.660/1.380	2.270	2.920	0.078	11/16	"7 feet	9 feet
1½"	1.900/1.610	2.720	3.603	0.106	11/16	"7 feet	9 feet
2"	2.375/2.067	3.650	5.108	0.175	3/4	"10 feet	13 feet

Sch. 40 steel pipe size	Approximate expansion allowance (inches) for each 10 feet of pipe to allow for water temperature of: (Note 2)				Offset distance and length for U-shaped offset expansion joint to absorb total expansion of:			
	150 °F	175 °F	200 °F	225 °F	½"	1"	1½"	2"
½"	0.08	0.10	0.12	0.14	19	27	34	39
¾"	0.08	0.10	0.12	0.14	22	31	37	43
1"	0.08	0.10	0.12	0.14	24	34	42	48
1¼"	0.08	0.10	0.12	0.14	27	38	47	54
1½"	0.08	0.10	0.12	0.14	29	41	50	58
2"	0.08	0.10	0.12	0.14	33	46	56	65

Note 1:   Support piping next to heavy components — immediately on either side of component.
Note 2:   Assumes steel piping installed at 50 °F or warmer. Use 225 °F column unless special controls will ensure temperature can go no higher than column selected.
Note 3:   Expansion data based on 2000 ASHRAE Handbook, "HVAC Systems and Equipment."

Hydronics Institute Section of AHRI

35 Russo Place

Berkeley Heights, NJ 07922-0218

part 5

Sizing
hydronic hot water heating systems

I=B=R Guide RHH

Residential Hydronic Heating . . .

Installation & Design

Hydronics Institute
Section of **AHRI**

I=B=R Guide RHH
Residential Hydronic Heating

Hydronics Institute Section of **AHRI**
35 Russo Place
Berkeley Heights, NJ 07922-0218

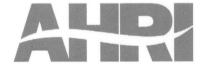

Contents – Part 5

Contents – continued

Contents _– continued_

● Appendix

Contents – *continued*

Illustrations

Contents *– continued*

Tables

Sizing methods

For a successful installation, you will need to correctly size and route the piping, and select or apply the right circulator. For most residential applications, you will be able to use the circulator packaged with the boiler. But some systems will require additional or larger circulators to do the job, and some residential boilers do not include a circulator.

Part 5 provides guidelines for selecting circulators and introduces four pipe-sizing methods:

Method 1
Quick-selector charts

The Quick-selector charts provide information for typical residential boiler packaged circulators (that is, circulators packaged with residential up to about 200,000 Btuh input). Your best use of the Quick-selector charts may be for a quick check whether you can apply series-loop piping to a system or branch. The charts include the maximum suggested sizes of series-loop systems using packaged circulators and the approximate head available at various flows for these circulators. You should always use Method 2 or Method 3 to verify the system design, using the Quick-selector charts only for roughing in the application.

Method 2
Milinch sizing method

The milinch method begins by selecting pipe sizing that will result in a known head loss per foot of pipe. (The head loss used is based on ASHRAE recommendations for limiting flow through piping to ensure quiet, reliable operation.) The next step is to multiply the loss per foot times the total length of the longest circuit. Finally, select a circulator that can deliver the required flow with this head loss.

Method 3
I=B=R head/ flow tables

The I=B=R tables simplify selection of pipe sizes for mains, trunk lines, and branch piping in series-loop, one-pipe and two-pipe systems. You only need measured length of piping. You do not need to calculate the equivalent length of piping (as in Method 3). You will need to use circulator curves to select (or verify) the correct circulator. Part 5 includes I=B=R tables for series-loop, one-pipe, and two-pipe applications.

Method 4
Equivalent length method

Equivalent length means the length of straight pipe that would cause the same pressure drop as the fittings and pipe in a circuit. Head loss (pressure drop) is proportional to the length of piping. By substituting an equivalent amount of straight pipe for each fitting, the calculation is simpler than dealing with each fitting separately. Method 4 provides tables of equivalent length for typical piping fittings plus tables and procedures for calculating head loss when you know equivalent length and flow rate in a piping system. We provide this method here to demonstrate a typical alternative pipe-sizing approach. (Method 4 may, by contrast, show how easy it is to work with the Method 3 tables.)

Heat/flow equations

To calculate flow rates and temperature drops when sizing circulators and piping, use equations 5-1 through 5-5. Throughout Part 5, temperature drop will be referred to as TD.

Btuh	gpm	TD	
			The equations below are approximate. The value, 500, changes with water temperature. The actual value varies from 500 for water at 60 °F to 486 for water at 180 °F. For most residential calculations, though, using 500 will be adequate and simplifies calculations.
			The number, 500, comes from: 8.33 (pounds/gallon) x 60 (minutes/hour) = 500.)
Find Btuh	(known)	(known)	**Btuh = 500 x gpm x TD** Equation (5-1) **Btuh** = heat transferred **gpm** = flow rate in gpm **TD** = temperature drop in °F
			Example: How much heat is transferred in a loop if the flow rate is 12 gpm and the temperature drops from 190 °F entering to 160 °F leaving?
			The temperature drop, TD, is: TD = 190 °F - 160 °F = 30 °F.
			Btuh = 500 x gpm x TD = 500 x 12 x 30 = 180,000 Btuh, or 180 MBH.
(known)	**Find gpm**	(known)	**gpm = Btuh ÷ (500 x TD)** Equation (5-2) **Btuh** = heat transferred **gpm** = flow rate in gpm **TD** = temperature drop in °F
			Example: You want to install a series loop system for a heat load of 80,000 Btuh. The temperature drop must not be more than 30 °F. What is the minimum flow rate?
			gpm = Btuh ÷ (500 x TD) = 80,000 ÷ (500 x 30) = 80,000 ÷ 15,000 = 5.3 gpm. The minimum flow rate for a 30 °F drop is 5.3 gpm.
			If the temperature drop had to be 20 °F:
			gpm = 80,000 ÷ (500 x 20) = 80,000 ÷ 10,000 = 8 gpm. The minimum flow rate for a 20 °F drop would be 8 gpm.
			Note: The higher the flow rate, the lower the temperature drop. The lower the flow rate, the higher the temperature drop.
(known)	(known)	**Find TD**	**TD = Btuh ÷ (500 x gpm)** Equation (5-3) **Btuh** = heat transferred **gpm** = flow rate in gpm **TD** = temperature drop in °F

Example: Suppose you install 40 feet of baseboard, with a heat output of 580 Btuh per foot (based on 4 gpm and an average temperature of 180 °F in the baseboard, 190 °F in/170 °F out, from Table 1, Page 5 – 10, repeated from Part 3). The circulator can deliver 4 gpm against the system head.

What is the temperature drop?

Btuh = 40 feet of baseboard x 580 Btuh per foot = 23,200 Btuh.

TD = Btuh ÷ (500 x gpm) = 22,000 ÷ (500 x 4) = 23,200 ÷ 2,000 = 11.6 °F. The temperature drop at 4 gpm would be 11.6 °F.

What if the flow was 1 gpm instead of 4 gpm?

The same baseboard (see Table 1, Page 5 – 10) would deliver 550 Btu per foot of baseboard at an average temperature of 180 °F and flow rate of 1 gpm, a slight reduction in output compared to the 4 gpm rating. So, what would the temperature drop be for a flow of 1 gpm if the baseboard put out 550 Btuh for each foot of baseboard?

Btuh = 40 feet of baseboard x 550 Btuh per foot = 22,000 Btuh.

TD = Btuh ÷ (500 x gpm) = 22,000 ÷ (500 x 1) = 22,000 ÷ 500 = 44 °F. The temperature drop at 1 gpm would be 44 °F if the baseboard output were 550 Btuh per foot throughout. But the baseboard output would not be 550 Btuh per foot in all the baseboard because the temperature drops too much before the water reaches the end of the loop. The temperature would drop from 190 °F to 170 °F by the time the water leaves the first 18 feet of baseboard! (Because a 20 °F drop at 1 gpm would be 20 °F x 500 x 1 gpm = 10,000 Btuh. Divide 10,000 Btuh by 550 Btuh per foot of baseboard, and the result is 18.2 feet.) The remaining 22 feet of baseboard would be supplied with water starting at 170 °F. The average output would be about 420 Btuh per foot instead of 550, a reduction of 24% in baseboard output.

This example should give an idea of the potential problem of an under-sized circulator on a series loop. If the flow is too low, the temperature will drop too much, causing reduced output in the heating units near the end of the loop. This translates to insufficient heat in these areas. You can avoid this problem by using a correctly-sized circulator or sizing and designing your piping for the circulator used (when using a packaged circulator, for example).

Heat flow equations (continued)

Simplified calculations: 20 °F temperature drop

The information below covers the heat/flow/temperature drop relationships. For most of your applications, you can design for a 20 °F temperature drop. For this case, the equations are:

Btuh = gpm x 10,000 (@ 20 °F TD) MBH = gpm x 10 (@ 20 °F TD)	**(5-1)**	gpm = Btuh ÷ 10,000 (@ 20 °F TD) gpm = MBH ÷ 10 (@ 20 °F TD)	**(5-2)**

Table 1   Typical ratings for finned-tube baseboard heaters

			(Hot water ratings, Btuh per linear foot, with 65°F entering air)									
Element	Water flow	Pressure drop (Millinches per foot)	140°F	150°F	160°F	170°F	180°F	190°F	200°F	210°F	215°F	220°F
Type 1 Baseboard with ¾" element	1 gpm	47	290	350	420	480	550	620	680	750	780	820
	4 gpm	525	310	370	440	510	580	660	720	790	820	870
Type 2 Baseboard with ½" element	1 gpm	260	310	370	430	490	550	610	680	740	770	800
	4 gpm	2880	330	390	450	520	580	640	720	780	810	850

NOTE: Ratings are for element installed with damper open, with expansion cradles. Ratings are based on active finned length (5" to 6" less than overall length) and include 15% heating effect factor. Use 4 gpm ratings only when flow is known to be equal to or greater than 4 gpm; otherwise, 1 gpm ratings must be used.

Sizing baseboard for lower temperature

You may want to apply a packaged circulator to a series loop longer than the maximum length given in the Quick-selector table. To determine whether it will work and how to size finned-tube baseboard for the loop, see the method on Page 5–50.

Selecting a circulator

Head

Head is usually given as "feet of head." As water moves through pipe and fittings, it loses energy to friction. Circulators provide the energy to make up for this loss, called the "head loss" through the system.

Head is an energy term, equal to the energy in foot-pounds required to push through each pound of fluid; i.e., head is really foot-pounds per pound, but is abbreviated to feet. The more water you push through the pipes, or the smaller the pipes are, the higher the head loss will be.

Circulator curves

To select a circulator, you need to know how much water it can move against how much resistance (head). Circulator graphs show head versus flow with a curve — with flow rate (gpm) along the horizontal axis and head (feet) along the vertical. The curve always slopes downward, showing that the head the circulator can provide reduces as flow increases. You use circulator flow curves to select a circulator or to check how a circulator will work on a given system.

Figure 1   Typical circulator performance curve

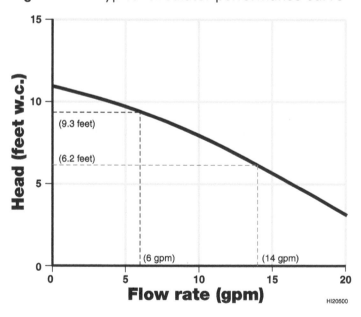

Look at Figure 1, a typical circulator curve.

• To read the curve, pick a flow rate and draw a line vertically until it strikes the curve.

• From this point, draw a line to the left and read the head on the vertical axis.

• The circulator can flow this many gpm against the head you read.

• For example, in Figure 1, two points are shown: at 6 gpm, this circulator can handle a head of 9.3 feet water column; at 14 gpm, it can handle a head of 6.2 feet water column.

Most circulator graphs include more than one curve (usually a family of circulators). Figure 2 includes the curves for three residential circulators, labeled A, B and C. These circulators have similar performance. Circulators B and C can provide slightly more head than circulator A for flows less than about 12 to 14 gpm. Circulator A provides more head above this flow rate.

Figure 2   Typical group of circulator curves

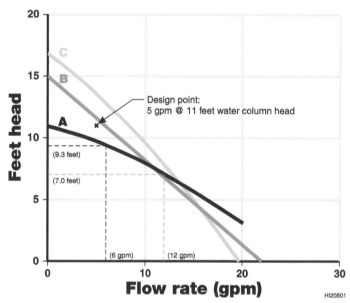

Design point and circulator selection

The design point is the flow rate and head loss you determine for your application. Place the design point on the circulator graph by drawing an x at the required flow rate and head loss. Then select a circulator whose curve is on or slightly above the design point, as shown in Figure 2 Here the design point is 5 gpm at 10 feet head. For this design point, the best circulator selection of the three shown would be Circulator B. You could use Circulator C as well, though it would deliver more flow. The curve for Circulator A falls below the design point, so it would not be able to deliver the required flow.

See Part 5 Appendix, Page 5–46, for procedure to predict actual performance of a circulator when its curve is not near the design point. You may also find this method useful to decide whether you can use a circulator whose curve is below the design point.

Method 1 . . . .

Follow Steps 1 through 6 as for Method 2, 3 or 4 (see following pages) to size the boiler, heating units and system components. Use the Method 1 tables for a quick check to see if a packaged circulator will work, to rough-in pipe sizes, or to estimate packaged circulator performance when a circulator performance curve is not available. You should design and/or verify your circuit using either Method 2, 3 or 4.

Table 2: Series-loop systems with packaged circulators

Use Table 2, Page 5–13 to check the likelihood of applying a packaged circulator on a series-loop application. You may be able to use a packaged circulator with a larger or longer system, but you should only do so if you verify the system and circulator using Method 2, 3 or 4 and the circulator curve. Do not use Table 2 if the circulator supplies more than a single series-loop.

The Table 2 columns are:

Pipe size	Size of type M copper tubing used in circuit.
Maximum load	Maximum suggested heating load handled by circulator in a single series loop.
Max GPM (20 °F)	The flow rate for Max load with a temperature drop of 20 °F.
Max loop length	The measured length of tubing and baseboard in the loop.
Baseboard size	Minimum size of baseboard. (These charts apply only if the pipe or tubing is the same size as the baseboard.)
Baseboard output	Baseboard output (Btu per foot) — used to determine maximum length of baseboard that can be supplied by the pipe size in column one.
Max feet baseboard	Maximum baseboard in loop for the Maximum load.

Table 3: Typical packaged residential circulator flow available

You can use the head/flow data of Table 3, Page 5–13 to check approximate packaged circulator performance for use in Method 2, 3 or 4 pipe sizing if you don't have a circulator curve available.

Table 4: Pipe flow capacities — flow limits for residential piping systems

Table 4, Page 5–15 shows the maximum recommended flow rates and corresponding heating loads for copper and steel piping.

Quick-selector charts

Table 2 Typical residential circulator capacity (use to check whether a typical circulator could handle a single series-loop circuit)

Pipe size (copper)	Maximum load (Btuh)	Maximum gpm (20 °F TD)	Maximum loop length (feet)	Baseboard size (Tubing diameter, inches)	Baseboard output (Btuh per foot at 180 °F)	Maximum baseboard (feet)
½	27,000	2.7	70 feet	½	580	46
¾	50,000	5.0	93 feet	¾	610	82
1	79,000	7.9	123 feet	1	760	104
1¼	125,000	12.5	164 feet	1¼	790	158

Note 1: Capacities above based on typical water-lubricated circulator as supplied with a residential boiler.
Note 2: This table and the I=B=R tables are based on head loss for water at 60 °F.

Table 3 Typical packaged residential circulator flow available

GPM	Head	Maximum measured length of copper or steel pipe & baseboard				Maximum equivalent length of copper or steel pipe & baseboard			
		½	¾	1	1¼	½	¾	1	1¼
1.0	10.7	635	-	-	-	815	-	-	-
1.5	10.6	268	-	-	-	386	-	-	-
2.0	10.5	144	714	-	-	226	910	-	-
2.5	10.4	90	445	-	-	149	594	-	-
3.0	10.3	62	295	-	-	105	418	-	-
3.5	10.1	-	208	-	-	-	310	-	-
4.0	10.0	-	154	607	-	-	239	783	-
4.5	9.9	-	118	469	-	-	189	622	-
5.0	9.7	-	93	368	-	-	153	505	-
6.0	9.4	-	62	239	-	-	106	350	-
7.0	9.1	-	-	165	751	-	-	254	954
8.0	8.7	-	-	119	554	-	-	191	720
9.0	8.4	-	-	89	413	-	-	147	557
10.0	8.0	-	-	69	312	-	-	115	438
11.0	7.5	-	-	-	239	-	-	-	349
12.0	7.1	-	-	-	185	-	-	-	281

Note 1: Capacities above based on typical water-lubricated circulator as supplied with a residential boiler.
Note 2: This table and the I=B=R tables are based on water at 60 °F.

Method 1: Quick-selector charts (continued)

Example 1:

You want to apply a packaged circulator on a series-loop system requiring 70,000 Btuh, with a measured length of copper tubing and baseboard of 120 feet. What size tubing and baseboard would be required?

Solution:

Read down the Maximum load column of Table 2, Page 5–13, for a load of at least 70,000 Btuh.

- You will find that 1" tubing and baseboard should work with a packaged circulator for a load of up to 79,000 Btuh and maximum measured length of 123 feet.

- You would not be able to use ¾" baseboard and tubing, because the limit is 50,000 Btuh.

Example 2:

If using a packaged residential circulator, what is the maximum loop length using ¾" baseboard and copper tubing?

Solution:

From Table 3, Page 5–13, the maximum measured length of tubing and baseboard is 124 feet, with a maximum heat load of 50,000 Btuh.

For longer lengths or loads:
- Use Method 2, 3 or 4 and the circulator curve to determine if the packaged circulator will work.

- If not, you could replace the circulator with a higher-capacity circulator, selected for the flow and head loss required as determined either by Method 2, 3 or 4.

- Or you might consider a larger temperature drop (lower flow rate), along with increased sizing of the baseboard heaters in the back half of the loop. (See Part 5 Appendix for this sizing procedure.)

NOTICE — Series-loop circuits that are too long will cause the living spaces near to the beginning of the circuit to heat more than those near to the end. This will cause uncomfortable living conditions — rooms either too warm at the front end or too cold at the back end of the system.

Prevent this by splitting long series loops and/or installing a circulator that can maintain a 20°F TD through the system. (DO NOT attempt to exceed the maximum recommended flow rate for the pipe size used.)

Table 4   Pipe flow capacities for steel pipe and copper tubing (based on maximum recommended flow rate)

Type M Copper tube size	Max flow (gpm)	Max load (MBH at temperature difference of:)			Maximum load (Total feet of baseboard in all circuits connected to pipe)			
		20 °F TD	30 °F TD	40 °F TD	½ (480 Btuh/ft)	¾ (610 Btuh/ft)	1 (760 Btuh/ft)	1¼ (790 Btuh/ft)
½	3.17	32	48	63	55	52	42	40
¾	6.44	64	97	129	111	106	85	82
1	10.9	109	163	218	188	179	143	138
1¼	16.3	163	245	326	281	268	215	207
1½	22.8	228	342	457	394	374	300	289
2	39.5	395	593	790	681	648	520	500
2½	80.9	809	1213	1617	1394	1326	1064	1024
3	129.7	1297	1946	2594	2236	2126	1707	1642

Steel pipe size	Max flow (gpm)	Max load (MBH at temperature difference of:)			Maximum load (Total feet of baseboard in all circuits connected to pipe)			
		20 °F TD	30 °F TD	40 °F TD	½ (480 Btuh/ft)	¾ (610 Btuh/ft)	1 (760 Btuh/ft)	1¼ (790 Btuh/ft)
½	3.79	38	57	76	65	62	50	48
¾	6.65	66	100	133	115	109	87	84
1	10.77	108	162	215	186	177	142	136
1¼	18.6	186	280	373	321	306	245	236
1½	25.4	254	381	508	438	416	334	321
2	41.8	418	627	837	721	686	550	530
2½	72.9	729	1093	1457	1256	1195	959	922
3	133.0	1330	1995	2660	2293	2180	1750	1684

Method 2 . . . . Milinch method

Rationale

All sizing methods require selecting pipe, circulator and component sizes that will provide good performance at reasonable cost. The milinch method can usually save time and simplify circulator and piping selections.

The milinch method begins by selecting pipe sizing that will result in a known head loss per foot of pipe. The head loss used is based on ASHRAE recommendations for limiting flow through piping to ensure quiet, reliable operation.

What is head loss?

Head loss due to friction of water flowing in pipes is determined in feet of head. As explained on Page 5–11, the term, "feet," really means energy lost through the piping in foot-pounds energy per pound of fluid.

Foot-pounds ÷ pounds = foot. So the term for head loss is simplified to just, "feet."

Feet of head may be confusing since we talk about feet of pipe, referring to the length of the piping system, and to feet head, referring to energy lost through flow.

You determine the head loss through a circuit by multiplying the head loss per foot of pipe times the total equivalent length of pipe in the circuit. That is, for each elbow, tee, and other fitting or component in the circuit, you find the length of straight pipe that would cause the same head loss. Add all of these equivalent lengths to the length of straight pipe in the circuit to find the "total equivalent length," often shortened to "TEL."

For residential systems, you won't need to do a lot of figuring to find a reasonable estimate for the TEL. Just multiple the actual length of pipe times 1.5. The extra 50% provides a reasonable estimated equivalent length for the fittings in a typical system.For example, if a circuit is 300 feet long, the TEL would be 1½ x 300 = 450 feet. Multiply 450 feet times the head loss per foot to find the head loss for the circuit.

What is a milinch?

A milinch is 1/1000 of an inch. Since a foot equals twelve inches, it takes 12,000 milinches to make one foot.

In hydronic systems, the friction loss through one foot of pipe is very low. To simplify the numbers, you can use the loss for a long length of pipe or use a smaller number than feet of head per foot of pipe. You will usually see head losses given in one of two ways:

Feet per hundred feet

This means feet head lost through each 100 feet of piping (typically 4.2 feet head per 100 feet of pipe or less). The notes in Table 5, Page 5–17 give equivalents for milinches per foot in feet head per foot, 10 feet head per foot, and 100 feet head per foot.

Milinches per foot

This means milinches head loss through each foot of piping (usually selected for 500 or 350 milinches head per foot of pipe).

Application

For residential hydronic systems, the milinch method begins with a maximum allowable head loss of 500 milinches head per foot of pipe. If this head loss results in too high a head for the circulator you want to use, try using a 350 milinch head per foot of pipe limit instead.

Table 5, Page 5–17 lists pipe capacities based on heating load (MBH) and flow rate (GPM). Heating load capacities are based on a temperature differential (T. D.) of 20°F. For other temperature differentials, find the flow rate needed and use the flow rate column to select pipe size.

Table 5   Pipe capacities in MBH and GPM for 500 or 350 milinches per foot head loss

Pipe capacity at **500** milinches restriction per foot of pipe ◆						Pipe capacity at **350** milinches restriction per foot of pipe ◆					
Pipe size	MBH	Friction head ◆	Flow rate	Velocity flow of water		Pipe size	MBH	Friction head ◆	Flow rate	Velocity flow of water	
Inches	Based on 20°F T. D.	Feet head per 100 feet of pipe	GPM at 20°F T. D.	Inches per second	Feet per minute	Inches	Based on 20°F T. D.	Feet head per 100 feet of pipe	GPM at 20°F T. D.	Inches per second	Feet per minute
½	17	4.2	1.7	23	115	½	15	2.9	1.5	18	90
¾	39	4.2	3.9	27	135	¾	31	2.9	3.1	17	85
1	71	4.2	7.1	34	170	1	59	2.9	5.9	23	117
1¼	160	4.2	16.0	40	200	1¼	130	2.9	13.0	26	132
1½	240	4.2	24.0	45 *	225	1½	185	2.9	18.5	27	134
2	450	4.2	45.0	54 *	270	2	360	2.9	36.0	35	173
2½	750	4.2	75.0	62 *	310						
3	1400	4.2	140.0	72 *	360						
4	2900	4.2	290.0	80 *	400						

✻ These are maximum recommended flow rates.

◆ **Friction loss in milinches:**
- In hot water systems, friction or pressure is measured in feet of water. Because friction loss in piping is small, it is measured in milinches. A milinch is 1/1000 of an inch, or 1/12,000 of a foot.
- 12,000 milinches (1 foot) represents the pressure exerted by a column of water one foot high. Convert milinches head per foot of pipe to feet head per foot of pipe by dividing by 12,000.

Friction loss		1 foot of pipe	10 feet of pipe	100 feet of pipe
500 milinches per foot	=	0.0416 feet head	0.416 feet head	4.16 feet head (rounds to 4.2)
350 milinches per foot	=	0.0292 feet head	0.292 feet head	2.92 feet head (rounds to 2.9)

Example

What is the required head and flow rate for a circulator supplying a 120,000 Btuh (120 MBH) series loop system? The installer wants to install ¾-inch finned-tube baseboard and obtain a 20°F temperature differential.

Solution

1   Start with the 500 milinch per foot portion of Table 5.

2   The first thing we need to look at is the ¾-inch baseboard. How much head load can be handled with a ¾-inch pipe?

   Look at the ¾-inch row in the 500 milinch table. The maximum heat load in a ¾-inch line should be 39 MBH.

3   This means the system has to be pipe in multiple series loop circuits, and the maximum heating load of each circuit must be 39 MBH.

4   For each loop, find the TEL by multiplying 1.5 times the length of the circuit from the boiler supply to boiler return.

5   Pick the longest TEL for any loop. Find the number of 100 feet of TEL by dividing by 100.

6   Multiply this number (TEL/100) times 4.2 feet head per hundred feet (500 milinches per foot).

   For example, if the longest loop is 200 feet, the TEL would be 1.5 times 200, or 300 feet of pipe. Then TEL/100 = 300/100 = 3.

Multiply 3 times 4.2 feet head per hundred feet of pipe = 12.6 feet head.

7   Size the trunk line by finding a pipe size in the 500 milinch head loss table large enough for 120 MBH. Use a 1¼-inch line since it can handle up to 160 MBH.

8   The flow rate at 20°F would be 120,000 / 500 / 20 = 12 GPM.

9   The circulator must be able to flow 12 GPM against a head loss of 12.6 feet.

10   If this head is too high for the circulator you want to use, try the 350 milinch table. At 350 milinch head loss per foot of pipe:

   a.   Maximum heat load per ¾-inch circuit is 31 MBH. So pipe circuits must not exceed 31 MBH each.

   b.   If the longest circuit is still 200 feet, the TEL is 300.

   c.   The head loss per 100 feet is 2.9, so the head total head loss is (300/100) x 2.9 = 8.7 feet.

   d.   Result: By limiting circuit size to no more than 31 MBH per circuit, the circulator only has to handle a head loss of 8.7 feet. The flow rate is the same as found in step 8, or 12 GPM.

   e.   The trunk line can still be 1¼ inch, since the 350 milinch table allows up to 130 MBH.

11   When piping multiple circuits, you may need to install balancing valves if the circuits are not similar lengths.

12   You can see from this example that the milinch method quickly tells you not only the trunk line size, but sets the size for each of the branch circuits as well.

Method 3 . . . . I=B=R tables

The I=B=R head/flow tables provide selection of pipe size and circulator capacity for the most common residential systems: series loop, one-pipe, and two-pipe. You can even use these tables for primary/secondary systems by considering each section of the system, based on its piping method (series, one-pipe, or two-pipe).

To use the tables, apply the steps listed below.

See "Example house" in Part 5 for sample applications of the I=B=R head/flow tables to the Guide H-22 house, showing heating unit selections, pipe sizing and circulator requirements for series-loop, one-pipe and two-pipe piping.

Table 6	Steps in using I=B=R tables — overview (see following pages for explanations of steps; Step 7 is explained for each of the 3 system types)
Step 1	**Calculate heat losses**
Step 2	**Select heat distributing units**
Step 3	**Select boiler size**
Step 4	**Select expansion tank**
Step 5	**Determine system type (series loop, one-pipe or two-pipe) and make piping layout — determine length of each circuit**
Step 6	**Determine flow rate (gpm)**
Step 7	**Use I=B=R tables for selected system type to determine circulator capacity and pipe size**

Step 1: Calculate heat losses

Use Guide H-22 or equivalent method. Record the heat loss of each space on a floor plan of the house (preferably to scale).

Step 2: Select heat distributing units

Select heating units. Then sketch in and identify each unit on the floor plan.

Notes on series-loop systems:

Use only baseboard heaters (cast iron or finned-tube). Do not mix high mass units (cast iron radiators) or higher pressure drop units (fan coil units, convectors, kick-space heaters, etc.).

The interconnecting pipe must be the same size or smaller than the baseboard to use these sizing tables. With ¾" baseboard, you can use ½" or ¾" interconnecting piping, depending on the system size and head loss. Cast iron baseboard uses ¾" connections, so the connecting piping cannot exceed ¾" when using these sizing tables. Use the size of the interconnecting piping, not the baseboard size, when applying the tables if the pipe size is smaller.

If the series-loop table does not provide a single-loop solution for the pipe size selected, you will have to use a multiple-loop arrangement and/or increase the baseboard and pipe size.

Notes on one-pipe systems:

The head loss through each heating unit also applies to the main, i.e., only Part of the flow goes through the branch, but the head loss at each branch is the same for the main and the branch.

When a heating unit is located below the main, use two fittings (unless not recommended by the fitting manufacturer). See Table 16, Page 5–47, for the approximate flow in one-pipe branches when the heating unit is located below the main.

Notes on two-pipe systems:

If using heating units with widely different head loss, you will have to install balancing valves to control flow. If the head losses of each branch are similar, you will not need balancing valves.

Notes on heating unit selection:

Sizing baseboard: Select baseboard based on manufacturer's ratings. Divide room heat loss by baseboard output per foot to determine minimum length of baseboard for that room.

Length = (Room Btuh) ÷ (Btuh per foot) **(5-6)**

Cast-iron radiators: Use Table 18, Page 5–54, to determine required radiation square feet based on Btuh and average water temperature in radiators. Use manufacturer's square feet ratings if available. If not, see Part 2 Appendix, for square feet (EDR) data of typical radiators.

Convectors: Use manufacturer's ratings for a 20 °F temperature drop. Select a convector (or convectors) for each space with total rating no less than space heating loss.

Step 3: Select boiler size

Select boiler with Net Load Rating no less than the building heat loss. See Part 2 for detailed discussion of boiler ratings and selection considerations. If using indirect water heating or other loads in addition to heat loss, see Part 6 for information on additional boiler capacity required, if any.

Step 4: Select expansion tank

See Part 3 for detailed discussion of expansion tank types and sizing. Use a diaphragm- or bladder-type tank when possible because of simpler system air control. Size per manufacturer's guidelines, preferably using a tank with an acceptance volume at least 5% of total system volume; or use rules of thumb:

- Conventional air-cushion tank: Volume no less than 1 gallon per 5,000 Btuh boiler input.

- Diaphragm- or bladder-type tank: Volume no less than 1 gallon per 7,000 Btuh boiler input for tanks precharged to at least 8 psig.

Step 5: Determine system type and make piping layout — determine length of each circuit

Sketch in boiler and all system components on the floor plan. Sketch the interconnecting piping on the plan and measure the length of each circuit. Estimate the length of vertical connections and add to the measured length. Include piping connections to heating units for series-loop and two-pipe applications, but not for one-pipe. For one-pipe systems, add 12 feet for each one-pipe fitting (diverter tee) used in the circuit.

Use only measured lengths. No additional allowance is required when using the I=B=R Head/flow tables.

Step 6: Determine flow rate (gpm)

For each circuit, apply equation 5-2 to calculate flow rate, assuming a 20 °F temperature drop (unless another temperature drop is required, such as for radiant heating).

$$gpm = Btuh \div 10,000$$
$$gpm = MBH \div 10 \qquad \textbf{(5-2)}$$

Method 3 . . . I=B=R tables
Step 7 – Series loop piping

Single-loop system:

Use Table 7, Page 5–21 to find the required head pressure of circulator for system pipe size chosen (same size or smaller than baseboard). If no selection is available for the circuit length or if the circulator head is higher than desired, you will need to try larger baseboard. If this does not work or is unacceptable, size the system for multiple series loops.

To use the table:

- Enter the upper section at the pipe size of the loop. Scan across to the first column with flow at least equal to the loop flow.
- Scan down this column to the lower section, to the measured length closest to the loop length.
- Follow this row left to read the minimum circulator head.

Multiple-loop system:

Determine the flow for each loop using Step 6 for the total heating load of the loop. For each loop, use Table 7 to find the required head of the circulator for the measured loop length and pipe size. The measured length of each loop must include the trunk line, if used, though the trunk line will probably be of a larger diameter.

Use the highest head of any of the loops to determine the required circulator head. Circulator flow equals the total of all connected loop flows.

Size the trunk line, if used, from Table 7. You use the table the reverse from previous procedures:

- Enter the lower portion of the table at the circulator head required.
- Scan across this row to the measured length closest to the longest loop length.
- Scan to the upper section, to the first row with gpm at least equal to the total circulator flow.
- Follow this row left to read the minimum trunk line size.

If you find no acceptable solution, consider another system design.

Table 7 I=B=R sizing table for series-loop baseboard systems

I=B=R Series-loop baseboard system sizing																
Circuit or trunk pipe size	Capacity of circuit (gpm)															
½	2.3	2.0	1.9	1.8	1.7	1.7	1.6	1.5	1.5	1.4	1.3	1.2	1.2	1.1	0.9	0.8
¾	5.0	4.3	4.1	3.8	3.7	3.6	3.4	3.2	3.1	2.9	2.8	2.6	2.4	2.2	2.0	1.8
1	9.6	8.3	7.7	7.3	7.0	6.8	6.5	6.3	5.9	5.7	5.5	5.0	4.6	4.3	3.8	3.4
1¼	-	18.0	17.0	16.0	15.0	15.0	14.0	14.0	13.0	12.0	11.0	11.0	9.7	9.0	8.3	7.3
Available head (feet of water)	Measured length of circuit, including baseboard (feet)															
4	35	45	50	60	65	70	75	80	90	100	110	130	150	180	220	290
5	45	60	65	70	80	90	95	100	120	130	140	160	190	230	290	360
6	55	70	80	90	100	110	120	130	140	160	180	200	240	290	350	450
7	65	90	100	110	120	130	140	150	170	190	210	240	290	340	420	540
8	75	100	110	130	140	150	160	180	200	220	250	290	330	400	490	620
9	85	110	130	150	160	170	190	200	230	250	290	330	380	450	560	710
10	100	130	140	170	180	190	210	230	260	290	320	370	430	510	620	790
11	110	140	160	190	200	220	240	260	290	320	360	410	480	570	690	880
12	120	160	180	200	220	240	260	290	320	350	400	450	540	620	760	960
14	150	190	210	250	260	290	310	340	380	420	470	540	620	730	900	1120
16	170	220	250	290	310	330	360	400	440	790	550	620	720	850	1020	-
18	190	250	290	330	350	380	420	450	500	560	620	710	830	950	1150	-
20	220	290	320	370	400	430	470	510	550	620	700	790	910	1060	-	-

Method 3 . . . I=B=R tables
Step 7 – One-pipe systems

Select circulator:

Select a circulator with flow capacity required for the system (using equation 5-2, Page 5–9, to find flow rate for entire system heating load). Note circulator head capacity.

Determine circuit, trunk and branch pipe sizing

You may need to reduce the branch circuit capacities as indicated later in this section (under "Adjustments to Table 8") if branch piping is unusually long or complex.

Single-circuit system:

Circuit length = Measured length of circuit from boiler and back to boiler, not including piping from circuit to heating units. Then add 12 feet to length for each one-pipe fitting (diverter tee) used in the system for total circuit length.

To use Table 8:

- Enter the upper table section at row with head closest to circulator capacity.

- Scan across to length closest to total circuit length.

- Drop down the column to lower section. First, find required circuit size from top line of each subsection. Find first circuit size with flow at least equal to total flow of circuit. Stay in this subsection to find the required branch pipe size for each branch, based on the flow in the branch.

Multi-circuit system:

For each circuit: Total circuit length = Measured length of circuit, not including branch piping, plus 12 feet for each one-pipe fitting (diverter tee) used in the circuit.

Trunk length = Total length of longest circuit.

Use Table 8, Page 5–23 as above to find the circuit and branch pipe sizes for each branch, based on only the flow in that circuit or branch. The circulator head is the head capacity of the circulator selected. Find the trunk pipe size using total system flow, with length equal to longest connected total circuit length.

Adjustments to Table 8:

- Decrease table branch flow capacities by 15% (multiply by 0.85) if the heating unit is below the circuit. No adjustment is needed, even for multiple-story locations, if the heating unit is above the circuit.

- The table allows for branch piping equivalent to 12 elbows and 15 feet of horizontal piping (no limit on vertical piping). For each additional elbow equivalent or foot of horizontal pipe, reduce table branch flow capacity by 1%.

- If gpm required by a heating unit exceeds capacity in table, consider:

  - Selecting a circulator with a higher head capacity.

  - Using two heating units instead of one, thus reducing the required flow to each. Remember to add 12 feet for each additional one-pipe fitting used.

Combination series-loop/one-pipe applications

The circulator must have a flow capacity equal to total system. Find the required head using the Series-loop Table based on the longest loop length in the system. Select the circulator. Then use the available head and the One-Pipe Table to size the one-pipe components.

Table 8   I=B=R sizing table for one-pipe systems

I=B=R One-pipe circuit sizing

Available head (feet of water)	Total length of circuit (feet) Measured length of circuit (not including branch piping) plus an allowance of 12 feet for each one-pipe fitting in the circuit															
4	35	45	50	60	65	70	75	80	90	100	110	130	150	180	220	290
5	45	60	65	70	80	90	95	100	120	130	140	160	190	230	290	360
6	55	70	80	90	100	110	120	130	140	160	180	200	240	290	350	150
7	65	90	100	110	120	130	140	150	170	190	210	240	290	340	420	540
8	75	100	110	130	140	150	160	180	200	220	250	290	330	400	490	620
9	85	110	130	150	160	170	190	200	230	250	290	330	380	450	560	710
10	100	130	140	170	180	190	210	230	260	290	320	370	430	510	620	790
11	110	140	160	190	200	220	240	260	290	320	360	410	480	570	690	880
12	120	160	180	200	220	240	260	290	320	350	400	450	530	620	760	960
14	150	190	210	250	260	290	310	340	380	420	470	540	620	730	900	1120

Pipe or tubing sizes		Trunk or circuit and branch capacity (gpm) (Steel pipe or copper tubing with one-pipe fittings, except for 3/8 branches*)															
Trunk or circuit	Branch																
3/4		5.0	4.3	4.1	3.8	3.7	3.6	3.4	3.2	3.1	2.9	2.8	2.6	2.4	2.2	2	1.8
	3/8 *	0.9	0.8	0.8	0.7	0.7	0.7	0.7	0.7	0.7	0.4	0.4	0.4	0.4	0.4	0.4	0.3
	1/2	1.8	1.5	1.5	1.2	1.2	1.2	1.2	1.2	1.2	0.8	0.8	0.8	0.8	0.8	0.8	0.6
1		9.6	8.3	7.7	7.3	7.0	6.8	6.5	6.3	5.9	5.7	5.5	5.0	4.6	4.3	3.8	3.4
	3/8 *	0.7	0.7	0.6	0.6	0.6	0.5	0.5	0.5	0.4	0.4	0.4	0.4	0.3	0.3	0.2	0.2
	1/2	1.4	1.3	1.1	1.1	1.1	1.0	1.0	1.0	0.8	0.8	0.8	0.8	0.7	0.7	0.5	0.5
	3/4	2.4	2.1	1.9	1.9	1.9	1.6	1.6	1.6	1.3	1.3	1.3	1.3	1.1	1.1	0.9	0.9
1 1/4		-	18.0	17.0	16.0	15.0	15.0	14.0	14.0	13.0	12.0	11.0	11.0	9.7	9.0	8.3	7.3
	3/8 *	-	0.9	0.9	0.8	0.8	0.8	0.7	0.7	0.7	0.7	0.6	0.9	0.4	0.4	0.4	0.3
	1/2	-	1.8	1.8	1.6	1.5	1.5	1.4	1.4	1.3	1.2	1.1	1.1	0.9	0.9	0.8	0.7
	3/4	-	2.9	2.9	2.6	2.4	2.4	2.3	2.3	2.1	2.0	1.8	1.8	1.5	1.5	1.3	1.2
1 1/2		-	-	25.0	24.0	23.0	22.0	21.0	20.0	19.0	18.0	17.0	16.0	15.0	13.0	12.0	11.0
	3/8 *	-	-	1.0	1.0	1.0	0.9	0.9	0.8	0.8	0.8	0.7	0.7	0.7	0.6	0.5	0.4
	1/2	-	-	2.0	2.0	1.9	1.8	1.7	1.6	1.6	1.5	1.4	1.3	1.2	1.1	1.0	0.9
	3/4	-	-	2.7	2.7	2.6	2.5	2.4	2.2	2.1	2.0	1.9	1.8	1.6	1.5	1.4	1.3

* 3/8" branch lines based only on type L soft copper, adapted to one-pipe fittings and baseboard using 1/2" x 3/8" bushings.

Method 3 . . . I=B=R tables
Step 7 – Two-pipe systems

Select circulator:

Calculate the required flow for each branch by dividing branch heating load by 10,000 (for 20 °F temperature drop). Record the information.

Select circulator with flow capacity required for the system (using equation 5-2, Page 5–10, to find flow rate for entire system heating load). The circulator requires a head capacity to handle the supply/return line flows plus the highest head loss of any branch circuit. When using higher pressure drop units, such as convectors or fan coil units, add their head loss to the required circulator head capacity. Also add for zone valves, mixing valves, check valves, or any other component with a head loss greater than 1 foot.

Note circulator head capacity.

Available head:

The head loss in a system is not only due to the piping and fittings, but to system components as well. The circuit sizing data of Table 9, Page 5–25 are based on the head available to overcome head loss through typical piping and fittings. It does not allow for high head-loss components.

The **available head** for use in sizing piping using Table 9 is NOT the head capacity of the circulator if high head-loss heating units, zone valves, etc., are used. The available head is the circulator head capacity MINUS the extra head loss caused by the high head-loss components.

To determine available head, find the branch with the highest component head loss (total of head losses of heating unit, zone valves, control valves, etc. in the branch). Then SUBTRACT this head from the circulator head capacity. Use only the remainder as the available head for pipe sizing.

Circuit length:

Find the longest distance through the supply main, the branch piping to and from the heating unit, and back to the boiler through the return main for any of the heating units in the system. Use this as the circuit length for sizing all main and branch piping.

Pipe sizing:

For each section of supply/return main, trunk or branch, use its flow and the circuit length from above in Table 9.

To use table:

- Enter the upper section of table at number closest to available head.

- Scan across table to the length closest to the circuit length.

- Drop down this column to lower section, to first gpm value at least equal to flow in pipe.

- Follow this row to the left to read minimum pipe size.

Trunk pipe sizing:

Size trunk piping, if used, as above. If using multiple circulators, size the trunk line based on lowest available head and longest circuit length connected to trunk.

Table 9   I=B=R sizing table for two-pipe reverse-return systems

I=B=R Two-pipe circuit sizing

Available head (feet of water)	Total measured length of circuit (feet)															
4	35	45	50	60	65	70	75	80	90	100	110	130	150	180	220	290
5	45	60	65	70	80	90	100	100	120	130	140	160	190	230	290	360
6	55	70	80	90	100	110	120	130	140	160	180	200	240	290	350	450
7	65	90	100	110	120	130	140	150	170	190	210	240	290	340	420	540
8	75	100	110	160	140	150	160	180	200	220	250	290	330	400	490	620
9	85	110	130	150	160	170	190	200	230	250	290	330	380	450	560	710
10	100	160	140	170	180	190	210	230	260	290	320	370	430	510	620	790
11	110	140	160	190	200	220	240	260	290	320	360	410	480	570	690	880
12	120	160	180	200	220	240	260	290	320	350	400	450	530	620	760	960
14	150	190	210	250	260	290	310	340	380	420	170	540	620	730	900	1120

Pipe size (Trunk, circuit or branch)	Capacity of trunk, circuit or branch (gpm)															
3/8	0.9	0.8	0.7	0.7	0.6	0.6	0.6	0.6	0.5	0.5	0.5	0.5	0.4	0.4	0.3	0.3
1/2	2.3	2.0	1.9	1.8	1.7	1.7	1.6	1.5	1.5	1.4	1.3	1.2	1.2	1.1	0.9	0.8
3/4	5.0	4.3	4.1	3.8	3.7	3.6	3.4	3.2	3.1	2.9	2.8	2.6	2.4	2.2	2.0	1.8
1	9.6	8.3	7.7	7.3	7.0	6.8	6.5	6.3	5.9	5.7	5.5	5.0	4.6	4.3	3.8	3.4
1¼	-	18	17	16	15.5	15	14.5	14	16	12	11.5	11	9.4	9.0	8.3	7.3
1½	-	27	25	24	23	22	21	20	19	18	17	16	15	13	12	11
2	-	-	-	-	-	42	40	39	38	36	34	32	29	27	24	21
2½	-	-	-	-	-	-	-	-	60	57	54	52	47	44	38	32

NOTICE

Use the Two-pipe sizing table, Table 9, only for reverse-return piping.

Method 4 . . . .

In Method 4, you measure the length of each circuit, then add the equivalent length for each fitting to determine the equivalent length of the circuit. (This was not required in Method 3 because the typical equivalent length for each circuit length is included in the table values.)

You will not need to use Method 4 for most residential applications. We provide this information here for larger residential and commercial systems and to give you some of the background behind the Method 3 tables.

The **equivalent length** of a fitting is the length of straight pipe in the same diameter that would cause the same head loss as the fitting.

- You will find the equivalent lengths of most common fittings in Table 11, Page 5–28.

- Also see Table 17, Page 5–48, to convert C_v values to equivalent feet of pipe or tubing. (C_v is the flow rate, in gpm, that will cause a pressure drop of 1 psig, or 2.309 feet water column.)

Head loss equation

You can calculate the head loss for each circuit using the equivalent length and the following equation:

$$\textbf{Head loss = (Equivalent length} \div \textbf{100) x a x (gpm)}^b$$

$$h = (Le/100) \times a \times (gpm)^b, \quad \text{(5-7)}$$

- **a** and **b** are the values from Table 12, Page 5–29, for the pipe size used,

- head loss, **h**, is the head loss in feet of water for the equivalent length, **Le**, at flow rate **gpm**.

Example:

A series circuit contains 150 feet of 1¼" copper tubing, (25) 1¼" 90-degree elbows and two ball valves (1¼"). What is the head loss for the circuit with 120 °F water flowing at 15 gpm?

From Table 11, the equivalent length of each 1¼" elbow is 3.0 feet and each 1¼" ball valve has an equivalent length of 7.0 feet. The circuit equivalent length is the sum of all of these, or:

$$Le = 150 + (25 \times 3.0) + (2 \times 7.0) = 150 + 75 + 14 = 239 \text{ feet}$$

From Table 12, a is .0338 for 1¼" copper tubing with water at 120 °F and b is 1.750. Use equation 5-7 to solve for head loss:

$$h = (239/100) \times .0338 \times 15^{1.75} = 9.2 \text{ feet}$$

So the head loss for this circuit with 120 °F water flowing at 15 gpm would be 9.2 feet. The head loss would be 11.0 feet if water were at 60 °F (using value of a = .0402 from Table 12 for water at 60 °F in 1¼" copper tubing).

Equivalent length

Table 10	Steps in using I=B=R tables — overview
Step 1	**Calculate heat losses**
Step 2	**Select heat distributing units**
Step 3	**Select boiler size**
Step 4	**Select expansion tank**
Step 5	**Determine system type (series loop, one-pipe or two-pipe) and make piping layout — determine length of each circuit**
Step 6	**Determine flow rate (gpm)**
Step 7	**Use I=B=R tables for selected system type to determine circulator capacity and pipe size**

Steps 1 to 6:
Same as Method 3 (Page 5–19)

Apply Steps 1 through 6 of Method 3 to determine flow rates and heating units for your application. Then proceed to Step 7, below.

Step 7:
Size piping, calculate head losses, select circulator(s)

Single-circuit system:

Calculate equivalent length of circuit

Find the equivalent length of each fitting in the circuit from Table 11, Page 5–28 For finned-tube baseboard, the equivalent length is the length of the element if the baseboard is the same size as the piping. For radiators or cast iron baseboard, add the equivalent length from the table plus the equivalent length of angle valves, if installed. Add the equivalent length for other system components (air separators, etc.) based on manufacturer's data. If data provides C_v instead of equivalent feet, use Table 17, Page 5–48, to convert. (C_v is the flow rate, in gpm, that will cause a pressure drop of 1 psig [2.309 feet water column].)

Calculate head loss

Use the flow rate required for the circuit (gpm) and the equivalent length (Le) from above in equation 5-7 to calculate head loss.

Select circulator

Select a circulator that will provide the required flow and head loss. If using a packaged circulator, verify using the circulator performance curve.

Multiple-circuit system:

Calculate equivalent length of each circuit

Multiple-circuit systems will usually include more than one pipe size. For each circuit and each pipe size in the circuit, calculate the equivalent length of pipe, fittings, and components for that segment.

Calculate head loss

Calculate the head loss using the flow for each segment of each circuit using equation 5-7. The total head loss for a circuit is the sum of the losses of all segments.

Select circulator

For circuits in parallel (two-pipe system or multiple-loop series circuit), select the highest head for any of the circuits. This is the required head for the circulator. The required flow is the sum of all circuit flows supplied by the circulator.

If using multiple circulators (zoned system, for example), the required flow is the flow for the circuit supplied. The required head is the highest head loss for any of the circuits in the system.

If you are interested in determining the actual flow that occurs in parallel circuits not equipped with balancing valves, see "Parallel resistances" in Part 5 Appendix.

Table 11   Equivalent length (feet) of common fittings and components

Copper fitting equivalent lengths (feet)

Copper size	½	¾	1	1¼	1½	2	2½	3	4
90 degree elbow	1.0	2.0	2.5	3.0	4.0	5.5	7.0	9.0	12.5
45 degree elbow	0.5	0.8	1.0	1.2	1.5	2.0	2.5	3.5	5.0
Tee (straight run)	0.3	0.4	0.5	0.6	0.8	1.0	0.5	1.0	1.0
Tee (side port)	2.0	3.0	4.5	5.5	7	9	12	15	21
Diverter tee (typical)		70	23.5	25	23	23			
Gate valve	0.2	0.3	0.3	0.4	0.5	0.7	1.0	1.5	2.0
Globe valve	15	20	25	36	46	56			
Angle valve	3.1	4.7	5.3	7.8	9.4	12.5			
Ball valve (standard port)	1.9	2.2	4.3	7.0	6.6	14			
Swing check valve	2.0	3.0	4.5	5.5	6.5	9	11.5	14.5	18.5
Flow-check valve (typical)		83	54	74	57	177			
Butterfly valve	1.1	2.0	2.7	2.0	2.7	4.5	10	15.5	15

Threaded fitting equivalent lengths (feet)

Pipe size		½	¾	1	1¼	1½	2	2½	3	4
90 degree elbow		1.6	2.1	2.6	3.5	4.0	5.2	6.2	7.7	10.1
Long radius ell (45° or 90°)		0.8	1.1	1.4	1.8	2.2	2.8	3.3	4.1	5.4
Standard tee, through flow		1.0	1.4	1.8	2.3	2.7	3.5	4.1	5.1	6.7
Standard tee, branch flow		3.1	4.1	5.3	6.9	8.1	10	12	15	20
Close return bend		2.6	3.4	4.4	5.8	6.7	8.6	10	13	17
Mitre bend	45°						2.6	3.1	3.8	5.0
	90°						10	12	15	20
Gate valve, full open		0.4	0.6	0.7	0.9	1.1	1.4	1.7	2.0	2.7
Globe valve, full open		18	23	30	39	46	59	70	87	114
Angle valve, full open		7.8	10	13	17	20	26	31	38	50
Swing check valve, full open		5.2	6.9	8.7	12	13	17	21	26	34
Butterfly valve							7.8	9.3	12	15
Flow/check valve, typical			27	42	60	63	83			

Typical component equivalent lengths (feet)

Cast iron radiator/baseboard	7.5
Finned-tube baseboard	Length of element
Other components	Consult manufacturer's data for equivalent length or C_v

Table 12 Head loss calculation factors for use in equation 5-7 Page 5–26

Head loss calculation factors
(feet head per 100 feet pipe = **a** x gpm^b)

Copper tube size	Minimum flow (gpm)(note 3)	Maximum flow) (gpm)(note 3)	**a** (notes 1 & 2)			**b** (note 2)	Minimum velocity (fps)(note 3)	Maximum velocity (fps)(note 3)	Pipe I.D. (inches)
			60 °F	120 °F	180 °F				
⅜ (Type L)	0.68	1.81	7.453	6.268	5.590	1.750	1.5	4.0	0.430
½ (Type M)	1.19	3.17	1.970	1.657	1.478	1.750	1.5	4.0	0.569
¾ (Type M)	2.41	6.44	0.366	0.308	0.274	1.750	1.5	4.0	0.811
1 (Type M)	4.09	10.90	0.1049	0.0882	0.0787	1.750	1.5	4.0	1.055
1¼ (Type M)	6.12	16.32	0.0402	0.0338	0.0302	1.750	1.5	4.0	1.291
1½ (Type M)	8.56	22.83	0.0181	0.0152	0.0136	1.750	1.5	4.0	1.527
2 (Type M)	14.8	39.5	0.00492	0.00414	0.00369	1.750	1.5	4.0	2.009
2½ (Type M)	32.1	80.9	0.00140	0.00118	0.00105	1.812	2.1	5.3	2.495
3 (Type M)	51.5	129.7	0.000594	0.000499	0.000445	1.812	2.4	6.0	2.981

Steel pipe size	Minimum flow (gpm)(note 3)	Maximum flow) (gpm)(note 3)	**a** (notes 1 & 2)			**b** (note 2)	Minimum velocity (fps)(note 3)	Maximum velocity (fps)(note 3)	Pipe I.D. (inches)
			60 °F	120 °F	180 °F				
½	1.42	3.79	1.317	1.163	1.073	1.822	1.5	4.0	0.622
¾	2.49	6.65	0.319	0.289	0.271	1.858	1.5	4.0	0.824
1	4.04	10.77	0.1000	0.0894	0.0830	1.838	1.5	4.0	1.049
1¼	6.99	18.6	0.0280	0.0245	0.0225	1.812	1.5	4.0	1.380
1½	9.52	25.4	0.0116	0.0105	0.0098	1.850	1.5	4.0	1.610
2	15.7	41.8	0.00329	0.00300	0.00281	1.863	1.5	4.0	2.067
2½	30.1	72.9	0.00121	0.00112	0.00106	1.890	2.0	4.9	2.469
3	53.9	133.0	0.000460	0.000415	0.000389	1.855	2.3	5.8	3.068

Note 1: The data above allow calculation for water temperature of 60 °F, 120 °F or 180 °F. Head loss decreases as temperature increases because of the decrease in water viscosity. The Quick-selector tables and I=B=R tables in this Guide are based on water at 60 °F. Calculating head loss for 60 °F water provides the most conservative results.

Note 2: To apply the data above, calculate the total equivalent length of the circuit or branch using the data for fitting equivalent lengths provided in the tables on the opposite page. Divide the total equivalent length by 100. Then multiply this amount times **a** x gpm^b to determine head loss at the given flow rate. The flow rate must be between the minimum and maximum values listed in the table.

Note 3: The minimum flow rates above represent the suggested minimum flow rate in the piping to ensure best removal of air from the system. The maximum flow rates are the recommended limits to avoid flow noise in the piping.

Note 4: **Glycol/water** applications. Glycol has a higher viscosity than water, causing a higher head loss. Glycol also reduces heat transfer. To compensate for a system filled with glycol/water, you can apply these correction factors:
To obtain approximately the same **heat transfer** as water in a heating unit, **increase the flow rate 15%** when using glycol up to a 50/50 concentration.
Calculate **head loss** for water only. Then multiply times **1.2 for 30%** glycol/water or **1.4 for 50%** glycol/water for water temperatures 100 °F and higher.

Example house

Overview

See Figure 1, Page 5–10. This is the house used in Hydronics Institute Guide H-22 for an example heat loss calculation. The following pages apply the Guide RHH sizing methods to various systems applied to this residence.

Table 13	Steps in using I=B=R tables — overview
Step 1	Determine heat loss
Step 2	Select heating units
Step 3	Select boiler
Step 4	Select expansion tank
Step 5	Determine system type (series loop, one-pipe or two-pipe) and make piping layout — determine length of each circuit
Step 6	Determine flow rate (gpm)
Step 7	Size piping and circulator(s)

Step 1: Determine heat loss

Page 5–52 and Page 5–53 are the Heat Loss Worksheets for this house (from Guide H-22). The worksheets show the heat loss for each room. The total heat loss for the house is 48,611 Btuh.

Table 14   Room-by-room heat losses for example house in Figure 3, Page 5–31

Room	Heat loss (Btuh)	Baseboard		
		Output (Btuh/foot)	Feet needed	Feet used
Bedroom 1	4,164	620	6.7	7
Bathroom 1	1,529	620	2.5	3
Family room	4,634	620	7.5	8
Kitchen	4,163	850	4.9	5
Dining room	3,080	620	5.0	5
Living room	4,272	620	6.9	7
Entry hall	1,992	620	3.2	4
Bedroom 3	2,531	620	4.1	5
Bedroom 2	3,860	620	6.2	7
Bathroom 2	235	None installed		
Basement	18,151	620	29.3	30
Total	48,611			

Figure 3   Example house from Guide H-22 used for Part 5 sizing examples

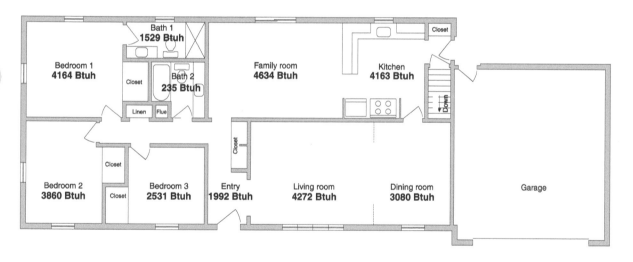

Bath 1
1529 Btuh

Bedroom 1
4164 Btuh

Closet

Bath 2
235 Btuh

Linen   Flue

Closet

Family room
4634 Btuh

Closet

Kitchen
4163 Btuh

Closet

Down

Bedroom 2
3860 Btuh

Closet

Closet

Bedroom 3
2531 Btuh

Entry
1992 Btuh

Living room
4272 Btuh

Dining room
3080 Btuh

Garage

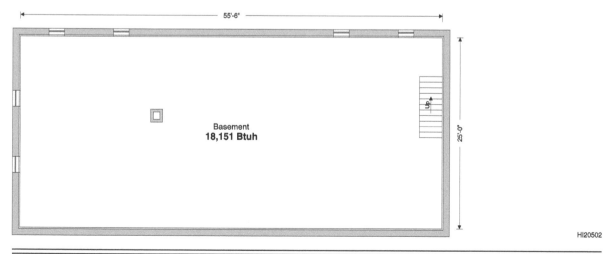

55'-6"

25'-0"

Up

Basement
18,151 Btuh

HI20502

See Hydronics Institute *Guide H-22* for detailed heat loss calculations. House is located in Billings, Montana. Outdoor design temp = -10 °F. Indoor design temp = 70 °F. Total heat loss (main floor only) — 30,460 Btuh; Total heat loss (including basement) — 48,611 Btuh.

Step 2:
Select heating units

We selected finned-tube baseboard heaters throughout the house because they meet the needs of the space and can be used well in all of the systems demonstrated in this section. This house would also be a candidate for radiant heating, convectors or other heating units. We use baseboard in all example cases for a clearer comparison of the sizing results.

Figure 7, Page 5–41 is a 1/8" = 1' scale floor plan of the home, with the selected baseboard heaters shown. We selected the baseboard heaters as follows.

All of the baseboard selected for the house is 3/4" copper finned-tube baseboard. The manufacturer's rating for the baseboard with 190 °F average water temperature is 620 Btuh per foot of element at 1 gpm and 660 Btuh per foot of element at 4 gpm. The exception is the kitchen, as explained by the following.

Main floor rooms except Kitchen and Bath 2: Each of these rooms has sufficient wall space for the required baseboard. Baseboard is sized using the 1 gpm rating so the sizing is adequate for each of the system sizing examples in this section.

Bath 2: The heat loss is only 235 Btuh, and the surrounding areas will provide some heat. To be sure the room is supplied, have the owner install a heat lamp, for use only as needed.

Kitchen: The kitchen only has about 5 feet of wall space available because of counters. This wall space is along the wall separating the kitchen from the stairway. By using high-output baseboard (rated 850 Btuh/foot at 1 gpm; 900 Btuh/foot at 4 gpm), the 5-foot length is adequate to handle the heat loss of 4,163 Btuh.

Basement: The heat loss from the basement is large — 18,151 Btuh. For series loop or one-pipe systems, the heat loss from the piping would not be enough to handle this load. For the example, we are assuming the owner wants the basement at 70 °F, for comfortable occupancy. We chose standard baseboard, selected for a flow of 1 gpm to be sure the baseboard is sized adequately for all applications discussed in this section. We place one baseboard on each wall, located for the best connection to the piping systems. The pipes will be insulated to prevent excessive heat loss to the basement.

Figure 4   Floor plan of example house, with room heat losses and baseboard selections shown

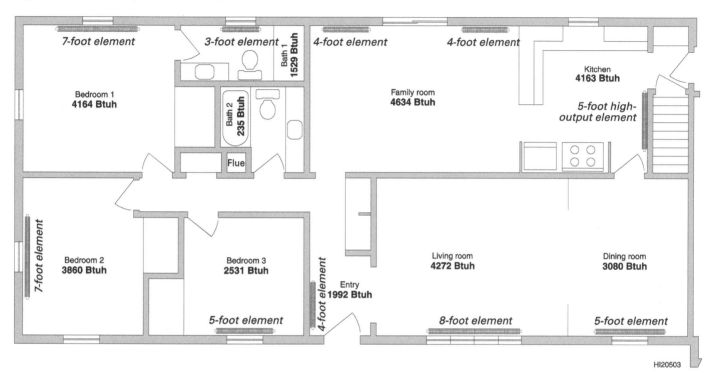

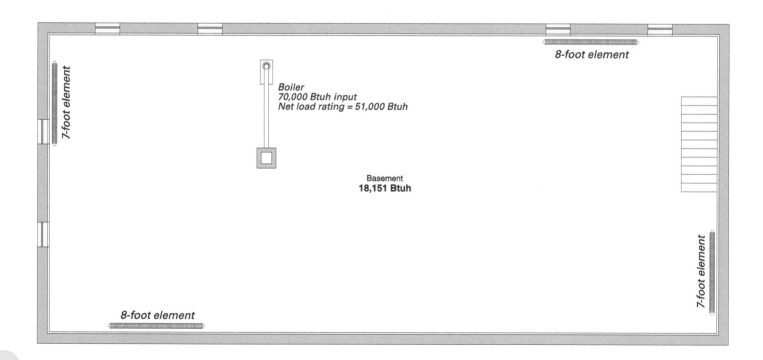

Scale: 1/8" = 1' 0"

Basement considerations

When you encounter a situation like this, consider:

- Recommending energy savings alternatives to the owner.

  - Add 2 x 4 studs, 3½" of insulation and drywall to the basement walls in this house, for instance. The result would be reduction in heat loss to 12,846 Btuh. You can check this by recalculating the basement heat loss with an exposed wall factor of 0.08 and a below-grade factor of 0.045 in place of the original values. This change would reduce the heat loss for the home 11% — meaning a substantial reduction in fuel cost.

  - Insulate the piping and installing baseboard in appropriate locations in the space rather than relying on boiler and piping heat loss to handle the basement load. This will assure comfortable, adequate heat, and will allow zoning if required.

Relying on bare pipes and the boiler to heat the basement can be risky, as explained in the following.

Heat loss from bare piping

Piping in basements is often left bare to provide heat to the space, assuming that the heat loss from the piping and boiler will keep the basement area warm.

That isn't necessarily the case. With the example house, even if the basement heat loss were reduced to 12,846 Btuh, the pipes and boiler wouldn't provide enough heat with some systems (single-loop series loop or one-pipe).

Look at Table 15, showing the heat loss from bare tubing and pipe at 180 °F temperature (45 Btuh per foot for ¾" horizontal copper or 55 Btuh per foot for 1" horizontal copper).

Split-loop or two-pipe systems would have enough tubing in the basement to provide this heat loss (though only if the walls were insulated as recommended before). But the heat would be near the ceiling and wouldn't comfortably heat the main occupied level.

The piping in a single-loop system, either series-loop or one-pipe, would give off about 7500 Btuh. The boiler would give off about 2% of its input (typical jacket loss from residential boilers), or about 1,400 Btuh, for a total of 8,900 Btuh. If the basement walls were covered with studs, insulation, and drywall, the heat loss would be 12,846 Btuh. The pipes and boiler would provide 8,900 Btuh of this, or about 70% of the loss. This would be enough for about

70% of the design temperature difference (80 °F), or 56 °F. That is, at design conditions (-10 °F outdoors), the basement temperature would be -10 °F + 56 °F = 46 °F. Some additional heat will also be lost to the basement through the floor, but you can see this would be a very uncomfortable temperature even if the basement were only used for laundry and occasional occupancy. And the floor temperature on the main level might become uncomfortable as well.

If you plan to zone, you have to insulate the basement piping. Otherwise, the basement would overheat because the pipes would give off heat any time any zone called for heat.

If you do plan to rely on the loss from bare pipes and the boiler to heat the basement, use Table 15 to make sure there will be enough heat. If the basement will be used for frequent occupancy, always recommend installing heating units for the most comfort.

Table 15   Heat loss from bare pipes at 180°F

Heat loss from bare pipes (Btuh per linear foot with water in pipes at 180 °F)		
Pipe or tube size (inches)	**Steel pipe**	**Copper tube**
½	59.3	34.0
¾	72.5	45.4
1	88.8	56.4
1¼	109.7	67.2
1½	123.9	77.6
2	151.8	98.0
2½	180.5	117.9
3	215.9	137.2

Source: 1997 ASHRAE Handbook, *Fundamentals*, pp 24.19 and 24.20,
Table 11A and Table 12 — based on tubing or pipe in still air at 80°F.

Step 3:
Select boiler

The chimney (flue) is in good condition and usable for natural-draft venting. We have chosen an atmospheric gas water boiler with an I=B=R net load rating of 51,000 Btuh (input of 70,000 Btuh). We chose this boiler because of its ease of installation into the existing chimney and the net load rating just larger than the house heat loss of 48,611 Btuh.

We have located the boiler near the chimney and the outside wall (to reduce piping runs to wall). See the floor plans for each circuit type on the following pages.

We would want to verify that:

1. The basement floor is in good condition and no foundation is required for the boiler.

2. There is no risk of water accumulation on the basement floor.

3. The house is not equipped with a whole-house fan or exhaust device which could cause a negative pressure.

4. No combustion air openings are required because the house is not of "unusually tight construction" as defined by the ANSI Z223.1 National Fuel Gas Code.

5. The area chosen for the boiler is not planned for any other future use.

6. The owner does not plan to use chemicals or products in the basement which could pose a contamination (corrosion) hazard to the boiler (since the boiler uses inside air for combustion).

7. The boiler will be directly mounted on the concrete floor - not on combustible flooring.

8. Review Part 2 of the Guide for other considerations in selecting and installing the boiler.

Step 4:
Select expansion tank

We have selected a diaphragm-type expansion tank. Our selection is based on the tank manufacturer's quick-selector guide, sized for a boiler with 70,000 Btuh input.

We will also be using an in-line air separator with 1" npt connections. (The air separator has an equivalent length rating of 18 feet of 1" copper tubing, or about 5 feet of ¾" tubing.)

Example house: Steps 5, 6 & 7 Single-loop series circuit (1 circuit)

Step 5:
Make piping layout and determine length of each circuit

Figure 5, Page 5–37 shows the piping for a single-loop series-loop circuit. The measured length of piping and baseboard is approximately 267 feet.

Step 6:
Determine flow rate (gpm)

The flow rate for a 20 °F TD would be:
gpm = 48,611 ÷ 10,000 = 4.9 gpm.

Step 7:
Size piping and circulator

Try Method 1 (Quick selector charts, Page 5–12)

For a quick check, take a look at Table 2, Page 5–13, "Typical packaged residential circulator capacity." Our circuit length of 267 feet exceeds the length limit for ¾" tubing. We might be able to use 1" baseboard and piping, but it would be much less costly to install a larger circulator. Size the circulator using one of the other methods as follows.

Try Method 2 (Milinch method, Page 5–16)

The heat load is 48,600 Btuh. Looking at Table 5, Page 5–17, you will see that the maximum load for ¾" pipe is 39 MBH at 500 milinches per foot, or 31 MBh for 350 milinches per foot. As for Method 1, we might be able to use 1" baseboard and piping, but it would be much less costly to install a larger circulator. Size the circulator using one of the other methods as follows.

Apply Method 3 (I=B=R tables, Page 5–20)

Enter Table 7, Page 5–21, upper section, at ¾" pipe size. Move across to 5 gpm (closest to required 4.9 gpm) in the first column.

Scan down this column to the measured length closest to our required 267 feet. You will find that there is no selection for a length of ¾" piping this long.

Your options would be:

• Use another circuit design (see split-loop and one-pipe discussions following)

• Use 1" piping and baseboard. [If you follow Table 7 for 1" pipe size across to the 5.0 gpm column, down to 290 feet (closest to 267), and read to the left column that the required circulator head is 8 feet. This would easily work for a packaged circulator (see Table 2, Page 5–13, which indicates a flow of 10 gpm at 8 feet head).]

System alternatives

Series-loop circuits are a good alternative and you should use them whenever they meet job requirements.

• This installation would work as a ¾" single-loop series loop *if we only had to distribute to baseboard on the first floor*.

• There would be no basement baseboard, and the total circuit length would be approximately 211 feet.

• The total heating load would be just the first floor load, or 48,611 – 18,151 = 30,460 Btuh.

• Flow rate for 20°F TD would be 30,460 ÷ 10,000 = 3.1 gpm.

• Follow Table 7, Page 5–21 for ¾" pipe to the third column, 4.1 gpm. Read down to 250 feet (closest to 211 feet), then left to a required circulator head of 16 feet.

• You can source a residential-sized circulator for 4.1 gpm and 16 feet head. (You would not be able to use the circulator packaged with a typical residential boiler.)

The best solution for this system if zoning is not required is a **split-loop series-loop system** (sizing discussion follows).

Figure 5 Piping arrangement for single-loop series circuit

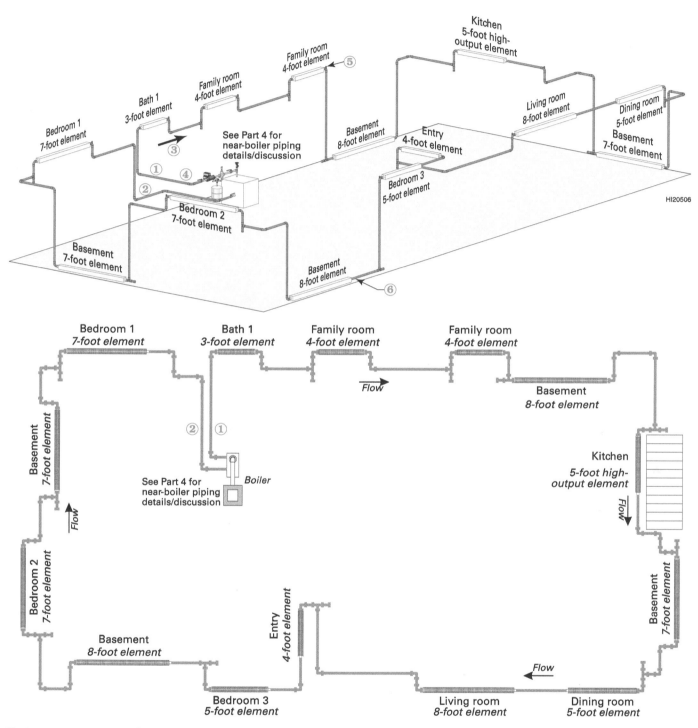

① Supply pipe

② Return pipe

③ Water flow direction

④ When all radiation is above the boiler and an indirect water heater is installed, provide a flow/check valve on the supply line to prevent gravity circulation. The system above would not require the flow/check valve. The basement branches provide thermal traps.

⑤ Air vents on outlet sides of main floor heating units

⑥ Install tee with nipple and cap or inverted baseboard tee with plug at low points to allow drainage.

Example house: Steps 5, 6 & 7 Single-loop series circuit (2 circuits)

Step 5:
Make piping layout and determine length of each circuit

Figure 6 shows the piping for a split-loop series-loop system. The measured lengths of piping and baseboard are:

- Left circuit — 179 feet
- Right circuit — 185 feet.

Step 6:
Determine flow rates (gpm)

The heat loads are:
- Left circuit — 23,153 Btuh
- Right circuit — 25,225 Btuh.

The flow rates for a 20 °F TD would be:
- Left circuit — 2.3 gpm
- Right circuit — 2.5 gpm.

The flow rate for the trunk would be the total of these, or 4.8 gpm.

Step 7: Size piping and circulator using Method 2 (Milinch method, Page 5–16)

Pipe sizes:

Check Table 5, Page 5–17.

- In the 500 milinch per foot section, note that the maximum flow rate for ½" copper is 1.7 gpm, and for ¾" copper is 3.9 gpm.
- This means both circuit 1 and circuit 2 must be ¾" copper.
- The trunk line carries 4.8 gpm, and must be 1" copper.

Circulator requirements

Use the length of the longest circuit, or 185 feet.

The TEL is estimated at 1.5 times measured length, or:

TEL = 1.5 x 185 = 278 feet

At 500 milinches per foot, the head loss is 4.2 feet per hundred feet. So head loss is:

Head loss = 4.2 feet/100 feet x 278 feet = 11.7 feet

Select a circulator that can handle 4.8 gpm at 11.7 feet (easily handled by a typical residential packaged circulator).

Step 7: Size piping and circulator using Method 3 (I=B=R tables, Page 5–20)

Circuit 1: Try ½" tubing first. Enter Table 7, Page 5–21, at ½" pipe size and scan to the first column, 2.3 gpm. Scan down to the closest length to 179 feet. This would be 170 feet. Read left to see the required head of 16 feet. This head is too high for a packaged circulator, so try ¾" tubing. Read across the ¾" row to the 2.4 gpm column. Scan down to 190 feet (closest to 179 feet). Read left to see the required head of 5 feet. Select ¾" tubing.

Circuit 2: Don't try ½" tubing because the flow is higher than circuit 1, and ½" didn't work for circuit 1. Read across the ¾" row to the 2.6 gpm column. Scan down to 200 feet (closest to 185 feet). Then read left to see the required head of 6 feet.

Trunk line: You must use the longest length of any of the circuits for sizing the trunk. This would be 185 feet, the length of the right circuit.

The circulator must supply 4.8 gpm at a head of at least 6 feet. A packaged circulator should be capable of 9.7 feet head at 4.8 gpm (per Table 2, Page 5–13).

Size the trunk line assuming a 10-foot head (packaged circulator capacity), and also assuming a 6-foot head (highest head loss of any of the circuits) just to see if the results differ.

For 10-foot head, enter Table 7 bottom section at 10 feet head. Read across to 190 (closest to length of 185 feet). Read up to the row with the flow rate just higher than the required 4.8 gpm. This would be the 6.8 gpm row. Read left to see a required trunk line size of 1".

If you enter the chart at 6 feet head, read across to 180 (closest to 185) and then up, you will find 5.5 gpm (1" pipe) or 11.0 gpm (1¼" pipe). The 1" pipe would be an acceptable selection even if available head was only 6 feet.

The trunk line must be 1" pipe.

The circulator must supply 4.8 gpm at a head ranging from 6 to 10 feet. A typical packaged circulator should be capable of this. In other words, you should be able to use the circulator supplied with the boiler (if included).

Figure 6   Piping arrangement for split-loop series-loop circuit

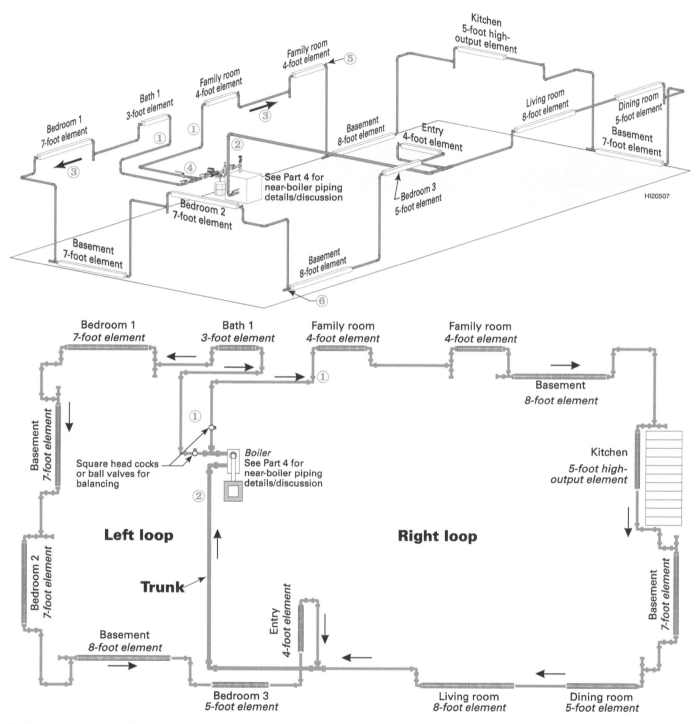

① Supply pipes
② Return pipe
③ Water flow direction

④ When all radiation is above the boiler and an indirect water heater is installed, provide a flow/check valve on the supply line to prevent gravity circulation. The system above would not require the flow/check valve. The basement branches provide thermal traps.

⑤ Air vents on outlet sides of main floor heating units
⑥ Install tee with nipple and cap or inverted baseboard tee with plug at low points to allow drainage.

Example house: Steps 5, 6 & 7 Single-loop one-pipe circuit

Step 5: Make piping layout and determine length of each circuit

Figure 7, Page 5–41 shows the piping for a single-loop one-pipe system.

We selected a single one-pipe fitting for each heating unit on the main floor (all above the main). We show two one-pipe fittings on each of the basement baseboard heaters because they are located below the main. Note that, when the main piping is ¾", some fitting manufacturers recommend only a single fitting, even when the unit is located below the main. Table 16, Page 5–47, shows the approximate flow in heating units located below the main. The table data for ¾" mains is based on a single fitting (with Cv of 4.9). Larger main sizes use two fittings.

The measured length of main circuit piping is 187 feet. The circuit contains (18) one-pipe fittings. Add 12 feet for each fitting to the measured length for a total circuit length of 403 feet.

Step 6: Determine flow rate (gpm)

The heat load is 48,611 Btuh. The main circuit flow at a 20 °F TD would be 4.9 gpm.

Each baseboard unit requires a flow of 1 gpm for the 1 gpm rating. Select branch piping, then, based on 1 gpm. (Note that no heating unit has a load in excess of 10,000 Btuh, so 1 gpm will always satisfy a maximum 20 °F TD.)

Step 7: Size piping and circulator using Method 3 (I=B=R tables, Page 5–22)

One-pipe circuits are best handled using Method 3, the I=B=R table for one-pipe systems, Page 5–22.

We want to use a packaged circulator with a head of 9.7 feet at 4.9 gpm. Use 10 feet for the available head.

Size the trunk by entering Table 8, Page 5–23, upper section at 10 feet head. Read across to the 430 column

(closest to circuit total length of 403 feet). Read down to 4.6 gpm (reasonably close to desired 4.8 gpm) and across to see this is for a 1" main. The branch flows are all to be 1 gpm. So read down from the 4.6 to find the 1.1 gpm row. Read across to find that all branches must be ¾" tubing. The table capacity for heating units below the main must be reduced to 85% or the value shown, or 85% of 1.1 = 0.94 gpm. This will be sufficient for the baseboard in the basement. As another check, look at Table 16, Page 5–47. Look for a 1" main with ¾" branch using two fittings and main flow of 4.6 gpm. This would be in between the flows of 4 and 5 gpm. The length of the 8-foot baseboard circuits is about 22 feet (7 feet below the main plus 8 feet of baseboard). Use the 40-foot branch circuit length column (this would provide enough margin to allow for a thermostatic valve if desired). At a flow of 4.6 gpm in the main, the approximate flow down through the baseboard would be 1.1 gpm (interpolating between the 4 and 5 gpm data). This indicates the application will work.

Summary

A packaged circulator providing 4.8 gpm at 9.7 feet head will suffice for this piping layout.

- The main will be 1" copper tubing.

- All branch piping will be ¾" copper tubing.

- The circuit requires (18) one-pipe fittings, 1" x 1" x ¾" (two for each basement baseboard connection and one for all other heating units).

- You can zone this system using a thermostatic valve on each heating unit.

- The sizing data in Table 8 allows for up to 15 feet of horizontal piping.

- If the thermostatic valves have an equivalent length no more than 15 feet, you won't need to double-check the sizing.

- If the valves have a higher pressure drop, make sure the added equivalent length won't lower the flow rate below 1 gpm.

Figure 7 Piping arrangement for single-loop one-pipe system

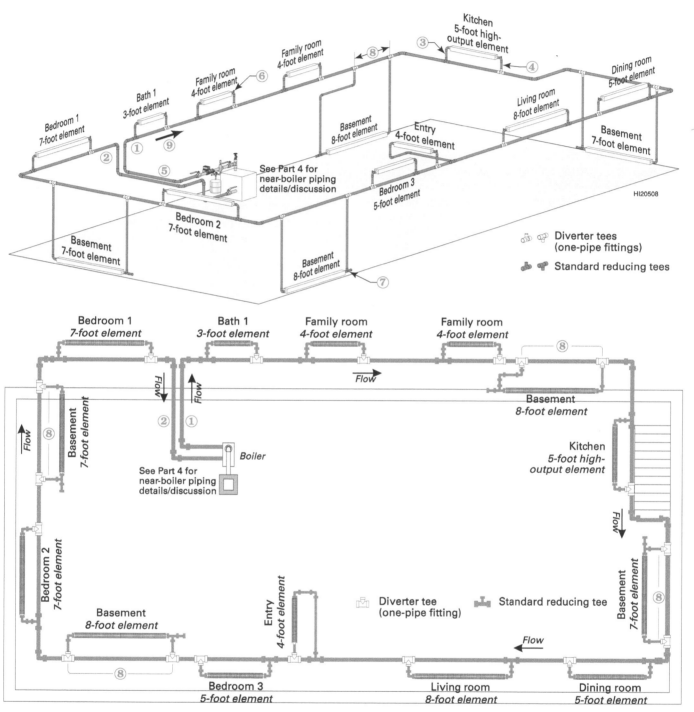

① Supply piping
② Return piping
③ Branch supply piping
④ Branch return piping

⑤ Install flow control valve on supply if an indirect water heater is used. Valve is optional otherwise.
⑥ Air vents on outlet sides of main floor heating units

⑦ Install tee with nipple and cap or inverted baseboard tee with plug at low points to allow drainage.
⑧ Install tees no closer than 6 inches.
⑨ Water flow direction

Example house: Steps 5, 6 & 7 Split-loop one-pipe system (2 circuits)

Step 5: Make piping layout and determine length of each circuit

Figure 8, Page 5–43 shows the piping for a split-loop one-pipe system.

We selected a single one-pipe fitting for each heating unit on the main floor (all above the main). We show two one-pipe fittings on each of the basement baseboard heaters because they are located below the main. Note that, when the main piping is ¾", some fitting manufacturers recommend only a single fitting, even when the unit is located below the main. Table 16, Page 5–47, shows the approximate flow in heating units located below the main. The table data for ¾" mains is based on a single fitting (with C_v of 4.9). Data for larger main sizes are based on two fittings.

The circuit lengths are: Left circuit — The measured length is 134 feet. The circuit contains (9) one-pipe fittings. Add 12 feet for each fitting to the measured length for a total circuit length of 242 feet. Right circuit — The measured length is 150 feet. The circuit contains (9) one-pipe fittings. Add 12 feet for each fitting to the measured length for a total circuit length of 258 feet.

Step 6: Determine flow rate (gpm)

Left circuit: Heat load of all rooms served is 23,151 Btuh. Flow at 20 °F TD is 2.3 gpm.

Right circuit: Heat load of all rooms served is 25,225 Btuh. Flow at 20 °F TD is 2.5 gpm.

Trunk: Flow is total, or 4.8 gpm.

Each baseboard unit requires a flow of 1 gpm for the 1 gpm rating. Select branch piping, then, based on 1 gpm. (Note that no heating unit has a load in excess of 10,000 Btuh, so 1 gpm will always satisfy a maximum 20 °F TD.)

Step 7: Size piping and circulator

We want to use a packaged circulator with a head of 9.7 feet at 4.9 gpm. Use 10 feet for the available head.

Left circuit: Enter Table 8, Page 5–23, upper section at 10 feet head. Read across to 230 (closest length to circuit length of 242 feet). Required flow is 2.3 gpm. Read down the 230 column to 3.2 gpm in the lower section of the table. Read left to see this is for a ¾" main.

For a 1 gpm branch capacity, read down the 3.2 gpm column in the ¾" main section. A ½" branch can provide up to 1.2 gpm. (For units below the main, reduce this capacity to 85%, = 1.02 gpm. This is satisfactory for the basement baseboard units.)

With a ¾" main, some one-pipe fitting manufacturers recommend only one fitting. Check Table 16, for the expected flow to the basement baseboard units if using a single fitting on the ¾" main. The length of the longest basement baseboard circuit is about 22 feet. Use the 40-foot column (which should allow for use of a thermostatic valve if desired). If the main flow is 2.3 gpm (use the 2.5 gpm row), the flow in the baseboard would only be 0.7 gpm. This would not be enough for the 1 gpm output rating. You might consider using two fittings (provided the manufacturer recommends this). You can source ¾" fittings which provide a C_v of 3.2 when two are used. This would be sufficient to ensure at least 1 gpm through the heating units. This is the choice we have taken in our application. Another alternative would be to increase the flow in the main. If you look at Table 16, you will see that, at a main circuit flow of 3.5 gpm, the flow in the branch would be 1.0 gpm. If you size a circulator to provide 7 gpm (3.5 in each circuit), you would have sufficient flow in each basement baseboard branch.

Right circuit: Enter Table 8, Page 5–23 at 10 feet head. Read across to 260 (closest length to 258 total circuit length). The required flow is 2.5 gpm. Read down the 260 column to the flow of 3.1 gpm (first section). Read left to see this is for a ¾" main as for the Left circuit. Use ½" branch piping, which will provide 1.2 gpm, read of the table below the 3.1 gpm value.

Trunk: The trunk flow rate is 4.8 gpm. Size the trunk using the longest circuit length, or 258 feet (length of Right circuit). Enter Table 8 at 10 feet head and read across to 260 (closest to 258 feet). Read down to the first value larger than 4.8 gpm, or 5.9 gpm. Read left to see this is for a 1" pipe size.

A packaged circulator should handle this circuit adequately (unless you use a single fitting on each basement baseboard unit, in which case you will need a circulator capable of 7 gpm at 11 feet head and a trunk size of 1¼").

In summary, a packaged circulator providing 4.8 gpm at 9.7 feet head will suffice for this piping layout. The trunk will be 1" copper tube, with ¾" main piping in each circuit. All branch piping will be ½" copper tubing. The circuit requires (18) one-pipe fittings, ¾" x ¾" x ¼" (two for each basement baseboard connection and one for all other heating units). You can zone this system using a thermostatic valve on each heating unit. The sizing data in Table 8 allows for up to 15 feet of horizontal piping. If the thermostatic valves have an equivalent length of no more than 15 feet, you won't need to double-check the sizing. If the valves have a higher pressure drop, make sure the added equivalent length won't lower the flow rate below 1 gpm.

Figure 8   Piping arrangement for split-loop one-pipe system

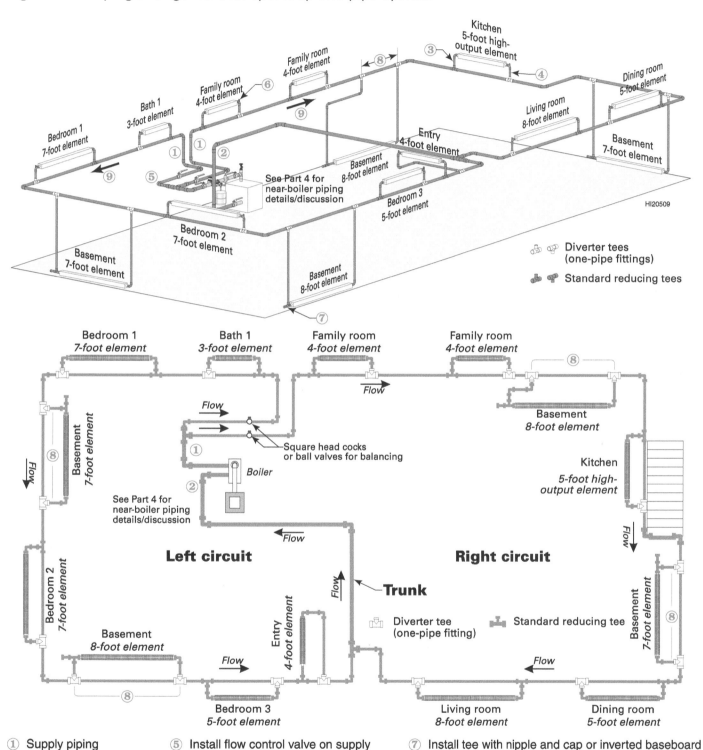

① Supply piping

② Return piping

③ Branch supply piping

④ Branch return piping

⑤ Install flow control valve on supply if an indirect water heater is used. Valve is optional otherwise.

⑥ Air vents on outlet sides of main floor heating units

⑦ Install tee with nipple and cap or inverted baseboard tee with plug at low points to allow drainage.

⑧ Install tees no closer than 6 inches.

⑨ Water flow direction

Example house: Steps 5, 6 & 7 Two-pipe reverse-return system

Step 5:
Make piping layout and determine length of each circuit

Figure 9, Page 5–45 shows the piping for a two-pipe reverse-return system.

The longest circuit in the system is approximately 222 feet. We combined some heating units in series to reduce the number of parallel branches, since each branch requires 1 gpm flow. Allow 15 feet equivalent length for a zone valve in each branch, giving a total length of 237 feet.

Step 6:
Determine flow rate (gpm)

To obtain the 1 gpm rating in each baseboard unit, each branch must have a flow rate of at least 1 gpm. The system contains (7) branches, for a total flow of 7 gpm. Note that all branches are less than 10,000 Btuh (1 gpm at 20 °F TD), so 1 gpm is adequate for every branch.

Step 7:
Size piping and circulator

Size all branches and the mains based on a length of 237 feet (longest branch).

- The flow in each branch is 1 gpm.
- The maximum flow in any portion of the mains is 7 gpm.

The supply and return takeoffs are labeled A through N in Figure 9, Page 5–45. At each supply takeoff, 1 gpm flows out to a branch. At each return, 1 gpm flows into the return from a branch. We want to use a packaged circulator, with a capacity of 7 gpm at 9 feet head. Size each portion of the supply and return mains using a length of 237 feet and an available head of 9 feet.

Enter Table 9, Page 5–25, at 9 feet head, scan across to the length closest to 237 feet (230-foot value). Drop down to find the first gpm larger than the flow in each section of the main.

Boiler to A:	7 gpm	(requires 1¼" pipe)
A - B:	6 gpm	(requires 1¼" pipe)
B - C:	5 gpm	(requires 1" pipe)
C - D:	4 gpm	(requires 1" pipe)
D - E:	3 gpm	(requires ¾" pipe)
E - F:	2 gpm	(requires ¾" pipe)
F - G:	1 gpm	(requires ½" pipe)
H - I:	1 gpm	(requires ½" pipe)
I - J:	2 gpm	(requires ¾" pipe)
J - K:	3 gpm	(requires ¾" pipe)
K - L:	4 gpm	(requires 1" pipe)
L - M:	5 gpm	(requires 1" pipe)
M - N:	6 gpm	(requires 1¼" pipe)
N - Boiler:	7 gpm	(requires 1¼" pipe)

All branches have a flow of 1 gpm. The available head is 9 feet. Enter Table 9, at 9 feet head. Scan across to 230 feet (the closest value to length of 237 feet). Scan down this column to the first value larger than 1 gpm. This value is 1.5 gpm. Read the left column to see the minimum branch pipe size is ½". You can use ½" tubing or pipe to connect between the ¾" baseboard units. The larger diameter of the baseboard will have little effect on the flow because the baseboard length is a small fraction of the total branch circuit length.

Figure 9 Piping arrangement for two-pipe reverse-return system

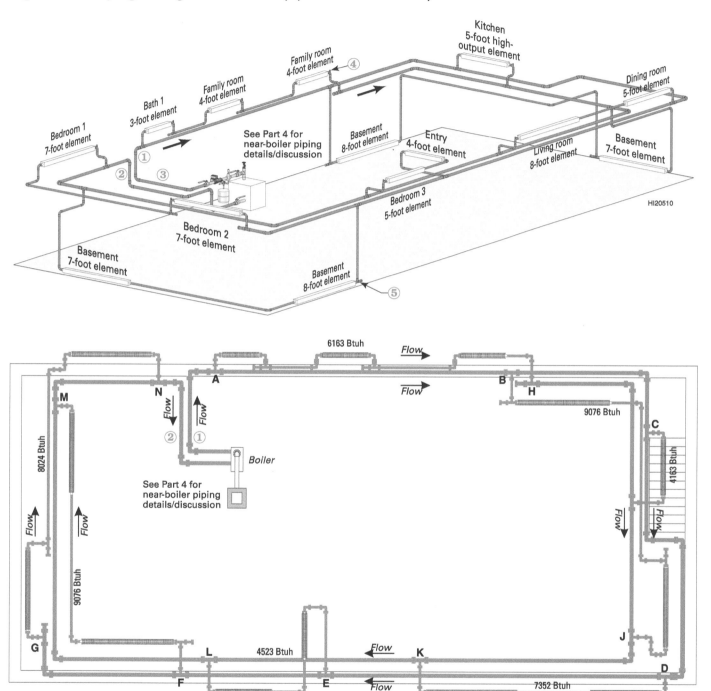

① Supply main

② Return main

③ Install flow control valve on supply if an indirect water heater is used. Valve is optional otherwise.

④ Air vents on outlet sides of main floor heating units

⑤ Install tee with nipple and cap or inverted baseboard tee with plug at low points to allow drainage.

Appendix

Approximate flow in heating units

Table 16   Approximate flow in heating units located below the main in one-pipe circuits

Approximate flow in one-pipe branch when heating unit is below main

Flow in main (gpm)		Heating unit 4 feet below main — Branch measured length (feet)				Heating unit 8 feet below main — Branch measured length (feet)				Heating unit 12 feet below main — Branch measured length (feet)			Heating unit 16 feet below main — Branch measured length (feet)		
		20	40	60	80	20	40	60	80	40	60	80	40	60	80
3/4" Main (3/8" branch) One fitting downstream (Cv = 4.9)	2	0.3													
	2.5	0.4	0.3	0.3	0.2	0.3	0.3								
	3	0.5	0.4	0.3	0.3	0.5	0.4	0.3	0.3	0.3	0.3				
	3.5	0.6	0.5	0.4	0.3	0.6	0.4	0.4	0.3	0.4	0.3	0.3	0.4	0.3	0.3
	4	0.7	0.6	0.5	0.4	0.7	0.5	0.4	0.4	0.5	0.4	0.4	0.5	0.4	0.3
	4.5	0.8	0.6	0.5	0.5	0.8	0.6	0.5	0.4	0.6	0.5	0.4	0.6	0.5	0.4
	5	0.9	0.7	0.6	0.5	0.9	0.7	0.6	0.5	0.7	0.5	0.5	0.6	0.5	0.5
3/4" Main (1/2" branch) One fitting downstream (Cv = 4.9)	1.5	0.4	0.3												
	2	0.7	0.5	0.4	0.4	0.4	0.3	0.3							
	2.5	0.9	0.7	0.6	0.5	0.7	0.6	0.5	0.4	0.4	0.3	0.3			
	3	1.1	0.9	0.7	0.6	1.0	0.8	0.6	0.5	0.7	0.5	0.5	0.5	0.4	0.4
	3.5	1.4	1.0	0.8	0.7	1.3	1.0	0.8	0.7	0.9	0.7	0.6	0.8	0.6	0.6
	4	1.6	1.2	1.0	0.9	1.5	1.1	0.9	0.8	1.1	0.9	0.8	1.0	0.8	0.7
	4.5	1.8	1.4	1.1	1.0	1.7	1.3	1.1	0.9	1.2	1.0	0.9	1.2	1.0	0.8
	5	2.0	1.5	1.3	1.1	2.0	1.5	1.2	1.1	1.4	1.2	1.0	1.4	1.1	1.0
1" Main (1/2" branch) Two fittings (Cv = 10)	3	0.3													
	4	0.6	0.5	0.4	0.3	0.4	0.3								
	5	0.9	0.7	0.5	0.5	0.7	0.5	0.4	0.4	0.4	0.3	0.3			
	6	1.1	0.8	0.7	0.6	1.0	0.7	0.6	0.5	0.6	0.5	0.5	0.5	0.4	0.4
	7	1.3	1.0	0.8	0.7	1.2	0.9	0.8	0.7	0.8	0.7	0.6	0.7	0.6	0.5
	8	1.6	1.2	1.0	0.8	1.5	1.1	0.9	0.8	1.0	0.9	0.7	1.0	0.8	0.7
	9	1.8	1.3	1.1	1.0	1.7	1.3	1.1	0.9	1.2	1.0	0.9	1.1	0.9	0.8
	10	2.0	1.5	1.2	1.1	1.9	1.4	1.2	1.0	1.4	1.1	1.0	1.3	1.1	1.0
1" Main (3/4" branch) Two fittings (Cv = 10)	3	0.9	0.7	0.5	0.5										
	4	1.7	1.2	1.0	0.9	1.0	0.7	0.6	0.5						
	5	2.3	1.7	1.4	1.2	1.9	1.4	1.2	1.0	1.0	0.8	0.7			
	6	2.9	2.2	1.8	1.6	2.6	1.9	1.6	1.4	1.7	1.4	1.2	1.3	1.1	0.9
	7	3.5	2.6	2.2	1.9	3.2	2.4	2.0	1.7	2.2	1.8	1.6	2.0	1.6	1.4
	8	4.1	3.1	2.5	2.2	3.8	2.9	2.4	2.1	2.7	2.2	1.9	2.5	2.1	1.8
	9	4.6	3.5	2.9	2.5	4.4	3.3	2.7	2.4	3.2	2.6	2.3	3.0	2.5	2.1
	10	5.2	3.9	3.2	2.8	5.0	3.8	3.1	2.7	3.6	3.0	2.6	3.5	2.9	2.5
1¼" Main (1/2" branch) Two fittings (Cv = 15)	6	0.6	0.5	0.4	0.3	0.4	0.3								
	8	1.0	0.7	0.6	0.5	0.8	0.6	0.5	0.4	0.5	0.4	0.3	0.3	0.2	0.2
	10	1.3	0.9	0.8	0.7	1.2	0.9	0.7	0.6	0.8	0.6	0.6	0.7	0.6	0.5
	12	1.6	1.2	1.0	0.8	S1.5	1.1	0.9	0.8	1.0	0.9	0.7	1.0	0.8	0.7
	14	1.8	1.4	1.1	1.0	1.8	1.3	1.1	1.0	1.3	1.0	0.9	1.2	1.0	0.9
	16	2.1	1.6	1.3	1.1	2.1	1.5	1.3	1.1	1.5	1.2	1.1	1.4	1.2	1.0
	18	2.4	1.8	1.5	1.3	2.3	1.8	1.5	1.3	1.7	1.4	1.2	1.7	1.4	1.2
	20	2.7	2.0	1.7	1.4	2.6	2.0	1.6	1.4	1.9	1.6	1.4	1.9	1.6	1.4
1¼" Main (3/4" branch) Two fittings (Cv = 15)	6	1.7	1.2	1.0	0.9	1.0	0.7	0.6	0.5						
	8	2.5	1.9	1.6	1.4	2.1	1.6	1.3	1.1	1.2	1.0	0.9	0.7	0.6	0.5
	10	3.3	2.5	2.1	1.8	3.0	2.3	1.9	1.6	2.0	1.7	1.5	1.8	1.4	1.3
	12	4.1	3.1	2.5	2.2	3.8	2.9	2.4	2.1	2.7	2.2	1.9	2.5	2.1	1.8
	14	4.8	3.6	3.0	2.6	4.6	3.5	2.9	2.5	3.3	2.7	2.4	3.2	2.6	2.3
	16	5.6	4.2	3.4	3.0	5.4	4.0	3.3	2.9	3.9	3.2	2.8	3.8	3.1	2.7
	18	6.3	4.7	3.9	3.4	6.1	4.6	3.8	3.3	4.5	3.7	3.2	4.4	3.6	3.1
	20	7.0	5.3	4.4	3.8	6.9	5.2	4.3	3.7	5.1	4.2	3.6	5.0	4.1	3.6

• Above flow rates are approximates only. They allow for a temperature difference of up to 150 °F between the main and heating unit to assure flow will occur properly on a cold start. The measured lengths are for a reasonable number of fittings only. They do not allow for insertion of thermostatic valves, zone valves or other devices in the branch circuit. If you know the equivalent feet of a device, you can add this amount to the measured length and use the appropriate measured length column for the total measured length of the branch plus the equivalent length of devices in the line.

Converting C_v values to equivelant lengths

Table 17   Converting C_v values to equivalent lengths

Converting C_v to equivalent feet of tubing or pipe										
Type M copper tube		**Equivalent feet of tubing for Cv of:**								
½	C_v	1	1.25	1.5	1.75	2	2.5	3	3.5	4
	Equiv. feet	117	79	58	44	35	24	17	13	10
¾	C_v	2	3	4	5	6	7	8	9	10
	Equiv. feet	188	92	56	38	27	21	17	13	11
1	C_v	4	6	8	10	12	14	16	18	20
	Equiv. feet	195	96	58	39	28	22	17	14	12
1¼	C_v	5	7.5	10	15	20	25	30	35	40
	Equiv. feet	344	169	102	50	30	21	15	11	9
1½	C_v	5	10	15	20	25	30	40	50	60
	Equiv. feet	763	227	112	67	46	33	20	14	10
2	C_v	10	20	30	40	50	60	70	80	100
	Equiv. feet	835	248	122	74	50	36	28	22	15
Sch 40 steel pipe		**Equivalent feet of pipe for Cv of:**								
½	C_v	1	1.25	1.5	1.75	2	2.5	3	3.5	4
	Equiv. feet	175	117	84	63	50	33	24	18	14
¾	C_v	2	3	4	5	6	7	8	9	10
	Equiv. feet	200	94	55	36	26	19	15	12	10
1	C_v	4	6	8	10	12	14	16	18	20
	Equiv. feet	181	86	51	34	24	18	14	11	9
1¼	C_v	5	7.5	10	15	20	25	30	35	40
	Equiv. feet	447	214	127	61	36	24	17	13	10
1½	C_v	5	10	15	20	25	30	40	50	60
	Equiv. feet	1014	281	133	78	52	37	22	14	10
2	C_v	10	20	30	40	50	60	70	80	100
	Equiv. feet	962	264	124	73	48	34	26	20	13

Circulator performance prediction — the system curve

Figure 10   Drawing a system curve

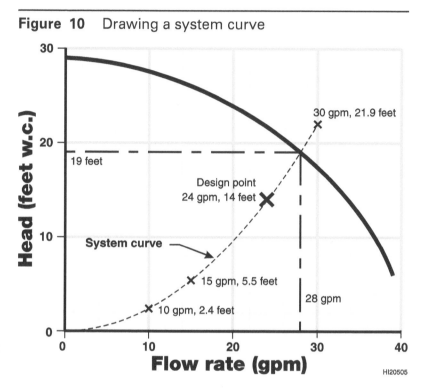

HI20505

For most residential circulator selections, you won't need to determine exactly how the circulator you select will perform on your system. But there may be a few instances when the design point is well off of the circulator performance curve.

Or you may want to use a circulator whose curve is below the design point. For these instances, you can determine what a circulator will do in your application by drawing a system curve on the circulator curve.

The system curve predicts the pressure drop at flows other than the design flow. To draw a system curve, calculate a few flow/head points in addition to the design point.

Apply the following formula to calculate other points:

head2 = design head x (flow2/design flow)^2

where head2 is the unknown head and flow2 is a selected flow rate.

Draw the points on the curve. Then connect them with a curved line. The point where the system curve crosses the circulator curve gives the performance point for the circulator on your system. Be sure to include at least one point above the curve to be sure your system curve line is accurate.

Example:

See Figure 10. The design point for this application is 24 gpm at a head of 14 feet water column. You can see the circulator curve is quite a bit above this design point. The additional points used to draw the system curve are:

10 gpm: head2 = design head x (flow2/design flow)^2 = 14 x (10/24)^2 = 14 x .174 = 2.4 feet

15 gpm: head2 = 14 x (15/24)^2 = 14 x 0.391 = 5.5 feet

30 gpm: head2 = 14 x (30/24)^2 = 14 x 1.563 = 21.9 feet

Connect these points with a curved line as shown on Figure 10. This line is the system curve. The system curve crosses the circulator curve at 28 gpm and 19 feet — the performance point for the circulator in this application.

Drawing a system curve is seldom necessary for residential applications. It can be useful, though, for commercial and large residential applications requiring large circulators, when circulators are piped in parallel or in series, or when you want to use a circulator whose curve is below the design point.

Sizing series-loop baseboard heaters for varying temperature

Baseboard rating tables are usually based on flow rate and average water temperature in the baseboard. The most common rating basis is for an average temperature of 180 °F. You may want to use a series loop with a packaged circulator, but the circulator cannot supply enough flow for a 20 °F temperature drop. You can ensure this system will work if you size all baseboard based on the water temperature supplied to it. Baseboard near the end of the loop will have a lower output per foot than that at the beginning of the loop.

Adding baseboard length shouldn't affect the measured length of most systems since the baseboard just replaces Part of the straight length of the connecting copper tubing.

Step 1:

Calculate the temperature drop for the series loop based on the flow rate the circulator can deliver: TD = Btuh ÷ (500 x gpm).

Step 2:

Break this temperature drop into 20°F stages. Start with a water supply temperature of 190 °F, with temperature drop of 20°F. The water temperature leaving this stage of the system is 170 °F. The average temperature is ½ x (190 + 170) = 180 °F. Find the output per foot of baseboard on the rating table for the baseboard you are using. In Table 2, Page 5–13, for example, the output per foot at 180 °F is 580 Btuh for a flow of 4 gpm or more, or 550 Btuh per foot if the flow is less than 4 gpm and more than 1 gpm.

Step 3:

Calculate how many feet of baseboard this first stage covers: Determine Btuh for this stage from: Btuh = 500 x gpm x 20 °F = 10,000 x gpm. Then determine how many feet of baseboard this would be by dividing Btuh by output per foot found in Step 2. Size this many feet at the beginning of the loop based on the 180 °F rating. If this length falls in the middle of a room, use the rating for the entire room.

Step 4:

Now check the next stage of the system (next 20 °F or less drop). The supply temperature to this stage is 170 °F. If the temperature drop is 20 °F, then the water will leave this stage at 150 °F, and the average water temperature will be ½ x (170 + 150) = 160 °F. Find the baseboard output rating for 160 °F. For the baseboard of Table 2, the output is 420 Btuh per foot for flow between 1 gpm and 4 gpm, for example. To determine the number of feet of baseboard in this stage: Determine Btuh from Btuh = 500 x gpm x 20 °F = 10,000 x gpm. The number of feet of baseboard is Btuh divided by output per foot. Size this length of baseboard based on the 160 °F rating.

Note: If the second stage drop is less than 20 °F, use the actual drop in the calculations. If the total drop is more than 40 °F, repeat the process for the next stage.

Example:

You want to use ¾" baseboard and copper tubing on a series loop with a total heat loss of 60,000 Btuh and a measured length of 195 feet. Find the flow rate for a ¾" loop 195-feet long with a packaged circulator and 60,000 Btuh load from Table 2, Page 5–13. For a loop length of 195 feet of ¾" copper, a packaged circulator should be able to flow about 3.5 gpm (where max length of tubing is 208 feet).

Step 1:

The temperature drop for 60,000 Btuh at 3.5 gpm would be:

TD = 60,000 ÷ (500 x 3.5) = 34 °F.

This would be the total temperature drop for the loop.

Step 2:

The first stage of the loop would have a 20 °F drop, from 190 °F to 170 °F, with average temperature of 180 °F. The flow rate is 5 gpm. So, for the baseboard of Table 2, Page 5–13, the output is 580 Btuh per foot of baseboard.

Step 3:

The output of stage 1 is: Btuh = 500 x 3.5 gpm x 20 °F = 35,000 Btuh. The length of baseboard at 580 Btuh per foot is: Length = 35,000 Btuh ÷ 580 Btuh per foot = 60 feet of baseboard. Size the first 60 feet of baseboard based on 580 Btuh per foot.

Step 4:

The temperature drop in stage 2 is 14 °F, the remainder of the 34 °F total drop for the loop. The output for this stage is the remainder of the 60,000 Btuh total, or 25,000 Btuh (60,000 minus 35,000 from stage 1). The supply temperature to stage 2 of the loop is 170 °F (leaving temperature of stage 1). The temperature leaving stage 2 is 170 °F - 14 °F = 156 °F. The average temperature of stage 2 is ½ x (170 + 156) = 163 °F. Use the 160 °F rating column for baseboard in stage 2. For the baseboard of Table 2, this is 440 Btuh per foot at 3.5 gpm (4 gpm rating). Size the remainder of the baseboard in the system based on 440 Btuh per foot.

Parallel resistances

When you pipe circuits in parallel, the head at the beginning of each circuit is the same because they are all connected; and the head at the end of each circuit is the same because the returns are all connected. If the flow resistance in the circuits is different, flow will favor the easiest path. This is why parallel circuits are usually fitted with balancing valves — to throttle the valves such that the resistance of each circuit is controlled.

If the balancing valves are set so the resistance of every circuit is the same, then every circuit will receive the same flow. If circuit resistances are different, then more flow will go through the circuits with lower resistance; less flow will go through circuits with higher resistance.

If no balancing valves are installed, how do you know what the flow will be in each circuit if resistances are different? The flow will balance itself such that the resulting head loss is the same in all circuits. To estimate the flow in parallel circuits, use the parallel resistance calculations in the following.

Circuit resistance

The "**resistance**" of a circuit is the equivalent length for that circuit times a constant dependent on the pipe diameter (the value, **a**, from equation 5-7, Page 5–26). When you pipe circuits in parallel, the equivalent length of the combination will determine the head loss for the system. Water flow will favor low head-loss circuits until the head loss in each circuit is the same (more flow means more head loss; less flow means less head loss). The formula for the effective resistance of the parallel system is approximately:

$$Ref = 1 \div (1/R1^{(1/b)} + 1/R2^{(1/b)} + 1/R3^{(1/b)} \ldots)^b \quad (5\text{-}8)$$

where R1, R2, R3, etc. are the resistances of each of the parallel circuits,

with $R = Leq \times a/100$, with **a** from Table 12, Page 5–29

b is also taken from Table 12, (If **b** varies between circuits, use the average of the values.)

If you substitute Ref from equation 5-8 into equation 5-7, Page 5–26, the head loss for the combined resistances is:

$$h = Ref \times (gpm)^b, \quad (5\text{-}9)$$

with **b** taken from Table 12. (If **b** varies between circuits, use the average of the values.)

Where you have to average the value of **b**, the resulting calculation will not be as close, but will be adequate for estimating the head loss and flows.

Parallel resistance rule:

The combined resistance of multiple circuits will always be less than the resistance of any of the connected circuits. This is because water can flow through the alternate paths as well. Use this rule as a check of your calculation.

Flow in parallel circuits

Equation 5-9 will give you the head loss for the combined flow of all circuits. This head loss applies across every individual circuit piped in parallel. So you can calculate the flow in each circuit using the reverse of equation 5-9, using **Ref** for the individual circuit once you have found the head loss, **h**, for the system:

$$gpm = (h/Ref)^{(1/b)}, \quad (5\text{-}10)$$

where **b** is taken from Table 12, Page 5–29. (For b = 1.75, 1/b = 0.571, for example).

Example: Two circuits in parallel:

For two circuits in parallel, the equivalent lengths and sizes are:

Circuit 1 —
1¼" copper (a = .0402 at 60 °F); b = 1.75; Leq = 720 feet

Circuit 2 —
1½" copper (a = .0181 at 60 °F); b = 1.75; Leq = 960 feet

Solution:
R1 = Leq x a/100 = 720 x .0402/100 = .289
R2 = Leq x a/100 = 960 x .0181/100 = .174
1/b = 1/1.75 = 0.571

Apply equation 5-8:
$$Ref = 1/(1/0.289^{0.571} + 1/0.174^{0.571})^{1.75}$$
$$Ref = 1/(1/0.4922 + 1/0.3684)^{1.75} = 1/(4.746)^{1.75} = 0.0655$$

The effective resistance for the two circuits piped in parallel is 0.0655, with a head loss of:

$$h = Ref \times (gpm)^b = 0.0655 \times (35)^{1.75} = 33.0 \text{ feet}$$

If the circulator will deliver 35 gpm against a head of 35.0 feet, then how much flow passes through each circuit?

Circuit 1: gpm = $(h/Ref)^{(1/b)}$ = $(33.0/.289)^{0.571}$ = 15.0 gpm

Circuit 2: gpm = $(h/Ref)^{(1/b)}$ = $(33.0/.174)^{0.571}$ = 20.0 gpm

HEAT LOSS WORKSHEET FOR GUIDE H-22

Page 1 **of** 2   **Total Btuh for Page** _17,805_

Total Btuh for Building _48,611_

Customer Name: _Residence R-1_   Telephone: _____

Example calculation

Address: _____   Heat loss calculated by _____

Billings, Montana   Date _____

Page 1 Rooms

Room	Length	Width	Height
Bedroom 1	12.75	12	8
Indoor/Outdoor desn. temp.	70 °F / -10 °F		
DTD (Indoor temp. - Outdoor temp.)	80 °F		
Exposed walls	# 1	# 1	

Room (Add 20% for bathrooms)

Bedroom 1 — Length 12.75, Width 12, Height 8 — 70 °F / -10 °F — DTD 80 °F — Exposed walls #1 / #1

	Volume (LxWxH)	X	Factor	X	DTD	=	Btuh
1 Infiltration (Table 2, sec. 1-3) — Heat loss = Use volume above grade only	1,224	X	0.018	X	80 °F		1,763
2 Ceiling (Table 3, sec. 22-24) — Heat loss = Area (LxW)	153	X	0.050	X	80 °F		612
3 Floor (Table 3, sec. 25-27) — Heat loss = Area (LxW)		X	none	X		=	
4 Slab (Table 3, sec. 27) — Heat loss = Exposed — Perimeter (feet)		X		X			
5 Windows (Table 3, sec. 4-5) — Heat loss = Sash Area (LxW)	16.25	X	0.480	X	80 °F		624
6 Doors (Table 3, sec. 6-9) — Heat loss = Door Area (LxW)		X	none	X			
Net Wall Area = Wall Area (LxH) − Window Area − Door Area =							182
7 Walls (Table 3, sec. 10-21) — Heat loss = Exposed — Net Wall Area	182	X	0.080	X	80 °F		1,165
Below grade more than 2 feet		X		X		=	
Cold partitions		X		X		=	
Baseboard length, feet —				Heat loss, Btuh —			4,164

Bathroom 1 — Length 10.25, Width 4.25, Height 8 — 70 °F / -10 °F — DTD 80 °F — Exposed walls #1 / #0

	Volume (LxWxH)	X	Factor	X	DTD	=	Btuh
Infiltration	348.5	X	0.012	X	80 °F		335
Ceiling — Area (LxW)	43.6	X	0.050	X	80 °F	=	174
Floor — Area (LxW)		X		X		=	
Slab — Perimeter (feet)		X		X		=	
Windows — Sash Area (LxW)	7.5	X	0.480	X	80 °F		288
Doors — Door Area (LxW)		X		X		=	
Net Wall Area = Wall Area (LxH) − Window Area − Door Area =							74.5
Walls — Net Wall Area	74.5	X	0.080	X	80 °F	=	477
Below grade		X		X		=	
Cold partitions		X		X		=	
Baseboard length, feet —				Heat loss, Btuh —			1,274
				Add 20% =			1,529

Bathroom 2 — Length 7, Width 7, Height 8 — 70 °F / -10 °F — DTD 80 °F — Exposed walls #0 / #0

	Volume (LxWxH)	X	Factor	X	DTD	=	Btuh
Infiltration		X	0	X			
Ceiling — Area (LxW)	49	X	0.050	X	80 °F		196
Floor — Area (LxW)		X	none	X		=	
Slab — Perimeter (feet)		X		X			
Windows — Sash Area (LxW)		X	none	X			
Doors — Door Area (LxW)		X	none	X			
Net Wall Area = Wall Area (LxH) − Window Area − Door Area =							
Walls — Net Wall Area		X		X		=	
Below grade		X		X		=	
Cold partitions		X		X		=	
Baseboard length, feet —				Heat loss, Btuh —			196
				Add 20% =			235

Family room — Length 17.25, Width 12, Height 8 — 70 °F / -10 °F — DTD 80 °F — Exposed walls #1 / #0

	Volume (LxWxH)	X	Factor	X	DTD	=	Btuh
Infiltration	1,656	X	0.012	X	80 °F		1,590
Ceiling — Area (LxW)	207	X	0.050	X	80 °F		828
Floor — Area (LxW)		X		X		=	
Slab — Perimeter (feet)		X		X		=	
Windows — Sash Area (LxW)		X		X		=	
Doors — Door Area (LxW)	37.13	X	0.530	X	80 °F		1574
Net Wall Area = Wall Area (LxH) − Window Area − Door Area =							100.9
Walls — Net Wall Area	100.9	X	0.080	X	80 °F		646
Below grade more than 2 feet		X		X		=	
Cold partitions		X		X		=	
Baseboard length, feet —				Heat loss, Btuh —			4,634

Kitchen — Length 14.25, Width 12, Height 8 — 70 °F / -10 °F — DTD 80 °F — Exposed walls #1 / #1

	Volume (LxWxH)	X	Factor	X	DTD	=	Btuh
Infiltration	1,368	X	0.018	X	80 °F		1,970
Ceiling — Area (LxW)	171	X	0.050	X	80 °F		684
Floor — Area (LxW)		X	none	X		=	
Slab — Perimeter (feet)		X		X		=	
Windows — Sash Area (LxW)	7.5	X	0.480	X	80 °F		288
Doors — Door Area (LxW)	16.88	X	0.290	X	80 °F		392
Net Wall Area = Wall Area (LxH) − Window Area − Door Area =							129.6
Walls — Net Wall Area	129.6	X	0.080	X	80 °F	=	829.4
Below grade		X		X		=	
Cold partitions		X		X		=	
Baseboard length, feet —				Heat loss, Btuh —			4,163

Dining room — Length 10, Width 13, Height 8 — 70 °F / -10 °F — DTD 80 °F — Exposed walls #1 / #1

	Volume (LxWxH)	X	Factor	X	DTD	=	Btuh
Infiltration	1,040	X	0.012	X	80 °F		998
Ceiling — Area (LxW)	130	X	0.050	X	80 °F		520
Floor — Area (LxW)		X		X		=	
Slab — Perimeter (feet)		X		X		=	
Windows — Sash Area (LxW)	12	X	0.480	X	80 °F		461
Doors — Door Area (LxW)		X		X		=	
Net Wall Area = Wall Area (LxH) − Window Area − Door Area =							172
Walls — Net Wall Area	172	X	0.080	X	80 °F		1,101
Below grade more than 2 feet		X		X		=	
Cold partitions		X		X		=	
Baseboard length, feet —				Heat loss, Btuh —			3,080

ROOM TOTALS

Baseboard information   Btuh per foot of baseboard _____   At average water temperature _____ °F

I=B=R Calculation Form 1504-98   Copyright 1998, The Hydronics Institute Division of GAMA   Berkeley Heights, NJ 07922-0218

HEAT LOSS WORKSHEET FOR GUIDE H-22

Page 2 **of** 2   **Total Btuh for Page** _30,806_

Total Btuh for Building _48,611_

Customer Name: **Residence R-1**   Telephone: _____
Example calculation

Address: _____   Heat loss calculated by: _____
Billings, Montana   Date: _____

Living room (Length 16, Width 13, Height 8)

70 °F / -10 °F   DTD 80 °F   Exposed walls #1

Calculation	Value	X	Factor	X	DTD	=	Btuh
Volume (LxWxH)	1,664	X	0.012	X	80 °F	=	1,597
Area (LxW)	208	X	0.050	X	80 °F	=	832
Area (LxW)		X		X		=	
Perimeter (feet)		X		X		=	
Sash Area (LxW)	32	X	0.480	X	80 °F	=	1,229
Door Area (LxW)		X		X		=	
Net Wall Area (Wall Area LxH – Window Area – Door Area)	96					=	96
Net Wall Area	96	X	0.080	X	80 °F	=	614
Below grade more than 2 feet		X		X		=	
Cold partitions		X		X		=	

Baseboard length, feet — ___   Heat loss, Btuh — **4,272**

Entry hall (Length 5, Width 13, Height 8)

70 °F / -10 °F   DTD 80 °F   Exposed walls #1, #0

Calculation	Value	X	Factor	X	DTD	=	Btuh
Volume (LxWxH)	520	X	0.027	X	80 °F	=	1,123
Area (LxW)	65	X	0.050	X	80 °F	=	260
Area (LxW)		X		X		=	
Perimeter (feet)		X		X		=	
Sash Area (LxW)		X		X		=	
Door Area (LxW)	21	X	0.290	X	80 °F	=	487
Net Wall Area (Wall Area LxH – Window Area – Door Area)	19					=	19
Net Wall Area	19	X	0.080	X	80 °F	=	122
Below grade more than 2 feet		X		X		=	
Cold partitions		X		X		=	

Baseboard length, feet — ___   Heat loss, Btuh — **1,992**

Bedroom 3 (Length 13, Width 9.75, Height 8)

70 °F / -10 °F   DTD 80 °F   Exposed walls #1, #0

Calculation	Value	X	Factor	X	DTD	=	Btuh
Volume (LxWxH)	1,014	X	0.012	X	80 °F	=	973
Area (LxW)	127	X	0.050	X	80 °F	=	508
Area (LxW)		X		X		=	
Perimeter (feet)		X		X		=	
Sash Area (LxW)	12	X	0.480	X	80 °F	=	461
Door Area (LxW)		X		X		=	
Net Wall Area (Wall Area LxH – Window Area – Door Area)	92					=	92
Net Wall Area	92	X	0.080	X	80 °F	=	589
Below grade more than 2 feet		X		X		=	
Cold partitions		X		X		=	

Baseboard length, feet — ___   Heat loss, Btuh — **2,531**

Bedroom 2 (Length 10, Width 13, Height 8)

70 °F / -10 °F   DTD 80 °F   Exposed walls #1, #1

Calculation	Value	X	Factor	X	DTD	=	Btuh
Volume (LxWxH)	1,040	X	0.018	X	80 °F	=	1,498
Area (LxW)	130	X	0.050	X	80 °F	=	520
Area (LxW)		X		X		=	
Perimeter (feet)		X		X		=	
Sash Area (LxW)	20.75	X	0.480	X	80 °F	=	797
Door Area (LxW)		X		X		=	
Net Wall Area (Wall Area LxH – Window Area – Door Area)	163.3					=	163.3
Net Wall Area	163.3	X	0.080	X	80 °F	=	1,045
Below grade more than 2 feet		X		X		=	
Cold partitions		X		X		=	

Baseboard length, feet — ___   Heat loss, Btuh — **3,860**

Basement (Length 55.5, Width 25, Height 7.5)

70 °F / -10 °F   DTD 80 °F   Use height of 1.5 feet for infiltration volume   Exposed walls #2, #2

Calculation	Value	X	Factor	X	DTD	=	Btuh
Volume (LxWxH)	2,081	X	0.018	X	80 °F	=	2,997
Area (LxW)		X		X		=	
Area (LxW)	1,388	X	0.040	X	80 °F	=	4,442
Perimeter (feet)		X		X		=	
Sash Area (LxW)	12	X	0.480	X	80 °F	=	461
Door Area (LxW)		X		X		=	
Net Wall Area (Wall Area LxH – Window Area – Door Area)	229.5					=	229.5
Net Wall Area	229.5	X	0.230	X	80 °F	=	4,223
Below grade more than 2 feet	966	X	0.078	X	80 °F	=	6,028
Cold partitions		X		X		=	

Baseboard length, feet — ___   Heat loss, Btuh — **18,151**

(blank room) (Length ___, Width ___, Height ___)

Calculation	Value	X	Factor	X	DTD	=	Btuh
Volume (LxWxH)		X		X		=	
Area (LxW)		X		X		=	
Area (LxW)		X		X		=	
Perimeter (feet)		X		X		=	
Sash Area (LxW)		X		X		=	
Door Area (LxW)		X		X		=	
Net Wall Area (Wall Area LxH – Window Area – Door Area)						=	
Net Wall Area		X		X		=	
Below grade more than 2 feet		X		X		=	
Cold partitions		X		X		=	

Baseboard length, feet — ___   Heat loss, Btuh — ___

Room table legend (left column):

- Room (Add 20% for bathrooms)
- Indoor/Outdoor desn. temp.
- DTD (Indoor temp. - Outdoor temp.)
- Exposed walls
- 1 Infiltration — Table 2, sec. 1 - 3 — Heat loss = Use volume above grade only
- 2 Ceiling — Table 3, sec. 22 - 24 — Heat loss =
- 3 Floor — Table 3, sec. 25 - 27 — Heat loss =
- 4 Slab — Table 3, sec. 27 — Heat loss = Exposed
- 5 Windows — Table 3, sec. 4 - 5 — Heat loss =
- 6 Doors — Table 3, sec. 6 - 9 — Heat loss =
- 7 Walls — Table 3, sec. 10 - 21 — Net Wall Area = Heat loss = Exposed / Below grade more than 2 feet / Cold partitions
- **ROOM TOTALS**

Baseboard information   Btuh per foot of baseboard ___   At average water temperature ___ °F

I=B=R Calculation Form 1504-98   Copyright 1998, The Hydronics Institute Division of GAMA   Berkeley Heights, NJ 07922-0218

Square feet of radiation required for heating loads

Table 18 Square feet of radiation required for heating loads from 1,000 to 12,900 Btuh

Square feet of cast iron radiation required for heating loads from 1,000 to 12,900 Btuh

Btuh	Average water temperature in radiator, °F										Btuh	Average water temperature in radiator, °F									
	170	175	180	185	190	195	200	205	210	215		170	175	180	185	190	195	200	205	210	215
	Btuh output per square foot EDR											Btuh output per square foot EDR									
	150	160	170	180	190	200	210	220	230	240		150	160	170	180	190	200	210	220	230	240
1000	6.7	6.3	5.9	5.6	5.3	5.0	4.8	4.5	4.3	4.2	7000	46.7	43.8	41.2	38.9	36.8	35.0	33.3	31.8	30.4	29.2
1100	7.3	6.9	6.5	6.1	5.8	5.5	5.2	5.0	4.8	4.6	7100	47.3	44.4	41.8	39.4	37.4	35.5	33.8	32.3	30.9	29.6
1200	8.0	7.5	7.1	6.7	6.3	6.0	5.7	5.5	5.2	5.0	7200	48.0	45.0	42.4	40.0	37.9	36.0	34.3	32.7	31.3	30.0
1300	8.7	8.1	7.6	7.2	6.8	6.5	6.2	5.9	5.7	5.4	7300	48.7	45.6	42.9	40.6	38.4	36.5	34.8	33.2	31.7	30.4
1400	9.3	8.8	8.2	7.8	7.4	7.0	6.7	6.4	6.1	5.8	7400	49.3	46.3	43.5	41.1	38.9	37.0	35.2	33.6	32.2	30.8
1500	10.0	9.4	8.8	8.3	7.9	7.5	7.1	6.8	6.5	6.3	7500	50.0	46.9	44.1	41.7	39.5	37.5	35.7	34.1	32.6	31.3
1600	10.7	10.0	9.4	8.9	8.4	8.0	7.6	7.3	7.0	6.7	7600	50.7	47.5	44.7	42.2	40.0	38.0	36.2	34.5	33.0	31.7
1700	11.3	10.6	10.0	9.4	8.9	8.5	8.1	7.7	7.4	7.1	7700	51.3	48.1	45.3	42.8	40.5	38.5	36.7	35.0	33.5	32.1
1800	12.0	11.3	10.6	10.0	9.5	9.0	8.6	8.2	7.8	7.5	7800	52.0	48.8	45.9	43.3	41.1	39.0	37.1	35.5	33.9	32.5
1900	12.7	11.9	11.2	10.6	10.0	9.5	9.0	8.6	8.3	7.9	7900	52.7	49.4	46.5	43.9	41.6	39.5	37.6	35.9	34.3	32.9
2000	13.3	12.5	11.8	11.1	10.5	10.0	9.5	9.1	8.7	8.3	8000	53.3	50.0	47.1	44.4	42.1	40.0	38.1	36.4	34.8	33.3
2100	14.0	13.1	12.4	11.7	11.1	10.5	10.0	9.5	9.1	8.8	8100	54.0	50.6	47.6	45.0	42.6	40.5	38.6	36.8	35.2	33.8
2200	14.7	13.8	12.9	12.2	11.6	11.0	10.5	10.0	9.6	9.2	8200	54.7	51.3	48.2	45.6	43.2	41.0	39.0	37.3	35.7	34.2
2300	15.3	14.4	13.5	12.8	12.1	11.5	11.0	10.5	10.0	9.6	8300	55.3	51.9	48.8	46.1	43.7	41.5	39.5	37.7	36.1	34.6
2400	16.0	15.0	14.1	13.3	12.6	12.0	11.4	10.9	10.4	10.0	8400	56.0	52.5	49.4	46.7	44.2	42.0	40.0	38.2	36.5	35.0
2500	16.7	15.6	14.7	13.9	13.2	12.5	11.9	11.4	10.9	10.4	8500	56.7	53.1	50.0	47.2	44.7	42.5	40.5	38.6	37.0	35.4
2600	17.3	16.3	15.3	14.4	13.7	13.0	12.4	11.8	11.3	10.8	8600	57.3	53.8	50.6	47.8	45.3	43.0	41.0	39.1	37.4	35.8
2700	18.0	16.9	15.9	15.0	14.2	13.5	12.9	12.3	11.7	11.3	8700	58.0	54.4	51.2	48.3	45.8	43.5	41.4	39.5	37.8	36.3
2800	18.7	17.5	16.5	15.6	14.7	14.0	13.3	12.7	12.2	11.7	8800	58.7	55.0	51.8	48.9	46.3	44.0	41.9	40.0	38.3	36.7
2900	19.3	18.1	17.1	16.1	15.3	14.5	13.8	13.2	12.6	12.1	8900	59.3	55.6	52.4	49.4	46.8	44.5	42.4	40.5	38.7	37.1
3000	20.0	18.8	17.6	16.7	15.8	15.0	14.3	13.6	13.0	12.5	9000	60.0	56.3	52.9	50.0	47.4	45.0	42.9	40.9	39.1	37.5
3100	20.7	19.4	18.2	17.2	16.3	15.5	14.8	14.1	13.5	12.9	9100	60.7	56.9	53.5	50.6	47.9	45.5	43.3	41.4	39.6	37.9
3200	21.3	20.0	18.8	17.8	16.8	16.0	15.2	14.5	13.9	13.3	9200	61.3	57.5	54.1	51.1	48.4	46.0	43.8	41.8	40.0	38.3
3300	22.0	20.6	19.4	18.3	17.4	16.5	15.7	15.0	14.3	13.8	9300	62.0	58.1	54.7	51.7	48.9	46.5	44.3	42.3	40.4	38.8
3400	22.7	21.3	20.0	18.9	17.9	17.0	16.2	15.5	14.8	14.2	9400	62.7	58.8	55.3	52.2	49.5	47.0	44.8	42.7	40.9	39.2
3500	23.3	21.9	20.6	19.4	18.4	17.5	16.7	15.9	15.2	14.6	9500	63.3	59.4	55.9	52.8	50.0	47.5	45.2	43.2	41.3	39.6
3600	24.0	22.5	21.2	20.0	18.9	18.0	17.1	16.4	15.7	15.0	9600	64.0	60.0	56.5	53.3	50.5	48.0	45.7	43.6	41.7	40.0
3700	24.7	23.1	21.8	20.6	19.5	18.5	17.6	16.8	16.1	15.4	9700	64.7	60.6	57.1	53.9	51.1	48.5	46.2	44.1	42.2	40.4
3800	25.3	23.8	22.4	21.1	20.0	19.0	18.1	17.3	16.5	15.8	9800	65.3	61.3	57.6	54.4	51.6	49.0	46.7	44.5	42.6	40.8
3900	26.0	24.4	22.9	21.7	20.5	19.5	18.6	17.7	17.0	16.3	9900	66.0	61.9	58.2	55.0	52.1	49.5	47.1	45.0	43.0	41.3
4000	26.7	25.0	23.5	22.2	21.1	20.0	19.0	18.2	17.4	16.7	10000	66.7	62.5	58.8	55.6	52.6	50.0	47.6	45.5	43.5	41.7
4100	27.3	25.6	24.1	22.8	21.6	20.5	19.5	18.6	17.8	17.1	10100	67.3	63.1	59.4	56.1	53.2	50.5	48.1	45.9	43.9	42.1
4200	28.0	26.3	24.7	23.3	22.1	21.0	20.0	19.1	18.3	17.5	10200	68.0	63.8	60.0	56.7	53.7	51.0	48.6	46.4	44.3	42.5
4300	28.7	26.9	25.3	23.9	22.6	21.5	20.5	19.5	18.7	17.9	10300	68.7	64.4	60.6	57.2	54.2	51.5	49.0	46.8	44.8	42.9
4400	29.3	27.5	25.9	24.4	23.2	22.0	21.0	20.0	19.1	18.3	10400	69.3	65.0	61.2	57.8	54.7	52.0	49.5	47.3	45.2	43.3
4500	30.0	28.1	26.5	25.0	23.7	22.5	21.4	20.5	19.6	18.8	10500	70.0	65.6	61.8	58.3	55.3	52.5	50.0	47.7	45.7	43.8
4600	30.7	28.8	27.1	25.6	24.2	23.0	21.9	20.9	20.0	19.2	10600	70.7	66.3	62.4	58.9	55.8	53.0	50.5	48.2	46.1	44.2
4700	31.3	29.4	27.6	26.1	24.7	23.5	22.4	21.4	20.4	19.6	10700	71.3	66.9	62.9	59.4	56.3	53.5	51.0	48.6	46.5	44.6
4800	32.0	30.0	28.2	26.7	25.3	24.0	22.9	21.8	20.9	20.0	10800	72.0	67.5	63.5	60.0	56.8	54.0	51.4	49.1	47.0	45.0
4900	32.7	30.6	28.8	27.2	25.8	24.5	23.3	22.3	21.3	20.4	10900	72.7	68.1	64.1	60.6	57.4	54.5	51.9	49.5	47.4	45.4
5000	33.3	31.3	29.4	27.8	26.3	25.0	23.8	22.7	21.7	20.8	11000	73.3	68.8	64.7	61.1	57.9	55.0	52.4	50.0	47.8	45.8
5100	34.0	31.9	30.0	28.3	26.8	25.5	24.3	23.2	22.2	21.3	11100	74.0	69.4	65.3	61.7	58.4	55.5	52.9	50.5	48.3	46.3
5200	34.7	32.5	30.6	28.9	27.4	26.0	24.8	23.6	22.6	21.7	11200	74.7	70.0	65.9	62.2	58.9	56.0	53.3	50.9	48.7	46.7
5300	35.3	33.1	31.2	29.4	27.9	26.5	25.2	24.1	23.0	22.1	11300	75.3	70.6	66.5	62.8	59.5	56.5	53.8	51.4	49.1	47.1
5400	36.0	33.8	31.8	30.0	28.4	27.0	25.7	24.5	23.5	22.5	11400	76.0	71.3	67.1	63.3	60.0	57.0	54.3	51.8	49.6	47.5
5500	36.7	34.4	32.4	30.6	28.9	27.5	26.2	25.0	23.9	22.9	11500	76.7	71.9	67.6	63.9	60.5	57.5	54.8	52.3	50.0	47.9
5600	37.3	35.0	32.9	31.1	29.5	28.0	26.7	25.5	24.3	23.3	11600	77.3	72.5	68.2	64.4	61.1	58.0	55.2	52.7	50.4	48.3
5700	38.0	35.6	33.5	31.7	30.0	28.5	27.1	25.9	24.8	23.8	11700	78.0	73.1	68.8	65.0	61.6	58.5	55.7	53.2	50.9	48.8
5800	38.7	36.3	34.1	32.2	30.5	29.0	27.6	26.4	25.2	24.2	11800	78.7	73.8	69.4	65.6	62.1	59.0	56.2	53.6	51.3	49.2
5900	39.3	36.9	34.7	32.8	31.1	29.5	28.1	26.8	25.7	24.6	11900	79.3	74.4	70.0	66.1	62.6	59.5	56.7	54.1	51.7	49.6
6000	40.0	37.5	35.3	33.3	31.6	30.0	28.6	27.3	26.1	25.0	12000	80.0	75.0	70.6	66.7	63.2	60.0	57.1	54.5	52.2	50.0
6100	40.7	38.1	35.9	33.9	32.1	30.5	29.0	27.7	26.5	25.4	12100	80.7	75.6	71.2	67.2	63.7	60.5	57.6	55.0	52.6	50.4
6200	41.3	38.8	36.5	34.4	32.6	31.0	29.5	28.2	27.0	25.8	12200	81.3	76.3	71.8	67.8	64.2	61.0	58.1	55.5	53.0	50.8
6300	42.0	39.4	37.1	35.0	33.2	31.5	30.0	28.6	27.4	26.3	12300	82.0	76.9	72.4	68.3	64.7	61.5	58.6	55.9	53.5	51.3
6400	42.7	40.0	37.6	35.6	33.7	32.0	30.5	29.1	27.8	26.7	12400	82.7	77.5	72.9	68.9	65.3	62.0	59.0	56.4	53.9	51.7
6500	43.3	40.6	38.2	36.1	34.2	32.5	31.0	29.5	28.3	27.1	12500	83.3	78.1	73.5	69.4	65.8	62.5	59.5	56.8	54.3	52.1
6600	44.0	41.3	38.8	36.7	34.7	33.0	31.4	30.0	28.7	27.5	12600	84.0	78.8	74.1	70.0	66.3	63.0	60.0	57.3	54.8	52.5
6700	44.7	41.9	39.4	37.2	35.3	33.5	31.9	30.5	29.1	27.9	12700	84.7	79.4	74.7	70.6	66.8	63.5	60.5	57.7	55.2	52.9
6800	45.3	42.5	40.0	37.8	35.8	34.0	32.4	30.9	29.6	28.3	12800	85.3	80.0	75.3	71.1	67.4	64.0	61.0	58.2	55.7	53.3
6900	46.0	43.1	40.6	38.3	36.3	34.5	32.9	31.4	30.0	28.8	12900	86.0	80.6	75.9	71.7	67.9	64.5	61.4	58.6	56.1	53.8

Notes

Hydronics Institute Section of AHRI

35 Russo Place

Berkeley Heights, NJ 07922-0218

part **6**

Auxiliary heating loads

Residential Hydronic Heating . . .

Installation & Design

Hydronics Institute
Section of **AHRI**

I=B=R Guide RHH
Residential Hydronic Heating

Hydronics Institute Section of **AHRI**
35 Russo Place
Berkeley Heights, NJ 07922-0218

Contents – Part 6

Contents *– continued*

Illustrations

Contents – continued

Tables

Domestic hot water heating

Overview

Hydronic heating systems can be adapted to many different loads as well as space heating — auxiliary heating loads such as domestic water heating, pool and spa heating, and snow melting, for example. When you combine these additional loads on a heating system, you will often have to increase boiler heating capacity to handle the additional load. If the additional load is large and used only for brief periods, you will want to use either multiple (or modular) boilers on the combined system (to allow staged heating) or use a dedicated boiler that only operates when the auxiliary load is required.

Part 6 provides information on required heating capacity and recommended piping for residential auxiliary heating applications. **Always use the recommended sizing information provided by the indirect water heater manufacturer**. If no sizing information is available from the manufacturer, you can apply the sizing methods in Part 6 for most of your residential domestic water heating, pool heating or spa heating applications.

For commercial or large residential DHW systems, refer to the information published by the water heater and/or boiler manufacturer.

For recommended piping and required heating capacity of a snowmelt load, you can use Hydronics Institute Guide S-40, "Snow Melting Calculation & Installation Guide," or the ASHRAE HVAC Applications Handbook.

The information in this Guide is intended as a general guideline only. Always consult the boiler and water heater manufacturers' instructions when applying their products.

Consider water conditions

For applications using hard water, recommend that the owner install a water softener. Hard water will cause lime deposits on heating surfaces, in lines and valves. The lime will cause reduction in performance and can lead to insufficient hot water supply.

Domestic water heating definitions

Indirect water heating

Indirect-fired water heater

A storage tank water heater with an internal heat exchanger. Indirect-fired water heaters include a thermostat that operates a valve or circulator to cause boiler water to flow through the heat exchanger to heat the domestic water in the tank. The heat capacity of an indirect-fired water heater depends on the size of its heat exchanger, the boiler water temperature and flow rate supplied to the tank, and the boiler heating capacity. Check the water heater manufacturer's requirements for boiler capacity, flow rate and temperature when selecting an indirect-fired water heater. Choose a water heater with a first-hour rating that meets the installation requirements. Indirect-fired water heaters generally allow using a smaller boiler than when using a tankless heater, because of the water stored in the tank.

Hot water storage tank

A tank used in conjunction with a tankless water heater. Insulate the piping between the tankless heater and the tank for maximum fuel efficiency. You can often add a storage tank to a tankless water heater installation if the user finds the need for additional hot water. The storage tank provides more hot water capacity for intermittent peak draws.

Tankless water heater

A heat exchanger inserted in a boiler, below the water line, to deliver domestic hot water. Tankless heaters require a flow regulating valve in the supply line to limit the flow so it doesn't exceed the heater capacity (unless the tankless heater is connected to a storage tank with a circulator). The hot boiler water heats domestic water as it flows through the coil. The boiler must be operated as a "hot boiler," with water maintained at 180°F to 200°F to be ready for a domestic water demand. Tankless heaters supply water on demand, at a flow rate set by boiler and heater capacity. (Each 1 gpm of hot water, heated from 40°F to 140°F, requires 50,000 Btuh. For 2 gpm, a boiler would have to have an output twice this much, or 100,000 Btuh, for example.)

Cold water supply line

The cold water inlet connection to an indirect-fired water heater. Size this line to at least the size of the tankless heater or cold water connection to the storage tank or indirect-fired water heater.

Domestic hot water

The heated water used for domestic or household purposes, such as washing clothes, dishes, bathing, etc.

Domestic priority

A control method that directs all available heating to domestic hot water on demand. Heat for space heating or other applications is interrupted during the domestic hot water heating cycle. This is accomplished by using a diverting zone valve or interrupting electrical power to the space heating zone valves or circulators.

High temperature domestic hot water

Domestic hot water for dishwashing, clothes washing or other uses requiring water hotter than used for showers and lavatories. (Most plumbing codes now require that water delivered to showers and lavatories be limited to 120°F.) Note that clothes washers use water at the temperature supplied. And most dishwashers include an electric element to heat the water to the minimum temperature required by NSF. You may not need to supply water higher than 120°F, even if the usage includes clothes and dish washing.

Hot boiler

This refers to operation that requires the boiler be kept hot in preparation for a domestic hot water demand. The boiler temperature is usually maintained at 180°F to 200°F.

Domestic water heating definitions

Valves

Flow regulating valve

A device for automatically controlling the rate of flow of water under varying pressure conditions. Use a flow regulating valve in the cold water supply to a tankless heater to limit the flow to the capacity of the heater. If the flow exceeds the heater capacity, the water won't be hot enough.

Thermostatic mixing valve

A self-contained device that mixes hot water with cold water to deliver heated water at a pre-defined temperature (usually adjustable).

Balanced-pressure valve

A flow/temperature regulating device that adapts to changes in pressure on the cold water line by reducing pressure to the valve on the hot water line, allowing close control of the valve temperature setting as line conditions vary.

Combination thermostatic/balanced-pressure valve

A flow/temperature regulating device that uses both thermostatic and balanced-pressure methods to maintain the outlet water temperature as set.

Anti-scald device (or temperature-actuated flow reduction valve

A device that stops or slows the flow when the water temperature exceeds 120°F. It may be a separate component or incorporated as part of a thermostatic mixing valve.

Pressure relief valve

A device for protecting a storage tank or system from excessive pressure. Use a relief valve with a pressure setting no higher than the maximum working pressure rating of the tank.

Pressure/temperature relief valve

A relief valve which opens on either a maximum pressure or temperature conditions. Pressure/temperature relief valves are used on indirect-fired water heaters. They may be mounted with stem either vertical or horizontal, installed in the tapping designated by the heater manufacturer.

Ratings

First-hour rating

The gallons of hot water that can be supplied from an indirect-fired water heater during the first hour of continuous operation, starting from a fully-recovered tank. This rating includes the usable storage in the tank.

The Indirect Water Heater Testing Standard, I=B=R IWH-TS-1, determines first hour rating for a 77°F domestic water temperature rise, 58°F to 135°F, with boiler water entering at 180°F at a boiler water flow rate specified by the manufacturer.

The boiler ratings must include the statement, "These ratings were obtained with a heat source output rate of _____ Btu/hr at a heat source flow rate of _____ gpm. Other results will be obtained under different conditions."

Continuous draw rating

This is the flow rate, gpm, that the tankless heater/boiler combination can supply on a continuous basis. [The Indirect Water Heater Testing Standard, I=B=R IWH-TS-1, determines continuous draw for a 77°F domestic water temperature rise, 58°F to 135°F. Boiler water conditions are the same as for first-hour rating, above.]

Intermittent draw rating

This is a tankless heater rating, giving the number of gpm the tankless heater/boiler combination can supply for short duration (5-minute draw). The boiler must be allowed to recover for at least 10 minutes between intermittent draws to achieve the rating again.

Three-hour rating

(Obsolete) — The gallons of hot water that can be supplied from an indirect-fired water heater during a three-hour period of continuous operation. This rating is no longer used for indirect-fired water heaters. The most common rating method is the first-hour rating.

Domestic hot water supply temperature

Scald hazard

The International Plumbing Code now requires either a master temperature-regulating device or a temperature-regulating device at each point of use for shower and tub use of hot water. The device(s) must limit supply temperature to 120°F. The IPC requires these devices to be thermostatic, balanced-pressure or combination mixing valves.

The reason for this limit is the severe damage that can occur to human skin for even short exposure to hotter water. It takes only 3 seconds for a child to develop third-degree burns if exposed to water at 140°F, for instance. The human threshold of pain is 118°F, but children and older adults may respond slowly to exposure. Even 120°F water can cause irreversible skin damage for exposure durations longer than about 8 minutes. Figure 1 is the warning label now required on

---

Figure 1   Water heater scald warning

Water temperature over 125°F can cause severe burns instantly or death from scalds.

Children, disabled and elderly are at highest risk of being scalded.

See instruction manual before setting temperature at water heater.

Feel water before bathing or showering.

Temperature limiting valves are available, see manual.

---

water heaters tested to the ANSI standard.

When you install any water heater, you need to consider whether the system provides the protection needed against scald hazard. New systems will most likely include point-of-use or master mixing valves because of code requirements. But existing systems may have no temperature regulation at all. For your highest assurance that you are protecting the user from hazards and yourself from liability, apply the requirements of the current plumbing codes when you install a water heater. Either install a master thermostatic mixing valve or combination thermostatic/pressure-balancing valve on the hot water line leading to the tub and shower fixtures. And recommend or install a temperature-actuated flow-reduction valve on each tap as well if not supplied from a master mixing valve. Serious scalds can occur quickly from tap water over 125°F.

Gas, oil and electric water heaters are either shipped with their thermostats set at 120°F or include instructions to do so. These thermostats are adjustable, however. And it has long been a practice to compensate for insufficient hot water by cranking up the setting on the water heater thermostat. If you put in tamper-proof mixing device(s), you can be more sure that water won't be fed to the fixtures above 120°F.

Legionnaires' disease

The following is from the 1999 ASHRAE Handbook: HVAC Applications, page 48.8. "Legionnaires' disease (a form of severe pneumonia) is caused by inhaling the bacteria Legionella pneumophila. It has been discovered in the service water heating systems of various buildings throughout the world. Infection has often been traced to Legionella pneumophila colonies in shower heads. Ciesielki et. (1984) determined that Legionella pneumophila can colonize in hot water maintained at 115°F or lower. Segments of service water systems in which the water stagnates (e.g., shower heads, faucet aerators, and certain sections of storage water heaters) provide ideal breeding locations.

Service water temperature in the 140°F range is recommended to limit the potential for Legionella pneumophila growth. This high temperature increases the potential for scalding, so care must be taken such as installing an anti scald or mixing valve. Supervised periodic flushing of fixture heads with 170°F water is recommended in hospitals and health care facilities because the already weakened patients are generally more susceptible to infection."

The Legionella biohazard has been associated with commercial and institutional installations more than with residential. It can occur, however, if water is maintained at warm temperatures below 115°F. As a safeguard against this problem, you may want to follow the ASHRAE guideline and set the indirect water heater thermostat for 140°F, but you must ensure the system provides adequate means to limit the water temperature to fixtures to no more than 120°F.

Estimating residential DHW required capacity

Sizing methods

ALWAYS use the recommended sizing information provided by the indirect water heater manufacturer when available. If sizing information is not available from the manufacturer, you can apply the *AHRI first-hour sizing method* (Table 1, Page 6–13) for residential domestic water heating applications. Sizing information is also available in the ASHRAE Handbooks.

Indirect water heater rating information — See information in the following pages about the I=B=R Testing Standard IWH-TS-1 water heater ratings. You will need this information to select the necessary water heater, boiler and circulator once you have determined the DHW heat demand.

NOTICE

Water heater ouput depends on operating conditions.

When selecting an indirect water heater, make sure your system will provide the required rating parameters — boiler output, boiler water temperature, boiler water flow rate, domestic water temperature rise and controls specified in the water heater rating information.

The water heater, boiler, circulator, system and controls must be selected to provide the peak and continuous DHW demands.

Standby loss ratings

The I=B=R IWH-TS-1 standby loss rating indicates the energy lost to the space during standby (no draw) conditions.

Storage-type indirect water heater applications

For storage-type indirect water heaters, boiler output can often be lowered by increasing water heater storage capacity and/or lengthening recovery time, provided the storage capacity and recovery times can handle the peak loads. The required storage varies with the application. Always apply the water heater manufacturer's recommendations and basic engineering principles.

Tankless heater applications

For indirect water heaters having no storage capacity, the connected boiler output must meet or exceed the load calculated from Equation 6-1 or 6-2, Page 6–14. In addition, the heat transfer capacity of the water heater must also meet or exceed this load. (Note that boilers with tankless water heaters are rated for intermittent draw as well as continuous draw. Intermittent capacities are higher than continuous draw capacities.)

Rating parameters

Indirect water heaters are tested and rated per the I=B=R standard, IWH-TS-1. For units tested/rated per this standard, the heater manufacture must show first hour rating, continuous draw rating and standby heat loss rating. First hour rating and continuous draw rating are given in GPH (gallons per hour), tested with DHW water entering at 58°F, leaving at 135°F, using boiler water at 180°F.

Indirect water heater output depends on boiler output, boiler water temperature, domestic water temperatures, flow rates and controls. Carefully read the requirements in the manufacturer's literrature and ensure your installation will provide the right conditions.

Boiler output

If boiler output is less than listed for the water heater rating, the water heater will not deliver the listed ratings. On the other hand, providing a boiler with an output larger than that shown for the heater rating may not increase heater ratings, because the maximum output from the heater is limited by the capability of its heat exchanger.

Boiler water supply temperature

Heat flows into the water heater because the boiler water temperature is higher than the domestic water temperature. The larger this difference, the greater the heat transfer. Heat transfer drops rapidly as boiler supply temperature drops (or as the domestic water temperature increases). Read the required boiler water supply temperature in the manufacturer's ratings. Make sure your system will provide the water temperature stated.

Boiler water flow rate

As the boiler water flows through the water heater and gives up heat to the domestic water, boiler water temperature drops. The lower the flow rate, the more the boiler temperature drops as it passes through the water heater (causing less heat transfer). Increasing boiler water flow rate reduces temperature drop, thus increasing heat transfer.

Read the required boiler water flow rate in the manufacturer's ratings. Also note the pressure drop through the heat exchanger. These flow rates and pressure drops are often higher than typical space heating zones. Make sure the circulator and piping you install will provide the required flow.

Domestic water temperature rise

The greater the temperature difference between boiler supply water temperature and domestic water temperature in the water heater, the greater the output. So the hotter the domestic water must leave the heater, the lower the heater output will be. Pay careful attention to the heater manufacturer's domestic water temperature rise. Typical ratings give FHR for domestic water temperature rises of 77 °F or 90 °F, for example. If the domestic water temperature rise for your application is greater than the rated rise, the heater won't deliver the rated capacity.

Control systems

The control system will sometimes affect the performance on an indirect water heater.

- Some water heater ratings specify **domestic priority** (deactivation of all space heating when the water heater calls for heat).

- The water heater thermostat differential can significantly affect how quickly the boiler responds to domestic water demand.

Be sure to follow all guidelines for controls and piping specified in the water heater ratings and instruction manual.

Domestic priority

Domestic priority is a control method that shuts off space heating or other applications when the DHW tank or controller calls for DHW heating. This dedicates all available boiler output to the DHW demand. Domestic priority is incorporated in many boiler controls, both integral controls and separate boiler/system electronic controllers. Domestic priority can also be accomplished using a relay or relays to control the space heating and DHW circulators or zone valves.

When utilizing domestic priority to direct all boiler heat output to domestic water heating on call from the DHW thermostat or controller, make sure there won't be substantial conflicts in demand. For example, consider the following situations.

- Domestic priority operation is satisfactory only if DHW demand will not be extensively long and will not disable space heating for long intervals during peak space heating periods.

- If using **setback operation** (reducing space temperature during the night and/or during unoccupied periods) — The setback times must be set to allow the space to come up to comfort temperature BEFORE the DHW demand begins. For example — a residence equipped with a setback thermostat takes 45 minutes for the house to heat up after setback, and the morning DHW peak begins at 6:30 AM. You must program the setback timer to return to space comfort temperature no later than 5:45 AM.

- Spa/whirlpool tub heating applications — Make sure the homeowner is aware that space heating will be shut down while the tub is being filled if domestic priority is used. The homeowner would need to fill the spa during off-peak space heating hours. If this isn't acceptable, it may be better to provide a separate heat source for the spa.

AHRI first-hour sizing method
(storage water heater systems only)

The AHRI first-hour method (Table 1, Page 6–13) determines the required capacity based on actual usage patterns.

Use this method only when sizing information is not available from the indirect water heater manufacturer.

When sizing the water heater, thoroughly discuss usage with the residents. Also instruct them regarding system capability and limitations.

> **NOTICE**: Indirect water heaters tend to have much higher recovery capacities than direct-fired water heaters. Consequently, they tend to have less storage capacity for similar first-hour ratings. You may find, for example, that an indirect water heater with only 20 gallons storage has a first-hour rating large enough to satisfy the domestic water requirements of a large home. Before making such an application, make sure the indirect water heater has adequate storage to handle expected peak demand. If not, the application might not meet user needs. For example, consider a residence with multiple showers. If two showers are taken during the peak hour, the required storage capacity would be much larger if these showers occurred at the same time instead of being staggered during the hour, with some time for recovery in between.

Use the AHRI first-hour sizing method when you can talk with the homeowner to determine the anticipated needs. You might use this opportunity to show the excellent capacity available with indirect-fired water heaters. First-hour ratings of indirect-fired water heaters are typically much higher than direct-fired heaters because of the higher input available from most boilers.

Ensure that the equipment selection and installation complies with applicable building codes (including HUD-FHA compliance, when required).

High-flow showers

When the installation will use high-flow showers or other unusual requirements, you will need to add extra capacity to handle the load. Find the hot water consumption rate from Table 2, Page 6–15. Enter the table at the flow rate of the shower head(s). Find the right section for the water heater storage temperature you will use (120, 130, 140, 150 or 160°F) and the cold water supply temperature (40, 50 or 60°F). Read across to the heater flow rate required for this shower flow (at 110°F). Multiply the table number times the length of shower expected, in minutes (typically 10 to 15 minutes per shower). This will give you the gallons of hot water consumed during each shower. Use the AHRI sizing method (Table 1), adding the special loads on the "Other" row in the table.

> Example: Shower flow rate is 6 gpm. The indirect-fired water heater thermostat will be set at 140°F. And the cold water temperature will be 50°F. Enter column 1 at 6 gpm in the 140°F "Heater outlet" section. In Table 2, read across to find the heater flow of 4.00 gpm with cold water at 50°F. The hot water flow from the heater would be 4 gpm during a shower. For a 15-minute shower, the consumption would be 4 x 15 = 60 gallons. Enter 60 gallons times the number of showers during a peak hour on the AHRI "Other" line. If two showers occur during the peak hour, enter 120 gallons, for instance.

Table 1   AHRI first-hour sizing method

AHRI Peak hourly demand estimate				
1. Determine the time of day when the peak usage of hot water is most likely to occur.				
2. Using the following table, determine the likely maximum usage during any one-hour period; this is the peak hour demand.				
NOTE — This table does not estimate total daily hot water usage. As an example, an average of 4 gallons of hot water is used each time dishes are washed by hand, but dishes washed by hand are usually done 3 times a day. The average daily hot water usage for hand dishwashing,12 gallons, is about the same as the average hot water usage for an automatic dishwasher used once a day.				
Use	**Average gallons of hot water per usage**		**Number of times used during one hour**	**Gallons used in one hour**
Shower	20	x		=
Bath	20	x		=
Shaving	2	x		=
Hands and face washing	4	x		=
Hair shampoo	4	x		=
Hand dishwashing	4	x		=
Automatic dishwasher	14	x		=
Food preparation	5	x		=
Residential clothes washer	32	x		=
Other	(specify) —	x		=
			Total (peak demand):	

Example: *A household uses the most hot water in the morning. During the busiest hour, the usage is:*				
(3) showers:	20	x	3	= 60 gallons
(1) shave:	1	x	2	= 2 gallons
(1) shampoo:	1	x	4	= 4 gallons
Handwashing dishes:	1	x	4	= 4 gallons
			Total (peak demand):	70 gallons
For this example, you would select a water heater with a first-hour rating of at least 70 gallons.				

Source: AHRI "Consumers' Directory of Certified Efficiency Ratings for Residential Heating and Water Heating Equipment". This excerpt has been reworded slightly from AHRI Directory.

NOTICE — Using I=B=R first hour ratings

The flow requirements in the AHRI table assume DHW water supplied at 140°F. The I=B=R first-hour and continuous draw ratings are based on DHW water supplied at 135°F. Supplying water at a lower temperature (135°F instead of 140°F) requires more flow for a given demand. To compensate, *increase* the peak demand result in the AHRI table by 6% to compensate for the lower DHW supply temperature when using I=B=R first-hour ratings.

Determining DHW flow rate

Calculating heat demand from DHW flow rate

If you want to know how much heat is required for a given domestic heating condition, you need to know the water temperature rise and flow rate. Calculate domestic water heating load using either equation 6-1 or 6-2:

$$\text{Btuh} = \text{gpm} \times 500 \times (T_{hot} - T_{cold}) \qquad \textbf{(6-1)}$$

$$\text{Btuh} = \text{gph} \times 8.33 \times (T_{hot} - T_{cold}) \qquad \textbf{(6-2)}$$

where T_{hot} is domestic hot water temperature and T_{cold} is the cold water temperature.

Table 2 Hot water flow required for a mixed temperature of 110 °F

Hot water flow required for a mixed temperature of 110 °F										
Use this table to find the required flow of water from the indirect heater										
Equation: **heater gpm = mixed gpm x [(110 - Tc)/(Th - Tc)]**, where **Th** = Heater outlet temperature, **Tc** = cold water temperature										
Flow at 110°F (gpm)	Heater outlet (°F)	Cold water (°F)	Heater flow (gpm)	Btuh	Cold water (°F)	Heater flow (gpm)	Btuh	Cold water (°F)	Heater flow (gpm)	Btuh
1	120	40	0.88	35,000	50	0.86	35,000	60	0.83	30,000
1.5	120	40	1.31	52,500	50	1.29	52,500	60	1.25	45,000
2	120	40	1.75	70,000	50	1.71	70,000	60	1.67	60,000
2.5	120	40	2.19	87,500	50	2.14	87,500	60	2.08	75,000
3	120	40	2.63	105,000	50	2.57	105,000	60	2.50	90,000
4	120	40	3.50	140,000	50	3.43	140,000	60	3.33	120,000
5	120	40	4.38	175,000	50	4.29	175,000	60	4.17	150,000
6	120	40	5.25	210,000	50	5.14	210,000	60	5.00	180,000
7	120	40	6.13	245,000	50	6.00	245,000	60	5.83	210,000
8	120	40	7.00	280,000	50	6.86	280,000	60	6.67	240,000
9	120	40	7.88	315,000	50	7.71	315,000	60	7.50	270,000
10	120	40	8.75	350,000	50	8.57	350,000	60	8.33	300,000
1	130	40	0.78	35,000	50	0.75	40,000	60	0.71	35,000
1.5	130	40	1.17	52,500	50	1.13	60,000	60	1.07	52,500
2	130	40	1.56	70,000	50	1.50	80,000	60	1.43	70,000
2.5	130	40	1.94	87,500	50	1.88	100,000	60	1.79	87,500
3	130	40	2.33	105,000	50	2.25	120,000	60	2.14	105,000
4	130	40	3.11	140,000	50	3.00	160,000	60	2.86	140,000
5	130	40	3.89	175,000	50	3.75	200,000	60	3.57	175,000
6	130	40	4.67	210,000	50	4.50	240,000	60	4.29	210,000
7	130	40	5.44	245,000	50	5.25	280,000	60	5.00	245,000
8	130	40	6.22	280,000	50	6.00	320,000	60	5.71	280,000
9	130	40	7.00	315,000	50	6.75	360,000	60	6.43	315,000
10	130	40	7.78	350,000	50	7.50	400,000	60	7.14	350,000
1	140	40	0.70	35,000	50	0.67	45,000	60	0.63	40,000
1.5	140	40	1.05	52,500	50	1.00	67,500	60	0.94	60,000
2	140	40	1.40	70,000	50	1.33	90,000	60	1.25	80,000
2.5	140	40	1.75	87,500	50	1.67	112,500	60	1.56	100,000
3	140	40	2.10	105,000	50	2.00	135,000	60	1.88	120,000
4	140	40	2.80	140,000	50	2.67	180,000	60	2.50	160,000
5	140	40	3.50	175,000	50	3.33	225,000	60	3.13	200,000
6	140	40	4.20	210,000	50	4.00	270,000	60	3.75	240,000
7	140	40	4.90	245,000	50	4.67	315,000	60	4.38	280,000
8	140	40	5.60	280,000	50	5.33	360,000	60	5.00	320,000
9	140	40	6.30	315,000	50	6.00	405,000	60	5.63	360,000
10	140	40	7.00	350,000	50	6.67	450,000	60	6.25	400,000

Is additional boiler capacity required?

General

For many residential installations, the boiler heating capacity is much greater than the maximum DHW capacity, and peak DHW demand doesn't typically coincide with peak space heating demand. Typically, on these installations, boiler capacity is not increased above that needed for the space heating load. But, for applications with high DHW demand, additional capacity may be necessary. The following information can be used to determine boiler sizing.

Storage-type system

No additional boiler capacity is needed unless the domestic water load is more than 25% of the space-heating load. When the domestic load is higher than 25% of the space-heating load, use Table 3, Page 6–17. Multiply the table factor times the space heating load to determine the total required boiler I=B=R net load rating in Btuh (data developed from 1999 ASHRAE Handbook: HVAC Applications, Figure 24, page 48.22).

- Example: A residence has a heat loss of 75,000 Btuh. The system requires an indirect-fired water heater with a first-hour rating of at least 90 gallons. You select an indirect-fired water heater with a first-hour rating (at 140°F) of 100 gallons. The heater manufacturer's literature shows a minimum boiler capacity of 60,000 Btuh to provide this first-hour rating. Use 60,000 Btuh as the required domestic water load.

- For this example, the space heating load is 75,000 Btuh and the domestic water load is 60,000 Btuh. Divide 60,000 by 75,000 to find the DHW ratio. The result is ratio = 0.8. For a DHW ratio of 0.8, Table 4, Page 6–17 gives a load factor of 1.69. Multiply this number times the space heating load to find the required boiler net load rating. The net load rating must be at least 1.69 x 75,000 = 126,750 Btuh.

Tankless (instantaneous) system

Find the required heating capacity for domestic water heating using Table 4, Page 6–17, using the required gpm of hot water determined from the HUD-FHA table (Table 1, page 6-13) or other recognized sizing method. Tankless coil ratings are usually based on intermittent draw testing. The performance is based on a 5-minute ON period followed by a 10-minute OFF period, averaged over time. Most tankless heater applications work satisfactorily using the intermittent draw ratings. Only use the continuous flow sizing information if you know the application will require continuous flow from the tankless heater.

The boiler must satisfy the space heating requirements and also be equipped with a tankless coil that can deliver the required flow. Select boiler with tankless coil:

Rule 1:

Boiler I=B=R net load rating must be at least equal to the space heating load for water boilers (or to total square feet of installed radiation if a steam boiler).

Rule 2:

The boiler/tankless coil combination must have a heating capacity (DOE Heating Capacity for boilers under 300,000 Btuh input, or Gross Output for larger boilers) at least equal to the DHW load.

- Example: Use same example as for the indirect-fired water heater above. The HUD-FHA requirement (Table 1, page 6-13) for an equivalent tankless heater is a 3.75 gpm capacity. From Table 4, 3.75 gpm (from 40 to 140°F) is a load of 187,500 Btuh.

- The space heating load is 75,000 Btuh. The boiler selected must have an I=B=R net load rating of at least 75,000 Btuh for rule 1 above. The boiler/tankless coil must also have a DOE Heating Capacity of at least 187,500 Btuh for rule 2.

Table 3   Load factors for combined space heating and domestic water heating

Load factors for combined space heating/DHW heating									
Multiply load factor below times space heating load to determine required boiler I=B=R net load rating.									
DHW Ratio = DHW load ÷ space heating load									
DHW ratio	Load factor	DHW ratio	Load factor	DHW ratio	Load factor	DHW ratio	Load factor	DHW ratio	Load factor
0.25	1.15	1.1	1.98	2.1	2.99	3.2	4.14	6	7.15
0.3	1.20	1.2	2.08	2.2	3.10	3.4	4.35	6.5	7.65
0.4	1.29	1.3	2.18	2.3	3.20	3.6	4.56	7	8.15
0.45	1.34	1.4	2.28	2.4	3.30	3.8	4.78	7.5	8.65
0.5	1.39	1.5	2.39	2.5	3.41	4	4.99	8	9.15
0.6	1.49	1.6	2.49	2.6	3.51	4.2	5.21	8.5	9.65
0.7	1.59	1.7	2.59	2.7	3.61	4.4	5.43	9	10.15
0.8	1.69	1.8	2.69	2.8	3.72	4.6	5.65	9.5	10.65
0.9	1.79	1.9	2.79	2.9	3.82	4.8	5.87	10	11.15
1	1.89	2	2.89	3	3.93	5	6.15		

Table 4   Required heating capacity, Btuh, for given tankless heater flow, gpm

Heating capacity vs tankless heater flow, gpm									
Use this table to find the required heating capacity for the given tankless heater flow rate Note that tankless coil ratings are usually based on **intermittent flow** (5 minutes ON, 10 minutes OFF)									
Equation: **Btuh = gpm x (140°F - 40°F) x 500** (divide by **3** for intermittent flow)									
Tankless flow (gpm)	Outlet temp (°F)	Cold water (°F)	Btuh required		Tankless flow (gpm)	Outlet temp (°F)	Cold water (°F)	Btuh required	
			Continuous flow	Intermittent flow 5 min ON, 10 min OFF				Continuous flow	Intermittent flow 5 min ON, 10 min OFF
1	140	40	50,000	16,600	5	140	40	250,000	83,300
1.25	140	40	62,500	20,800	5.5	140	40	275,000	91,600
1.5	140	40	75,000	25,000	6	140	40	300,000	100,000
1.75	140	40	87,500	29,100	6.5	140	40	325,000	108,300
2	140	40	100,000	33,300	7	140	40	350,000	116,600
2.25	140	40	112,500	37,500	7.5	140	40	375,000	125,000
2.5	140	40	125,000	41,600	8	140	40	400,000	133,300
2.75	140	40	137,500	45,800	8.5	140	40	425,000	141,600
3	140	40	150,000	50,000	9	140	40	450,000	150,000
3.25	140	40	162,500	54,100	9.5	140	40	475,000	158,300
3.5	140	40	175,000	58,300	10	140	40	500,000	166,600
3.75	140	40	187,500	62,500	10.5	140	40	525,000	175,000
4	140	40	200,000	66,600	11	140	40	550,000	183,300
4.25	140	40	212,500	70,800	11.5	140	40	575,000	191,600
4.5	140	40	225,000	75,000	12	140	40	600,000	200,000
4.75	140	40	237,500	79,100					

DHW applications

Domestic water connections

See Figure 2, Page 6–19, showing domestic water connections to and from a typical indirect water heater. The items noted with a star are optional on many installations, depending on applicable code requirements for the area. Though item 13, a thermostatic mixing valve, is optional on many installations, you should use it. No device is fail-safe, but providing a thermostatic mixing valve is added protection for the owner against scald hazards.

The function of each item in Figure 2, Page 6–19 is:

1   Gate or globe valve for isolation

2   Backflow preventer, when required

3   Pressure-reducing valve, when required on high supply-pressure systems

4   Expansion tank, always recommended when using a backflow preventer, to prevent pressure from rising as water in the lines is heated

5   Vacuum breaker, when required

6   Thermal trap, when required, prevents thermal migration in the piping

7   Domestic water supply to heater

8   Dip tube (Never solder fitting attached to water heater supply connection — the heat could damage the plastic dip tube. Always solder the connection before screwing the fitting to the tank connection.)

9   Domestic hot water outlet from heater

10 Unions as needed to simplify piping disconnection, if needed

11 Globe valve or square-head cock to allow throttling flow

12 Shock arrester, when required, to prevent damage from hydraulic shock due to quick closure of valves

13 Thermostatic mixing valve, required on all tankless heater installations, highly recommended on all applications, as added protection against scald hazard

14 Flow regulator valve, used with tankless heater installations to limit the flow of water to the tankless heater

Figure 2 Domestic cold and hot water connections to water heater

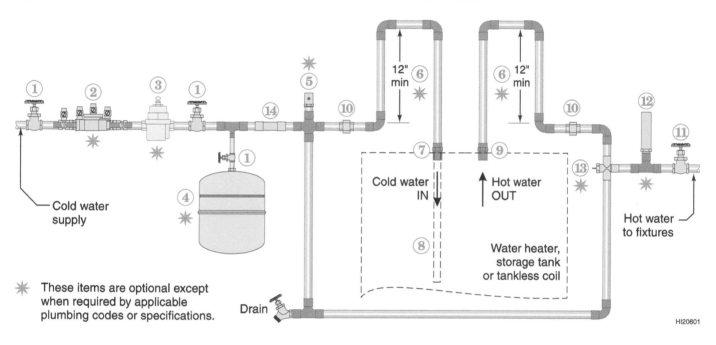

These items are optional except when required by applicable plumbing codes or specifications.

Figure 3   Domestic water connections to tankless heater, typical (refer to boiler/heater manufacturer's instructions for piping and control connections)

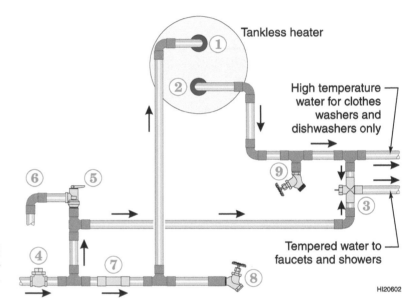

NOTICE: Provide all required components in the cold water supply line. See Figure 2, Page 6-19.

Tankless heater

See Figure 3, typical piping to a tankless heater. Always connect the cold water piping from below the mixing valve (item 3). This puts cold water below the valve, thus resisting gravity circulation and heat migration into the cold water line.

Refer to the boiler manual for suggested piping and instructions on application of tankless heaters.

The function of each item is:

1 Cold water inlet to heater, connected at top

2 Hot water outlet from heater

3 Thermostatic mixing valve, to limit supply temperature to shower, tub and faucets. (If supplying a higher-temperature, unmixed, water supply, connect only for dishwashing and clothes washing applications.)

4 Square-head cock or globe valve to allow flow adjustment, if needed

5 Relief valve, to limit pressure in line

6 Connect relief valve outlet full size, using only metal pipe, to a safe discharge location

7 Flow restrictor valve, sized to the flow capacity of the tankless heater

8 Hose bibb valve

9 Hose bibb valve, used in conjunction with valve 8 to allow flushing the heater. Flush heater periodically to eliminate sediment that accumulates from the fresh water supply. Connect water supply hose to valve 9. Connect drain hose to valve 8. Open valves 8 and 9 to allow water to flow through heater to flush.

Figure 4 Domestic water connections to tankless heater with storage tank, typical (refer to boiler/heater manufacturer's instructions for piping and control connections)

NOTICE: Provide all required components in the cold water supply line. See Figure 2, Page 6–19.

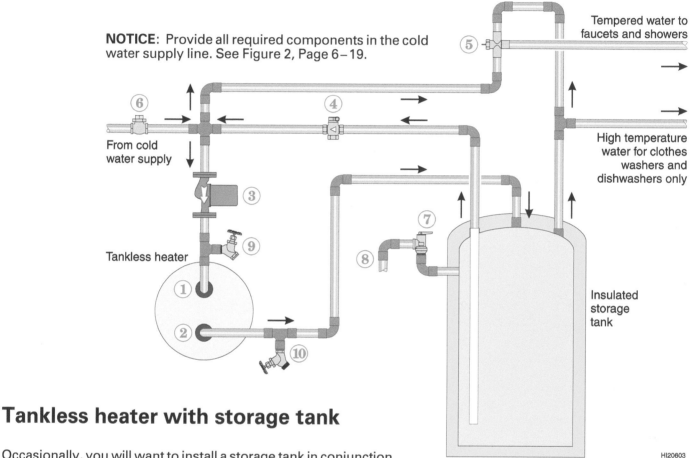

Tankless heater with storage tank

Occasionally, you will want to install a storage tank in conjunction with a tankless heater to provide additional hot water capacity for the owner. You may also find this a helpful remedy on some tankless heater installations that encounter insufficient hot water supply at times.

Refer to the boiler manual for suggested piping and instructions on application of tankless heaters.

See Figure 4, showing a typical application of a tankless heater with a storage tank. Note that a circulator circulates tank water to and from the tankless heater. Cold domestic water enters the tank, not the tankless heater.

The function of the piping components is:

1 Tankless heater supply tapping, in top

2 Tankless heater outlet water tapping

3 Bronze-body circulator, to circulate cold water from bottom of tank, through tankless heater, and back to top of tank

4 Check valve, to prevent cold water from entering top of tank

5 Thermostatic mixing valve, recommended for regulating maximum water temperature available to showers/tubs/faucets. (If supplying a higher-temperature, unmixed, water supply, connect

only for dishwashing and clothes washing applications.)

6 Square-head cock or globe valve for adjusting cold water flow rate, if needed

7 Pressure or pressure/temperature relief valve (pressure/temperature valves may be mounted with stem horizontally)

8 Connect relief valve outlet full size, using only metal pipe, to a safe discharge location

9 Hose bibb valve

10 Hose bibb valve, used in conjunction with valve 9 to allow flushing the heater. Flush heater periodically to eliminate sediment that accumulates from the fresh water supply. Connect water supply hose to valve 9. Connect drain hose to valve 10. Open valves 9 and 10 to allow water to flow through heater to flush.

Indirect-fired water heaters

All tank-type heaters (indirect-fired water heaters) require circulation of boiler water through the tank heat exchanger. Thus, these heaters have two connections for domestic water (cold and hot) plus a boiler water supply and boiler water return connection. The following discussion deals with typical piping arrangements to control boiler water flow to the indirect water heater. For information on the domestic water connections, see the previous discussion and Figure 2, Page 6–19.

Indirect-fired water heater with zone valve zoning (with or without domestic priority)

See Figure 5 for a typical application of an indirect-fired water heater to a zone-valve-zoned system. The water heater zone valve is operated by the water heater thermostat, opening the valve on a call for tank heating. This system can be arranged by domestic priority by using a zone controller or adding relays to interrupt power to the space heating zone valves when the tank thermostat calls for heat.

Indirect-fired water heater with zone valve zoning and 3-way priority zone valve

See Figure 6, a zone-valve-zoned system using a 3-way zone valve to divert flow to either the water heater or the space heating circuits. The 3-way (priority) valve, item 1, is normally-open to the space heating system, allowing boiler water to flow to the heating circuits on call from their zone valves. When the water heater thermostat calls for heat, the 3-way valve opens to the water heater and closes to the space heating circuits. This diverts all available heat to the water heater.

Figure 5 Indirect water heater with zone-valve zoning

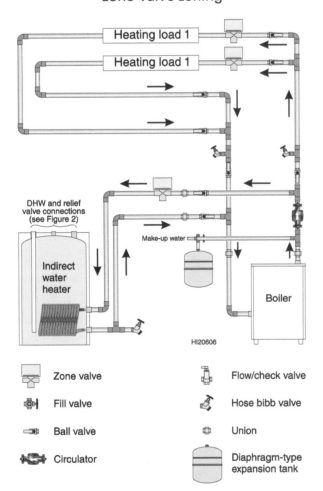

Figure 6 Indirect water heater with 3-way priority zone valve

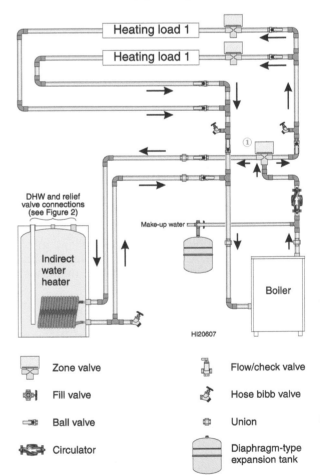

Indirect-fired water heater with circulator zoning (with or without domestic priority)

See Figure 7, a circulator-zoned system. On a call for heat from the water heater thermostat, the water heater circulator operates, delivering hot boiler water to the heater internal heat exchanger. This system can be controlled to provide domestic priority by using a zoning controller with priority feature or installing relays to deactivate the relays operating the space heating circulators when the water thermostat calls for heat. (Always use flow/check valves on circulator-zoned systems, as shown in Figure 7, to prevent gravity flow or flow induced in idle zones due to operation of other circulators.)

Indirect-fired water heater (or pool or spa heater) with dedicated boiler

In Figure 8 a boiler is piped directly to an indirect-fired water heater. The boiler and its circulator operate on a call for heat from the water heater thermostat. Use a dedicated boiler for domestic water heating or pool/spa heating applications that have a high heating load compared to the space-heating load in order to reduce excess boiler cycling due to over-sizing. Alternatively, you can use multiple (or modular) boilers, with boilers staged on as load increases. Refer to the boiler manufacturer's instructions for recommended piping.

Figure 7 Indirect water heater with circulator zoning

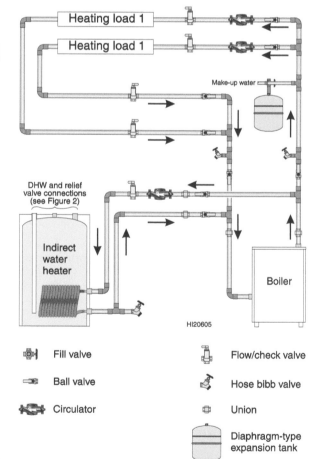

Figure 8 Indirect water heater with dedicated boiler

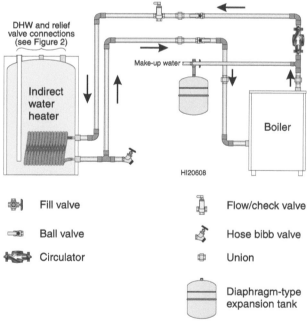

Spa & pool heating

NOTICE: When applying indirect water heaters to pool or recirculating spa heating, use only heaters or heat exchangers rated for pool heating application. Consult boiler and heater manufacturers' instructions for application, sizing and piping information.

Estimating spa heating requirements

Once-through spa heating

When filling a spa with hot water, with once-through heating [using an indirect-fired water heater, heat exchanger or tankless coil] the heating capacity required will be:

$$\text{Btuh} = \text{gpm} \times 500 \times \text{TD}, \quad \textbf{(6-3)}$$

where gpm is the fill rate and TD is the temperature rise. If you don't know the fill rate, find the spa water volume from manufacturer's literature or calculate approximate volume (see Figure 9, Page 6–26). Divide volume by the number of minutes over which you want to fill the tub to find the gpm fill rate. (Example, a 200-gallon spa filled in 30 minutes would require 6.7 gpm (200 divided by 30). This method can require a very large boiler or heat exchanger if fill time is short or volume large. Use as much storage as possible to allow using a smaller boiler, heating the tank over a longer period.

Recirculating spa heating

In recirculating heating, part of the filter pump flow is routed through a pool heater or pool heat exchanger to heat the spa. Consult the heater and spa manufacturers' instructions for piping requirements.

Find the spa volume from the manufacturer's literature or calculate the approximate volume (see Figure 9). Determine the allowable time to bring the spa up to temperature. Then use the bottom section of Table 6, Page 6–29, to find the required heater capacity in Btuh. Note that the heating capacities are based on heating spa water from room ambient temperature (70°F) to 105°F. Heating on initial cold water fill would take up to twice as long, depending on cold water fill temperature.

> **Example**: Your application has a 600-gallon spa to be heated with recirculating piping through a pool heat exchanger. The owner wants to be able to raise the spa water from room temperature in 2 hours. From Table 6, the heater capacity must be at least 87,400 Btuh.

Additional boiler heating capacity for spa heating

If the spa heat exchanger will be installed on a combination space heating system, use Figure 3, Page 6–17, to find the required boiler net load rating. Add the domestic hot water load and spa heating load together to determine the load ratio. Divide the total of the spa heating and domestic water heating loads by the space heating load to find the "DHW ratio."

If the spa load is large compared to the space heating load (over 100% of space heating load), you should use multiple (or modular) boilers or provide a separate boiler for the spa. Otherwise, the boiler will cycle often because it will be too large for the system load most of the time.

Estimating pool heating requirements

Residential indoor-pool heating

Find indoor-pool heating capacity based on heat-up time, using Table 6, Page 6–29. This table assumes an average depth in the pool of 5.5 feet. See Figure 9, Page 6–26, for assistance in calculating pool heating surface. Table 6 assumes the heat-up time begins with the pool at room ambient temperature of 70°F. Heat-up time at initial fill will take longer because the water must be heated from ground water temperature to 80°F. This could take 2 to 3 times longer than the Table 6 values if ground water is at 50°F, for example.

If pool-heating load is large compared to space heating load (more than 50%), use multiple (or modular) boilers or apply a separate boiler for the pool-heating load.

- Example: You want to heat a kidney-shaped indoor pool. The owner is satisfied with a 36-hour heat-up time (from room temperature). The pool dimensions are (see Figure 9, Page 6–26): A=10 feet; B=18 feet; L=24 feet; average depth = 5.5 feet. From Figure 9, the approximate pool surface area = 0.45 x L x (A + B) = 0.45 x 24 x (10 + 18) = 302 square feet. Use the 300-square feet line in Table 6, Page 6–29. The required heater capacity is 38,300 Btuh. See Equation 6-3, Page 6–24, to determine total boiler capacity required with a combination space heating system.

Residential outdoor-pool heating

You will need to determine whether the pool heat-up time will be specified. If the pool must be heated from ambient temperature water to 80°F in a given time period, use Table 5, Page 6–28, to find the required heater capacity. If the heat-up time is not specified and not considered important, you can use the upper section of Table 6, Page 6–29, based on the heat capacity required to maintain the pool at 80°F. Be sure to find out the lowest outdoor temperature at which the pool temperature must be maintained when applying either Table 5 or Table 6.

To determine the pool surface area, you can use Figure 9, Page 6–26. Calculate pool surface and volume from the equations provided.

Note that the data in Table 5 and Table 6 are for a wind speed of 3.5 mph. Use the correction factors given in the table for other wind speeds. Select whichever of the two required heating capacities is the largest. If the wind speed will exceed 3.5 mph, see the notes in the tables for instructions. Make sure the owner is aware that he can reduce pool heat loss (saving energy) by using a solar cover (providing heat gain during the day and preventing evaporative loss when the pool is not in use). Installing a hedge or fence for a windbreak will also reduce heat loss.

- Example: What is the heater capacity required for an oval-shaped outdoor pool 24 feet wide and 20 feet along the straight sides to be heated at an outdoor temperature of 50°F? The average depth is 5.5 feet. The wind speed over the pool will not exceed 3.5 mph. The pool can be heated up over a 48-hour period. Size both for time-rise and maintaining-temperature.

- The pool surface area is (from Figure 9, Page 6–26): Area = (0.785 x D2) + (L x D) = (0.785 x 242) + (20 x 24) = 932 square feet. Volume = 7.48 x Area x Average depth = 7.48 x 932 x 5.5 = 38,300 gallons.

- Time-rise method: Enter Table 5, Page 6–28, at volume = 40,000 gallons. Find a required heating capacity under the 48-hour section and 50°F ambient temperature of 361,300 Btuh.

- Maintaining temperature method: Enter Table 6, Page 6–29, at surface area = 1000 square feet. Scan across to the 50°F ambient temperature column to find a required heating capacity of 315,000 Btuh. The required capacity could be reduced from 361,300 to 315,000 if the heat-up time was not important.

Additional boiler capacity for pool heating

Pool heating is a sustained load. Pool heating demand, then, will occur at peak heating load times. Therefore, the pool heating capacity must be added to the space heating load without applying a load factor (as for domestic hot water heating). You do not need to add a piping and pick-up allowance for the pool load — just for the space heating load. Use the following formula to determine total boiler net load rating:

Net load required = (Pool heating load) + (1.15 x Space heating load) (6-3)

If the system includes domestic water and/or spa loads:

Net load required = (Pool heating load) + (Load factor x Space heating load) (6-4)

where "Load factor" is found from Figure 3, Page 6–17, as described previously for domestic water or spa loads.

If the pool-heating load is large compared to the space-heating load, use multiple (modular) boilers or a separate boiler dedicated to pool heating. Otherwise, the boiler will cycle too frequently, causing substantial off-cycle losses.

Figure 9   Estimating area and volume of pools and spas

Kidney-shaped pools

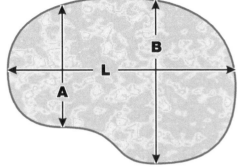

Area (square feet) = $0.45 \times L \times (A + B)$ *(approximately)*

Volume (gallons) = $7.48 \times$ Area $\times$ Average depth

Circular pools

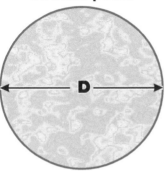

Area (square feet) = $0.785 \times D^2$

Volume (gallons) = $7.48 \times$ Area $\times$ Average depth

Oval pools

HI20609

Area (square feet) = $(0.785 \times D^2) + (L \times D)$

Volume (gallons) = $7.48 \times$ Area $\times$ Average depth

Rectangular pools

Area (square feet) = $L \times W$

Volume (gallons) = $7.48 \times$ Area $\times$ Average depth

Elliptical pools

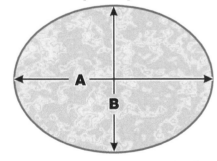

Area (square feet) = $0.785 \times A \times B$

Volume (gallons) = $7.48 \times$ Area $\times$ Average depth

Figure 10 Typical piping with pool heater exchanger and combination space heating

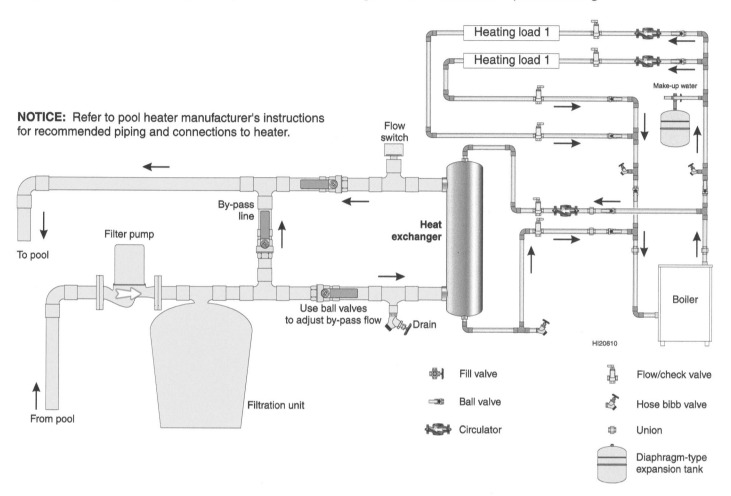

NOTICE: Refer to pool heater manufacturer's instructions for recommended piping and connections to heater.

Fill valve		Flow/check valve
Ball valve		Hose bibb valve
Circulator		Union
		Diaphragm-type expansion tank

Table 5   Residential outdoor in-ground pool heating requirements — Time-rise method

Residential outdoor in-ground pool heating requirements in Btuh — Time-rise method

Heat capacity (Btuh) required to heat pool water from ambient air temperature to 80°F
at given ambient temperatures in the amount of time shown (@ 3½ mph wind speed — see Note 1)

Pool volume (gallons)	48 hours heat-up time				36 hours heat-up time			
	40°F Ambient	50°F Ambient	60°F Ambient	70°F Ambient	40°F Ambient	50°F Ambient	60°F Ambient	70°F Ambient
500	5,900	4,500	2,900	1,400	7,100	5,300	3,500	1,700
750	9,000	6,700	4,500	2,200	10,700	8,000	5,300	2,600
1,000	12,000	9,000	5,900	2,900	14,300	10,700	7,100	3,500
2,500	30,000	22,500	14,900	7,400	35,800	26,800	17,800	8,800
5,000	60,200	45,100	30,000	14,900	71,700	53,800	35,800	17,800
7,500	90,200	67,700	45,100	22,500	107,600	80,700	53,800	26,800
10,000	120,400	90,200	60,200	30,000	143,500	107,600	71,700	35,800
15,000	180,600	135,400	90,200	45,100	215,300	161,500	107,600	53,800
20,000	240,800	180,600	120,400	60,200	287,100	215,300	143,500	71,700
25,000	301,100	225,800	150,500	75,200	358,900	269,200	179,400	89,700
30,000	361,300	270,900	180,600	90,200	430,700	323,000	215,300	107,600
35,000	421,500	316,100	210,700	105,300	502,500	376,800	251,200	125,500
40,000	481,700	361,300	240,800	120,400	574,300	430,700	287,100	143,500
50,000	602,200	451,700	301,100	150,500	717,900	538,400	358,900	179,400

Pool volume (gallons)	24 hours heat-up time				12 hours heat-up time			
	40°F Ambient	50°F Ambient	60°F Ambient	70°F Ambient	40°F Ambient	50°F Ambient	60°F Ambient	70°F Ambient
500	9,400	7,100	4,600	2,300	16,300	12,300	8,100	4,000
750	14,200	10,600	7,100	3,500	24,600	18,400	12,300	6,100
1,000	18,900	14,200	9,400	4,600	32,800	24,600	16,300	8,100
2,500	47,400	35,500	23,600	11,700	82,100	61,500	41,000	20,400
5,000	94,900	71,100	47,400	23,600	164,300	123,200	82,100	41,000
7,500	142,300	106,700	71,100	35,500	246,400	184,800	123,200	61,500
10,000	189,800	142,300	94,900	47,400	328,600	246,400	164,300	82,100
15,000	284,700	213,500	142,300	71,100	493,000	369,700	246,400	123,200
20,000	379,600	284,700	189,800	94,900	657,300	493,000	328,600	164,300
25,000	474,600	356,000	237,300	118,600	821,700	616,300	410,800	205,400
30,000	569,600	427,100	284,700	142,300	986,100	739,500	493,000	246,400
35,000	664,500	498,300	332,200	166,000	1,150,400	862,700	575,200	287,500
40,000	759,400	569,600	379,600	189,800	1,314,700	986,100	657,300	328,600
50,000	949,300	712,000	474,600	237,300	1,643,500	1,232,600	821,700	410,800

Note 1: Above heat capacities based on 3½ mph wind speed. For higher wind speeds, consult pool heater manufacturer's recommendations.

Note 2: Above calculations based on average pool depth of 5.5 feet to estimate square feet of surface typical.

Source: Developed from ASHRAE 1999 HVAC Applications Handbook, Chapter 48. Sizing accounts for surface heat loss during heat-up.

Table 6 Residential pool and spa heating requirements

Residential outdoor in-ground pool heating requirements in Btuh — maintaining temperature only

Heat capacity required to **maintain** pool at 80°F (at wind speed 3½ mph — see Note 1)

Surface area (square feet)	Ambient temperature					
	40°F	50°F	55°F	60°F	65°F	70°F
100	42,000	31,500	26,250	21,000	15,750	10,500
150	63,000	47,250	39,375	31,500	23,625	15,750
200	84,000	63,000	52,500	42,000	31,500	21,000
250	105,000	78,750	65,625	52,500	39,375	26,250
300	126,000	94,500	78,750	63,000	47,250	31,500
350	147,000	110,250	91,875	73,500	55,125	36,750
400	168,000	126,000	105,000	84,000	63,000	42,000
450	189,000	141,750	118,125	94,500	70,875	47,250
500	210,000	157,500	131,250	105,000	78,750	52,500
600	252,000	189,000	157,500	126,000	94,500	63,000
700	294,000	220,500	183,750	147,000	110,250	73,500
800	336,000	252,000	210,000	168,000	126,000	84,000
900	378,000	283,500	236,250	189,000	141,750	94,500
1000	420,000	315,000	262,500	210,000	157,500	105,000
1100	462,000	346,500	288,750	231,000	173,250	115,500
1200	504,000	378,000	315,000	252,000	189,000	126,000

Note 1: Adjustment for different wind speeds: Less than 3½ mph: multiply by 0.75; Over 3½ to 5 mph: multiply times 1.25; Over 5 to 10 mph: multiply times 2.0

Residential indoor pool heating requirements in Btuh (assumes 70°F ambient temperature @ 50% relative humidity)

Surface area (sqare feet)	Pool volume (gallons)	To heat pool from 70°F to 80°F in given time:				To maintain 80°F only
		48 hours	36 hours	24 hours	12 hours	
100	4,100	10,400	12,700	17,500	31,700	6,630
150	6,100	15,400	19,000	26,000	47,200	9,945
200	8,200	20,800	25,500	35,000	63,500	13,260
250	10,200	25,900	31,800	43,600	79,000	16,575
300	12,300	31,200	38,300	52,500	95,200	19,890
350	14,300	36,400	44,600	61,200	110,800	23,205
400	16,400	41,600	51,100	70,100	127,000	26,520
450	18,500	47,000	57,700	79,100	143,300	29,835
500	20,500	52,000	63,900	87,600	158,800	33,150
600	24,600	62,400	76,700	105,100	190,500	39,780
700	28,700	73,000	89,600	122,800	222,400	46,410
800	32,900	83,500	102,600	140,600	254,800	53,040
900	37,000	94,000	115,400	158,200	286,600	59,670
1000	41,100	104,400	128,200	175,700	318,400	66,300
1100	45,200	114,800	140,900	193,200	350,100	72,930
1200	49,300	125,200	153,700	210,800	381,900	79,560

Note 2: This table assumes pool water has come to ambient temperature before being heated to 80°F. Heat-up time at initial fill would be up to 3 times longer than above if starting with water at 40 or 50°F.

Residential indoor spa heating — Btuh required to raise spa water temperature 35°F (70 to 105 F) in time shown

Volume (gallons)	30 minutes	45 minutes	1 hour	1½ hours	2 hours	3 hours	4 hours
200	116,600	77,700	58,300	38,800	29,100	19,400	14,500
300	174,900	116,600	87,400	58,300	43,700	29,100	21,800
400	233,200	155,400	116,600	77,700	58,300	38,800	29,100
500	291,500	194,300	145,700	97,100	72,800	48,500	36,400
600	349,800	233,200	174,900	116,600	87,400	58,300	43,700
700	408,100	272,100	204,000	136,000	102,000	68,000	51,000
800	466,400	310,900	233,200	155,400	116,600	77,700	58,300
900	524,700	349,800	262,300	174,900	131,100	87,400	65,500
1000	583,100	388,700	291,500	194,300	145,700	97,100	72,800

Hydronics Institute Section of AHRI

35 Russo Place

Berkeley Heights, NJ 07922-0218

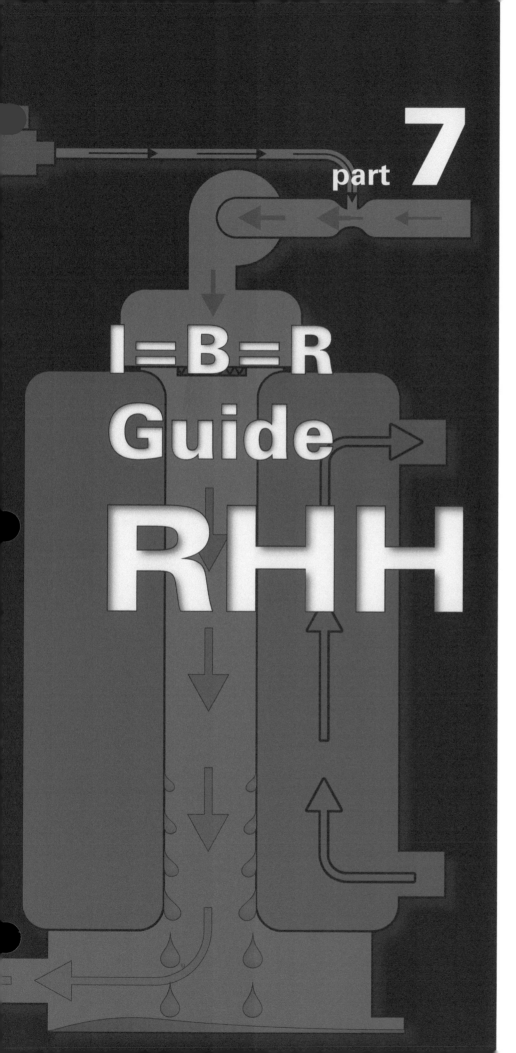

part 7

Condensing boilers . . .

- Application
- Optimizing for maximum efficiency

Residential Hydronic Heating . . .

Installation & Design

Hydronics Institute
Section of **AHRI**

I=B=R Guide RHH
Residential Hydronic Heating

Hydronics Institute Section of **AHRI**
35 Russo Place
Berkeley Heights, NJ   07922-0218

Contents - Part 7

Contents *– continued*

Contents *– continued*

● Reset controls

● Application examples

Contents – continued

Illustrations

Contents *– continued*

Tables

Introduction

What is a condensing boiler?

Figure 1, Page 7–9, shows a schematic of a typical gas-fired condensing boiler. Every condensing boiler includes the basic elements shown in the illustration: Burner (item 1 — a premix burner in this example); Heat exchanger (item 2); Flue gas condensate collection/removal section (item 3); and Vent system (item 4). The heat exchanger is designed to promote condensation, and to be resistant to the corrosive potential of flue gas condensate. Some condensing boilers use a two-stage heat exchanger — the first stage for non-condensing operation and the second stage for condensing operation.

Typical condensing boilers feed boiler RETURN water through the condensing portion of the exchanger, as in Figure 1. This provides the coolest possible surfaces for condensation. Note that condensation ONLY occurs where the water temperature (and heat exchanger surface temperature) is below the flue gas dewpoint temperature. *If the return water temperature is higher than the flue gas dewpoint temperature, NO boiler can condense, regardless of its design.*

Residential condensing boilers typically have AFUE ratings from 90% to 95%. (The DOE test for AFUE is done with 120°F return water and 140 °F outlet water.)

WARNING	Remember that most ***conventional*** boilers are not capable of sustained condensing operation. Flue gas condensate can seriously damage the heat exchanger, insulation, burners and other components, causing early failure. System piping must be designed to prevent condensation in conventional boilers if low water temperatures are likely. Never install a conventional boiler for low-temperature operation unless recommended by the boiler manufacturer.

Why condense?

Flue gas condensation is needed for high efficiency. Every pound of water vapor that condenses gives off about 1,000 Btu. Nearly 10% of the heat in natural gas flue products is in the water vapor. For propane and oil, it is about 8% and 6½%, respectively. To recover this heat, the boiler must condense.

Consider a natural gas boiler for example. If no condensation occurs, then the combustion efficiency cannot get above 90%. But, lower the boiler water temperature enough to cause condensation, and the combustion efficiency can reach 98%, or even higher.

What causes condensation?

Natural gas, propane and fuel oil contain carbon and hydrogen. Carbon burns to form carbon dioxide. Hydrogen burns to form water vapor. The water vapor in the flue products condenses when exposed to a surface that is colder than the water vapor dewpoint — the same way that water beads on a cold glass.

Gases can hold more water vapor at high temperature than at low temperature. So, as moist gases cool, they reach a temperature at which they cannot hold all of the water vapor, and the vapor begins to condense. This is the dewpoint temperature (also called the saturation temperature).

Notice in Figure 1, Page 7–9 that water condenses on the boiler heat exchanger only where the surface temperature is cooler than the flue gas dewpoint temperature. If the return water temperature is ABOVE the flue gas dewpoint, NO condensate can occur.

Figure 1 Schematic — typical condensing boiler

Typical operation with return water temperature below flue gas dewpoint temperature

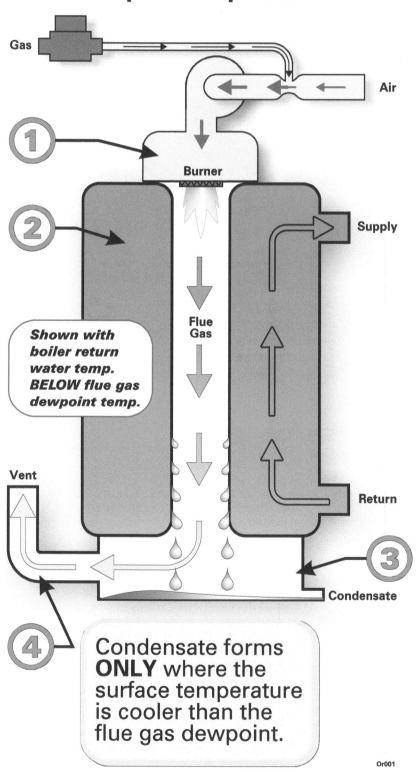

Gas

Air

① Burner

②

Shown with boiler return water temp. BELOW flue gas dewpoint temp.

Flue Gas

Supply

Vent

Return

③

Condensate

④ Condensate forms **ONLY** where the surface temperature is cooler than the flue gas dewpoint.

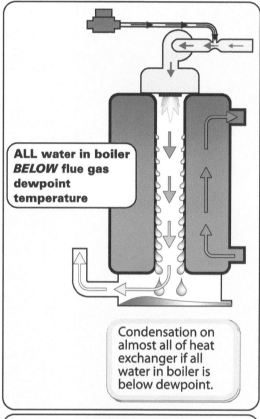

ALL water in boiler *BELOW* flue gas dewpoint temperature

Condensation on almost all of heat exchanger if all water in boiler is below dewpoint.

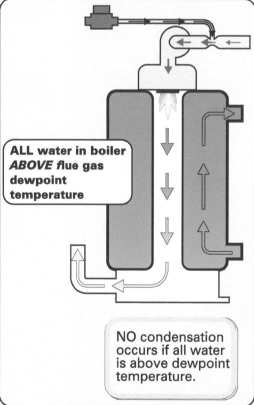

ALL water in boiler *ABOVE* flue gas dewpoint temperature

NO condensation occurs if all water is above dewpoint temperature.

Or001

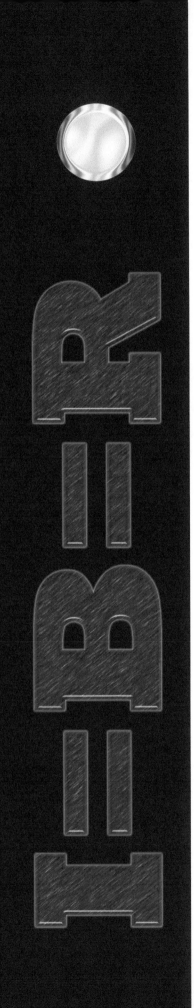

Advantages

Overview

Typical condensing boilers utilize boiler and control designs that result in substantial improvement in efficiency and flexibility compared to conventional boilers. Advantages of condensing boilers include:

- ❏ **Higher efficiency**

- ❏ **Improved comfort**

- ❏ **Improved vent and air control**

- ❏ **Simpler piping**

- ❏ **Lower emissions**

❏ Higher efficiency

Condensing boilers provide higher combustion efficiency and seasonal efficiency due to: heat exchanger design, condensing operation, lower-excess-air combustion, and reduced cyclic and jacket losses.

Condensing operation

Condensing boilers are designed to operate with low water temperatures. Like any heat exchanger, boiler heat transfer depends on the difference in temperature between the boiler water (heated medium) and the flue gases (heating source). The lower the water temperature in the boiler, the higher the efficiency.

To avoid harmful corrosion, conventional boilers must operate with water temperature above the flue gas dewpoint (greater than 130°F usually). This means the flue gases should not be cooled as much as with condensing boilers.

Figure 2, Page 7–11 shows the effect on boiler efficiency as boiler return water temperature drops. Below the dewpoint temperature, heat is taken from the flue gases by cooling them and by condensing flue gas water vapor. The jump in efficiency below the dewpoint temperature is due to the heat gained from condensation.

Figure 2 Typical natural gas-fired condensing boiler combustion efficiency vs return water temperature

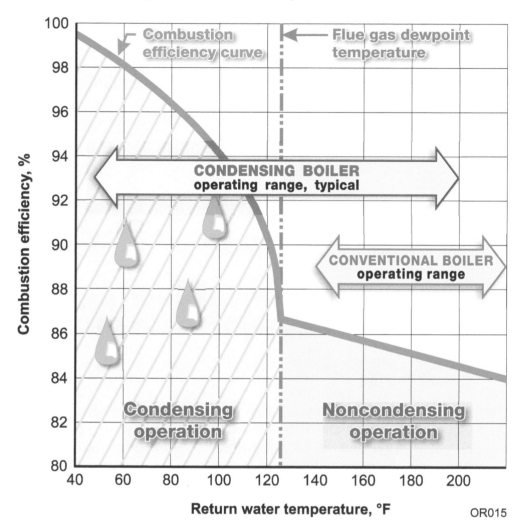

Condensing boilers often have a maximum operating temperature of 200°F or less, as reflected in the operating limit band shown in Figure 2.

Lower excess air combustion

Condensing boilers typically use technologies that allow reliable combustion with low excess air. For example, a premix combustion boiler might operate at 10% CO_2 (19% excess air), compared to 8% CO_2 (46% excess air) for common atmospheric-gas boilers

Reduced excess air means lower flue gas volume, resulting in higher combustion efficiency and lower pressure loss through the boiler flueways.

Reduced cyclic and jacket losses

Typical on/off-fired boilers produce the same amount of heat regardless of the actual demand from the system. When low loads occur, such as during spring and fall, or when only a few zones call for heat, the water temperature rises because the system cannot absorb the heat. This causes the boiler to shut down on high limit. This is actually the most common way for on/off boilers to cycle.

Efficiency definitions

Figure 2 shows how boiler combustion efficiency changes with return water temperature for a typical condensing boiler.

Combustion efficiency reflects how much heat is taken from the flue gases. It is equal to 100% minus the % heat remaining in the flue gases as they leave the boiler. It applies to steady-state operation of the boiler, and does not reflect jacket or cyclic losses.

Thermal efficiency differs from combustion efficiency in that it accounts for jacket losses. Thermal efficiency is determined by measuring the heat gained by water flowing through the boiler and dividing by the energy available in the fuel used.

AFUE (annual fuel utilization efficiency) is a computer-calculated value that estimates how the boiler will perform over a typical heating season. It accounts for cyclic losses, jacket losses and any other factors that cause heat to be lost during the heating season. AFUE is defined by the U. S. Department of Energy, and is determined for each boiler using prescribed test methods.

For a given boiler, combustion efficiency directly affects both thermal efficiency and AFUE. The higher the combustion efficiency, the higher the AFUE or thermal efficiency.

Energy is wasted each time the boiler cycles, and the unnecessarily high water temperatures cause wasteful swings in room temperature.

Many condensing boilers feature modulating firing, allowing them to reduce firing rate during times of light demand. With turndowns of 4 or 5-to-one on these boilers, cycling is significantly reduced, because the boiler can closely match heat input to demand.

The advanced controls used on condensing boilers provide intelligent anticipation of system needs, providing closer temperature control of the supply water and, consequently, reduced swings in room temperature.

Condensing boilers usually have much lower heat loss from the jackets, particularly those with modulating firing. The reduced jacket loss is due to typically smaller boiler size and fully water-jacketed combustion chambers.

❑ Improved comfort

Typical on/off boilers frequently supply too much heat to the system, causing supply water temperature to rise much higher than needed. The high temperature water causes baseboard heaters and other units to put slugs of heat into the room instead of a uniform heat flow. The heat rises to the ceiling, causing stratification in the room. By the time the room thermostat can react, the room temperature has already taken a large rise.

Condensing boilers with modulated firing typically feature electronic controls that monitor how the system responds to heat supplied. The control modulates boiler firing rate and cycling to match heat input to demand. Baseboard heaters and other units receive water at the temperature needed. So heat output to the room is smooth and steady. The room temperature changes gradually, and the room thermostat can more easily control the room.

The result is a steadier room temperature, for greater comfort.

❑ Improved vent and air control

Most condensing boilers allow sealed combustion (direct vent), with air piped from outside to the boiler. The vent system is assembled using stainless steel, pressure-tight vent pipe. (Some boilers allow use of PVC or CPVC pipe.) The result is a brand new vent system rather than an application relying on an old chimney and vent piping.

The vent and air piping can be terminated vertically or through a side wall.

Because the air is piped to the boiler, there is no need to rely on combustion air supplied from inside the building. There is less likelihood of contamination of the air from laundry products and other potentially corrosive household materials.

Flue products are forced through the vent system by the boiler combustion blower. Venting doesn't rely on a chimney to produce draft.

❏ Eliminates conventional boiler low-temp piping

Conventional boilers require by-pass piping or a temperature-regulating valve to protect against low temperature operation. But condensing boilers are made for low-temperature operation. It is not necessary to add piping components and controls to protect the boiler, and would reduce a condensing boiler's efficiency.

Many systems work best with low temperature water (outdoor reset-operated finned-tube baseboard systems, for example). Some radiant systems (in-slab radiant floor, for example) require constant low-temperature water. The water temperature can be controlled directly by the condensing boiler in most cases. No additional components or mixing valves are needed.

The only time additional components are required is when the system requires multiple supply water temperatures. This applies when a radiant floor zone is used on a system that also uses finned-tube baseboard. This situation is easily handled by installing a mixing valve or injection pump for the low-temperature zones.

❏ Reduced emissions

Condensing boilers are environmentally friendly. They reduce the generation of greenhouse gases (CO_2) and typically produce less nitrogen oxides (NO_x).

A side benefit of the higher seasonal efficiency of condensing boilers is reduced emissions of CO_2. Because less fuel is burned, the boiler makes less CO_2 during the year. If the application maximizes the condensing time for the boiler, this can be a substantial reduction.

Condensing boilers typically use advanced combustion designs, such as premix. These burners operate with much lower excess air than conventional boilers, and provide exceptional fuel/air mixing. The result is a major reduction in nitrogen oxide (NO_x) emissions.

Application considerations

Overview

Condensing boiler designs involve new technologies and usually require different or additional installation procedures compared to conventional boilers. Pay close attention to the boiler manufacturer's instructions to ensure a successful, trouble-free installation. Consider the following guidelines.

Replacement boiler installations

Investigate first — Make sure to investigate why the original boiler failed. For example, if it failed due to oxygen corrosion or lime accumulation in the boiler heat exchanger resulting from too much make-up water, you need to find the leak. If you don't, the new boiler will have a very short life. Use the System Survey in Part 13 of the I=B=R Residential Hydronic Heating Guide.

Clean the system piping — Existing systems usually contain large amounts of sludge, iron deposits and lime. If you don't clean the piping thoroughly, this sediment will find its way to the new boiler, causing early failure. Even new piping needs to be cleaned to remove oil and debris before starting the boiler.

Provide water treatment

Most condensing boiler manufacturers require careful attention to the system water chemistry. Read the boiler manual thoroughly, and follow all guidelines for cleaning the system and providing water treatment and/or water softening. Pay attention to the needs of the boiler.

Special venting

Condensing boilers ALWAYS require a special vent system. The vent piping must be pressure-tight and resistant to corrosion from flue gas condensate. Use only the vent components listed in the boiler manuals.

	WARNING Condensing boiler vent systems are pressurized. If you do not completely seal the vent, following all boiler instructions, and using only the materials approved by the boiler manufacturer, flue products can leak into the building, possibly leading to carbon monoxide poisoning.

The only exceptions to pressurized venting are those applications approved by the boiler manufacturer as ANSI Vent category II. This means a vent that relies on gravity flue gas flow, with vent materials and construction suitable for condensing operation. These applications *still require special vent materials*. DO NOT use conventional vent materials, such as Type B vent.

Pay attention to the code requirements (and boiler instructions) for placement of the vent termination. It must be located a minimum distance from windows, doors and building air intakes. And it must not terminate over a public walkway. If the air and vent piping are terminated separately, follow the instructions to place the air termination in the correct orientation and distance from the vent termination.

Combustion air piping

Typical condensing boilers allow either installation using inside air or air piped from outside to the boiler (sealed combustion, or direct vent). Some require only direct vent (sealed combustion).

Combustion air piping must be installed according to the boiler manufacturer's instructions. Pay close attention to the placement of the air intake fitting. It must be located with the correct orientation and spacing from the vent termination.

Ensure clean combustion air

All boilers need clean air for combustion. Air contaminants that contain chlorine or fluorine, in particular, create hydrochloric or hydrofluoric acid in the boiler and vent system, causing serious failures. These contaminants come from laundry or dry cleaning products, paints and other common materials. Follow the boiler manufacturer's instructions to provide uncontaminated air.

Even if combustion air is piped to the boiler (sealed combustion, or direct vent), pay attention to the location of the air intake. It must not be near a swimming pool, dryer exhaust, beauty salon vent, or any other location that could contain corrosion-causing contaminants.

Circulator sizing

Many condensing boilers have a high pressure drop through the boiler compared to conventional boilers. Read the boiler manual for the recommended piping and circulator requirements.

Make sure not to oversize circulators. Larger motors consume more electricity, and will reduce energy savings. What is more, applying a high-head circulator in the wrong application can result in damage to components and operating noise.

Condensate drainage system

All condensing boilers require a condensate drainage system, with at least the installation of condensate tubing. You will need to install a condensate pump if condensate cannot be eliminated by gravity flow.

Install a condensate neutralization system when required by codes or job conditions.

WARNING

Flue gas condensate is acidic. It should not be drained into metallic piping without first being treated with a condensate neutralization system.

Dewpoint

Flue gas dewpoint temperature depends on:

❑ **The percentage of water vapor in combustion products.**

Compared to propane and number 2 fuel oil, *natural gas* has the highest ratio of hydrogen to carbon. It is mostly methane (CH_4), which has a ratio of 4 to 1. So natural gas flue gases contain more water vapor.

Propane (C_3H_8) is the next highest, with a 2.7 to 1 H/C ratio.

Number 2 fuel oil has a lower ratio yet, at about 1.7 to 1 H/C ratio.

❑ **The amount of excess air used in combustion.**

The more excess air, the lower the dewpoint temperature, because the extra air in the flue products dilutes the water vapor. Notice in Figure 3 and Figure 4 how dewpoint temperature drops as excess air increases.

❑ **The amount of moisture in the combustion air.**

Water vapor in the combustion air adds to the water vapor from combustion, slightly increasing the total water vapor in the flue products.

Dewpoint variation between fuels

The higher the amount of hydrogen in the fuel, the more water vapor is made in combustion. Table 1, Page 7–17, and Figure 3, Page 7–17, show how dewpoint temperature varies with fuel and excess air. The dewpoint curve is highest for natural gas, then propane and then fuel oil, the same order as the hydrogen ratio in the fuels. And the higher the excess air, the lower the dewpoint for any fuel.

Dewpoint variation with combustion air humidity

Air brought in for combustion contains water vapor. The higher the humidity, the more water vapor. So dewpoint varies with the amount of moisture brought in with the combustion air. Combustion air moisture can raise the dewpoint temperature as much as 10°F, depending on which fuel and the amount of excess air in combustion.

(NOTICE: A traditional boiler with a draft hood or barometric damper has a lot of excess air in the chimney because of the dilution air added by the draft hood or damper. All the excess air lowers the dewpoint temperature of the flue gases, reducing the likelihood of condensation in the vent system.)

Figure 4, Page 7–17, shows the change in dewpoint temperature with combustion air humidity for propane combustion. Notice that the dewpoint temperature rises as the humidity of the air increases.

Keep condensing boiler return water temperature below the dewpoint as much as possible

Notice that, for a typical natural gas boiler, the dewpoint temperature is about 130 °F. For a typical oil boiler, the dewpoint temperature is about 110 °F. The return water to the boiler has to be lower than these temperatures if condensation is to occur.

Table 1 Dewpoint versus excess air and CO2

Natural gas

CO$_2$ %	Excess air @ 70% relative humidity	Dewpoint, approximate °F
7	65	126
8	46	130
9	31	133
10	19	136
11	9	139

Propane

CO$_2$ %	Excess air @ 70% relative humidity	Dewpoint, approximate °F
8	66	120
9	49	123
10	35	126
11	23	129
12	14	131

No. 2 Fuel oil

CO$_2$ %	Excess air @ 70% relative humidity	Dewpoint, approximate °F
9	69	113
10	53	116
11	39	118
12	28	120
13	19	122
14	11	124

Figure 3 Dewpoint (saturation temperature) versus excess air and CO$_2$

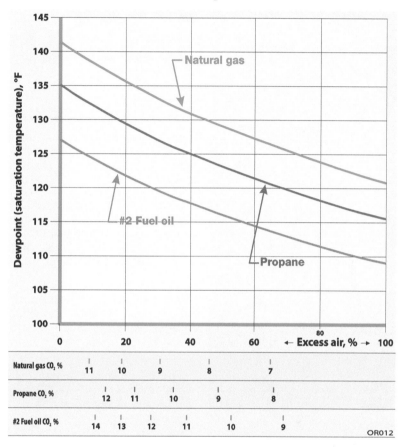

Figure 4 Propane combustion — effect of combustion air humidity on dewpoint temperature

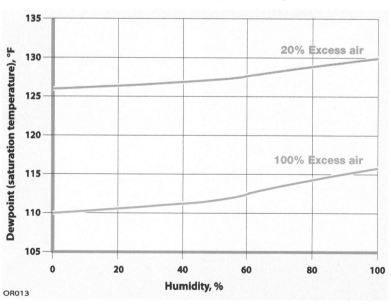

Making the boiler condense

Making a condensing boiler condense

Installing a condensing boiler is just the first step toward fuel savings. To make the most of the condensing boiler, you need to keep the water temperature as low as possible, for as long as possible. For that you need good system design and good controls.

Radiant heating systems, as noted in the following, usually operate at temperatures low enough to ensure sustained condensing-mode operation. For radiant applications such as staple-up, though, the design heating load supply temperature may be as high as 140°F. Outdoor reset may be helpful in these applications to increase condensation time.

The discussion that follows in this Addendum concentrates on how to design and control finned-tube baseboard (or fan-coil and cast iron radiator) systems to ensure the longest possible condensing-mode operation while still meeting the water temperature needs of the terminal units.

> A condensing boiler cannot condense if the boiler return water temperature is above the flue gas dewpoint temperature.
>
> To keep the boiler water temperature low, apply controls that match heat input to heat loss — reducing water temperature as demand reduces. Outdoor reset and indoor reset controls do just that.

Dewpoint temperature (or saturation temperature) depends on fuel, boiler design, excess air and other factors. But a reasonable rule-of-thumb for considering whether flue gas condensation is likely is that the condensing boiler return water temperature must be below:

- Natural gas   130°F.
- Propane   125°F.
- No. 2 fuel oil 115°F.

Radiant heating systems

Radiant heating systems typically operate constantly at temperatures low enough for condensation, particularly in-slab systems, where the water temperature is usually around 100°F.

Radiant heating systems are a natural application for condensing boilers because they operate with sustained low water temperatures, providing maximum condensation, and the greatest possible efficiency.

Some radiant heating systems, such as staple-up installations, require hotter water temperatures. Use outdoor reset to reduce water temperature based on heat demand. This will increase the amount of condensation in the boiler.

Finned-tube baseboard or fan-coil systems

What about finned tube baseboard, fan-coil or cast-iron radiator systems — usually designed for supply water at or above 180°F?

If a system always operates with 180°F supply and a 20 or 40-degree temperature drop, the boiler return water temperature never drops below 140°F, except momentarily during cold-starts. A condensing boiler operated this way would seldom condense, and combustion efficiency would never be very high.

Solution: Use water-temperature reset

Finned-tube baseboard systems only need the hottest water during the coldest weather. To make best use of a condensing boiler, *use a reset control, adjusted so the water temperature is always just hot enough to meet system demand*. This can be an outdoor reset control or a control that senses inside temperature and adjusts boiler firing rate based on system response (indoor reset).

Brookhaven National Laboratory's October, 2004, study[1] estimates that a condensing boiler with outdoor reset could operate in the condensing range 96% of the season for natural gas, 90% of the season for propane, or 80% of the season for #2 fuel oil.

The following pages discuss finned-tube baseboard applications. The same approaches apply to other systems, such as fan-coil and cast-iron radiation systems.

Combined heating/DHW systems

Indirect-fired water heaters usually require water at 180°F or higher. Provide boiler and/or system controls that automatically switch to supply this higher-temperature water to the DHW loop on a call for heat from the water heater.

Reset controls

Outdoor reset

Outdoor reset controls have been available for a long time, but condensing boilers can really take advantage of them.

The idea behind outdoor reset is to lower the supply water temperature as outdoor temperature rises. Lowering supply water temperature reduces terminal unit output — so the heat input to the building drops as the outdoor temperature rises. In other words, the heat input to the building is matched to the heat loss from the building.

This addendum will provide suggestions on how to select and adjust outdoor reset controls. Some condensing boilers include an electronic control that provides an outdoor reset option, usually just by adding an outdoor temperature sensor and adjusting control settings.

If the boiler control does not provide a reset method, obtain an electronic control designed for outdoor reset operation. Controls are now available to not only reset water temperature, but to control system components, such as mixing valves and injection pumps.

Keep in mind that outdoor reset controls can be used with conventional boilers (within operating limits of the boilers), but such applications have limited, if any, effects on boiler efficiency. Outdoor reset on condensing boiler applications optimizes boiler efficiency by ensuring the coolest possible return water temperatures, thus the greatest possible amount of flue gas condensation.

Indoor reset

Some electronic controls now monitor indoor response to heating and adjust water supply water temperature and boiler firing rate accordingly. The controls look at response of the system as heat is applied to determine how high to fire the boiler (or boilers, if connected as a multiple boiler system).

---

[1] Brookhaven National Laboratories, BNL-73314-2004-IR, "Hydronic Baseboard Thermal Distribution System with Outdoor Reset control to enable the Use of a Condensing Boiler"

Finned-tube baseboard

Baseboard sizing

See Part 3 of the Residential Hydronic Heating Guide for recommended sizing procedures for new installations. Even if you aren't installing new baseboard, you'll want to consider the existing baseboard sizing in order to make the best decisions when installing a condensing boiler with reset controls. See "How oversized is the existing finned-tube baseboard?," page 22.

Some condensing boilers have a maximum outlet water temperature of 200°F or less. For applications like this, use a design supply water temperature no higher than the boiler maximum to be sure the boiler and system can supply hot enough water to the baseboard to meet peak demand. *Use the lowest possible design water temperature to ensure the most possible condensing-mode operation.*

See Table 2, reproduced from a typical baseboard rating chart. Notice that the output depends heavily on the water temperature. Figure 5, Page 7–21 shows how quickly output of finned-tube baseboard drops as average water temperature is reduced. Be sure to consider the water temperature available to the heater when sizing.

Increasing flow rate above 1 gpm will slightly increase output. So ratings are given in ratings tables for both 1 gpm and 4 gpm flow rates. Use the 1-gpm ratings unless you are sure the flow will be at least 4 gpm. Then use the 4 gpm data.

The temperatures in the ratings chart are average water temperatures in the baseboard, based on a 20°F temperature drop. Add 10°F to the temperature shown to determine the required entering water temperature (assuming a 20°F temperature

Table 2 Output ratings of typical finned-tube baseboard, ¾" and ¼"

Btuh per foot of baseboard for average water temperatures of: (Performance based on 65°F entering air)										
Element	Flow	140°F	150°F	160°F	170°F	180°F	190°F	200°F	210°F	220°F
Type A ¾" element	1 gpm	290	350	420	480	550	620	680	750	820
	4 gpm	310	370	440	510	580	660	720	790	870
Type B ½" element	1 gpm	310	370	430	490	550	610	680	740	800
	4 gpm	330	390	450	520	580	640	720	780	850

NOTE: Ratings are for element installed with damper open, with expansion cradles. Ratings are based on active finned length (5" to 6" less than overall length) and include 15% heating effect factor. Use 4 gpm ratings only when flow is known to be equal to or greater than 4 gpm; otherwise, 1 gpm ratings must be used.

drop through the heater).

Output versus water temperature

See the finned-tube baseboard cutaway on the opposite page. When hot water runs through the copper tube, it heats the tube and the attached fins. Air between the fins heats up. Because it is lighter when heated, the air rises through the fins and out the top of the baseboard enclosure. The hotter the air, the faster it rises, like a hot-air balloon. The hotter the water in the tube, the hotter the fins and the hotter the surrounding air.

The heat transferred to the air moving through the finned-tube baseboard depends on the temperature difference between the fins and the incoming room air. But it also depends on how fast the air moves.

When the water temperature is lower, the temperature difference between the fins and the air is lower. So less heat is transferred because of temperature difference. In addition, the air rises more slowly because it is not as hot. The slower air movement reduces heat transfer even more. The double effects of temperature difference and air flow cause the baseboard output line (Figure 5) to have a downward hump.

The straight line in Figure 5 is what the output of the baseboard would be if the only factor was the temperature difference between the room air and the fins. The red line is the actual output.

This is why you need to be careful when setting the operation of an outdoor reset control. Traditional outdoor reset controls assume heating unit output changes in a straight line. But the output of finned-tube baseboard drops faster than a straight line as water temperature is lowered. Apply a reset control that either:

• Adjusts supply temperature so the output from the baseboard is a straight line (i.e., a characterized reset curve), . . .

• Has an automatic adjustment (such as temperature boost, to increase supply temperature if the building isn't heating quickly enough), or . . .

• Monitors the system response to heat input and adjusts supply temperature accordingly (such as indoor reset or outdoor reset with an indoor sensor).

Condensing boilers with integral electronic reset controls generally provide one or more of these options. You just need to set the control accordingly.

See the following pages for more details about reset controls.

Output versus flow rate

Make sure your piping and components will always provide at least 1 gpm. As you can see in Figure 5, flow rates below 1 gpm cause rapid fall-off in performance. Consider this carefully if the system will include variable-speed pumping.

Consider also that pressure drops through the water side of some condensing boilers are substantially higher than traditional boilers. Select your circulators carefully, considering this pressure drop where applicable.

Figure 5 Finned-tube baseboard output vs entering temperature (compared to ratings for 210°F entering water)

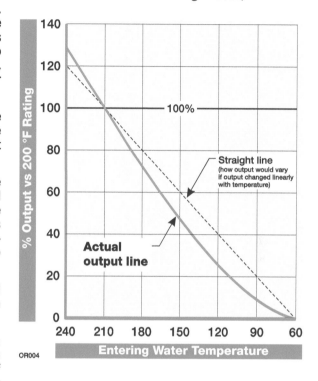

OR004

Figure 6 Finned-tube baseboard output vs flow rate changes (compared to 1 GPM ratings)

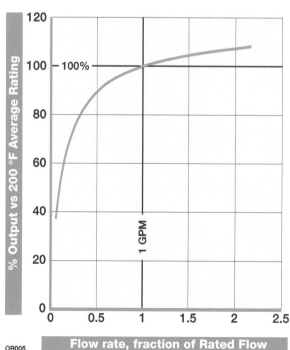

OR005

How oversized is the existing finned-tube baseboard?

You can determine the required supply water temperature to meet the heating demands at the outdoor design temperature (ODT) by measuring the length of the existing baseboard and using the baseboard sizing table backwards, from output to temperature.

❏ Do a heat loss

Knowing the heat loss of each room will allow you to make the best decision on the required supply water temperature at peak load.

Room-by-room heat loss

Determine the heat loss for each room, using available software packages, I=B=R Guide H-22, or other means.

Whole-house heat loss

Estimate room-by-room heat loss by first doing a whole-house heat loss (Hydronics Institute Form 1504WH), then determining room losses as a fraction of the whole house loss.

❏ Determine output per foot

Record room-by-room losses

Write down the heat loss for each room.

Measure lengths of baseboard

Measure and write down the finned lengths of baseboard (feet) in each room next to the heat loss for the room.

Use the highest number

Find the highest output per foot required for any room. You will use this rating to determine the water temperature required at peak load.

What if there is a big difference?

If one or more of the rooms has a much higher output per foot required than the other rooms, consider adding additional baseboard or replacing existing baseboard with new baseboard, and more feet, so the requirements for each room are similar.

Calculate output required

For each room, divide the heat loss by the total feet of finned baseboard in the room. The answer is the required output in Btuh per foot.

☐ Find required temperature

Use the ratings table supplied by the baseboard manufacturer, if available. If not, use Table 3.

For the size of baseboard used (½" or ¾"), scan across the 1 gpm row until you reach an output per foot just higher than what is required. (If you are sure the flow will be at least 4 gpm, you can use the 4 gpm row instead of the 1 gpm row.)

Table 3   Default output ratings

Btuh per foot of finned-tube baseboard for average water temperatures, of:								
¾" Element								
Flow	**140 °F**	**150 °F**	**160 °F**	**170 °F**	**180 °F**	**190 °F**	**200 °F**	**210 °F**	**220 °F**
1 gpm	290	350	420	480	550	620	680	750	820
4 gpm	310	370	440	510	580	660	720	790	870

½" Element								
Flow	**140 °F**	**150 °F**	**160 °F**	**170 °F**	**180 °F**	**190 °F**	**200 °F**	**210 °F**	**220 °F**
1 gpm	310	370	430	490	550	610	680	740	800
4 gpm	330	390	450	520	580	640	720	780	850

Example

Room	Heat loss, Btuh	Feet of ¾" baseboard	Output per foot
Living room	4,272	10	427
Family room	4,400	10	440
Dining room	3,080	7	440
Bath 1	1,529	4	382
Bath 2	235	2	118
Kitchen	4,500	10	450
Bedroom 1	4,164	9	463
Bedroom 2	3,860	9	429
Bedroom 3	2,531	6	422
Basement	18,151	40	454

The highest output per foot of baseboard required is 463 Btuh per foot (Bedroom 1).

In Table 3, read across the 1 gpm row for ¾" baseboard. The first number just higher than 463 is 480, in the 170°F average temperature column. Supply water temperature would be 180°F (170 + 20/2 for a 20°F temperature drop).

Set the reset control to provide 180°F supply water to the baseboard when the outside temperature equals the ODT for the area.

Reset controls

Reset strategy

Heat loss from a room is proportional to the difference between outside temperature and inside temperature. Heat loss or heat gain is also affected by wind speed, solar exposure and other factors. But, for most purposes of estimating heat loss from the house, heat loss is proportional to the difference in temperature between inside and outside. (See I=B=R Guide H-22 for more details.)

Reset controls attempt to match the heat input to the space with the heat loss from the space. Because the heat loss from the building is proportional to the difference between inside and outside temperatures, traditional reset controls change the supply water temperature proportional to the outdoor temperature change.

Another method of matching heat input to heat loss is to reduce the maximum boiler firing rate based on the demand sensed by the control. The terminal units (finned-tube baseboard or other) use available heat from the water, causing the water temperature to drop.

Whether the water temperature is controlled or the boiler input is limited, these strategies reduce heat input to the system as outdoor temperature rises, or increase input as outdoor temperature decreases.

Advantages of matching heat input to heat loss

More constant room temperature

Matching the room heat loss with the heat input by finned-tube baseboard units reduces temperature swings in the room.

- Putting heat out of the baseboard too fast causes the wall above the baseboard to heat more than it needs to. This actually increases heat loss to outside because the wall temperature is hotter.

- Bursts of heat cause stacking of hot air near the ceiling. Room temperature swings above the thermostat setting before the thermostat can shut off the heat.

- With cooler water in the baseboard, the air is warmed less and rises less rapidly. The wall temperatures remains cooler and stacking of hot air is less likely.

- Room temperature changes are slower, and the thermostat can react more effectively.

The circulator will run longer

- Room thermostats will call for heat longer because the room heats more slowly. Most boilers are designed to run the circulator as long as there is a call for heat. The boiler will cycle the burner on an off to control outlet water temperature, but continue to run the circulator as long as the thermostat contact is closed.

- Most boilers today cycle more often on the limit control than on the thermostat, because they put out more heat than one zone or a few zones can use during a heat call.

- Because the circulator runs longer, the heat in the pipes and the boiler spreads more evenly throughout the building.

Less cycling means higher efficiency

- Studies of conventional heating systems operated with outdoor reset have shown improved seasonal efficiency due to reduced boiler cycling and better use of residual heat in the boiler and piping.

Outdoor reset

Figure 7 is a typical outdoor reset curve. It shows how supply water temperature is changed as the outdoor temperature changes. Most reset controls are set by adjusting the four temperature values shown in Figure 7.

- **T1** — Supply water temperature at design outdoor temperature.

   This is the water supply temperature required when the outside temperature is at the ODT (outside design temperature). In other words, when water at this temperature is supplied to the heating units (finned-tube baseboard, for example), the heat output will meet the heat loss at the design minimum outside temperature.

- **T2** — Supply water temperature for zero heat output.

   This is usually the design room temperature. If water is supplied at room temperature, no heat is output from the heating units. In the example of Figure 7, with a supply temperature of 70°F, there would be no heat transfer to the space because the room temperature is also 70°F.

- **T3** — This is the ODT, or outdoor design temperature.

   Heat loss calculations are done for this outdoor temperature, and the boiler and terminal units are sized as required to meet the heat loss.

- **T4** — The outside temperature at which no heat loss occurs.

   This is usually the balance point temperature for the building. At this outside temperature, heat gain (solar, etc.) is enough to match the heat loss from the building. In the example of Figure 7, T4 is 60°F for a room temperature of 70°F.

Reset ratio — This is the slope of the reset curve

Reset ratio gives the required rise in supply water temperature for a 1°F drop in outside temperature. See Figure 7 inset. Some reset controls prompt for reset ratio instead of inputting all four temperatures above. To calculate reset ratio:

- Find the *water temperature change* by subtracting the minimum water temperature (usually 70°F) from the water temperature at outdoor design temperature.

- The *outdoor temperature change* is the difference between the outdoor design temperature (ODT) and the zero-heat outdoor temperature.

- The reset ratio is the *water temperature change* divided by the *outdoor temperature change*.

The reset ratio for Figure 7 is 1.8, as shown. In this example, the baseboard is sized for a supply water temperature of 180°F. At a 20°F temperature drop the average temperature in the baseboard

Figure 7   Outdoor reset curve, typical

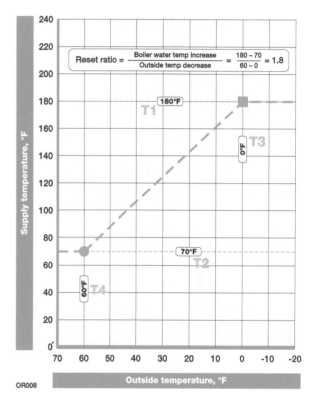

$$\text{Reset ratio} = \frac{\text{Boiler water temp increase}}{\text{Outside temp decrease}} = \frac{180 - 70}{60 - 0} = 1.8$$

OR008

Figure 8   Linear reset curve examples

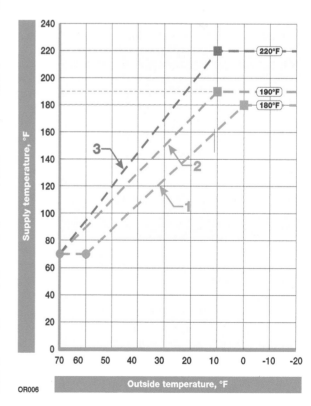

OR006

would be 170°F (180-20/2). You would use the 170°F column to size the finned-tube baseboard using the baseboard manufacturer's ratings, or Table 3, page 23. If the baseboard sizing requires hotter water, set the reset curve to supply at least this water temperature at the ODT.

Figure 8, page 25, shows other example reset curves. The top curve requires 200°F water at the ODT, which would be an average temperature in the baseboard of 190°F (220-20/2). The average water temperatures for the other curves would be 180°F for the center curve, and 170°F for the lower curve.

Reset with finned-tube baseboard

In "Output versus water temperature," page 7–21, you will see that finned-tube baseboard does not have a straight-line curve of heat output vs water temperature.

Linear reset issues

Figure 9 shows the problem of applying conventional linear reset to finned-tube baseboard. The linear reset curve (purple dotted line) would always supply water colder than required, because the required water temperature line (red) is always above the purple line.

For the example shown, a linear reset curve would have to be as in the blue dotted line in order to always supply hot enough water. The reset control would never attempt to supply water colder than 140°F.

Compensation for linear curve issues

A linear reset curve will probably pose problems with heat output from the baseboard unless the reset control includes compensating features. One option is called "temperature boost." With temperature boost, the reset control increases the supply water temperature if the call for heat sustains for a long period, say 15 minutes for example. At the end of this interval, the reset control increase supply water temperature by a specified value, and continues to apply this increase for each additional interval the call for heat continues. This way, the reset control automatically adapts the needed water temperature.

Characterized reset curves

The solution is for the reset control to adapt the reset curve to the needs of the heating unit. This is done by control to a curve shaped accordingly, that is a characterized curve. Figure 10 shows examples of characterized curves. The control is still set up by inputting the four temperature values shown in Figure 7, Page 7–25, but the control resets water temperature following a curved line instead of a straight one. Reset controls with characterized curves provide a set-up selection for the type of heating unit to ensure the curve shape is appropriate.

Figure 9 Issues affecting outdoor reset with finned-tube baseboard

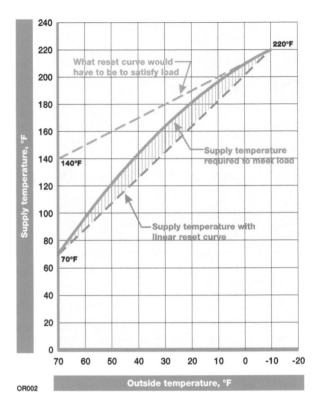

Figure 10 Characterized reset curves

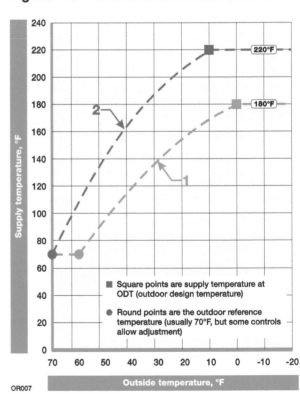

Indoor reset

Another method of matching heat input at the heating units to heat loss from the building is indoor reset. In this method, the boiler firing rate is limited based on the reset control's monitoring of indoor temperature and the response of indoor temperature to changes in the heat supplied. Water temperature is inherently regulated because the heating units "soak off" available heat, automatically reducing water temperature.

Summary

When applying reset to finned-tube baseboard, or any heating unit with a non-linear heat output curve (radiators, fan-coil units, etc.), select a reset control that compensates for the heat response of the baseboard. If operating with outdoor reset, the control should include automatic compensation or characterization.

Common terms

Reset — Supply water temperature, boiler firing rate and/or flow rate are adjusted to reduce heat input to the system based on demand.

Outdoor reset — The supply water temperature is lowered as the outside air temperature increases, as in Figure 7, Page 7–25.

Indoor reset — Indoor temperature is monitored (directly or via thermostats) and supply temperature, flow rate and/ or boiler firing rate are adjusted based on how much and how fast the system changes as heat is added to the space.

Proportional reset — This is applied in a system requiring multiple supply temperatures (such as a mixed system with finned-tube baseboard in some zones and radiant floor heating in others). It means the use of a fixed-position mixing valve supplying the low-temperature portion of the system, in lieu of installing a more expensive automatic mixing valve. The reset is proportional because the temperature drop in the low-temperature system is proportional to the drop in the high-temperature system.

Offset or parallel shift — The control's reset curve is adjusted upward or downward to account for building heat gain (downward shift) or increased heat loss due to wind or other factors (upward shift).

Temperature boost — The control increases supply water temperature if the call for heat extends longer than a given time.

Characterized curve — The reset curve is shaped in the same way as the heat output curve for the terminal units rather than resetting on a straight-line curve. See Figure 10, Page 7–26.

Boiler reset and mixing reset — Boiler reset means the supply temperature is controlled directly by the boiler. Mixing reset means the supply temperature is controlled by a mixing device, such as a mixing valve or injection pumping system.

Outdoor design temperature — The outside temperature assumed when calculating the required heating load of the building.

Balance point temperature — The outside temperature at which the heat loss from the building matches the heat gain (from solar or other sources of heat).

Application examples

> **WARNING** Follow all instructions provided by the boiler manufacturer when installing and piping the boiler. Failure to comply with the instructions could result in severe personal injury or premature failure of the boiler and/or system components.

Application to fan-coil and other systems

The application examples shown in the following pages for finned-tube baseboard can be applied to fan-coil and other systems as well.

Piping for high pressure-drop condensing boilers

Some condensing boilers utilize heat exchangers that may require larger pumps to handle the pressure loss. For these applications, boiler manufacturers often recommend using primary/secondary connections for the boiler. This isolates the high pressure drop component from the system. The suggested piping diagrams on the following pages show the use of primary/secondary connections for boilers rated as "high pressure drop." For other boilers, you can choose either of the piping alternatives shown for each application. Always follow the boiler manufacturer's instructions for piping.

Radiant space heating only (no DHW)

For most applications using only one type of radiant heating, no special piping should be necessary. Follow the boiler manufacturer's or tubing manufacturer's design guidelines. See Figure 11, Page 7–29 for applications other than high pressure drop boilers. For high pressure drop boilers, use the method of Figure 12, Page 7–29 instead. Notice that no special piping is needed (and should not be used) to regulate boiler return water, because the boilers are designed for condensing operation.

Radiant space heating with DHW

Figure 13, Page 7–29 shows piping for a radiant space heating with DHW heating. The DHW tank and circulator connect directly off of the boiler to minimize heat loss from the piping during summertime operation. The radiant heating circulators should be disabled automatically on a call for DHW, because the operating temperature will be too high for most radiant applications (typically 180°F or higher). Alternatively, you would need to provide automatic temperature regulation to the radiant loops, such as with 3-way or 4-way valves, or injection mixing.

Figure 14, Page 7–29 shows DHW connection for a high pressure drop boiler (or when primary/secondary is desired). The DHW tank is piped as a secondary loop off of the boiler piping so the DHW circulator does not have to overcome the boiler pressure drop. (If pressure drop is not an issue, you can connect the DHW tank across the boiler as in Figure 13, Page 7–29.) The boiler loop circulator must operate during DHW operation. The radiant heating circulators should be disabled automatically on a call for DHW, because the operating temperature will be too high for most radiant applications (typically 180°F or higher). Alternatively, you would need to provide automatic temperature regulation to the radiant loops, such as with 3-way or 4-way valves, or injection mixing.

Figure 11 Radiant heating system
Multiple manifolds
Space heating only
Low pressure drop boiler

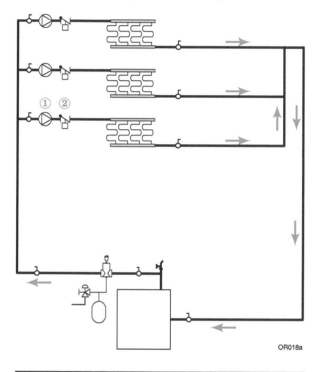

OR018a

Figure 12 Radiant heating system
Multiple manifolds
Space heating + DHW
High pressure drop boiler

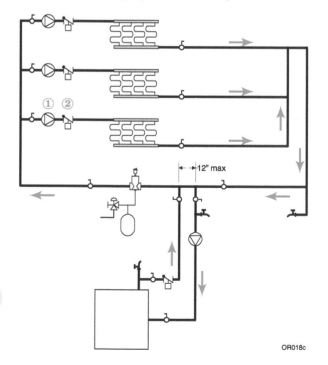

12" max

OR018c

Figure 13 Radiant heating system
Multiple manifolds
Space heating + DHW loop
Low pressure drop boiler

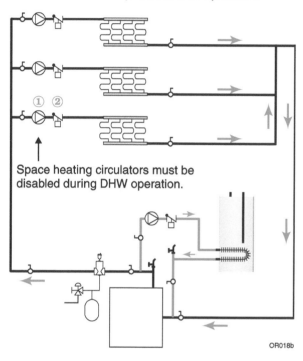

Space heating circulators must be
disabled during DHW operation.

OR018b

Figure 14 Radiant heating system
Multiple manifolds
Space heating + DHW loop
High pressure drop boiler

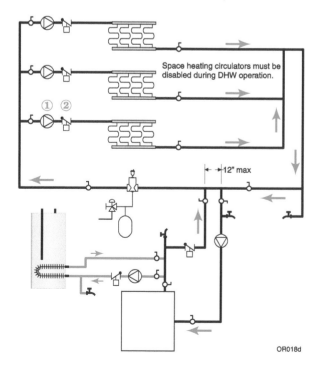

Space heating circulators must be
disabled during DHW operation.

12" max

OR018d

Baseboard space heating & DHW — zone valve zoning

For combined space heating and DHW systems, follow the suggested piping in Figure 15, Page 7–31. For high pressure drop boilers, or when primary/secondary piping is desired, apply Figure 16, page 31. (Disregard the DHW piping for space heating-only applications.)

- Low pressure drop boilers (Figure 15, Page 7–31) — The DHW circuit should be piped directly across the boiler, as shown, so the DHW system can operate in summer months without having to circulate the space heating system.

- High pressure drop boilers (Figure 16, Page 7–31) — The DHW tank is piped as a secondary loop off of the boiler piping so the DHW circulator does not have to overcome the boiler pressure drop. (If pressure drop is not an issue, you can connect the DHW tank across the boiler as in Figure 15, Page 7–31.) The boiler loop circulator must operate during DHW operation.

- Install flow/check valves (item 2) as shown to prevent unwanted circulation in the DHW or space heating loops.

- If the space heating system is operating with reset, the best results are likely by enabling domestic priority when setting the controls. With domestic priority, the space heating circuits are deactivated during a DHW call (unless the duration exceeds a preset time limit for some controls).

Adding a radiant space heating zone

When connecting a radiant heating zone to a finned-tube baseboard system, you can apply the piping options shown in Figure 17, Page 7–31. For high pressure drop boilers, or when primary/secondary piping is desired, apply Figure 18, Page 7–31. (Disregard the DHW piping for space heating-only applications.)

- Low pressure drop boilers (Figure 17, Page 7–31) — The DHW circuit should be piped directly across the boiler, as shown, so the DHW system can operate in summer months without having to circulate the space heating system.

- High pressure drop boilers (Figure 18, Page 7–31) — The DHW tank is piped as a secondary loop off of the boiler piping so the DHW circulator does not have to overcome the boiler pressure drop. (If pressure drop is not an issue, you can connect the DHW tank across the boiler as in Figure 17, Page 7–31.) The boiler loop circulator must operate during DHW operation.

- Install flow/check valves (item 2) as shown to prevent unwanted circulation in the DHW or space heating loops.

- If the space heating system is operating with reset, the best results are likely by enabling domestic priority when setting the controls. With domestic priority, the space heating circuits are deactivated during a DHW call (unless the duration exceeds a preset time limit for some controls).

- The radiant zone (item 3) is fed through a mixing valve (item 5). Operate the mixing valve using an electronic controller or self-contained regulation. Item 4 is the radiant zone circulator. Place the radiant zone take-off on a parallel circuit, as shown. Use a primary/secondary type take-off (item 6) particularly if the radiant zone circulator is a high-head type. A high-head circulator piped directly into the baseboard piping, without the secondary isolation, could cause flow check valve lifting in the baseboard zones.

Figure 15 Finned-tube baseboard
Space heating + DHW
Low pressure drop boiler

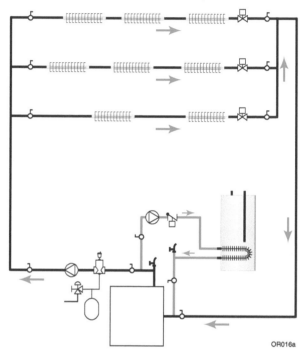

OR016a

Figure 17 Finned-tube baseboard
Space htg. + DHW + radiant
Low pressure drop boiler

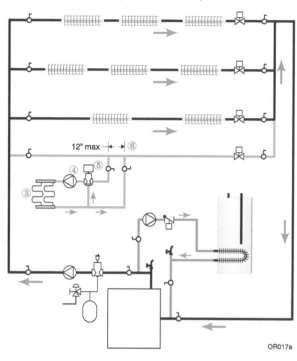

OR017a

Figure 16 Finned-tube baseboard
Space heating + DHW
High pressure drop boiler

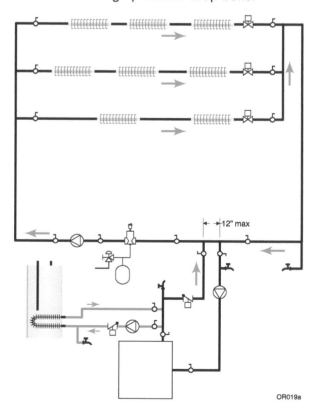

OR019a

Figure 18 Finned-tube baseboard
Space htg. + DHW + radiant
High pressure drop boiler

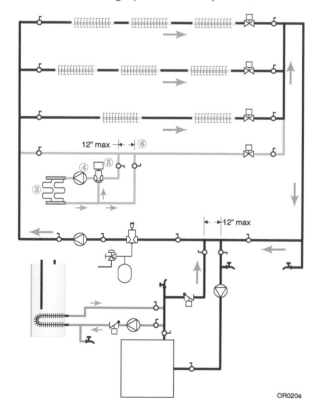

OR020a

Baseboard space heating & DHW — circulator zoning

For combined space heating and DHW systems, follow the suggested piping in Figure 19, Page 7–33. For high pressure drop boilers, or when primary/secondary piping is desired, apply Figure 20, Page 7–33. (Disregard the DHW piping for space heating-only applications.)

- Low pressure drop boilers (Figure 19, Page 7–33) — The DHW circuit should be piped directly across the boiler, as shown, so the DHW system can operate in summer months without having to circulate the space heating system.

- High pressure drop boilers (Figure 20, Page 7–33) — The DHW tank is piped as a secondary loop off of the boiler piping so the DHW circulator does not have to overcome the boiler pressure drop. (If pressure drop is not an issue, you can connect the DHW tank across the boiler as in Figure 19, Page 7–33.) The boiler loop circulator must operate during DHW operation.

- Zone circulators in the illustrations are identified as item 1.

- Install flow/check valves (item 2) as shown to prevent unwanted circulation in the DHW or space heating loops.

- If the space heating system is operating with reset, the best results are likely by enabling domestic priority when setting the controls. With domestic priority, the space heating circuits are deactivated during a DHW call (unless the duration exceeds a preset time limit for some controls).

Adding a radiant space heating zone

When connecting a radiant heating zone to a finned-tube baseboard system, you can apply the piping options shown in Figure 21, Page 7–33. For high pressure drop boilers, or when primary/secondary piping is desired, apply Figure 22, Page 7–33. (Disregard the DHW piping for space heating-only applications.)

- Low pressure drop boilers (Figure 21, Page 7–33) — The DHW circuit should be piped directly across the boiler, as shown, so the DHW system can operate in summer months without having to circulate the space heating system.
- High pressure drop boilers (Figure 22, Page 7–33) — The DHW tank is piped as a secondary loop off of the boiler piping so the DHW circulator does not have to overcome the boiler pressure drop. (If pressure drop is not an issue, you can connect the DHW tank across the boiler as in Figure 21, Page 7–33.) The boiler loop circulator must operate during DHW operation.
- Zone circulators in the illustrations are identified as item 1.
- Install flow/check valves (item 2) as shown to prevent unwanted circulation in the DHW or space heating loops.
- If the space heating system is operating with reset, the best results are likely by enabling domestic priority when setting the controls. With domestic priority, the space heating circuits are deactivated during a DHW call (unless the duration exceeds a preset time limit for some controls).
- The radiant zone (item 3) is fed through a mixing valve (item 5). Operate the mixing valve using an electronic controller or self-contained regulation. Item 4 is the radiant zone circulator. Place the radiant zone take-off on a parallel circuit, as shown. Use a primary/secondary type take-off (item 6) particularly if the radiant zone circulator is a high-head type. A high-head circulator piped directly into the baseboard piping, without the secondary isolation, could cause flow check valve lifting in the baseboard zones.

Figure 19   Finned-tube baseboard
Space heating + DHW
Low pressure drop boiler

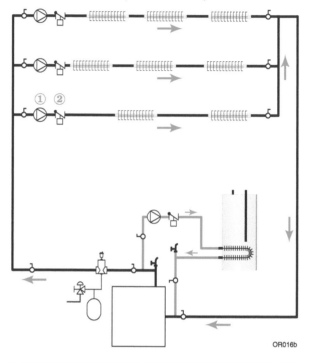

OR016b

Figure 20   Finned-tube baseboard
Space heating + DHW
High pressure drop boiler

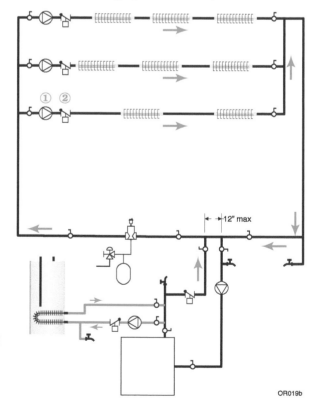

OR019b

Figure 21   Finned-tube baseboard
Space htg. + DHW + radiant
Low pressure drop boiler

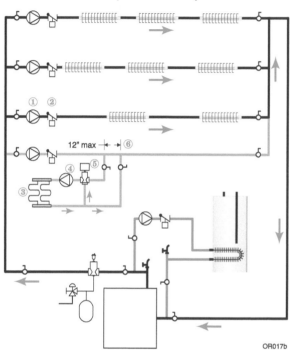

OR017b

Figure 22   Finned-tube baseboard
Space htg. + DHW + radiant
High pressure drop boiler

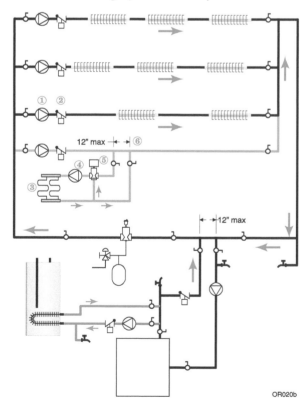

OR020b

Hydronics Institute Section of AHRI

35 Russo Place

Berkeley Heights, NJ 07922-0218

part 8

Radiant heating basics

I=B=R
Guide
RHH

Residential Hydronic Heating . . .

Installation & Design

Hydronics Institute
Section of AHRI

I=B=R Guide RHH
Residential Hydronic Heating

Hydronics Institute Section of **AHRI**
35 Russo Place
Berkeley Heights, NJ 07922-0218

Contents – Part 8

Contents – *continued*

Contents – *continued*

Illustrations

Background

Overview

Part 8 provides an introduction to space heating with radiant panels. For more detailed information on radiant heating system design and application, see Hydronics Institute Guide 400, "Radiant Floor Heating," and John Siegenthaler's book, "Modern Hydronic Heating." You will also find publications available from the Radiant Panel Association, available for order "from their website." Many manufacturers of radiant heating equipment and controls provide comprehensive information, both in hard copy and as software.

How radiant panels work

What is a radiant panel?

A radiant panel is any surface (floor, wall, ceiling or other surface) heated above room temperature that relies on heat transfer mostly due to radiant heat.

How radiant heating works

In Part 1, we discussed the three ways heat moves — conduction, convection, and radiation. Conventional heating systems rely mostly on convection (moving air to distribute heat). Radiant heating systems rely mostly on radiant heat transfer.

Radiant heat transfer occurs when radiant energy leaves a warm surface and strikes a cooler surface. For radiant heat transfer to be effective, the warm surface needs to be very large or very hot. Radiant panel heating uses a floor, wall or ceiling as the radiator — hence, a large surface. (An example of a radiating surface that is small, but very hot, is an overhead-mounted radiant heater.)

With the room temperature at 70°F, a floor surface at 85°F will emit about 30 Btuh per square foot of surface, most of this (50% to 75%) by radiating heat to the walls and objects in the room. Radiant heat doesn't heat air. It heats surfaces. Radiant heat passes through the air almost without touching it.

The design temperature for radiant floors is usually 85°F in most areas. This is ideal for the human body since the body surface temperature is 85°F. Figure 1, Page 8–7 shows the "ideal" room temperature distribution curve. Our bodies are most comfortable when the lower third of the body is in a warmer environment than the upper portion, preferring 85°F at our feet and 68°F at our heads.

Figure 1, **B**, is a typical temperature profile in a room heated with a radiant floor. This curve is very close to the "ideal" profile of Figure 1, **A**. To see why radiant heating is so effective, compare the radiant floor heating profile of Figure 1, **B**, with the profile typical of residential warm-air heating in Figure 1, **C**. The warm-air heating system tends to stack heat near the ceiling because warm air rises. Radiant floors don't rely as much on air movement, so the stack effect is much less.

Is the floor always at 85°F?

No.

The floor is the heating unit in a radiant heating system. The hotter the floor surface, the greater the heat it gives off. The heat given off by the floor is about proportional to the temperature difference between the room air and the floor (at a rate of about 2 Btuh per square foot per each 1°F difference in temperature). At 85°F floor temperature with

Figure 1 Comparison of ideal floor-to-ceiling temperature profile with profile resulting from radiant floor heating and forced air heating

A Ideal profile

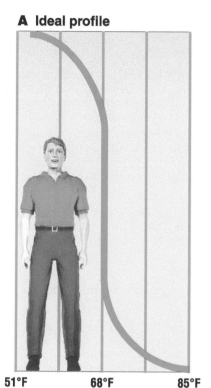

51°F   68°F   85°F

B Radiant floor profile

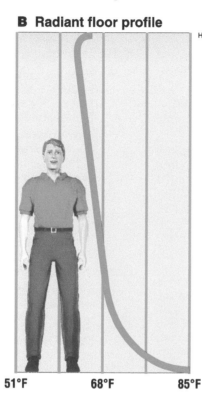

HI21101

51°F   68°F   85°F

C Forced air heating profile

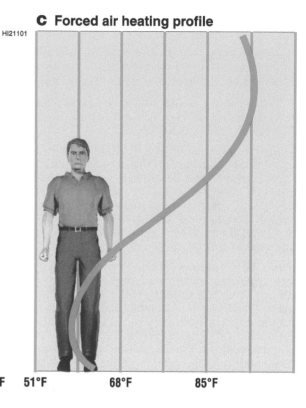

51°F   68°F   85°F

70°F in the room, the temperature difference is 15°F. The heat output is 30 Btuh per square foot of floor surface.

You want the heat output from the heat exchanger (floor) to match the heat lost from the space. The only time you need 30 Btuh per square foot is for design conditions — when the outdoor temperature is at the design low temperature. The average floor temperature will change with the outdoor temperature if the heating system is working correctly. Even if only a room thermostat is used, the time average heat output from the floor will match the time average heat loss from the space.

Mean radiant temperature

So, if the floor temperature gets cooler as the outdoor temperature rises, why is the occupant still comfortable? The occupant stays comfortable because of "mean radiant temperature" — the average temperature of all surfaces in the space.

In January, when outdoor temperature is at or near design point, the average floor temperature will be 85°F, but the outside wall of the space will be cooler than room temperature because of the large heat loss to the outside. The warm floor will offset this cool outside wall so the mean radiant temperature for the room is comfortable. The occupant will lose heat by radiation to the outside wall, but won't lose heat to the floor, even if bare-footed.

In Fall and Spring, the heat loss from the space is low, so the average temperature of the floor will be closer to 70°F (if the inside design temperature is 70°F). But the outside wall be will warmer than in January because the outside temperature is higher. The mean radiant temperature in the room will still be comfortable.

Our bodies give off radiant heat (making us feel chilled) to any surface colder than our body surface temperature of 85°F — even if we are in a warm room. The key to comfortable heating is to warm the surfaces of the room (floor, walls, ceiling and objects) to minimize the radiant heat loss from our bodies.

Our bodies need to give off heat, typically about 500 Btuh. If we can't shed the heat our bodies make, we become hot. If we shed more heat than our bodies make, we feel cold. The key to a comfortable heating system is to control the space temperatures (wall, floor, ceiling and air) so we only give off the right amount of heat.

Warming the radiant panel

Radiant panel systems currently use mostly plastic or elastomeric tubing embedded in or attached to the radiant panel. Copper tubing can be used as well. The system circulates hot water from a boiler or other heat source through the tubing. The water in the tubing gives off its heat to the surrounding flooring materials.

Thermal mass

When you read about radiant heating, you will encounter the term, "thermal mass." The thermal mass of a system refers to how much mass has to be heated in order to heat the finished flooring surface to the desired temperature.

- If the tubing is embedded in a concrete slab floor, the system has high thermal mass because of the mass of all the concrete plus any flooring on top.

- If the tubing is attached to a suspended wood floor, the system will be low thermal mass because the only mass is that of the flooring.

Thermal mass acts like a flywheel. The more mass, the more heat it stores, and the longer it takes to change the floor temperature. High thermal mass tends to cause smooth changes in inside temperature. On the other hand, high thermal mass makes it difficult for the heating system to respond quickly when the user changes the thermostat setting or when the heat gain or loss in the space changes quickly.

If you want a slab-on-grade home heating system to provide quick response, for example, consider mounting the tubing on top of the slab rather than in the slab, with at least ½ inch of polystyrene placed on the slab first. The tubing can be installed this way using either a thin slab (concrete or gypsum underlayment) or an above-slab sleeper system. The insulation on the slab prevents the main slab from being Part of the thermal mass, limiting the thermal mass of the installation mostly to the material from the tubing up. The higher the insulation R-factor, the better this will work. Remember when using an above-subfloor or above-slab system that the building design must include allowance for this extra thickness at stair risers and rough openings (for doors and windows).

Radiant heat advantages

- Radiant heating is almost a perfect match to human comfort needs:
  - Room temperature profile and mean radiant temperature are nearly ideal.
  - Each space can be individually controlled if desired, providing a perfect match to homeowner needs and improving efficiency by allowing setback in unoccupied spaces — using heat only where and how much it is needed.
- Radiant heating can be more efficient than other systems:
  - Operating temperatures are lower.
  - Room air isn't stratified.
  - There are no duct losses.
- Radiant heating is unseen and unheard:
  - Tubing is embedded in floor (or walls/ceilings), so furniture placement is unimpaired.
  - Heating units are unseen.
  - The system is quiet — no noisy fans.

- Radiant heating is clean:
  - System doesn't require air movement through the house that scatters dust, odors, and germs.
  - System doesn't require periodic cleaning of the distribution system or filters.
- Radiant heating is versatile:
  - Systems can be combined with domestic water heating, snow melting, pool heating, and space heating, for example, just by piping and controlling the flow of water from the heat source.
  - Low-temperature radiant heating systems can use low-temperature heat sources, such as geothermal or solar.
- Radiant heating encourages architectural variety in the home:
  - Tile and hardwood floors are warm with radiant floor heating, and can be left uncovered. Floors don't have to be carpeted for comfort.

Radiant heat limitations

- High mass systems may have slow response to thermostat changes.
- Radiant floor heating may not be usable where a large portion of floor is blocked by furniture, where heavy carpeting or padding is used, or where very high loading is required, though some situations can be handled by using supplemental heat or radiant wall/ceiling rather than radiant floor.
- Unlikely as it is, a leak in embedded tubing can be difficult and expensive to repair.
- Radiant heating systems do not provide ventilation or air cleaning, and usually require a separate air conditioning system.
- Radiant heating must be compatible with many building materials and methods, and suitable for exposure over long periods.

Every radiant installation must:

- Regulate **three** temperatures:
  - Space (room).
  - Supply water to tubing.
  - Return water to boiler.
- Regulate supply water temperature to the tubing circuits to:
  - Properly control space temperatures.
  - Prevent overheating tubing.
  - Protect slab and/or flooring from excessive temperature.
- Regulate return water temperature to boiler or heater
  - If the boiler or heater has a minimum allowable return water temperature to prevent damage due to condensation.

Components

Tubing

Tubing types

Radiant tubing can be steel, copper, plastic or elastomeric. Most installations currently use plastic or elastomeric tubing. Plastic tubing is usually cross-linked polyethylene (PEX). Elastomeric tubing is usually synthetic rubber.

Plastic tubing is available with or without an oxygen barrier. Most tubing vendors recommend using tubing with an oxygen barrier if the heating system contains ferrous components (iron or steel). This is because plastic tubing allows oxygen to move through the tubing wall. Oxygen will pass into the water through the tubing wall attempting to balance the oxygen inside the tubing to atmospheric oxygen. Oxygen barriers are either embedded metallic liners or specially formulated materials that resist oxygen penetration (similar to the materials used in food packaging such as potato chip bags).

PEX, polybutylene, and EPDM rubber are compatible with Portland cement, gypsum underlayments, and most common building materials. PEX tubing has a slight advantage over synthetic rubber tubing because it is resistant to oils as well.

Tubing sizes

You will find plastic or rubber (elastomeric) tubing available in sizes starting at about 3/8 inch inside diameter up to 1 inch inside diameter or larger. Most installations will use 3/8, 1/2, or 5/8-inch tubing.

Follow the tubing supplier's recommendations for tubing selection. Smaller tubing is less expensive per foot and easier to manipulate, but has a much higher pressure drop than larger diameter tubing. Tubing circuit length will depend on design and layout. You will find recommended maximum lengths per circuit at about:

- 3/8-inch — 250 feet
- 1/2-inch — 350 feet
- 5/8-inch — 500 feet
- 3/4-inch — 600 feet

Tubing is supplied in rolls, with roll length dependent on tube diameter. Tubing lengths range from 300 to 1000 feet per roll. (Doing an accurate tubing layout drawing allows you to select the best roll sizes to minimize leftover tubing for your installation.)

Applications

Once you select a tubing vendor, check its recommendations for fittings, manifolds, and any special application requirements. Fitting designs and attachment methods may vary between tubing manufacturers.

Use the information available from your vendor when doing your tubing layout drawing and preparing the purchasing list.

You can use larger-diameter tubing (5/8-inch inside diameter and larger) for distribution of hot water to remote manifolds as well as for radiant panel installation. You may find this considerably faster than using conventional sweat-fitted copper piping. Find the required flow for the distribution line. Then use the tubing manufacturer's design guidelines to find the head loss for the line.

Fittings

Plastic and rubber tubing fittings are designed to adapt tubing ends to common pipe thread sizes, male and female, and to connect to tubing manifolds. Tubing is usually attached to the fitting using compression methods or band clamps. Follow fitting and tubing manufacturers' instructions for application and installation of fittings.

Verify that the fittings, tubing, and manifolds have compatible connections.

Manifolds

Tubing circuits are piped in parallel — never in series. Tubing manufacturers offer piping manifolds that provide for multiple tubing circuits. The manifold take-offs use connections designed for tube fittings, and manifold ends are usually tapped for standard pipe threads.

Flow balancing — If the lengths of all tubing circuits connected to a manifold assembly are within 10% of each other, you probably won't need to balance flow. If the lengths differ by more than 10%, however, provide some means to adjust flow (a ball valve or integral manifold valve for each tubing circuit, for example) or you may encounter performance and temperature distribution problems.

Manifolds include, or are available with, integral valves and flow indicators. Always use integral valves when possible. These not only allow isolation of circuits for air purging and troubleshooting, but can often be used to manually balance flow in circuits.

To simplify flow balancing, install a flow indicator on each circuit. These will probably pay for themselves in the time you can save making flow adjustments.

Most manifolds are available with zone valves (electric valve actuators). When possible, include zone valves in your quotation and installation. You probably won't zone each circuit separately. You can combine several circuits in a single zone, operated by that zone thermostat, by wiring the zone valves in parallel. Changing the zoning plan is usually just a matter of installing or relocating thermostats, then changing the wiring to the zone valves. (You can also zone systems with zone valves installed in the piping to the manifolds, but won't have the versatility provided by installing manifold zone valves, providing separate regulation of each circuit.)

Controls

You can install a successful radiant heating system for a small home using a single zone (one room thermostat), one circulator and no additional controls. Below-floor tubing installations lend themselves better to a single-zone approach because the thermal mass is relatively small and all rooms (usually) have the same radiant heating method. These systems usually operate with warm enough water in the tubing to prevent condensate problems at the boiler. If necessary, you can include a manual by-pass arrangement in the near-boiler piping to raise the average boiler water temperature. Many boiler manufacturers provide suggested by-pass piping in their instruction manuals. Most radiant systems will probably need electronic controls and some type of zoning in order to match the temperature and flow to the needs of different spaces. See the tubing manufacturer's guidelines and the references listed at the beginning of Part 8.

Temperature limit control

In every installation, regardless of the control complexity, install a control that will prevent water being delivered to the tubing above its rated maximum operating temperature. For slab installations, include a temperature sensor to protect the slab against overheating.

Multiple zones and/or circulators

Single-zone, single-circulator methods probably won't work for large homes or for systems having varying tubing installation methods (such as slab-on-grade embedded tubing on the first floor and below-floor staple-up tubing on the second floor) or widely different flooring conditions (carpet in some areas, tile and hardwood exposed flooring in others). Use the information from the tubing supplier and the references listed at the beginning of Part 8 to determine the best method for your installation.

Electronic controls

For multi-zone or high-mass (embedded slab) installations, you will want to use electronic controls, probably with outdoor reset, when possible. Controls available for radiant heating provide better system response. Outdoor reset, in particular, offers the advantage of reducing cyclic expansion and contraction of the tubing and piping.

Circulators

Proper flow to each circuit is key to a successful radiant heating installation. You need to consider the special needs of radiant heating circuits:

• The pressure drop through radiant tubing circuits is considerably higher than pressure drops through typical hydronic circuits, mostly due to the unusually long length of each circuit. Each circuit may be up to 300 feet long, though some may be longer with larger-diameter tubing. Radiant circuits are long because the tubing is the heat exchanger — there are no terminal units, unless added for supplemental heating.

• Radiant heating systems consist of multiple flow loops (tubing circuits), each of which must be provided with the correct flow.

• Radiant heating systems are usually designed especially for the space. The heating system designer will plan the flow and water temperature in each circuit, appropriate for the type of flooring (or wall/ceiling), floor coverings, and required heat output.

• The circulators you supply must provide the total required flow at the correct pressure drop. Packaged residential boiler circulators are unlikely to provide the correct performance.

The number and specifications of circulators needed will depend on the system piping and control design.

You will find sizing information in the design guides available from many tubing manufacturers. You can apply the general guidelines provided in Guide RHH, Parts 3 through 5, to radiant heating as well, as long as you use the pressure drop (head loss) data for the tubing you are using (available from the tubing supplier).

Heat sources

A boiler is the usual heat source for radiant heating systems. Boilers have been developed over many years specifically for space heating applications. And the testing standards used to verify boiler performance and efficiency have evolved from extensive experience in space heating applications.

Some radiant heating systems have used other heat sources. But note that the heat source must provide carefully-controlled water temperatures. Check with local codes before using any heat source other than a boiler. For some small systems, water heaters have been used where permitted by local codes. But their limited capacity usually makes them impractical, less effective and less efficient.

If you plan to install a system for combined service water heating and space heating, using the same water for both, be aware that many local and national authorities are concerned about the potential health hazards due to bacterial growth in stagnant regions of heating systems.

Condensing and non-condensing boilers

The controls and piping of a radiant system have to meet the needs of not only the system, but all of the equipment used in the system. If the boiler is a conventional (non-condensing) design, the piping and controls must control the return water temperature to the boiler in addition to controlling the water temperature supplied to the radiant panels. The return water temperature must be maintained at or above the minimum temperature specified by the boiler manufacturer.

Some boilers are intended for condensing operation, with no limit on return water temperature. Condensing boilers can be operated without return water temperature control because flue gas condensation is accounted for in the boiler design.

Do not confuse the ANSI vent category rating of a boiler with its ability to operate in condensing mode. A boiler rated as Category II or IV is classified by ANSI standards as "condensing" because condensation in the vent system is likely. Most boilers rated as ANSI Category II or IV are actually high-efficiency conventional boilers, not condensing boilers. Consult the boiler manufacturer's literature and instructions carefully to determine whether return water temperature control is required.

Applications

Tubing application methods

Figure 2, Page 8–13 through Figure 5, Page 8–16 show typical installation techniques for radiant floor applications of plastic or rubber tubing. These drawings are for general information only. Always refer to the tubing manufacturer's instructions and local building codes when designing and installing a radiant heating system. You will find more detailed information on these system options, as well as typical wall and ceiling installations, in the references listed at the beginning of Part 8.

Installation considerations

Working with plastic and rubber tubing
Read and follow the tubing manufacturer's instructions for applying and installing its tubing. Use necessary precautions when handling and installing the tubing to prevent kinking, collapsing, or puncturing of the tubing. Some general guidelines to keep in mind are:

- Use a tubing uncoiler to ensure smooth unrolling of the tubing as you apply it. Pull tubing straight off of the coil; don't allow kinking or twisting.

- Some tube fittings require special equipment to correctly compress the fitting. Use the equipment specified by the manufacturer.

- Don't install tube-fitting joints in slabs or inaccessible locations. Tubing circuits should be continuous, with no fitting joints.

- Protect tubing at slab penetrations — use PVC sweep elbows or manufactured pipe bends to sleeve tubing.

- Fasten tubing securely for slab installations (tubing can float if not held down).

- Tag tubing at manifolds same as on layout plan.

- Clearly mark all tubing locations on slab or subflooring and studs.

- Pressure test tubing and circuits at installation and monitor throughout remainder of building construction. Use air not water (it eliminates freeze potential when installed in cold conditions).

- Leave copy of layout drawing with owner. Document job with photos.

- Protect against freezing and water damage if water will remain in tubing where it could freeze (use antifreeze if necessary).

Figure 2 Typical construction with tubing embedded in a slab on grade

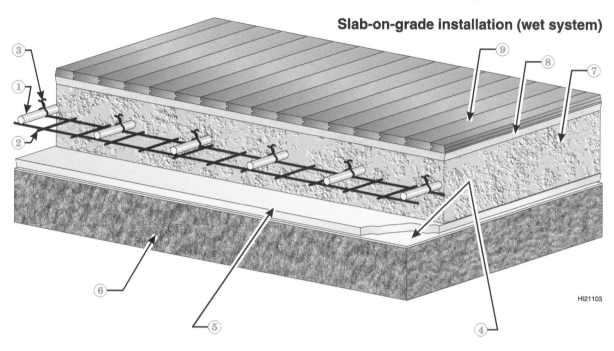

Slab-on-grade installation (wet system)

HI21103

① Plastic or rubber radiant tubing

② Wire mesh or rebar slab reinforcement

③ Plastic ties (typically) used to secure tubing to wire mesh or rebar when routing tubing. These also prevent tubing from floating when the slab is poured. Wire mesh or rebar should be lifted during pour to suspend tubing as close as possible to center of slab, but no closer than 1 inch from top of slab.

④ Polyethylene sheet moisture barrier, placed on compacted fill before placing polystyrene insulation board.

⑤ Polystyrene insulation board, preferably 25 psig or higher rating. Place under slab and around perimeter.

⑥ Compacted fill must be smooth and level with no gaps or rocks. If the fill causes insulation board to bridge gaps, cut insulation board to 1-foot strips to allow it to fit more closely to fill surface.

⑦ Concrete slab

⑧ Subflooring (if used)

⑨ Finished flooring (Some laminated flooring may not require a subfloor. Consult flooring supplier.)

Figure 3 Typical construction with tubing installed under the subfloor

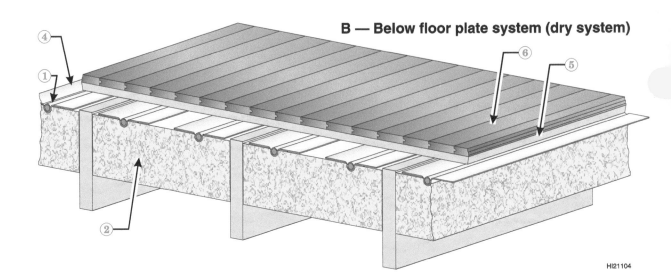

A — Below floor "staple-up" (dry system)

B — Below floor plate system (dry system)

HI21104

① Plastic or rubber radiant tubing — usually stapled to subfloor using special staples for method A, secured to floor with plates in method B.

② Fiberglass insulation — Insulate under tubing even if application is interior second story. Insulation prevents heat from going down instead of up through flooring. Use foil-face insulation for method A. Use plastic-wrapped insulation (preferred) for method B.

③ Air space between insulation and subfloor — Method A uses the air space between insulation and subfloor to distribute heat on flooring — Air is heated and transfers heat to subfloor.

④ Aluminum plates include a groove sized for the tubing. Tubing is usually lightly stapled to subfloor. Then plates are pressed over the tubing and stapled securely in place. Aluminum plates help distribute heat across flooring in method B.

⑤ Subflooring

⑥ Finished flooring (Some laminated flooring may not require a subfloor. Consult flooring supplier.)

Figure 4   Typical construction with tubing embedded in a thin slab

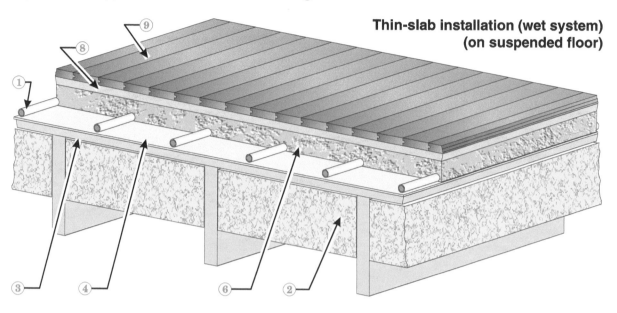

**Thin-slab installation (wet system)
(on suspended floor)**

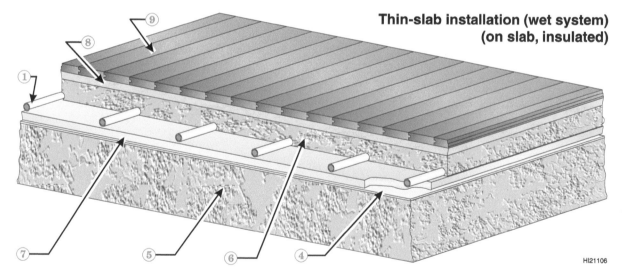

**Thin-slab installation (wet system)
(on slab, insulated)**

HI21106

① Plastic or rubber radiant tubing

② Fiberglass insulation — Insulate under floor even if application is interior second story. Insulation prevents heat from going down instead of up through flooring.

③ Subfloor

④ Moisture barrier. When using concrete thin slab, use a polyethylene sheet. When using gypsum underlayment, use moisture barrier specified by underlayment supplier.

⑤ Concrete slab on grade

⑥ Concrete or gypsum thin slab (typically 1½ to 2 inches)

⑦ Place minimum ½-inch polystyrene insulation on top of moisture barrier slab (unless designer wants a high-thermal-mass system that includes the main slab).

⑧ Plywood or OSB sheeting (when used) to provide a nailing board for flooring.

⑨ Finished flooring (Some laminated flooring may not require a subfloor. Consult flooring supplier.)

Figure 5 Typical construction with tubing installed above the subfloor or slab with sleepers

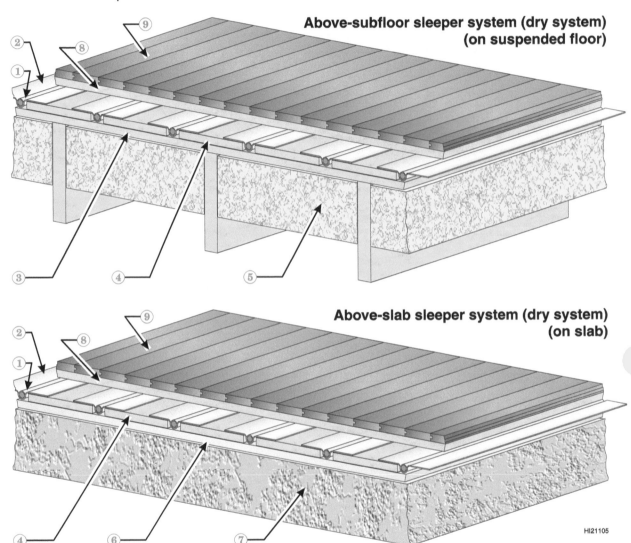

Above-subfloor sleeper system (dry system)
(on suspended floor)

Above-slab sleeper system (dry system)
(on slab)

HI21105

① Plastic or rubber radiant tubing

② Aluminum plates with groove for tubing (typical)

③ Subfloor

④ Sleepers (typically ¾" lumber nailed to subfloor)

⑤ Fiberglass insulation — Insulate under tubing even if application is interior second story. Insulation prevents heat from going down instead of up through flooring.

⑥ Place a polyethylene sheet moisture barrier on slab. Then place minimum ½-inch thick polystyrene insulating board on moisture barrier.

⑦ Concrete slab

⑧ Plywood or OSB sheeting to provide a smooth, regular surface and nailing board for flooring.

⑨ Finished flooring (Some laminated flooring may not require a subfloor. Consult flooring supplier.)

Layout/design

Design and layout considerations

Radiant heating installations require some design work, regardless of how simple the system may seem. You need to do a complete heat loss for the home first. Then do a scaled tubing layout drawing (as in the example shown in Figure 7, Page 8–19. Notice the table included with the layout to show the tubing circuit lengths, used to determine the best coil lengths for minimum scrap.

The tubing layout drawing is essential because:

- The layout drawing provides accurate lengths of circuits and feasibility of paths. This will avoid running out of tubing when you install the circuits. You don't want joints in the circuits.

- You can use the layout to measure for tubing locations when marking locations for slab and above-subfloor or above-slab installations.

- You will be able to order tubing in lengths that will avoid excessive scrap.

- You can determine how to pipe and control the system based on the required locations of manifolds and circuits.

When doing the layout drawing, consider:

- Avoid where possible: slab control or expansion joints; routing under walls to reduce risk of fastener damage.

- Locate manifold stations centrally, away from outside walls and accessible.

- Identify each circuit on plan.

- Plan circuits so spaces can be individually controlled, regardless of the initial control design.

- Limit circuit lengths: Each circuit within 10% of all others connected to a manifold unless balancing valves (and preferably flow indicators) are used; no circuit longer than tubing manufacturer's recommendation.

See Figure 8, Page 8–20, for common tubing layout patterns. These examples show the usual procedure of routing the hottest supply water to tubing run along the outside walls. Notice the use of these patterns in Figure 7, Page 8–19.

For detailed information on radiant heating system design, see the tubing supplier's design guides and references listed at the beginning of Part 8.

Radiant heating installations require some design work, regardless of how simple the system may seem. You need to do a complete heat loss for the home first. Then do a scaled tubing layout drawing.

Why do a layout drawing?

- The layout drawing provides accurate lengths of circuits and feasibility of paths. This will avoid running out of tubing when you install the circuits. You don't want joints in the circuits.

- You can use the layout to measure for tubing locations when marking locations for slab and above-subfloor or above-slab installations.

- You will be able to order tubing in lengths that will avoid scrap.

- You can determine how to pipe and control the system based on the required locations of manifolds and circuits.

Tube spacing and water temperature

The following is general information on how tube spacing and water temperature are determined. Refer to Hydronics Institute Guide 400 and the tubing manufacturer's design procedures for details.

- Tube spacing and water temperature in the tubes control the floor surface temperature.

- As the insulation effect (R-factor) of flooring materials above the tubing increases, the tube spacing needs to decrease and/or water temperature needs to increase to compensate.

- Wide tubing spacing can cause noticeable "striping," meaning the temperature of the floor will vary enough to be noticed by the occupant. Flooring materials above the tubing cause the heat to spread as it rises through the materials, reducing thermal striping. Striping will be most noticeable on bare wood or tile floors because they are thin and have low thermal resistance. For these surfaces, particularly in baths, closer tube spacing will reduce noticeable striping.

- Tube spacing is commonly made closer near walls with high heat losses (due to patio doors or large windows, for example) to increase the heat output. The floor temperature will often be above 85°F in these areas. This won't cause an uncomfortable condition for occupants because the perimeter areas are usually not occupied.

- As a general guideline, you will find that floor temperature (and heat output per square foot) increases with closer tube spacing or higher water temperature, and decreases with increasing R-factor of floor covering.

- Tubing design/layout uses heat loss for the space and space geometry to determine tube spacing and water temperatures, depending on the R-factor of the flooring and floor coverings to be used.

Figure 6 Example house from Guide H-22, used for tubing layout, opposite page

HI21109

Figure 7   Tubing layout drawing example — Guide H-22 house, thin-slab on suspended floor

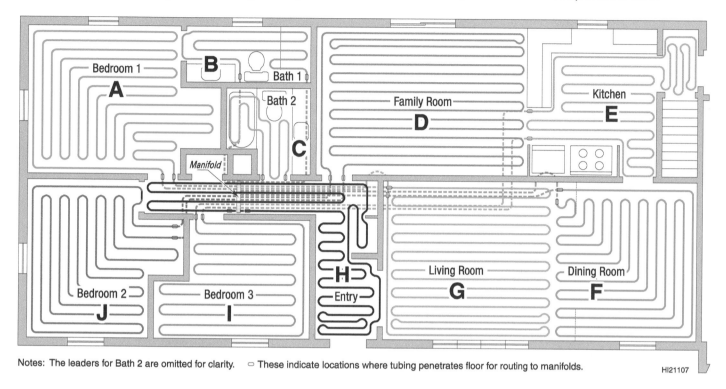

Notes: The leaders for Bath 2 are omitted for clarity.   ○ These indicate locations where tubing penetrates floor for routing to manifolds.   HI21107

Tubing circuit	Finished flooring	Tube spacing (center to center)		Circuit length (including leader — see Note)	Length allowance	Total tube length
		Normal	Close			
		Inches	Inches	Feet	Feet	Feet
A	Low carpet w/thermal pad	9	--	307	10	317
B	Ceramic tile	9	--	71	10	81
C	Ceramic tile	9	--	41	10	51
D	Low carpet w/thermal pad	9	6	314	10	324
E	Ceramic tile	9	6	215	10	225
F	Laminated hardwood	9	--	280	10	290
G	Laminated hardwood	9	6	306	10	316
H	Ceramic tile	9	6	191	10	201
I	Low carpet w/thermal pad	10	--	150	10	160
J	Low carpet w/thermal pad	9	--	194	10	204

Note: The leader is the tubing connecting from the heated space tubing to the manifolds.

Coil number	Coil length	Tube size	Use for circuits listed	Total length (feet)
1	1000	½"	A, D, G	957
2	1000	½"	B, C, E, H, I, J	925
3	300	½"	F	290

Figure 8 Tubing layouts usually route hottest water along outside walls

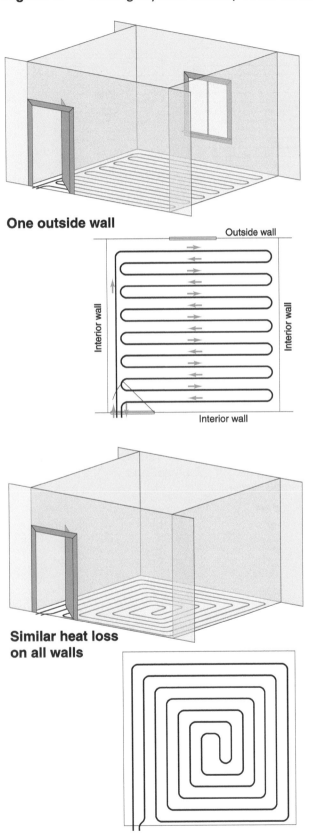

One outside wall

Similar heat loss on all walls

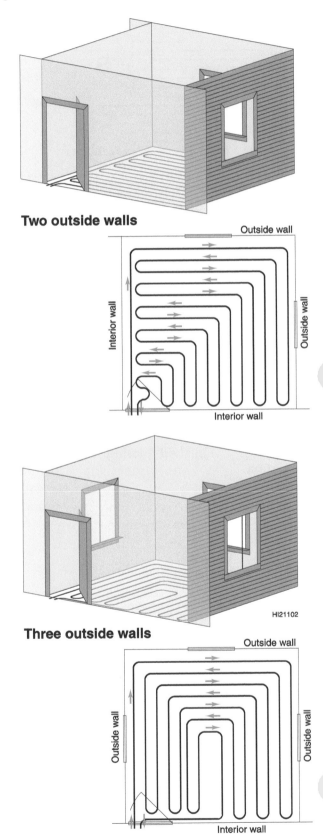

Two outside walls

Three outside walls

Coordination

Coordinating installation with other trades

You will need to work with the general contractor and other trades to coordinate your heating system installation with masons, rough carpenters, finish carpenters, and flooring installers, etc. You also need to notify the general contractor and/or other trades about the location of tubing in the structure and avoidance of damage due to accidental penetration or abuse.

The following pages provide suggestions for trade coordination depending on radiant heating application (slab-on-grade, thin slab, etc.). You may find this information helpful when planning the installation with the general contractor and/or homeowner.

Inspect and pressure test

Inspect the tubing installation:

Before the slab is poured, thoroughly examine all of the installed tubing to verify it has not been penetrated or crushed. During the pour, have someone watch to encourage the masons to lift the tubing, if needed, so it is located near the center of the slab and to visually verify no tubing has been penetrated or crushed.

Pressure test the tubing after installation, preferably using air since it isn't susceptible to freezing as is water. Temporarily install a pressure gauge located for easy accessibility by all trades and pressurize the tubing for the duration of construction. Ask others to monitor the gauge, making sure there is no loss of pressure that would indicate tube penetration. Pressure testing won't show kinked or crushed tubing, but it will notify of any tube penetrations.

EMBEDDED TUBING: SLAB-ON-GRADE

Design/materials/special requirements

1. Tubing may affect structural rating of slab.
2. Compacted fill must be smooth, level and gap-free.
3. Polyethylene moisture barrier required.
4. Slab should be insulated with polystyrene board.
5. Coordinate scheduling with other trades.

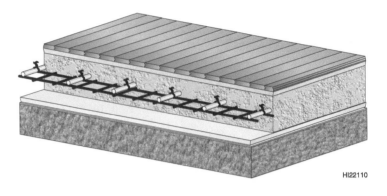

HI22110

Flooring

6. No heavy carpeting or pads over heated floors.
7. Material changes could affect heating performance.
8. Must be rated for floor heating.
9. Cure slab thoroughly before finished flooring installation.
10. Follow hardwood industry guidelines for hardwood flooring.

Construction

11. Limit traffic across tubing.
12. Compacted fill must be smooth and flat, with no gaps.
13. Pressurize tubing and monitor during construction.
14. Moisture barrier on fill, then polystyrene board insulation.
15. Place tubing in center of slab.
16. Tubing must be protected at all slab joints.
17. Protect tubing during construction.
18. Mark location of all tubing.

Explanation of checklist items:

1. The tubing could affect the structural rating of the slab, and should be considered by the architect/designer. If tubing cannot be installed in the main slab, it will have to be installed in a thin slab (1½ to 2 inches) applied on top of the slab or installed as an above-slab system using wood sleepers.

2. The compacted fill must be smooth and level with no gaps. If the polystyrene insulation bridges any gaps, there is a risk of later slab failure.

3. New construction: Place polyethylene moisture barrier on compacted fill before placement of tubing or slab reinforcement. Retrofit: Place polyethylene moisture barrier on slab before placing tubing or flooring.

4. Provide insulation to prevent heat loss from the heated slab.

   • The most important insulation is around the perimeter of the slab and under the slab near the perimeter. Locate insulation as shown in Figure 2, Page 8–13, (or as recommended by tubing manufacturer). Use 2 inches of polystyrene board around the perimeter and under the slab near the perimeter, with 1 inch of polystyrene under the remainder of the slab. Exception: For installations in moderate climates with dry soil or where the ground is considered Part of the thermal mass, the under-slab insulation can be omitted (use perimeter insulation only).

   • The slab load itself is minor (only about 0.4 psig for a 6-inch concrete slab). For assurance of a generous safety factor to account for concentrated loads, use 25 psig-rated polystyrene board insulation, though 15-psig board is probably adequate for most residential applications. Verify the compressive load on the polystyrene insulation with architect/designer.

   • Do not install insulation under load-bearing areas of the slab (perimeter and other locations carrying load-bearing walls or supports).

5. The heating system is installed in stages:

   Stage 1: Before slab is poured — Heating contractor will measure and mark tubing layout on insulation board. Heating contractor will install tubing and tie to wire mesh or rebar. Temporarily mount tubing manifold(s), supported by temporary posts (rebar or metal posts driven in ground). Heating contractor will inspect and pressure test tubing.

   Stage 2: After concrete pour and before rough carpentry, heating contractor should mark the slab showing locations of all tubing, particularly where tubing passes under interior walls.

   Stage 3: After rough carpentry and electrical installations, and before drywall installation, heating contractor may inspect installation to verify all tubing in good condition. Heating contractor may want to install wiring to thermostats and verify wiring is available for controls, pumps and boiler.

Stage 4: After drywall completion, heating contractor will complete heating system installation, including installation of boiler, controls, pumps and any supplementary heating units (radiant baseboard, for example). NOTE: If the flooring contractor needs the heating system to operate for acclimatization of the flooring materials, the heating contractor may need to install the system for temporary heating, then finish the installation after finished flooring has been installed.

6. Carpet and padding are insulators (particularly wool carpet). Using too great a thickness of either can limit the heat available from the floor, possibly requiring addition of perimeter baseboard heating. The thicker the carpet and pad, the more expensive the heating installation may be because of the amount of tubing required. Preferably, use a rubber or slab foam rubber pad 1/4 or 3/8 inch thick, with carpeting no thicker than 1/2 inch. Explain the reasoning to the owner when determining final flooring selections. Try not to exceed a total R-value of 2.0 for the complete flooring system above the tubing.

7. The heating system designer determines tubing layout, piping and control design based on the flooring materials to be used. If materials are changed, the system performance could be affected. If the owner wants flexibility to change in the future, the heating system designer will want to provide for this in design.

8. All materials used in flooring must be rated for heated-floor application to ensure a successful installation. Plywood or OSB (oriented strand board) is preferred for subflooring (particle board may contain adhesives that could outgas). For carpeted floors, rubber or foam rubber padding is preferred. Do not use urethane foam padding. Finished flooring should be rated by the manufacturer for heated-floor application and should be installed following the manufacturer's guidelines.

9. Moisture left in the slab will damage finished flooring. Cure the slab completely and test for moisture content. Apply a 4-foot by 4-foot sheet of polyethylene over a section of the slab. Turn on the heating system. If moisture appears under the plastic, the slab must be cured at least another day. Continue this process until no moisture appears. If the heating system is not operational, leave the polyethylene barrier in place for two or three days. The slab is cured only if there is no discoloration of the slab or formation of condensation under the plastic sheeting.

10. Sawn-wood flooring shrinks and expands with variations in moisture content. The flooring must be acclimatized to the environment it will experience to ensure the least possible movement. The Hardwood Council recommends (as of July, 2000):

• Operate the heating system at least 72 hours to balance the house's moisture content before installing the flooring. Don't have the hardwood flooring delivered until the house humidity is at normal operating level. Additional heating time may be needed if the subflooring moisture content is high.

• Hardwood flooring moisture content should be suitable for the climate in which it will be used. Moisture content should typically be between 6 and 9%, but should be based on normal inside relative humidity. The flooring installer should check with the hardwood supplier to determine the appropriate moisture content for the area. If the flooring moisture content is lower when received, return it. If the flooring moisture content is higher than desired, spread the flooring out and allow it to dry with the heating system in normal operation, checking the moisture content daily (keeping written records) until the appropriate level is reached. DO NOT install the flooring until

the hardwood flooring and the subflooring are at the correct moisture level.

• Use quartersawn tongue-and-groove or beveled-edge hardwood flooring strips no more than 3" wide. Wider planks could possibly cause noticeable separation at joints.

11. Only cross tubing with pneumatic tire wheelbarrows — no heavy equipment, such as trucks or power buggies. Never nose down a wheelbarrow on tubing. Notify heating contractor immediately (before pouring slab) if any damage should occur.

12. If the polystyrene insulation bridges a gap in the fill, the slab may fill to the insulation surface only, leaving a void between the insulation and compacted fill. This can lead to slab failure later. At the minimum, remove all rocks. If the surface is not smooth and flat, cut the polystyrene to 1-foot strips to allow better fit to the fill contour.

13. The tubing should be kept under air pressure throughout construction so any puncture damage will show immediately. Mount a pressure gauge at an accessible location and ask construction personnel to check pressure gauge periodically, notifying heating contractor (or general contractor) if pressure drops excessively. The pressure will fluctuate slightly due to temperature changes.

14. Place polyethylene moisture barrier on compacted fill. Then place polystyrene insulation over entire slab area and around slab perimeter. If insulation does not conform closely to fill surface, cut insulation to 1-foot strips to ensure better fit.

15. Lift wire mesh or rebar onto chairs or pull up during pour to place tubing in center of slab. Tubing should be positioned as close as possible to center of slab, but never higher than 1 inch below the slab surface.

16. Tubing installer must protect tubing at all control and expansion joints. Mason needs to advise of the location of all joints. Tubing could fail if not protected because the slab can shear at a joint. Most plastic and rubber tubing can be affected by extended exposure to ultraviolet light. Tubing must be protected from excessive direct sunlight. Store tubing rolls inside, out of direct sunlight, and limit the time between tubing placement and pour to no more than about two weeks.

17. Protect tubing where it penetrates the slab (using PVC sweep elbows or manufactured pipe bends to sleeve the tubing — installed by heating contractor). Masons should avoid striking the tubing with power trowels. Protect tubing where it passes through studs by installing nail stops on both sides. Finish carpenters and drywall installers will need to be aware of tubing locations.

18. Mark the location of all tubing on the slab (particularly where tubing passes under interior walls). Inform other trades to avoid these areas when driving fasteners.

BELOW-FLOOR STAPLE-UP OR PLATE METHOD

Design/materials/special requirements
1. Insulate below tubing.
2. Insulate perimeter trusses.

Flooring
3. No heavy carpeting or pads over heated floors.
4. Material changes could affect heating performance.
5. Must be rated for floor heating.
6. Allow all subflooring to dry before installing finished flooring.
7. Follow hardwood industry guidelines for hardwood flooring.

Construction
8. Pressurize tubing and monitor during construction.
9. Protect tubing during construction.
10. Advise other trades of tubing locations.

Explanation of checklist items:

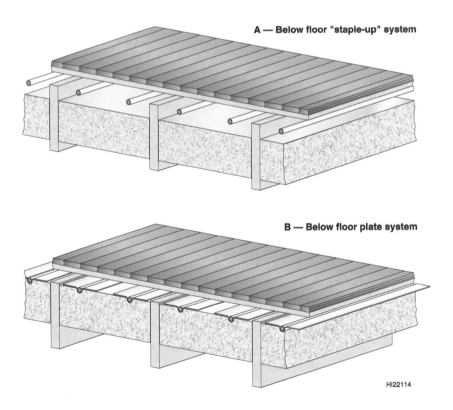

A — Below floor "staple-up" system

B — Below floor plate system

HI22114

1. Insulate below the tubing to prevent downward heat loss.

2. The truss cavities are heated with a below-floor installation. Insulate the perimeters to prevent heat loss as well as to protect against the potential of tubing freeze-up during no-flow or non-heating periods.

3. Carpet and padding are insulators (particularly wool carpet). Using too great a thickness of either can limit the heat available from the floor, possibly requiring addition of perimeter baseboard heating. The thicker the carpet and pad, the more expensive the heating installation may be because of the amount of tubing required. Preferably, use a rubber or slab foam rubber pad 1/4 or 3/8 inch thick, with carpeting no thicker than 1/2 inch. Explain the reasoning to the owner when determining final flooring selections. Try not to exceed a total R-value of 2.0 for the complete flooring system above the tubing.

4. The heating system designer determines tubing layout, piping and control design based on the flooring materials to be used. If materials are changed, the system performance could be affected. If the owner wants flexibility to change in the future, the heating system designer will want to provide for this in design.

5. All materials used in flooring must be rated for heated-floor application to ensure a successful installation. Plywood or OSB (oriented strand board) is preferred for subflooring (particle board may contain adhesives that could outgas). For carpeted floors, rubber or foam rubber padding is preferred. Do not use urethane foam padding. Finished flooring should be rated by the manufacturer for heated-floor application and should be installed following the manufacturer's guidelines.

6. Moisture left in the subflooring will damage finished flooring. Cure the subflooring completely and test for moisture content. Have the finished floor installer verify acceptable moisture content of subflooring. It may be necessary to operate the heating system for two to three days to aid in drying of the flooring materials.

7. Sawn-wood flooring shrinks and expands with variations in moisture content. The flooring must be acclimatized to the environment it will experience to ensure the least possible movement. The Hardwood Council recommends (as of July, 2000):

- Operate the heating system at least 72 hours to balance the house's moisture content before installing the flooring. Don't have the hardwood flooring delivered until the house humidity is at normal operating level. Additional heating time may be needed if the subflooring moisture content is high.

- Hardwood flooring moisture content should be suitable for the climate in which it will be used. Moisture content should typically be between 6 and 9%, but should be based on normal inside relative humidity. The flooring installer should check with the hardwood supplier to determine the appropriate moisture content for the area. If the flooring moisture content is lower when received, return it. If the flooring moisture content is higher than desired, spread the flooring out and allow it to dry with the heating system in normal operation, checking the moisture content daily (keeping written records) until the appropriate level is reached. DO NOT install the flooring until the hardwood flooring and the subflooring are at the correct moisture level.

- Use quartersawn tongue-and-groove or beveled-edge hardwood flooring strips no more than 3" wide. Wider planks could possibly cause noticeable separation at joints.

8. The tubing should be kept under air pressure throughout construction so any puncture damage will show immediately. Mount a pressure gauge at an accessible location and ask construction personnel to check pressure gauge periodically, notifying heating contractor (or general contractor) if pressure drops excessively. The pressure will fluctuate slightly due to temperature changes.

9. Most plastic and rubber tubing can be affected by extended exposure to ultraviolet light. Tubing must be protected from excessive direct sunlight. Store tubing rolls inside, out of direct sunlight. Protect tubing where it passes through studs by installing nail stops on both sides. Finish carpenters and drywall installers will need to be aware of tubing locations. Finished flooring installers should use care when nailing not to penetrate tubing.

10. Mark tubing locations where possible. Inform other trades by providing copy of tubing layout drawing.

EMBEDDED TUBING: THIN SLAB (SUSPENDED FLOOR)

Design/materials/special requirements

1. Provide moisture barrier on subfloor.
2. Insulate under floors.
3. Include slab weight in structural design.
4. Account for slab/flooring height at stair risers and rough openings.
5. Heating installer and other trades need to coordinate scheduling.

Flooring

6. No heavy carpeting or pads over heated floors.
7. Material changes could affect heating performance.
8. Must be rated for floor heating.
9. Cure slab thoroughly before finished flooring installation.
10. Follow hardwood industry guidelines for hardwood flooring.
11. Apply plywood or OSB sheeting on slab for hardwood strip flooring.
12. Use anti-fracture membrane for tile floors.

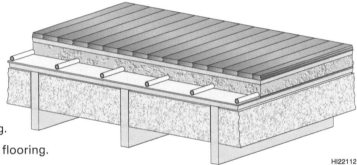

HI22112

Construction

13. Pre-install sole plates before slab pour.
14. Limit traffic across tubing.
15. Pressurize tubing and monitor during construction.
16. Tubing must be protected at all slab joints.
17. Protect tubing during construction.
18. Mark location of all tubing.

Explanation of checklist items:

1. Apply a polyethylene moisture barrier to the subfloor before pour for concrete floors. Use moisture barrier specified by gypsum flooring manufacturer if using gypsum slab. Prevent unwanted downward heat flow by insulating between floor trusses. For exposed-beam construction, install polystyrene board insulation on subfloor before pouring slab (include height in allowance for stair risers and rough openings).

2. If the tubing will be secured with tracks, the minimum thickness of the slab is 2 inches (unless specified otherwise by tubing supplier). For tubing secured with staples, the minimum thickness is 1½ inches (unless specified otherwise by tubing supplier).

3. Verify that the slab weight is acceptable for the structural design of the trusses and flooring.

4. Verify that the height of the slab (and any under-slab insulation board used) is accounted for in stair risers and rough opening (doors and windows) locations.

5. The heating system is installed in stages:

   Stage 1: Before slab is poured — Heating contractor will measure and mark tubing layout on subfloor or insulation board. Heating contractor will install tubing and secure it to the subfloor, and may complete the heating system piping. Heating contractor will inspect and pressure test tubing (and system, if installed).

   Stage 2: After slab pour and before rough carpentry, heating contractor should mark the slab showing locations of all tubing, particularly where tubing passes under interior walls.

   Stage 3: After rough carpentry and electrical installations, and before drywall installation, heating contractor may inspect installation to verify all tubing in good condition. Heating contractor may

want to install wiring to thermostats and verify wiring is available for controls, pumps, and boiler.

Stage 4: After drywall completion, heating contractor will complete heating system installation, including installation of boiler, controls, pumps and any supplementary heating units (radiant baseboard, for example). NOTE: If the flooring contractor needs the heating system to operate for acclimatization of the flooring materials, the heating contractor may need to install the system for temporary heating, then finish the installation after finished flooring has been installed.

6. Carpet and padding are insulators (particularly wool carpet). Using too great a thickness of either can limit the heat available from the floor, possibly requiring addition of perimeter baseboard heating. The thicker the carpet and pad, the more expensive the heating installation may be because of the amount of tubing required. Preferably, use a rubber or slab foam rubber pad 1/4 or 3/8 inch thick, with carpeting no thicker than 1/2 inch. Explain the reasoning to the owner when determining final flooring selections. Try not to exceed a total R-value of 2.0 for the complete flooring system above the tubing.

7. The heating system designer determines tubing layout, piping and control design based on the flooring materials to be used. If materials are changed, the system performance could be affected. If the owner wants flexibility to change in the future, the heating system designer will want to provide for this in design.

8. All materials used in flooring, including the slab material, must be rated for heated-floor application to ensure a successful installation. Plywood or OSB (oriented strand board) is preferred for subflooring (particle board may contain adhesives that could outgas). For carpeted floors, rubber or foam rubber padding is preferred. Do not use urethane foam padding. Finished flooring should be rated by the manufacturer for heated-floor application and should be installed following the manufacturer's guidelines.

9. Moisture left in the slab will damage finished flooring. Cure the slab completely and test for moisture content. Apply a 4-foot by 4-foot sheet of polyethylene over a section of the slab. Turn on the heating system. If moisture appears under the plastic, the slab must be cured at least another day. Continue this process until no moisture appears. If the heating system is not operational, leave the polyethylene barrier in place for two or three days. The slab is cured only if there is no discoloration of the slab or formation of condensation under the plastic sheeting.

10. Sawn-wood flooring shrinks and expands with variations in moisture content. The flooring must be acclimatized to the environment it will experience to ensure the least possible movement. The Hardwood Council recommends (as of July, 2000):

    • Operate the heating system at least 72 hours to balance the house's moisture content before installing the flooring. Don't have the hardwood flooring delivered until the house humidity is at normal operating level. Additional heating time may be needed if the sub-flooring moisture content is high.

    • Hardwood flooring moisture content should be suitable for the climate in which it will be used. Moisture content should typically be between 6 and 9%, but should be based on normal inside relative humidity. The flooring installer should check with the hardwood supplier to determine the appropriate moisture content for the area. If the flooring moisture content is lower when received, return it. If the flooring moisture content is higher than desired, spread the flooring out and allow it to dry with the heating system in normal operation, checking the moisture content daily (keeping written records) until the appropriate level is reached. DO NOT install the flooring until the hardwood flooring and the subflooring are at the correct moisture level.

    • Use quartersawn tongue-and-groove or beveled-edge hardwood flooring strips no more than 3" wide. Wider planks could possibly cause noticeable separation at joints.

11. If a good nailing surface is needed for strip hardwood flooring, apply a 3/4-inch plywood or OSB sheet on top of slab before installing flooring. Limit nail penetration to the 3/4-inch sheet to avoid tubing penetration.

12. If installing tile on the slab, apply an anti-fracture membrane on the slab before tile installation. Otherwise, movement of the slab could damage tile.

13. Have sole plates of same height as slab installed before slab pour. These can be used for a screwed surface, provide a nailer for wall framing, and allow accurate placement of tubing by heating contractor.

14. Only cross tubing with pneumatic tire wheelbarrows — no heavy equipment, such as trucks or power buggies. Never nose down a wheelbarrow on tubing. Notify heating contractor immediately (before pouring slab) if any damage should occur.

15. The tubing should be kept under air pressure throughout construction so any puncture damage will show immediately. Mount a pressure gauge at an accessible location and ask construction personnel to check pressure gauge periodically, notifying heating contractor (or general contractor) if pressure drops excessively. The pressure will fluctuate slightly due to temperature changes.

16. Tubing installer must protect tubing at all control and expansion joints. Mason needs to advise of the location of all joints. Tubing could fail if not protected because the slab can shear at a joint. Most plastic and rubber tubing can be affected by extended exposure to ultraviolet light. Tubing must be protected from excessive direct sunlight. Store tubing rolls inside, out of direct sunlight.

17. Masons should avoid striking the tubing with power trowels. Protect tubing where it passes through studs by installing nail stops on both sides. Finish carpenters and drywall installers will need to be aware of tubing locations.

18. Mark the location of all tubing on the slab (particularly where tubing passes under interior walls). Inform other trades to avoid these areas when driving fasteners.

EMBEDDED TUBING: THIN SLAB (SLAB ON SLAB)

Design/materials/special requirements

1. Provide moisture barrier.

2. Insulate under the thin slab.

3. Verify thin slab depth.

4. Account for slab/flooring height at stair risers and rough openings.

5. Heating installer and other trades need to coordinate scheduling.

Flooring

6. No heavy carpeting or pads over heated floors.

7. Material changes could affect heating performance.

8. Must be rated for floor heating.

9. Cure slab thoroughly before finished flooring installation.

10. Follow hardwood industry guidelines for hardwood flooring.

11. Apply plywood or OSB sheeting on slab for hardwood strip flooring.

12. Use anti-fracture membrane for tile floors.

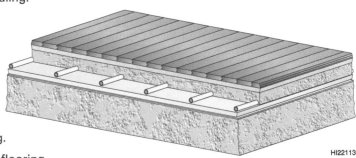

HI22113

Construction

13. Pre-install sole plates before slab pour.

14. Limit traffic across tubing.

15. Pressurize tubing and monitor during construction.

16. Tubing must be protected at all slab joints.

17. Protect tubing during construction.

18. Mark location of all tubing.

Explanation of checklist items:

1. If the main slab does not include a moisture barrier between the slab and fill, apply a layer of polyethylene below insulating board.

2. Apply a minimum of ½ inch of polystyrene insulating board on top of the main slab (or moisture barrier), being sure this height is included in the allowance for stair risers and rough openings (doors and windows). This insulation prevents heat loss to the slab below. Exception: If the main slab is intended as Part of the thermal mass for the heating system, omit the insulation if directed by the heating system designer.

3. If the tubing will be secured with tracks, the minimum thickness of the slab is 2 inches (unless specified otherwise by tubing supplier). For tubing secured with staples, the minimum thickness is 1½ inches (unless specified otherwise by tubing supplier).

4. Verify that the height of the slab (and any under-slab insulation board

used) is accounted for in stair risers and rough opening (doors and windows) locations.

5. The heating system is installed in stages:

Stage 1: Before slab is poured — Heating contractor will measure and mark tubing layout on subfloor or insulation board. Heating contractor will install tubing and secure it to the subfloor, and may complete the heating system piping. Heating contractor will inspect and pressure test tubing (and system, if installed).

Stage 2: After slab pour and before rough carpentry, heating contractor should mark the slab showing locations of all tubing, particularly where tubing passes under interior walls.

Stage 3: After rough carpentry and electrical installations, and before drywall installation, heating contractor may inspect installation to verify all tubing is in good condition. Heating contractor may want to install wiring to thermostats and verify wiring available for controls, pumps and boiler.

Stage 4: After drywall completion, heating contractor will complete heating system installation, including installation of boiler, controls, pumps, and any supplementary heating units (radiant baseboard, for example). NOTE: If the flooring contractor needs the heating system to operate for acclimatization of the flooring materials, the heating contractor may need to install the system for temporary heating, then finish the installation after finished flooring has been installed.

6. Carpet and padding are insulators (particularly wool carpet). Using too great a thickness of either can limit the heat available from the floor, possibly requiring addition of perimeter baseboard heating. The thicker the carpet and pad, the more expensive the heating installation may be because of the amount of tubing required. Preferably, use a rubber or slab foam rubber pad 1/4 or 3/8 inch thick, with carpeting no thicker than 1/2 inch. Explain the reasoning to the owner when determining final flooring selections. Try not to exceed a total R-value of 2.0 for the complete flooring system above the tubing.

7. The heating system designer determines tubing layout, piping and control design based on the flooring materials to be used. If materials are changed, the system performance could be affected. If the owner wants flexibility to change in the future, the heating system designer will want to provide for this in design.

8. All materials used in flooring, including the thin slab material, must be rated for heated-floor application to ensure a successful installation. Plywood or OSB (oriented strand board) is preferred for subflooring (particle board may contain adhesives that could outgas). For carpeted floors, rubber or foam rubber padding is preferred. Do not use urethane foam padding. Finished flooring should be rated by the manufacturer for heated-floor application and should be installed following the manufacturer's guidelines.

9. Moisture left in the slab will damage finished flooring. Cure the slab completely and test for moisture content. Apply a 4-foot by 4-foot sheet of polyethylene over a section of the slab. Turn on the heating system. If moisture appears under the plastic, the slab must be cured at least another day. Continue this process until no moisture appears. If the heating system is not operational, leave the polyethylene barrier in place for two or three days. The slab is cured only if there is no discoloration of the slab or formation of condensation under the plastic sheeting.

10. Sawn-wood flooring shrinks and expands with variations in moisture content. The flooring must be acclimatized to the environment it will experience to ensure the least possible movement. The Hardwood Council recommends (as of July, 2000):

• Operate the heating system at least 72 hours to balance the house's moisture content before installing the flooring. Don't have the hardwood flooring delivered until the house humidity is at normal operating level. Additional heating time may be needed if the subflooring moisture content is high.

• Hardwood flooring moisture content should be suitable for the climate in which it will be used. Moisture content should typically be between 6 and 9%, but should be based on normal inside relative humidity. The flooring installer should check with the hardwood supplier to determine the appropriate moisture content for the area. If the flooring moisture content is lower when received, return it. If the flooring moisture content is higher than desired, spread the flooring out and allow it to dry with the heating system in normal

operation, checking the moisture content daily (keeping written records) until the appropriate level is reached. DO NOT install the flooring until the hardwood flooring and the subflooring are at the correct moisture level.

• Use quartersawn tongue-and-groove or beveled-edge hardwood flooring strips no more than 3" wide. Wider planks could possibly cause noticeable separation at joints.

11. If a good nailing surface is needed for strip hardwood flooring, apply a ¾-inch plywood or OSB sheet on top of slab before installing flooring. Limit nail penetration to the ¾-inch sheet to avoid tubing penetration.

12. If installing tile on the slab, apply an anti-fracture membrane on the slab before tile installation. Otherwise, movement of the slab could damage tile.

13. Have sole plates of same height as slab installed before slab pour. These can be used for a screwed surface, provide a nailer for wall framing, and allow accurate placement of tubing by heating contractor.

14. Only cross tubing with pneumatic tire wheelbarrows — no heavy equipment, such as trucks or power buggies. Never nose down a wheelbarrow on tubing. Notify heating contractor immediately (before pouring slab) if any damage should occur.

15. The tubing should be kept under air pressure throughout construction so any puncture damage will show immediately. Mount a pressure gauge at an accessible location and ask construction personnel to check pressure gauge periodically, notifying heating contractor (or general contractor) if pressure drops excessively. The pressure will fluctuate slightly due to temperature changes.

16. Tubing installer must protect tubing at all control and expansion joints. Mason needs to advise of the location of all joints. Tubing could fail if not protected because the slab can shear at a joint. Most plastic and rubber tubing can be affected by extended exposure to ultraviolet light. Tubing must be protected from excessive direct sunlight. Store tubing rolls inside, out of direct sunlight.

17. Masons should avoid striking the tubing with power trowels. Protect tubing where it passes through studs by installing nail stops on both sides. Finish carpenters and drywall installers will need to be aware of tubing locations.

18. Mark the location of all tubing on the slab (particularly where tubing passes under interior walls). Inform other trades to avoid these areas when driving fasteners.

ABOVE-SUBFLOOR or ABOVE-SLAB SLEEPER METHOD

Design/materials/special requirements

1. Provide subfloor on top of sleeper/tubing layer.

2. Insulate under the tubing or floor.

3. Account for total tubing/flooring height at stair risers and rough openings.

4. Heating installer and other trades need to coordinate scheduling.

Flooring

5. No heavy carpeting or pads over heated floors.

6. Material changes could affect heating performance.

7. Must be rated for floor heating.

8. Allow all subflooring and sleepers to dry before installing finished flooring.

9. Follow hardwood industry guidelines for hardwood flooring.

10. Apply plywood or OSB sheeting on top of sleepers.

Construction

11. Limit traffic across tubing.

12. Pressurize tubing and monitor during construction.

13. Protect tubing during construction.

14. Mark location of all tubing on upper subfloor.

Above-subfloor sleeper system (on suspended floor)

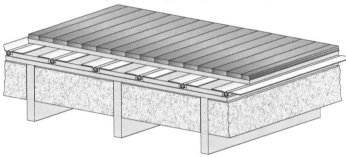

Above-slab sleeper system (on slab)

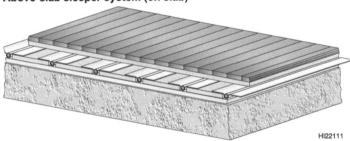

HI22111

Explanation of checklist items:

1. Apply wood sheeting (preferably plywood or OSB) to provide a smooth, flat surface for finished flooring.

2. If applying over a concrete slab, install a polyethylene moisture barrier on the slab before installing tubing and sleepers. Provide at least 1/2 inch of polystyrene insulating board on the barrier before installing sleepers and tubing. If applying on a suspended floor, install insulation between the trusses to prevent unwanted downward heat flow. If the trusses are intended as exposed beams, apply a minimum of 1/2 inch of polystyrene insulating board on the subfloor before installing sleepers and tubing.

3. Verify that the height of the slab (and any under-slab insulation board used) is accounted for in stair risers and rough opening (doors and windows) locations.

4. The heating contractor may need to coordinate timing of heating system and tubing layout with other trades (rough carpenters, electricians, finish carpenters, and drywall installers). The heating contractor may mark the upper subfloor for tube locations to assist other trades in protecting tubing from accidental penetration.

5. Carpet and padding are insulators (particularly wool carpet). Using too great a thickness of either can limit the heat available from the floor, possibly requiring addition of perimeter baseboard heating. The thicker the carpet and pad, the more expensive the heating installation may be because of the amount of tubing required. Preferably, use a rubber or slab foam rubber pad 1/4 or 3/8 inch thick, with carpeting no thicker than 1/2 inch. Explain the reasoning to the owner when determining final flooring selections. Try not to exceed a total R-value of 2.0 for the complete flooring system above the tubing.

6. The heating system designer determines tubing layout, piping and control design based on the flooring materials to be used. If materials are changed, the system performance could be affected. If the owner wants flexibility to change in the future, the heating system designer will want to provide for this in design.

7. All materials used in flooring, including the thin slab material, must be rated for heated-floor application to ensure a successful installation. Plywood or OSB (oriented strand board) is preferred for subflooring (particle board may contain adhesives that could outgas). For carpeted floors, rubber or foam rubber padding is preferred. Do not use urethane foam padding. Finished flooring should be rated by the manufacturer for heated-floor application and should be installed following the manufacturer's guidelines.

8. Moisture left in the slab or subflooring will damage finished flooring. Cure the slab and/or subflooring completely and test for moisture content. (For a slab: Apply a 4-foot by 4-foot sheet of polyethylene over a section of the slab. Turn on the heating system. If moisture appears under the plastic, the slab must be cured at least another day. Continue this process until no moisture appears. If the heating system is not operational, leave the polyethylene barrier in place for two or three days. The slab is cured only if there is no discoloration of the slab or formation of condensation under the plastic sheeting.) Have the finished floor installer verify acceptable moisture content of subflooring. It may be necessary to operate the heating system for two to three days to aid in drying of the flooring materials.

9. Sawn-wood flooring shrinks and expands with variations in moisture content. The flooring must be acclimatized to the environment it will experience to ensure the least possible movement. The Hardwood Council recommends (as of July, 2000):

    • Operate the heating system at least 72 hours to balance the house's moisture content before installing the flooring. Don't have the hardwood flooring delivered until the house humidity is at normal operating level. Additional heating time may be needed if the subflooring moisture content is high.

    • Hardwood flooring moisture content should be suitable for the climate in which it will be used. Moisture content should typically be between 6 and 9%, but should be based on normal inside relative humidity. The flooring installer should check with the hardwood supplier to determine the appropriate moisture content for the area. If the flooring moisture content is lower when received, return it. If the flooring moisture content is higher than desired, spread the flooring out and allow it to dry with the heating system in normal operation, checking the moisture content daily (keeping written records) until the appropriate level is reached. DO NOT install the flooring until the hardwood flooring and the subflooring are at the correct moisture level.

    • Use quartersawn tongue-and-groove or beveled-edge hardwood flooring strips no more than 3" wide. Wider planks could possibly cause noticeable separation at joints.

10. Apply a 3/4-inch plywood or OSB sheet on top of sleepers before installing finished flooring. If applying wood flooring, limit nail penetration to the 3/4-inch sheet to avoid tubing penetration.

11. Only cross tubing with pneumatic tire wheelbarrows — no heavy equipment. Never nose down a wheelbarrow on tubing. Notify heating contractor immediately if any damage should occur.

12. The tubing should be kept under air pressure throughout construction so any puncture damage will show immediately. Mount a pressure gauge at an accessible location and ask construction personnel to check pressure gauge periodically, notifying heating contractor (or general contractor) if pressure drops excessively. The pressure will fluctuate slightly due to temperature changes.

13. Most plastic and rubber tubing can be affected by extended exposure to ultraviolet light. Tubing must be protected from excessive direct sunlight. Store tubing rolls inside, out of direct sunlight. Protect tubing where it passes through studs by installing nail stops on both sides. Finish carpenters and drywall installers will need to be aware of tubing locations. Finished flooring installers should use care when nailing not to penetrate to tubing depth.

14. Mark the location of all tubing on upper subfloor (particularly where tubing passes under interior walls). Inform other trades to avoid these areas when driving fasteners.

Controls/piping

General guidelines

- Provide flow control means for each circuit, at least with adjustable isolation valves; flow metering preferred.

- Provide means to zone circuits in the future.

- Select circulators for the actual flow requirements. Residential boiler packaged circulators may not provide proper flow.

- Slab-on-grade systems will respond slowly. Plan for this in control design.

- Use outdoor reset and continuous circulation when possible — it provides smoother response and reduces expansion cycling.

- Use electronic controls to regulate space and supply water temperatures. Controls using PI (proportional-integral) or PID (proportional-integral-derivative) regulation will anticipate effects, reducing override. See discussion below.

- Do not use night setback thermostats — this will likely result in comfort problems due to slow response and overshoot, and not gains in operating efficiency overall.

- Group and establish zones considering:

  - client wishes
  - physical separation of spaces, floor levels, frequency of use, type of use, heat loss characteristics
  - solar heat gain
  - internal heat gain
  - types of floor coverings
  - types of floor construction
  - excessive zoning increases system cost, complexity, and owner interaction, and may be of little benefit
  - typical circuit design
  - Figure 9, Page 8–34, compares two circuit designs for a radiant heating system, both of which accomplish the requirement of controlling all three temperatures - space, supply water, and boiler return water. Both systems are piped primary/secondary, with the boiler circuit serving as the primary.
  - for multi-function circuits and other piping design alternatives, see the tubing manufacturer's design guides and the references given at the beginning of Part 8.

PI and PID control operation

- Proportional controls react based on how far the actual temperature (water or space temperature) is from the desired temperature.
- PI (proportional/integral) controls react to how much the actual temperature has changed as a result of a control response as well as how far the actual temperature is from the desired temperature.
- PID (proportional/integral/differential) controls react to the amount of change, how far the actual temperature is from desired temperature and also how fast the actual temperature changes as a result of control action.

Figure 9   Typical piping arrangements for radiant heating

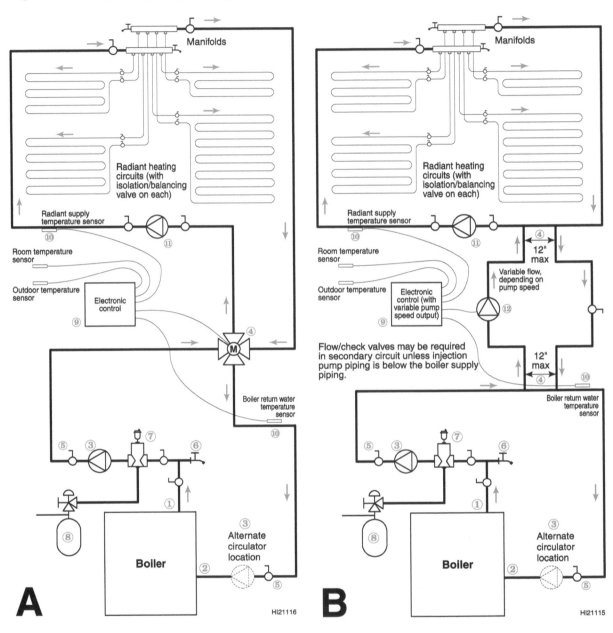

①	Boiler outlet (supply)	⑧	Diaphragm expansion tank with fill valve
②	Boiler inlet (return)	⑨	Electronic control to operate mixing valve
③	Boiler loop circulator		
④	Motorized 4-way mixing valve	⑩	Radiant supply and boiler return sensors
⑤	Isolation valves	⑪	Radiant circuit circulator
⑥	Purge cocks		
⑦	Air separator with automatic air vent		

①	Boiler outlet (supply)	⑧	Diaphragm expansion tank with fill valve
②	Boiler inlet (return)	⑨	Electronic control with variable pump speed
③	Boiler loop circulator		
④	Secondary connection	⑩	Radiant supply and boiler return sensors
⑤	Isolation valves	⑪	Radiant circuit circulator
⑥	Purge cocks	⑫	Injection mixing pump
⑦	Air separator with automatic air vent		

Motorized mixing valve method — Figure 9 – A

- System A uses a motorized 4-way mixing valve. The valve responds to signals from the electronic control to regulate supply water temperature to the radiant heating circuit. The 4-way valve porting allows hot water from the boiler loop to be injected into the radiant heating loop. For more information on 4-way valves, see Part 3, Figure 12, page 3-20.

- The electronic control senses all three required temperatures (room, supply water, and boiler return water). It can also be equipped with an outdoor temperature sensor to provide outdoor reset operation.

- The electronic control can be equipped with a boiler return water temperature sensor to control the 4-way mixing valve so the return water temperature to the boiler will remain above the minimum specified by the boiler manufacturer.

- Size the radiant circuit circulator for the head loss and flow requirements of the radiant system. Size the 4-way mixing valve for a head loss just large enough to allow good control.

- If the operating temperature of the radiant system will always be above the boiler minimum, or if there is no minimum return temperature to the boiler, the 4-way valve can be replaced by a 3-way valve that only controls supply water temperature to the radiant system.

- For boilers that do not require a minimum flow, the boiler-loop circulator is optional. If using a boiler packaged with a circulator, use the boiler circulator in all cases.

Injection mixing pump method — Figure 9 – B

- System B uses a method called "injection mixing." Pump 12 is a small circulator that only has to inject the required amount of hot water from the primary circuit (boiler side) into the secondary circuit (radiant).

- Injection mixing is sometimes done with an on/off injection pump. But this should be limited to high thermal mass applications to prevent frequent cycling. Use an electronic control that provides a variable speed output circuit for the injection pump when possible.

- The electronic control senses all three required temperatures (room, supply water, and boiler return water). It can also be equipped with an outdoor temperature sensor to provide outdoor reset operation. The control uses the injection pump to add heat as needed in the radiant circuit to obtain the design supply water temperature. If the boiler return-water temperature starts to drops too low, the control reduces the amount of heat taken from the boiler circuit enough to keep the return water temperature above the minimum you set on the control (per boiler manufacturer's requirement).

Thermostatic mixing valve method — Figure 9 – C

- System A uses two thermostatic mixing valves. Valve 9 controls supply water temperature to the system. Valve 10 controls return water temperature to the boiler. (This system assumes the boiler and/or zone valves are controlled by room thermostat(s).)

- The circulator supplied with a typical residential boiler should be sufficient as circulator 3 (boiler loop circulator) for most applications when piped as shown.

- You will need a circulator sized specifically for the radiant heating circuit, circulator 11 in the illustration.

- Notice that the piping connection between the boiler loop and the radiant loop must be spaced no further than 12 inches apart. This is a primary/secondary connection. The close spacing ensures that the pressure drop through this section of tubing won't induce unwanted flow in the other circuit.

- Obtain mixing valves with ranges suitable for the required temperatures. Set the boiler-circuit mixing valve at or above the minimum return-water temperature specified by the boiler manufacturer. (For condensing boiler installations, you can omit mixing valve 10. You will still need mixing valve 9 to regulate radiant supply water temperature.)

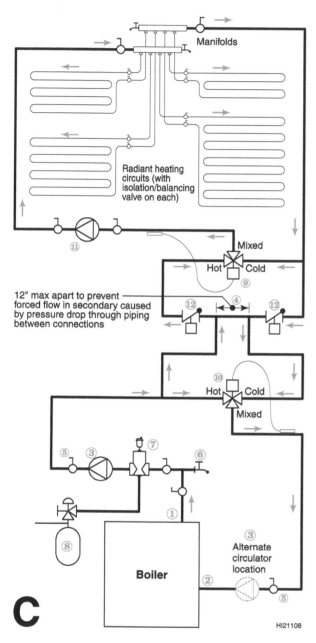

C

HI21108

① Boiler outlet (supply)	⑧ Diaphragm expansion tank with fill valve
② Boiler inlet (return)	⑨ Radiant supply water thermostatic mixing valve
③ Boiler loop circulator	⑩ Boiler return water thermostatic mixing valve
④ Secondary connection	
⑤ Isolation valves	⑪ Radiant circuit circulator
⑥ Purge cocks	⑫ Flow/check valves (to prevent gravity- or flow-induced circulation)
⑦ Air separator with automatic air vent	

Start-up and troubleshooting

After completing a radiant installation, we suggest you ust the "Start-Up Check List" below. This check list is from the Radiant Panel Association publication, "Standard Guidelines for the Design and Installation of Residential Radiant Panel Heating Systems."

Also see the "Troubleshooting Tool Kit", excerpted from the same publication for suggestions on what you'll need when troubleshooting radiant heating systems.

Start-Up Check List

(from R.P.A. publication, "Standard Guidelines for the Design and Installation of Residential Radiant Panel Heating Systems")

System Design

❏ Heat loss analysis for the building done.

❏ Building construction complete as shown in the same plans used to design the system.

❏ Acceptable procedure or design process used for the radiant panel system design.

❏ Installed system meets the design criteria and/or other accepted standards.

❏ Met customers' expectations and level of understanding of how the system will operate, particularly with respect to the differences between radiant heating and more traditional forms of heating systems.

Installation

A. Circuit Installation

❏ Circuit Spacing (on center distance between each pipe) as specified by the system design.

❏ Installed length of circuits as specified by the system design +/-3%.

❏ Installation methods comply with the manufacturers recommendations and RPA Guidelines.

❏ Installed circuits subjected to an acceptable pressure test following completion of the tube installation.

❏ All circuits and manifolds accurately labeled for length, location, and zone.

❏ Proper manifold location and height for accessibility and ease of service.

B. Mechanical and Piping

❏ Boiler or heat source capacity or output as specified in system design.

Start-Up Check List (continued)

(from Radiant Panel Association publication, "Standard Guidelines for the Design and Installation of Residential Radiant Panel Heating Systems")

❏ Boiler installation, piping, venting, safety equipment, combustion air supply, condensate drainage in compliance with the system design, manufacturer's guidelines, and local code requirements. When in question contact the equipment manufacturer.

❏ Thermal expansion tank pre-charge pressure and system fill valve setting match the system design pressure.

❏ Circulating pumps in the proper location and orientation. Recommended location is on the supply pipe after the system expansion tank.

❏ Circulation pump size, pipe size, valves, and accessories as specified in the system design requirements.

❏ Required flow control valves present and proper position.

❏ Boiler or heat source operating temperature as specified.

C. Controls

❏ Electrical schematic provided for the system's control? If not, try each zone control individually, observe operation and compare to expected results.

❏ Control settings match the system design specifications, or accepted industry standards.

❏ Room temperature controls or thermostats located and installed and wired properly.

❏ Thermostat anticipators set to match the current draw of the controls they are connected to.

D. System Start-Up

❏ Fill the boiler and boiler room piping; check for and repair any leaks.

❏ Fill and purge distribution system. This is usually done one zone or manifold at a time. Some installers prefer to fill and purge circuits individually.

❏ System cleaned and flushed.

❏ Suitability of system fluid verified: acceptable ph, freeze protection, potability, etc.

❏ System control settings checked:

1. Reset control heating curve or slope, heating curve shift or parallel displacement, maximum and minimum supply temperatures.

2. Set thermostatic or manual mixing valve settings to achieve design supply water temperature for each zone, if applicable.

3. Thermostat anticipator settings must correspond to control current draw.

❏ Start and operate the boiler or heat source.

❏ Proper and safe operation of boiler verified, (refer to the boiler manual) venting system, controls and safety devices.

❏ Operate one zone at a time and evaluate flow to each zone.

❏ Check operation of other control valves installed.

❏ With system zones ON verify that there is a normal temperature difference between the system supply and return. Typical 10-20°F.

❏ With zones running under load the boiler should cycle. If not: repurge, check circulator operation, check control valves, check boiler operating temperature.

❏ Owner has received proper instruction on system operation and use.

❏ Owner has been provided with the system design specification, equipment and component documents, warranties, and operation.

Troubleshooting Tool Kit

(from Radiant Panel Association publication, "Standard Guidelines for the Design and Installation of Residential Radiant Panel Heating Systems")

1. Hand tools including screwdriver with interchangeable bit set, pocket screwdriver, special tools or wrenches for manifolds and tube fittings, adjustable wrench, ratchet and sockets, needle nose pliers, "channel lock" pliers, etc.

2. Pressure gauge (0-100#) with ½", ¾" bushings and couplings and female hose thread adapter.

3. Two or three double ended (washing machine) hoses.

4. Utility/charging pump for filling, purging, and flushing systems. Essential for use with antifreeze and chemical additives.

5. Thermometers: pocket, gauge, and infrared; and thermocouples for use with multimeters to check water, pipe, and surface temperatures.

6. Manometer: U-tube, inclined or magnehelic gauge to measure and monitor fuel gas pressure.

7. Multimeter/tester: AC/DC volts, resistance, amperage/current with test leads and alligator clips and jumper wires.

8. Combustion test kit or flue gas analyzer with draft gauge, flue gas thermometer, smoke tester, and CO_2 and O_2 tester.

9. Antifreeze test strips or spectrometer.

10. Ultrasonic flow detector.

11. Spare parts assortment appropriate for system; i.e. relays, transformers, circulators, zone valves, sensors, copper and tube/hose fittings and adapters, etc.

12. Manufacturers equipment manuals.

• Start-up and troubleshooting

Hydronics Institute Section of AHRI
35 Russo Place
Berkeley Heights, NJ 07922-0218

part 9

Basic electricity

I=B=R Guide RHH

Residential Hydronic Heating . . .

Installation & Design

Hydronics Institute
Section of **AHRI**

I=B=R Guide RHH
Residential Hydronic Heating

Hydronics Institute Section of **AHRI**
35 Russo Place
Berkeley Heights, NJ   07922-0218

 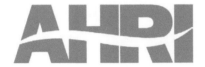

Contents – Part 9

Contents – *continued*

Contents *– continued*

Illustrations

Background

Electrons

Atoms are the basic building blocks of all matter. An atom behaves like our solar system, with the nucleus in the center. **Electrons** orbit around the nucleus like the planets. The nucleus contains two kinds of particles — **protons** and **neutrons**. See Figure 1.

What keeps the electrons in their orbits? Electrons are attracted to protons because of their electrical charge. Protons have a positive charge. Electrons have a negative charge. This electrical attraction is a strong force, and holds the electrons in place. Each atom contains an equal number of electrons and protons. So the negative charges balance the positive charges. The atom has no net charge. (You might think that the protons in the nucleus would want to push apart because they are charged the same, and would repel one another. But there is a force stronger than the electrical repulsion, called the nuclear force, that holds the protons and neutrons together in the nucleus.)

Figure 1 The atom

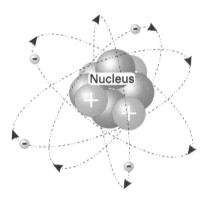

- ⊖ Electrons (negative charge)
- ⊕ Protons (positive charge)
- ⚫ Neutrons (no charge)

HI20709

The electrons in some materials are held more tightly than in others. The electrons in most **metals** can be dislodged easily. Those in electrical **insulators**, such as glass, fiberglass, most plastics, and rubber, are held tightly, preventing their movement. This is why metals make good electrical conductors — because it is easy for electrons to move through the material. Electrical insulators, on the other hand, resist electron movement, making them effective for separating electrical components from one another, from ground, and from us.

To be a good conductor, a material has to have something that will carry a charge, such as the electrons in a metallic conductor. Water is actually an insulator if it is pure. But dissolve a chemical, such as table salt (sodium chloride) in the water and it becomes a good conductor. See page 9-8 for more information.

Voltage

The force exerted by charged particles on other charged particles is called an **electric field**. An electric field develops around any charged object. See Figure 2. We can cause objects to develop a net charge through:

- chemical reactions (as in a battery)

- interactions with magnetic fields, as in an electric generator or motor

- interaction between dissimilar metals, as in a thermocouple.

Figure 2 Electric field

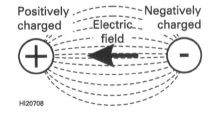

Positively charged — Electric field — Negatively charged

HI20708

Voltage (continued)

If you expose a conductive material, such as copper, to an electric field, electrons in the material will move toward the positively-charged object (or away from negatively charged objects). We usually measure electric fields in **volts**. Other common measurements are kilovolts (1,000 volts — used for power lines, for instance) and millivolts (one thousandth of a volt — created by thermocouples, for example). Batteries typically generate voltages between their terminals of 1.5 volts for an AA battery, up to 12 volts for a dry cell battery or automotive battery. The voltage at a wall outlet is 120 volts. The voltage in a boiler low-voltage circuit is usually 24 volts.

Current

Current is the movement of charged particles (usually electrons) in an electric field. We measure current in amperes most often (usually abbreviated to amp or amps). Other measurements are milliamperes or milliamps (one thousandth of an amp — used in thermostat circuits, for example) and microamperes or microamps (one millionth of an amp — used to measure the current flowing in flame rod sensors monitoring gas pilot flames, for example).

Resistance

When current (electrons) flows in a wire or any device, the electrons collide with other objects in their path, causing them to lose energy, like water loses energy to friction when flowing in a pipe. In electricity, we call this effect **resistance**. The longer or smaller a wire, the greater its resistance. Resistance also depends on the conducting material. Copper has much lower electrical resistance than steel, for example. Copper and aluminum are the most common metals used for distribution wire. Conductivity is the opposite (inverse) of resistance. Copper and aluminum have high conductivity (low resistance).

Voltage, current and resistance affect one another, as shown in **Ohm's Law**:

   Voltage = Current x Resistance, or **V = I x R**   (Eq 7-1),

   with current in amperes, voltage in volts, and resistance in ohms, typically.

This can be written three ways, depending on which of the values you want to find:

   V = I x R   (Eq 7-1)

   I = V/R   or I = V ÷ R   (Eq 7-2)

   R = V/I   or R = V ÷ I   (Eq 7-3)

Equation 7-3 shows that resistance controls current. Increase resistance and current decreases. Decrease resistance and current increases.

Examples:

• A current of 20 amps flows through a resistance of 30 ohms. What is the voltage required to cause this current?

   Use Equation 7-1: V = I x R = 20 x 30 = 600 volts.

• A voltage of 200 volts is applied across a 40-ohm resistance. What is the current?

   Use Equation 7-2: I = V/R = 200/40 = 5 amps.

• When a voltage of 120 volts is applied across an unknown resistance, the current is 10 amps. What is the resistance?

   Use Equation 7-3: R = V/I = 120/10 = 12 ohms.

Batteries

When chemicals dissolve in water, they separate into **ions**. Ions are positively or negatively charged particles, depending on the chemical elements or combinations of elements they are made of. Dissolve table salt in water, for instance, and the salt (sodium chloride) separates into positively-charged sodium ions (Na+) and negatively-charged chloride ions (Cl-). This is because there is an electron in sodium that easily separates when it is combined with water, leaving the ion positive. Chloride, on the other hand, encourages available electrons to attach when it combines with water. You wouldn't detect a net charge on the solution unless you applied an electric field. The electric field would cause negative ions to move toward the positive direction, and positive ions to move toward the negative direction.

A battery consists of metallic rods or plates immersed in a chemical solution. The rods are of different metals. The ions are attracted to the metal rods, with the positive ions preferring one metal, the negative ions preferring the other. This causes one rod to become positively-charged, the other to be negative. Attach a wire between these metal rods and electrons will flow through the wire, from the negative side to the positive side because of their attraction.

HI20718

Thermocouples

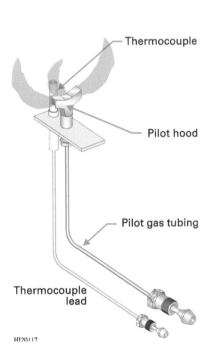

Thermocouple

Pilot hood

Pilot gas tubing

Thermocouple lead

HI20117

Another way to create an electric field is to fuse two metals together. Certain combinations of metals create a condition that encourages electrons to shift from one metal to the other. As the temperature increases, more electrons use the energy available from the heat to make the jump to the other metal. The voltage difference that occurs because of this charge separation is directly related to the temperature. For each combination of the right metals, the voltage difference (in millivolts) will always be the same for any given temperature. This is how thermocouples are used to measure temperature. Common thermocouple metal combinations are iron/constantan, copper/constantan, and chromel/alumel. Each combination has its own temperature curve.

Because thermocouples respond to temperature by developing a voltage across the junction, they

have been used for many years to prove pilot burner flames. A metal tube containing a thermocouple is immersed in the pilot flame. Wires attached to each side of the thermocouple connect to a gas valve. The gas valve can only stay open if there is a high enough voltage on the thermocouple, proving that a flame is present. (The thermocouple voltage powers a small solenoid, only strong enough to hold the valve open once it has been manually reset.)

The voltage across a thermocouple is approximately 250 millivolts. A higher-output device is the "thermopile," a group of thermocouples in a single body. The output of a thermopile is from 500 to 750 millivolts.

NOTE: Excessive heat on a thermocouple will cause the junction to fail.

Direct current circuit

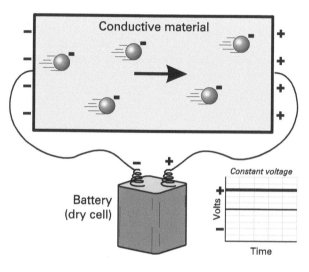

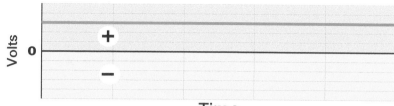

Alternating current

Batteries generate what is called direct current (or DC) — always flowing in the same direction. Most power sources, though, generate alternating current (or AC). This means that the current flow direction alternates. Alternating current means the electrical field (voltage) changes alternately from positive to negative. Alternating voltage is created using electrical generators — devices that create voltage using the way magnetic fields can create electrical fields.

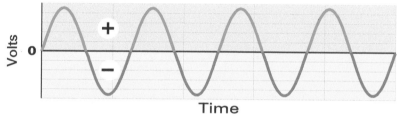

Notice the difference in the figures on this page between direct current and alternating current.

- Direct current voltage (battery source) always stays positive, so the plot of voltage versus time shows a straight horizontal line, always in the positive part of the graph.

- Alternating current is caused by a voltage source that alternates between positive and negative. When the voltage is positive (above zero voltage line), electrons (current) flow in one direction in the metal. When the voltage changes to negative, the electrons (current) flows in the opposite direction.

Alternating current (AC) circuit

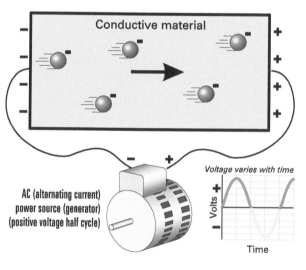

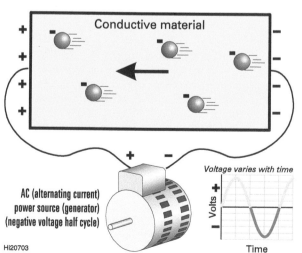

Most electrical applications in hydronic heating involve AC power and components. As we have discussed, the voltage and current in an AC circuit change constantly. So, how do we apply Ohm's law to an AC application?

- We discuss AC voltage and current using what is called the **rms** (root mean square) value — the equivalent direct current (constant) voltage and/or current that would have the same effect as the AC voltage and current.

- When your multimeter measures the voltage on a wall outlet, for instance, it will read 120 volts AC. This is actually the rms voltage.

- When we talk about AC voltage and current in this Guide, we will always be talking about rms values.

- Using rms values, you can apply Ohm's Law as if voltage and current were DC (constant).

HI20703

Power

As electrons move through wiring and components, they lose energy due to collisions with other objects. This energy consumption is called **electrical power**. Power is most often given in watts or kilowatts. You can calculate power from the equation:

- Power = Volts x Amps, or P = V x I,   (Eq 7-4)

- Current = Power ÷ Volts, or I = P ÷ V,   (Eq 7-5)

  where V and I are rms values of voltage and current for an AC circuit.

- *Note: Equations 7-4 and 7-5 assume voltage and current are in phase with one another; i.e., phase angle = 0.*

A common application of electrical power is to use the heat created when electrons lose energy to resistance. An electric water heater element consists of a nickel-chromium alloy wire coiled and suspended in electrically-insulating magnesium oxide powder in a metal tube. Rolling the tube decreases the tube diameter, causing the powder to become solid. As current flows through the nickel-chromium wire, the electron collisions in the nickel-chromium wire cause heat. This heat transfers through the magnesium oxide, through the tube and into the water. You can see the same concept in electric baseboard heaters, electric irons, hair dryers, and toasters.

Another way we apply electrical power is what makes it possible for you to read this page indoors — electric lights, which use the energy of electrons (or ions) to create visible light. Incandescent bulbs (ordinary light bulbs) use a small-diameter wire capable of being heated to a very high temperature (tungsten, for instance). When a material gets hot enough, its electrons begin to use the energy to jump in and out of orbits. Each jump takes a certain amount of energy. When the electrons fall back to their regular orbits, they give off this energy. You see it as visible light. The hotter the filament, the brighter the light will be. Thomas Edison's original electrical light didn't use a filament. It used a carbon arc — two carbon rods placed close together, with a high voltage between them. With the voltage high enough, the gas between them is pulled apart, into charged ions. They rush between the electrodes, giving off visible light when they join back together. This is the same effect that causes lightning to be visible. Neon and fluorescent lights work because voltage applied to the gas inside the bulb causes the gas to ionize. As the gas recombines, it gives off visible light.

How much current does it take to operate a light bulb? Use Equation 7-5. A 60-watt bulb operated on 120 volts would take ½ amp (60 watts ÷ 120 volts = 0.5 amp).

Power distribution

Measure the voltage at a wall outlet, and you will read between 115 and 120 VAC. How does it get there? Electrical power originates at a power plant, is transmitted through electrical lines, and finally reaches the residence or point of use. See Figure 3 for a brief description of electrical power distribution. Notice that power is generated as three-phase and distributed at very high voltage (up to 750,000 volts (750 kilovolts). Even the power lines running through residential neighborhoods will typically be 7,200 volts.

Why is the voltage from a power plant so high? Take a look at Equation 7-5. If you increase voltage, you decrease current, because current is power divided by voltage — the larger the voltage, the smaller the current. Smaller current means you can use smaller wires to do the same job. This makes it practical, then, to deliver large quantities of electrical power over long distances. Power companies keep the voltage high for as long as possible. Then they reduce the voltage near the point of use with transformers.

Figure 3   Power distribution to residences

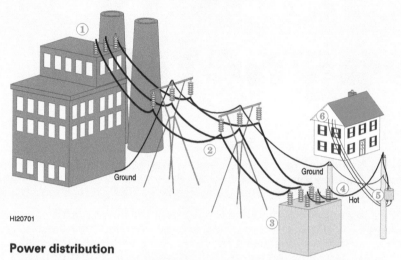

HI20701

Power distribution

1   **Power plants** generate power and increase voltage with transformers to reduce the current and resultant energy loss in transmission. Power plants deliver power to long distance transmission lines at as high as 750,000 volts. Power is generated and transmitted as 3-phase. Transformers in the distribution system adapt this to single phase for residential use. Industrial and commercial installations use 3-phase power at voltages such as 480 or 208 volts.

2   Long-distance **transmission lines** (up to 300 miles) connect to transmission stations, where voltage is lowered using transformers.

3   **Transformers** at transmission stations and substations reduce voltage for intended use — typically 7,200 volts for residential areas.

4   **Local distribution lines** typically carry power at 7,200 volts. These lines provide power to transformers for each residence. Three hot lines and a ground feed the network, but only one hot wire and a ground are connected to each residential drum transformer.

5   **Drum transformers** on power poles drop the voltage to 240 volts for residential use. Three wires enter the house — two insulated "hot" wires, with 240 volts difference, and a bare ground wire, from the earth ground wire on the pole. The voltage between each "hot" wire and the ground wire is 120 volts. This ground wire is connected to the neutral of the breaker box. The wires feed the breaker box, where wires are fed to the house from the breakers, providing both 120-volt and 240-volt circuits as needed.

What do we mean by phase when we say three-phase? Most residential motors are single phase, for example. Power is generated at power plants as three-phase. Phase means how voltage differences vary with time. The AC voltage/time curve on page 9-9 shows a single curve. In single-phase circuits, this single curve represents how voltage varies with time. Three-phase means there are three different curves — one for each power lead. Each curve is offset 1/3 of a cycle from the other. There are three power leads and a neutral. You won't encounter three-phase power in residential applications. But three-phase power (particularly 208 volts/3-phase and 480 volts/3-phase) is common for commercial installations.

Power in the United States and Canada is generated at 60 cycles per second (one cycle per second is also called a hertz, abbreviated hz). This means there would be 60 complete up and down curves like those on page 9-9 every second. We call this the frequency of the voltage. Many areas of the world generate at 50 cycles per second, or 50 hz. Be careful applying any motorized appliance designed for one frequency on another frequency. If you apply 50-hertz power to a 60-hertz motor it will run slower (about 50/60 times, or 5/6). Apply 60 hertz to a 50-hertz motor and it will run faster, probably causing it to overheat and quickly fail.

Electricity & magnetism

Overview

When charged particles move, they create a magnetic field. Scientist Hans Christian Oersted detected this in 1819. Michael Faraday later quantified the effect and discovered magnetic induction (creation of an electric field by a magnetic field).

Flow current in a coiled wire and a magnetic field will flow in and around the coil as shown in Figure 4. (Permanent magnets apparently have their constant magnetic field because of the way electrons move naturally within the material.) Likewise, magnetic fields can cause an electrical field (**magnetic induction**). Pass a changing magnetic field through a wire coil and a voltage develops across the coil (Figure 4).

These principles make our electrical power systems possible today. Magnetic induction in power generators creates the voltage that runs the world.

We also apply this electro-magnetic effect to make transformers, relays (using electromagnets), solenoids, and motors — components used in heating circuits and control systems.

Transformers

Transformers consist of two coils (primary and secondary) wound around an iron core, as in Figure 4. Iron interacts with magnetic fields, and concentrates the fields. So iron is used in transformers to concentrate the magnetic field created when current runs through the **primary coil**, and routes it through the **secondary coil**. The varying magnetic field flowing through the secondary coil creates (induces) a voltage in the coil. Connect the coil to an electrical circuit and current will flow.

We apply transformers throughout the electrical power distribution system to regulate the voltage suitably for the application.

- Power companies transform voltage from high line voltage to usage voltage at the power pole (to 240 volts, single-phase, 3-wire for residential applications).

- 240-volts, single-phase, 3-wire means two of the wires differ in voltage by 240 volts. Each wire differs from ground (third wire feeding house) by 120 volts.

- This allows you to apply higher voltage to heavy loads, such as electric stoves, electric baseboard and air conditioners, for example. You apply 120 volts to smaller for wall outlets and lighting.

Another way we apply transformers is to reduce voltage for control circuits.

- Lower voltage requires less insulation on the wiring and easier routing of the wires.

- Reducing voltage to 24 volts eliminates injury hazard from electrical shock, as opposed to 120 volts, which can cause lethal shock.

- You will find 24 volts used in thermostat circuits, zone valves and residential appliance gas valves, for instance.

Figure 4   Electricity & magnetism

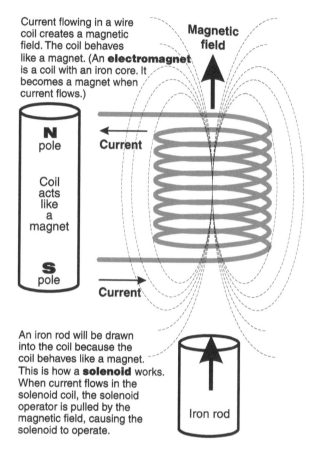

Current flowing in a wire coil creates a magnetic field. The coil behaves like a magnet. (An **electromagnet** is a coil with an iron core. It becomes a magnet when current flows.)

An iron rod will be drawn into the coil because the coil behaves like a magnet. This is how a **solenoid** works. When current flows in the solenoid coil, the solenoid operator is pulled by the magnetic field, causing the solenoid to operate.

Changing magnetic fields can cause current to flow in a coil. When a varying magnetic field flows through a wire coil, it causes an AC voltage across the coil.

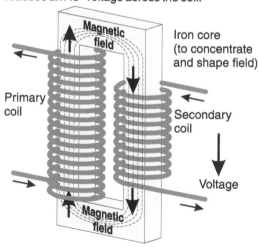

This is how a **transformer** works. Alternating current (AC) flowing in the primary coil causes a varying magnetic field. The iron core of the transformer concentrates and shapes the field. The magnetic field flowing through the secondary coil causes a voltage across the coil. If the circuit is closed, current will flow in the secondary circuit. If the secondary has fewer windings than the primary, the voltage will be lower than the primary. The secondary voltage is approximately equal to the primary voltage times the ratio of windings. So, a 120 VAC/24 VAC transformer would have about 5 times (120 ÷ 24) as many windings in the primary as in the secondary coil.   HI20705

Electromagnets & relays

Place an iron rod in the center of a wire coil and you have an electromagnet. When current flows in the wire coil, the magnetic field concentrates in the iron core, and the iron core behaves like a magnet. We apply this principle to relays (Figure 5). (Large electromagnets are often used to move metal, such as in industrial plant material handling and automotive scrap yards.)

Relays use electromagnets, as shown in Figure 5. When the relay coil is supplied with voltage, current flows through the coil, creating a magnetic field and turning the iron core into a magnet. The magnet attracts the metal relay arm, causing the contacts to be activated as shown.

Figure 5   Relay operation

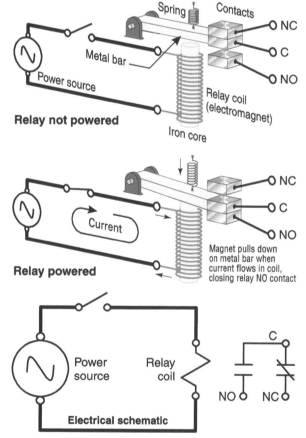

Typical relay, with voltage source and switch. This relay is called single-pole, double throw. The number of poles is the number of sets of contacts. Double throw means the relay contacts provide a normally-open (NO) and a normally-closed (NC) contact. The common terminal (C) is common to both. When the relay is not powered, there is a connection from common to normally-closed. When the relay is powered, the normally-closed contact opens, while the normally-open contact closes.   HI20710

Solenoids

Iron is attracted to a magnet. Place an iron rod in a magnetic field and it will be drawn into it (as in Figure 4, Page 9–13). We apply this principle in the operation of solenoid valves (Figure 6).

Figure 6   Solenoid valve operation

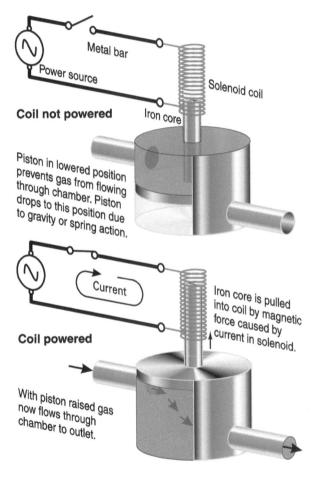

Typical operation of a solenoid gas valve. Piston rests at lower position with no power on coil due to gravity or spring action.

When coil is powered, current flows through coil. This creates a magnetic field through coil.

The iron coil is pulled into the magnetic field, just as it would be if a permanent magnet were placed where the coil is.

With piston above gas inlet opening, gas flows through chamber to outlet port, as shown.

When power to the coil is interrupted, piston drops back to closed position.   HI20713

Motors

Magnets attract when their poles are aligned opposite, and repel when they are aligned the same. We apply this principle in electric motors. Motors use wire coils distributed to take advantage of the magnetic attraction and repulsion of electromagnets. Apply current to the motor windings and the rotor (moving part of motor) begins to spin because of this interaction.

DC motors

The simplest example is the basic DC motor, shown schematically in Figure 7. DC motors usually use a permanent magnet to interact with the magnetic field caused by the windings.

Single-phase AC motors

AC motors also use the interaction between magnetic fields. But they don't use permanent magnets. AC motors use wire coils oriented so their fields cause a current to flow in the rotor, usually a squirrel-cage-shaped component made with conductive bars that allow current to flow back and forth from end to end of the rotor. (The rotor is the moving part of the motor, while the stator is the stationary part that contains the primary windings.)

What is most notable about AC motors is that, by nature, they don't want to start. They won't turn unless the magnetic field changes in the right way. The difference between types of AC motors is how they overcome the resistance to starting.

Split-phase AC motors

Split-phase motors overcome the stubbornness to start by adding a set of coils (called "windings") perpendicular to the primary windings. These are called the starter windings because they are only used to start the motor. The starter windings are shorter than the primary windings, so their magnetic field behaves differently from the primary windings. This causes an upset in the fields enough to cause the rotor to start turning. Once the rotor begins to turn, the field it creates interacts with the primary winding field, causing the motor to keep turning. The starter windings are then electrically disconnected. You may find split-phase motors used on two-piece residential and small commercial circulators.

Capacitor start AC motors

These motors also use a starter winding located perpendicular to the primary windings. In order to make the starter winding magnetic field different from the primary, they put a capacitor in series with the starter windings. The capacitor makes the starter winding field different from the primary winding field. This provides a stronger starting force than the split-phase motor, giving this type of motor a higher starting torque. The starter windings are electrically disconnected once the motor begins to turn. Typical applications are for pumps, tolls, and air conditioning motors.

Permanent split capacitor AC motors

The motors also use a capacitor in series with the starter windings, but the capacitor is smaller than a capacitor start motor, and the starter windings remain on during motor operation. The starting torque is lower than capacitor start motors, but the running efficiency is higher. These motors can also be run at variable speeds because the starter windings are on continuously to keep the rotor moving. You will find these motors applied to water-lubricated circulators and large fan motors such as used in unit heaters.

Shaded pole AC motors

Shaded pole motors unbalance the magnetic field by using "shading coils" — strategically placed small windings that disturb the magnetic field enough to cause rotor movement. The shading coils remain on during motor operation, making these motors suitable for variable speed uses. They have lower starting torque than capacitor motors. Typical applications include light motor loads, such as kickspace heaters and hot water unit heaters.

Synchronous AC motors

These motors, used on small loads such as clocks and motorized zone valves, operate similarly to shaded pole motors, but the speed is synchronized to a moving magnetic field. They are often referred to as "clock motors."

Three-phase AC motors

Three-phase motors don't need special starter or shading windings. The magnetic fields created by the three windings being out of phase with one another provide the rotational force.

Figure 7 Typical DC motor operation

DC motor

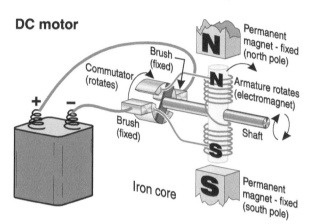

Schematic construction of a DC motor.

Applying voltage to the brushes feeds power to the commutator (split in two sections as shown). The armature windings connect to the commutator as shown.

Power fed to the armature coil causes a current in the windings, resulting in a magnetic field — turning the armature core into a magnet. With the north pole of the armature lined up with the permanent magnet north, as above, the armature is pushed away, causing it to rotate, trying to align its poles opposite to the electromagnet.

When the armature rotates about 180°, the commutator now gets its voltage in reverse because the commutator halves are in contact with different brushes. This reverses the current in the armature, causing the magnetic field to reverse as well. Once again the magnetic fields line up north to north and south to south, repelling the armature, sending it through another 180° rotation. When it reaches the top again, the commutator voltage reverses, and the cycle continues.

The opposing magnetic fields of the armature electromagnet and the permanent magnet cause the armature and shaft to rotate as long as DC voltage is applied to the brushes.

HI20711

Semiconductor devices

Our discussion in Part 9 deals with **electromechanical** devices. Electromechanical means using electrical and magnetic properties to operate mechanisms. But many of the controls we use today contain semiconductor components and even micro-processors, so we will discuss them briefly.

Silicon is an abundant mineral, capable of being formed in a highly pure condition. In its pure state, silicon is an excellent electrical insulator, but it has a peculiar behavior when it contains certain impurities. If silicon contains traces of aluminum, the silicon/aluminum contains "gaps," called holes, that attract electrons if made available. This is called a "p-type" semiconductor, because it behaves as if positively charged. Silicon/phosphorous, on the other hand, contains electrons that are easily excited to move. This is called an "n-type" semiconductor because it has negative particles available.

Place a p-type semiconductor next to an n-type semiconductor and you have the makings of a semiconductor device. Nothing happens unless you apply an electric field to the junction. Then current starts to flow — electrons jumping from the n-type material to the "holes" in the p-type material. These materials are called semicon-ductors because they only conduct when excited by the right conditions (electric field, light, or heat, for examples).

So, how is this semiconductor behavior used? Place the right silicon mixtures next to one another, and you can create a "device" which operates when you apply an electrical voltage to it. You can make semiconductor switches (triacs), amplifiers (transistors), or rectifiers (diodes — devices that allow current only to flow in one direction, causing an AC current to become DC) by carefully controlling the amount of impurity and location of impurities on a silicon wafer. Semiconductor manufac-turers do this by exposing the silicon to the right gases in selected areas, using chemical masking materials to protect the areas not to be affected. This process of adding impurities to the silicon is called doping. This is how microprocessors are made — with the high precision of the process, thousands of semiconductor devices are created on a small silicon wafer.

Today's electronic controls use semiconductor devices to provide advanced func-tionality cost effectively and in small spaces. Programmable thermostats, for ex-ample, would be practically impossible if not made with electronic components.

Circuits

What is a circuit?

An electrical circuit contains:

* a **power source** (battery, generator, thermocouple, etc.)
* a **conductor** (wire, for example)
* a **load** (motor, relay coil, resistor, solenoid, or light bulb, for examples)
* a **switch** (usually).

Wiring diagrams

To simplify design, installation, and troubleshooting of electrical circuits, wiring layouts are usually drawn using electrical schematic symbols, as shown in Figure 9, Page 9–18.

You will find both point-to-point wiring diagrams (that show actual connection points for wiring) and ladder wiring diagrams (that show wiring connections as if they were steps on a ladder). See Figure 10, Page 9–18 for a typical example comparing wiring diagram types.

Wire functions

Most of the wiring diagrams you will work with include wires with one of three functions:

"Hot" wires carry line voltage.

"Neutral" wires carry current back to the source, but have no measurable voltage.

"Ground" wires attach to the appliance chassis, connecting it to earth ground. Ground wires protect against the appliance chassis becoming electrically charged should a component or wire inside short to the enclosure.

Figure 8   Voltage splitting with resistors

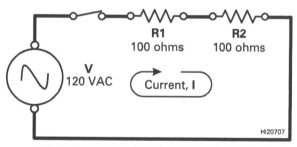

Find current from Ohm's Law: **I = V/R**
    Total resistance, R = R1 + R2 = 200 ohms
    Voltage, V = 120 volts
    Current, **I = V/R** = 120/200 = 0.6 amps
Find voltage drops:
    V1 = voltage drop across R1 = **I x R1** = 0.6 x 100 = 60 volts
    V2 = voltage drop across R2 = **I x R2** = 0.6 x 100 = 60 volts

Loads

A **load** is any mechanism operated by electricity, such as a motor, relay coil, solenoid, light bulb, valve operator, or a resistor. The resistor, an electronic device, is used to regulate current. It can also be used to distribute voltage. Voltage drop is the change in voltage as current flows through a device. See Figure 8 for an example of voltage distribution with two resistors in series. The voltage drop across each resistor is 60 volts, as shown. Electronic circuits use resistors extensively to distribute voltage and control current.

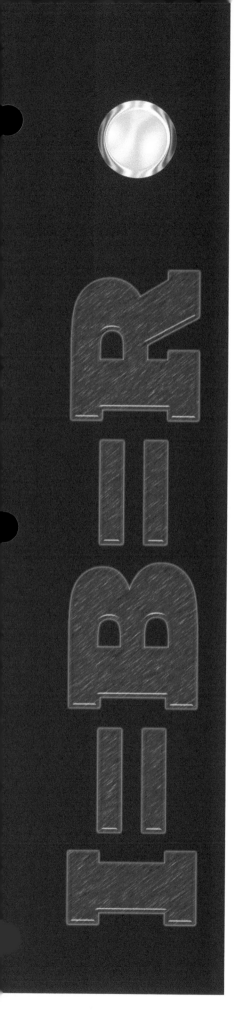

Figure 9 Electrical schematic symbols

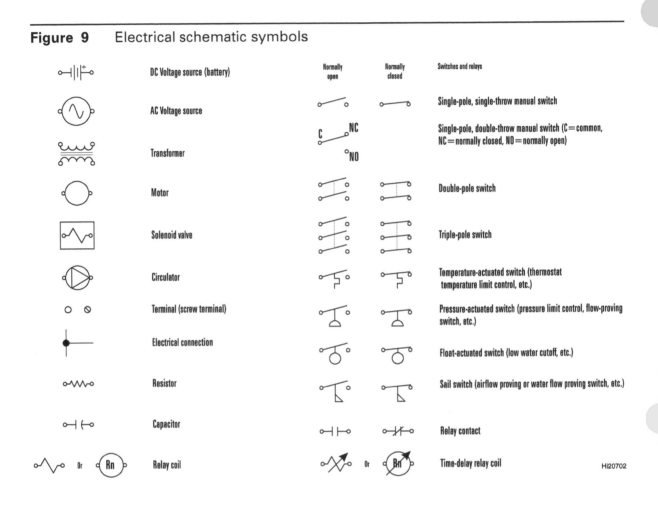

Figure 10 Comparison of point-to-point and ladder wiring diagrams

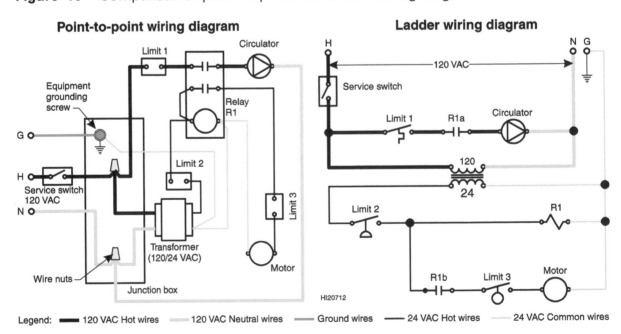

Load ratings

When loads are wired in series, each takes part of the available voltage in the circuit, as in Figure 8, Page 9–17, showing resistors in series. Most hydronic applications, however, involve loads in parallel, such as a damper motor and gas valve on a gas boiler, the ignition transformer and burner motor on an oil boiler, or multiple circulators powered from a common 120 VAC circuit on a circulator-zoned system. Always keep in mind that the power source has to supply enough current to satisfy all of these devices. Check the load rating of each connected load:

• Motors will usually be rated in full load amps. You will need to verify that the fuse and/or breaker in the circuit has sufficient amp capacity for all connected motors and loads.

• Relay coils, solenoids, and valves will usually have a "va" rating, giving the required volts times amps for the device. These devices are usually supplied from a transformer. Transformer load capacity is usually given in va or volt-amps. Make sure the transformer va capacity meets or exceeds the total connected load in va. To estimate the required amperage to the transformer, divide the transformer va rating by the transformer primary voltage.

• Calculate the combined amp load of all connected loads to be sure the breaker or fuse and wire are rated properly.

NOTICE: For control circuits (zone valves, etc.), always make sure the total va of all connected devices is **LESS** than the control transformer rating. A 38-va transformer will burn up (fail) if you connect a load larger than 38 va, for example.

Wire ratings

The higher the current, the bigger the wire must be to handle the load. Verify that the wire size you are using complies with the National Electrical Code and local code requirements. Also check equipment manufacturer's instructions for any special requirements regarding wire size or type.

Series/parallel connections

Electric circuits are constructed with components either in series (one after the other) or parallel (Figure 11).

A typical boiler limit circuit may contain a temperature limit switch, a low water cutoff switch and other limit devices. These are generally wired in series so the boiler will shut down if any one of the limit controls should open.

On a zoned system using zone valves, each zone valve usually has a switch that closes when the zone calls for heat (called an end switch). The zone switches are wired in parallel in the thermostat connection to the boiler so the boiler will come on when any one of the switches closes.

See Part 8 of this Guide for more information on application of controls.

Figure 11 Comparison of series and parallel wiring

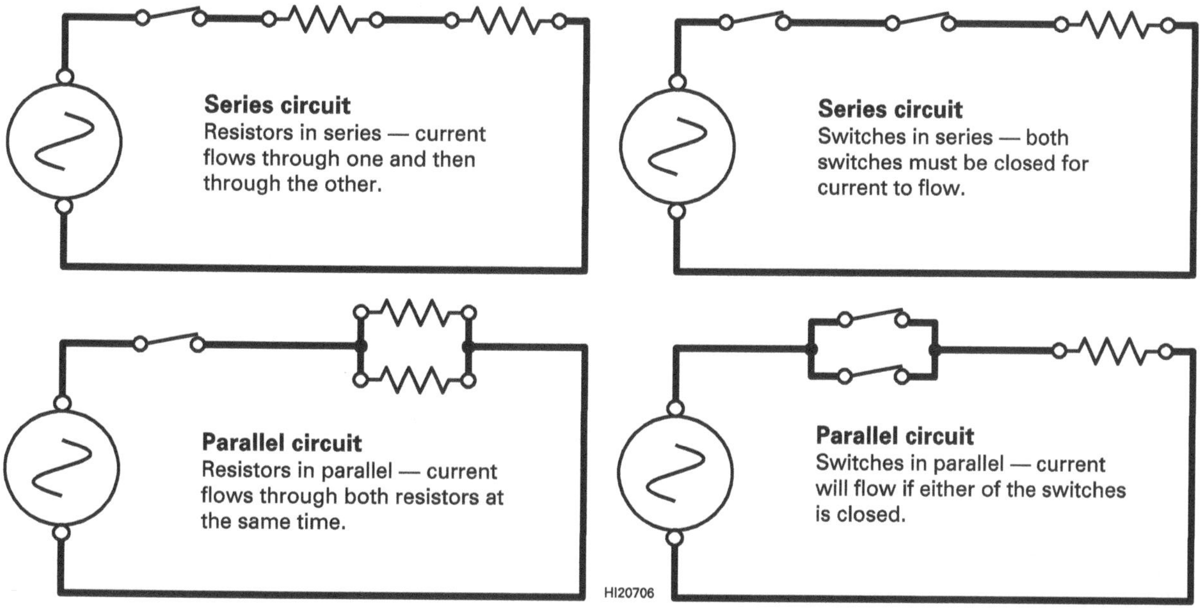

Series circuit
Resistors in series — current flows through one and then through the other.

Series circuit
Switches in series — both switches must be closed for current to flow.

Parallel circuit
Resistors in parallel — current flows through both resistors at the same time.

Parallel circuit
Switches in parallel — current will flow if either of the switches is closed.

HI20706

Electrical test equipment

Voltage indicators

Many styles of electrical indicators are available. The pen-style indicators are very portable and simple to use. Always check the device on a known live circuit before using it to check other circuits.

Troubleshooting with a voltage test light or indicator is much faster than using a continuity tester (below). You can check for voltage compared to neutral and also check directly across a load to see if voltage is available.

Continuity testers

You connect a continuity tester to a circuit to check whether there are any interruptions in the circuit (such as open switches or limits, broken or disconnected wires, for examples).

The example in Figure 12, Page 9–21 shows a multi-function voltage indicator and continuity tester. This device is easily held with one hand — one probe is attached to the meter in the probe holder slot. The fixed probe can be touched to a device, leaving the other hand free to use the loose probe to check for voltage or continuity.

Most continuity testers cannot be used with live voltage. You must disconnect power to use one unless it is specifically designed to automatically detect voltage.

Multimeters (VOMs and DVMs)

VOM means a volt-ohmmeter. DVM means a digital multimeter. Digital meters have higher input resistance, making them more accurate when measuring resistance.

Multimeters have the advantage of allowing you to measure the value of current, resistance and voltage — particularly useful if checking for low voltage conditions on the line. Many are available with current transformer-type amperage probe. This probe consists of a coil that opens like a pair of pliers and springs closed. By placing this coil around a wire, you can measure the current with the meter without having to remove any wires. The coil uses the magnetic induction principle used in transformers. When current flows through the coil wires, it causes a voltage to develop in the coil. The meter measures the value of this voltage to determine the amount of current.

Figure 12, Page 9–21 shows a typical multimeter, showing the options available with the front dial. Always remember to zero the scale on these meters before use.

Invest in a set of alligator clips (that press onto the probe ends). They allow you to attach one probe while working with the other, or to attach both probes and work hands free from the meter.

Figure 12 Typical continuity tester (multi-function) and multimeter

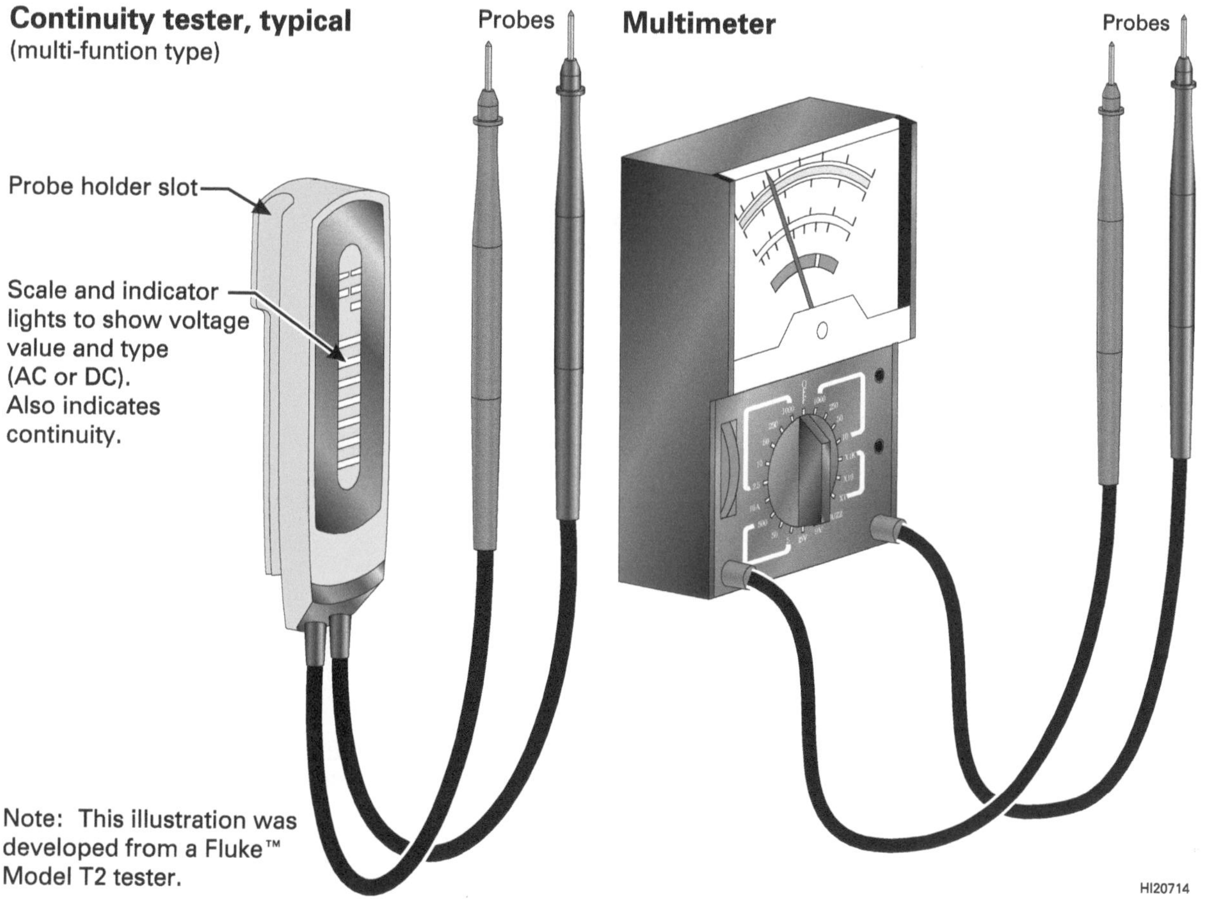

Continuity tester, typical
(multi-funtion type)

Probes

Multimeter

Probes

Probe holder slot—

Scale and indicator —
lights to show voltage
value and type
(AC or DC).
Also indicates
continuity.

Note: This illustration was
developed from a Fluke™
Model T2 tester.

HI20714

Measuring DC voltage and current

DC voltage and current have polarity. That is, they have a definite positive and negative direction.
When measuring DC current, then, you will need to make sure the positive probe is on the positive
side of the circuit to obtain a correct reading. Most multimeters won't register if the probes are re-
versed. If you don't get a reading, try reversing the probe positions.

NOTICE: ALL voltmeters —Always make sure to set the meter scale for the correct range and voltage type (AC or DC). If you are unsure of the voltage you will find, set the meter for the
highest scale. Change the scale to a lower range only after you are sure of the voltage.

Troubleshooting with a ladder schematic

Using the ladder diagram

To understand how a circuit operates, study the ladder wiring diagram. With the ladder diagram, you can trace the power source for any load. If a load isn't coming on when it should, use the ladder wiring diagram and a continuity tester or voltmeter.

To troubleshoot with a continuity tester, be sure to disconnect power to the appliance. Also make sure there is only one way to get continuity. That is, check the ladder diagram to see if you could get continuity through a backwards path.

For example, look at Figure 13. Suppose you want to check continuity from point A to point B. If you didn't make provision before checking, there are two ways current from the continuity tester could flow backwards, making it look like you had continuity. Notice the two backwards paths the current could take. To prevent these paths, what should you do? Remove one of the wires connected to the motor. Now, no current could flow backwards. So your continuity test will work. This is a valuable use of the ladder wiring diagram. You can easily check for false signal (backwards) paths. The general rule to make sure you won't get false readings is to disconnect the circuit on at least one end of the section you want to test.

Figure 13 Disconnect a wire on one side of the section to be tested to prevent backwards paths of continuity

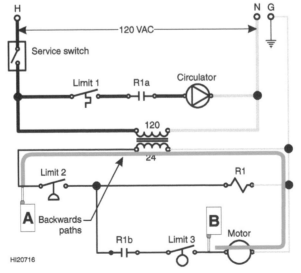

If you use a multimeter to check continuity, be sure to set it on the lowest resistance range (usually ohms X1 or lower). Never connect a multimeter set for resistance or a continuity tester to a powered circuit.

You could also check continuity from A to B in Figure 13, Page 9–22, if power is on using a voltmeter or voltage indicator. To do this, check voltage from point A to ground. It should be 24 VAC if the transformer is powered and functional. If there is continuity from A to B, then, you should also read 24 VAC at point B when you probe with a voltmeter. If you cannot power the appliance, you will have to use a continuity tester instead.

WARNING: When checking for voltage in a circuit, do not contact wires or electrical terminals. Even a wire or terminal that should be neutral can have live voltage if the circuit is open. Contact with a live contact or wire could result in severe personal injury or death.

Troubleshooting example

We will use the circuit discussed above to troubleshoot a problem. The symptom is: Power is on (we think), and the limits (1, 2, and 3) should be satisfied and closed, but the motor isn't operating. What is wrong?

Figure 14, below, shows what the ladder diagram looks like if you don't know whether switches and limits are open or closed. We have shown the voltage readings found as we check through the circuit. Which switch or device is open, preventing voltage from getting to the motor? Two problems are illustrated in the figure.

NOTICE: Using a test light or voltmeter, you can check across a load (such as relay coil R1 in Case 2, Figure 14). If you measure voltage across the load, you know that there is voltage available AND there are no breaks in the circuit to and from the load.

Figure 14 Troubleshooting with a ladder wiring diagram

Case 1

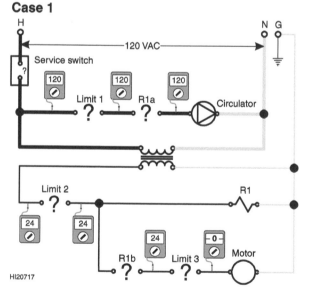

With this situation, checking with a voltmeter, we find voltage at all points up to the motor. There is voltage before Limit 3, but no voltage after limit 3. Limit 3 is not closed. All other limits and relay R1 seem to be functional, and closed.

Check conditions. If Limit 3 should be closed, but is not, check wires to and from limit to be sure they are securely attached. Replace limit if required and retest.

Always disconnect power before attempting to remove any component or perform continuity testing.

Case 2

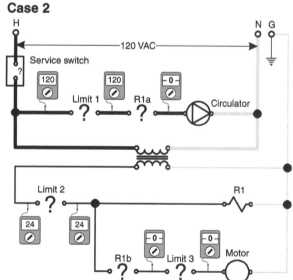

When checking this circuit with a voltmeter, we find the conditions shown. We show voltage up to relay contact R1a, but no voltage to the circulator. Relay contact R1a is not closed for some reason. Take a look at the low voltage (24 VAC) circuit. We find voltage up to the relay R1 coil, so the relay should be operating.

Turn off power and check wires to the relay and its contacts. If they are intact and securely connected, replace the relay and retest.

If replacing the relay doesn't solve the problem, turn off power and check wires to the relay and its contacts using a continuity tester.

Be sure to disconnect one side of the section you are testing to avoid false readings.

Hydronics Institute Section of AHRI

35 Russo Place

Berkeley Heights, NJ 07922-0218

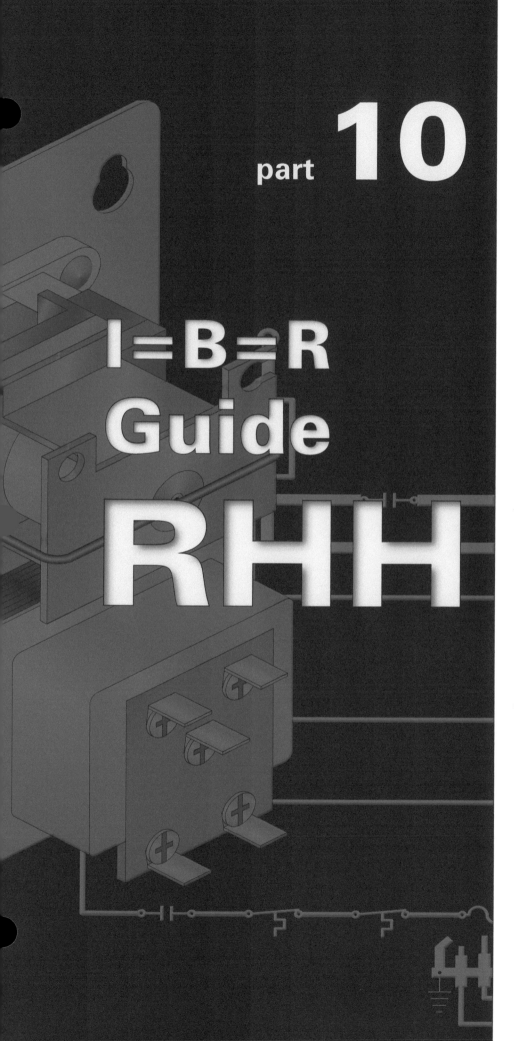

part **10**

Controlling and wiring hydronic systems

Residential Hydronic Heating . . .

Installation & Design

Hydronics Institute
Section of **AHRI**

I=B=R Guide RHH
Residential Hydronic Heating

Hydronics Institute Section of **AHRI**
35 Russo Place
Berkeley Heights, NJ   07922-0218

Contents – Part 10

Contents

• About controls

• How boilers operate

Contents *– continued*

- ## Wiring zone controls

- ## Outdoor reset controls

Contents _– continued_

Illustrations

About controls

Overview

For many of your hydronic installations, you will only need to wire from a thermostat or thermostats to the boiler. For others, you will install zoning devices (controls and zone valves or circulators). For these installations, you will have to connect more wiring and make the right decisions on components and capacities. Part 10 will explain the functions and wiring requirements of the primary components you will need to understand in order to electrically connect your installation.

The conclusion of Part 10 includes an introduction to outdoor reset controls. Remember when applying outdoor reset or any control that will affect return water temperature to the boiler — pipe the boiler as discussed in Part 3 for low temperature systems. You must do this to protect the boiler from potential damage due to flue gas condensation unless the boiler manufacturer's instructions say there is no minimum return water temperature.

What controls do

Hydronic system controls do one or all of the following: monitor conditions, sequence devices, or annunciate.

Controls monitor condition(s)

Controls "monitor" conditions such as temperature, pressure and flame presence. They monitor using an "input," such as the spiral bi-metal band in a thermostat. The spiral band expands as its temperature rises. This movement opens a switch. An electronic control might take input from temperature sensors mounted in the room or attached or inserted in the water. An ignition control or primary control monitors flame using a flame rod (flame rectification) or cad cell (sensing light from flame). Here the input is the current that flows through the flame sensor.

Controls sequence devices

Controls use the input they monitor to cause a response — closing a switch or relay to activate a burner or circulator, for example. They can also send a "signal" to a device to cause a proportional response — that is, allowing just the right amount of reaction. For example, a proportional control can drive the motor on a 3-way valve to position the valve for just the right mix of hot and cold water needed. Controls often sequence multiple devices, such as a pilot gas valve followed by a main gas valve on a gas ignition control.

Controls annunciate

Controls can signal what is happening, that is "annunciate." This can be as simple as closing an alarm contact or as elaborate as an electronic display providing a readout of system conditions.

Recognize how each control in your system monitors, sequences, and annunciates. Your installations will be simpler and your troubleshooting faster.

Avoid common errors

☐ Know how system controls work and how to set up for the system.

☐ Know limitations and requirements: voltage, current, accuracy, availability of replacement parts, etc. Be sure to size transformers large enough to handle the total connected va load.

☐ Make control systems as simple as possible to meet job requirements.

☐ Install components in a neat, well-organized layout.

☐ Label all components and maintain all documentation and instructions.

☐ Separate low-voltage circuits from high-voltage circuits where possible.

☐ Use multiple wire colors to ensure accuracy and traceability.

Figure 1   Typical room thermostat, mechanical type

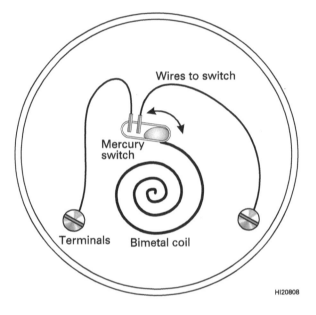

HI20808

Room thermostats

Mechanical thermostats

Mechanical thermostats usually use a bi-metal coil, such as shown in Figure 1. "Bi-metal" means two dissimilar metals bonded together. The idea is to use metals that expand differently with temperature changes. Because one metal tries to move more than the other, the bonded bi-metal bends. Form a bi-metal strip in a coil and the coil expands as its temperature rises or contracts when its temperature drops.

The typical thermostat shown in Figure 1 uses a bi-metal spiral band with a mercury switch attached to the end. The switch is shown open, indicating the room temperature is above the setting of the control. When the room temperature drops, the coil cools off, causing the spiral to contract. The switch tube becomes level, causing the mercury inside to level out as well. The switch contacts will then be immersed in the mercury, causing the circuit to close because the mercury is conductive. The boiler, zone valve or circulator for the zone will start.

NOTICE: You must be sure mercury-switch controls are installed level. Follow the manufacturer's guidelines. If you don't, the control will not regulate temperature (or pressure) as shown on the setting dial. On steam boiler applications, controls are often piped to a siphon tube. Siphon tubes expand and contract with pressure and temperature. So always mount mercury-switch pressure controls so the control will move front-

to-back and not side-to-side as the siphon moves. The mercury switch is mounted parallel to the front, and its position would change if the control were rotated by side-to-side siphon movement.

Electronic thermostats

Electronic thermostats use electronic temperature sensors. They usually contain a microprocessor and electronic components to allow programming their operation. When applying electronic thermostats, check to see whether they require a constant voltage source in order to charge their batteries. If so, make sure they wire correctly to the system to avoid battery rundown. Instruct the owner how to adjust and program the thermostat. This could avoid unnecessary tampering that could result in poor system performance.

Anticipator setting

Thermostats usually include an anticipator. Its purpose is to anticipate room temperature change as heat is added by the heating system, shutting off the heat before room temperature reaches the thermostat setting. The anticipator assumes there will be a continued rise in temperature after the heating unit is shut off. This continued rise is called the override, and is due to heat left in the terminal units after the heat cycle is stopped and to delays in room air temperature mixing.

The anticipator usually includes a resistor circuit that heats up due to the current flowing through the thermostat. You have to set the anticipator to the amount of current that will actually flow. If the anticipator setting doesn't match actual current, the thermostat will cycle too often or stay on too long.

Determine the anticipator setting:

• Thermostat wired directly to boiler: Read boiler manufacturer's wiring diagram label or instructions for the required anticipator setting.

• Thermostat wired to zone valve: Read valve current from manufacturer's literature or label.

• Thermostat wired to circulator relay: Read relay current from manufacturer's literature or label.

• Thermostat wired to zoning panel or electronic control: Control literature or label will state required thermostat anticipator setting.

Selection and application tips

Check the voltage of the application:

Most hydronic thermostat circuits are low voltage (24 volts), but be sure. Some control systems require line-voltage thermostats (particularly electric baseboard installations).

Check the amp draw:

You need to know the amp draw to set the anticipator and to be sure the load (amps) is within the limits of the thermostat you use. When using line voltage thermostats to directly control heating (electric baseboard, for example), don't connect more heaters than the thermostat can support.

Determine customer's needs:

You may need to use electronic thermostats if the customer wants to program them for night setback or occupancy changes.

Verify thermostat specifications:

Check the specifications of the thermostats you use to make sure they meet the above requirements.

Mounting/locating thermostats

Mount (and level) thermostats according to the manufacturer's instructions. Locate thermostats on interior walls away from drafts, direct sunlight, and heating/cooling effects of heating units or appliances. Why?

Mounted on an outside wall, the thermostat would always sense a temperature below the average in the room.

Mounted above a heating unit or near a stove or oven, for example, the thermostat would sense a temperature higher than the average in the room.

Exposed to drafts or direct sunlight, the thermostat would respond

Figure 2 Using 2-stage room thermostat for extended circulator operation

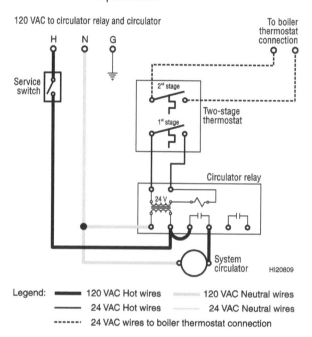

Legend:
— 120 VAC Hot wires   — 120 VAC Neutral wires
— 24 VAC Hot wires   — 24 VAC Neutral wires
------- 24 VAC wires to boiler thermostat connection

incorrectly, causing poor control of the average room temperature.

Continuous/extended circulation

Operating the circulator continuously improves temperature control throughout the system and reduces potential noises due to rapid expansion of piping. You can wire the circulator to run continuously (with a manual switch to turn it off during non-heating periods). The better method is to use two-stage thermostat(s), resulting in extended circulator operation.

You can upgrade most zoned systems by using two-stage room thermostats, with the first stage used to operate the circulator (or zone valve), the second stage to operate the boiler. This provides the smooth temperature control advantage of continuous circulation, but turns off the circulator(s) when heating isn't needed. The boiler will cycle much less often than when cycling the circulator(s) with the boiler because the zones use heat available in the system water before starting the boiler.

See Figure 2 for a typical wiring diagram. You will need to install a circulator relay as shown and electrically disconnect the circulator from the boiler controls if using a boiler-mounted circulator. This method starts zone circulation first. The zone only turns on the boiler if the heat available in the system isn't enough to bring the room up to temperature.

Limit and operating controls

Boilers use limit and operating controls to control water tempera- ture (water boilers) or steam pressure (steam boilers). The controls monitor temperature (or pressure) and shut off the boiler/burner if the temperature (pressure) exceeds the control setting. An op- erating control regulates the boiler temperature or pressure at a given setting. A limit control limits the maximum temperature or pressure. If there is only one control, it serves both as operating control and limit. If multiple controls are used, the limit is the one with the highest setting.

The boiler "limit circuit" means the portion of the wiring includ- ing limit controls (high temperature or pressure, low water, high or low gas pressure, etc.). Be sure to test all limit controls at least annually (or more often if directed by the control and/or boiler manufacturer) to verify they are operating correctly.

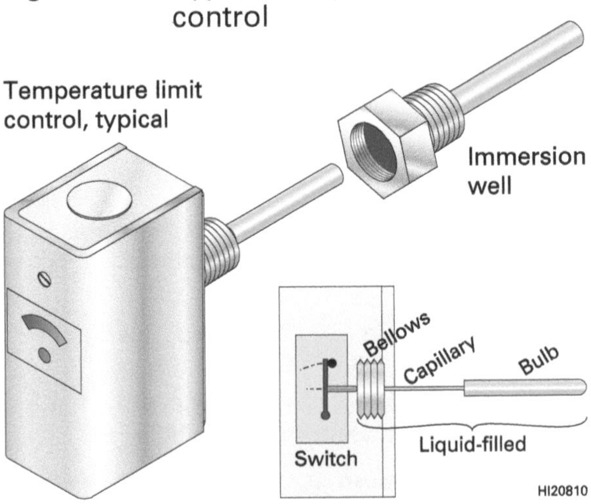

Temperature controls

See Figure 3, a typical liquid-filled temperature control. The con- trol sensing bulb, capillary, and bellows are filled with a liquid that expands a known amount as temperature increases. As the liquid expands, the bellows flexes open, pushing the switch linkage until the switch opens (high limit) or closes (low limit).

To simplify replacement and servicing, the bulb is often inserted in a water-tight well, as shown. This allows removal of the control with pressure on the system.

Other temperature controls include:

Bi-metal devices that use a bi-metal disc. The disc snaps inside out because of the bi-metal action, closing or opening a switch from the movement. Most bi-metal disc controls are direct mount (with the disc in contact with a pipe or surface). You will find bi-metal temperature limit controls used for spill switches (mounted on draft diverters to shut off the boiler in the event of downdraft).

Thermal fuse elements. Many gas boilers use a thermal fuse ele- ment (TFE) on which the fuse melts, opening the circuit, if the tem- perature exceeds its limit. These fuse-style limit controls are most often used for rollout switches (to shut down boiler if temperature in burner area rises too high).

Pressure controls

Steam boilers use pressure limit and operating controls to turn the boiler off when the steam pressure reaches the control setting. Low pressure boilers can never operate above 15 psig steam.

Most pressure limit controls use a diaphragm operator as shown in Figure 4. As the pressure increases, the diaphragm is pushed upward, eventually opening the switch.

Another common application of diaphragm-operated pressure switches is for airflow proving. Induced draft gas boilers use pres- sure switches to ensure there is a negative pressure upstream of the inducer blower, indicating the blower is operating.

Figure 3 Typical temperature limit control

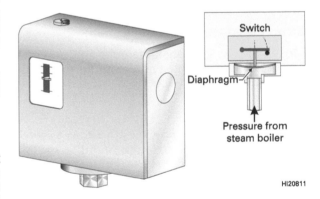

Figure 4 Typical pressure limit control

Differential

A control's differential is the difference between the cut-in and cut-out values. The cut-in value is the temperature (pressure) at which the control closes, and the cut-out value is the temperature (pressure) at which the control opens. The dif- ferential is usually subtractive; that is, the control cut-out value is the control dial setting, and the cut-in value is the cut-out value minus the differ- ential. For example, if a limit control with a 20°F differential is set at 180°F, the control switch will open at 180°F and close again if the temperature drops 20°F below that setting, or 160°F. A pres- sure switch with a 5 psig differential set at 8 psig will open at 8 psig and close again when pressure drops below 3 psig.

If a limit control has an adjustable differential, set the differential so the control comes back in within a reasonable time without excessive cycling. A wide differential will reduce cycling, but can cause large room temperature control swings by failing to react quickly enough. Be particularly careful of wide differential settings on residential steam boiler controls. Most hydronic steam heating systems should operate with a cut-out pressure of about 2 psig. On one-pipe systems, be sure the cut-in pressure is lower than the dropout pressure of the air vents, to allow them to open between cycles.

Manual reset

Limit controls will often include a manual reset feature, requiring human intervention to restart the boiler after they open. Always instruct owners not to use the reset button excessively. If a limit control shuts down often, something may be wrong with the system. The owner should have the system checked/tested by a qualified technician. Instruct the owner never to attempt replacing a thermal fuse element rollout switch. A qualified technician should investigate why the control opened and correct the condition before replacing the control and restarting the boiler.

Multifunction controls

Transformer relays (Figure 5)

Most residential boiler control systems cycle the circulator with the boiler. The circulator starts when the thermostat closes (calls for heat). The circulator continues to operate so long as the thermostat circuit is closed. The burner shuts down any time the limit control opens, but the circulator continues to operate.

To simplify wiring and service, many boiler manufacturers provide a circulator relay with the boiler. It is combined with the transformer as in the transformer relay (Figure 5) or in a combination control (see discussion following). See Figure 16, Page 10–21, and discussion for a typical boiler application.

The relay is usually a plug-in type, allowing its replacement without replacing the entire assembly. Carefully match the replacement to the original relay to avoid possible serious problems.

Combination limit/relay controls (Figure 6)

Combination limit/relay controls provide the functions of a temperature limit control with the transformer/relay function. The control includes its own transformer, a temperature limit switch and a relay to operate a circulator and the boiler. See Figure 16, Page 10–21, and discussion for a typical application.

Boilers equipped with tankless heater coils often use a tri-function control (sometimes called a triple aquastat). These controls include a low limit setting as well as a high limit setting. When the boiler water temperature drops below the low limit setting, the control turns on the burner (without starting the circulator) in order to bring the boiler water temperature up to the minimum needed for quick heating response for the tankless coil. See Figure 17, Page 10–23, and discussion for typical applications.

Figure 5   Typical transformer/relay

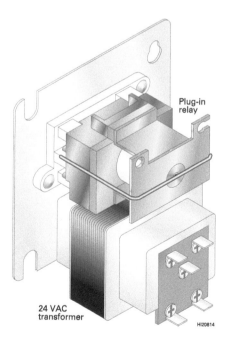

Plug-in relay

24 VAC transformer

HI20814

Figure 6   Typical combination limit/circulator relay

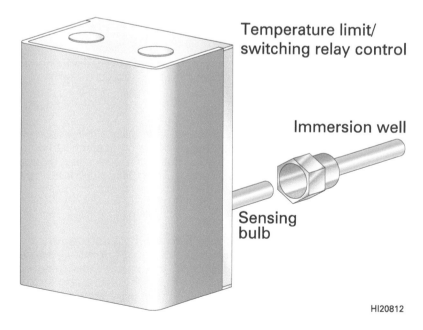

Temperature limit/ switching relay control

Immersion well

Sensing bulb

HI20812

Low water cutoffs

Probe-type low water cutoffs

Probe-type low water cutoffs (Figure 7) use electricity to verify water level. They include a probe inserted into the boiler at the lowest permissible water level. The probe is usually wired into the low water cutoff's relay coil circuit. The coil will only activate if current flows through the probe to ground (boiler sections or pressure vessel). Current will flow from the probe to ground if it is covered with water and the water is conductive.

In some situations, if the water is too pure, it may not conduct enough electricity to satisfy the probe low water cutoff. Consult probe low water cutoff manufacturer's instructions and a chemical treatment technician for water treatment.

Many probes now incorporate a special timing feature intended to rule out the possibility of fooling the probe circuit with foam (instead of water). The problem is that foam can build up inside steam boilers due to the violent movement inside combined with contaminants in the water that form foam bubbles. In some cases, the water level can drop below the probe, but enough foam forms to provide a ground path for the probe. Timed probe controls turn off power to the boiler periodically and check to see if the probe still has a ground path. The rationale is that foam disperses when the boiler stops firing. If the probe was being fooled by foam, the ground path would be eliminated when the foam dispersed.

Care and operation:

Pay close attention to the control and boiler manufacturers' instructions regarding service and maintenance. Most probe controls require the probe be removed annually, inspected, and cleaned or replaced if necessary. This is to avoid the possibility of sediment accumulation on the probe creating a permanent ground path.

Test the control to verify operation as instructed by low water cutoff and boiler manufacturers.

Never install a low water cutoff probe with teflon tape. The tape can insulate the probe from ground, causing nuisance problems.

Float-type low water cutoffs

Float-type controls include a water chamber piped to the boiler. Inside the chamber is a float linked to a switch in the electrical box. When the water level in the chamber drops too low, the float drops, opening the limit circuit switch. See Figure 8 for typical float-type low water cutoff construction.

Care and operation:

Pay careful attention to the control and boiler manufacturers' instructions regarding operation and blowdown (periodically opening bottom valve to blow out sediment). If sediment accumulates in the float chamber it will prevent the float from dropping as the water level drops, causing the control to stop protecting from low water. In addition to periodic blowdown, you should also unpipe the control periodically and inspect and rod out the pipes connecting the control to the boiler. If the pipes become plugged with sediment, the control will not sense the correct water level. Test the control to verify operation as instructed by low water cutoff and boiler manufacturers.

Figure 7 Typical probe-type cutoff

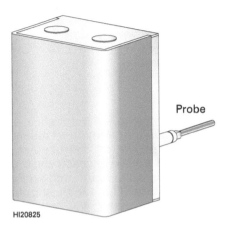

Probe

HI20825

Figure 8 Typical float-type cutoff

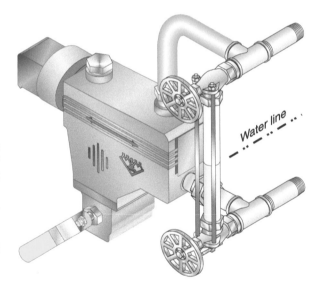

Water line

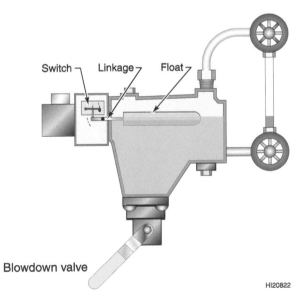

Switch   Linkage   Float

Blowdown valve

HI20822

Vent dampers

Modern atmospheric gas boilers usually include a vent damper that must be installed directly on the appliance vent outlet or near the appliance in the vent piping. Some boilers may use a flue damper instead — a damper that mounts between the appliance and its draft hood. Never install a damper unless approved by the boiler manufacturer, and only in strict accordance with the boiler installation instructions. Locating the damper in the wrong place could result in serious injury or death from carbon monoxide caused by improper combustion.

Vent dampers keep the damper closed except during calls for heat from the thermostat. This reduces the flow of warm room air up the vent system, saving energy (increasing seasonal efficiency of boiler). Vent dampers used on standing pilot gas boilers are made with a hole in the damper to allow removal of flue products from the pilot during the off cycle.

When the thermostat circuit closes, the vent damper rotates the damper open. When the damper reaches open position, an end switch closes, proving the damper is open. When this end switch closes, the burner is activated. When the thermostat is satisfied, the thermostat circuit opens. The damper motor then drives the damper closed or allows it to spring-load closed on some dampers. The damper remains closed until the next call for heat.

Pay close attention to the boiler manufacturer's instructions regarding the installation and wiring of the damper. Inspect and test the damper periodically to verify it is operating correctly.

Sequence of operation (typical):

The damper consists primarily of a motor, a damper blade and vent section (See Figure 9), a relay (TR) and end switches (S1, S2, and S3) operated by the rotating damper shaft.

See Figure 10. This series of schematic wiring diagrams shows three conditions of the vent damper components as the damper cycles. (The wiring diagrams are a simplified version of the Effikal RVGP vent damper.)

- The top frame shows the damper in standby mode — no call for heat. This is the position the damper returns to after a heating cycle.
- The center frame shows the condition when the boiler limit/thermostat circuit has called for heat, applying 24 VAC on the TR relay coil line. Current flows through the coil to 24 VAC common, activating the relay. Relay normally-open contacts and now closed, normally-closed contacts are now open. This provides a current path for the damper motor through TRb and S1 and as well as the current path for relay TR coil as shown. The damper motor now begins to rotate the damper open.
- The bottom frame shows the damper in the full open position. When the damper reaches this position, end switches S1, S2, and S3 trip as shown in the wiring. When S1 trips, it breaks the current path for the motor, so the motor stops rotating. Relay TR coil remains activated as long as the boiler limit/thermostat circuit calls for heat. The damper now completes the boiler limit circuit to the gas valve (or ignition control) through relay contact TRa and switches S2 and S3.
- When the call for heat is completed (no 24 VAC to TR coil), relay TR drops out. Relay contact TRc now closes, providing 24 VAC to the motor. The motor rotates until it reaches the closed position. The cam that operates the switches drops out S1, S2, and S3 by the time the damper reaches the closed position. When S1 drops out, the motor stops rotating. The damper is now in the condition shown in the top frame.

Figure 9   Typical vent damper

Motor or actuator

HI20826

Figure 10   Typical vent damper operation

No call for heat — Damper closed

Relay TR not activated; end switches S1, S2, S3 not tripped yet; motor stopped

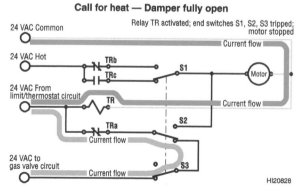

Call for heat — Damper opening

Relay TR activated; end switches S1, S2, S3 not tripped yet; motor operating

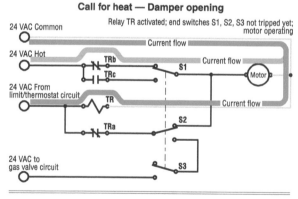

Call for heat — Damper fully open

Relay TR activated; end switches S1, S2, S3 tripped; motor stopped

HI20828

How boilers operate

Overview

In this section, we will discuss wiring and operation for typical boilers.

Gas boilers

The gas boiler discussions are all based on atmospheric gas (gravity vented) to simplify the wiring. Induced-draft gas boilers are very similar in operation of controls, but they include a blower motor and airflow-proving switch in place of a vent damper.

Oil boilers

The oil-fired boiler discussions all assume the simplest oil primary control. Other oil primary controls will wire and operate similarly, but may include prepurge (blowing air through boiler before oil flows to nozzle) and/or postpurge (blowing air through boiler after firing cycle is completed) and wiring to oil valves.

Boiler operation

For information on operation of any boiler, study the ladder wiring diagram and installation instructions.

Once you understand the symbols and control functions, you will find ladder diagrams very helpful in understanding, installing and troubleshooting boilers.

Condensing boilers

Condensing boilers typically include integral electronic controls. The controls function to regulate boiler operation and often include programmable features that allow outdoor reset, multiple boiler, etc. Consult the boiler manufacturer's literature for operating characteristics and adjustment options. Refer to Part 7 of Guide RHH for discussion of condensing boilers.

Case 1:

Water boiler, atmospheric gas, constant-burning (standing) gas pilot burner (Figure 11)

This boiler is equipped with a circulator and a transformer/relay assembly. The transformer provides 24 VAC to the control circuit. The relay operates the circulator and provides a contact to start the burner. Note that the relay coil **CR** is powered by 24 VAC when the thermostat circuit closes.

The boiler limit circuit includes:

- Spill switch **SW** (mounted on draft diverter, protects against blocked flue conditions).

- High limit **HL** (shuts off burner if temperature rises above setting on switch).

- Thermal fuse element **TFE** (fuse element mounted near burner, protects against excessive heat from the burner area).

- Vent damper (operated by motor **D**, provides contact **Da** to close when damper is open).

Burner:

The burner is atmospheric type with a standing pilot gas valve. The thermocouple in the pilot flame must be hot for the valve to open.

Sequence of operation:

1. Pilot must be manually lighted (per boiler lighting instructions in manual).

2. Wiring diagram shows service switch open. Close service switch to allow boiler to fire.

3. Call for heat (thermostat circuit closed through thermostat and/or zone controls).

4. When thermostat circuit closes, coil **CR** receives 24 VAC and activates. This closes relay contact **CRa**, starting the circulator. Contact **CRb** also closes, providing power to the limit circuit.

5. If all limits are closed (**SW**, **HL** and **TFE**), vent damper motor receives 24 VAC and damper begins to open. When damper is fully open, damper switch **Da** closes. This provides 24 VAC to the gas valve. NOTE: If any limit is open, the boiler will not fire until the limit is closed.

6. The gas valve receives 24 VAC. If the thermocouple senses pilot flame, the gas valve

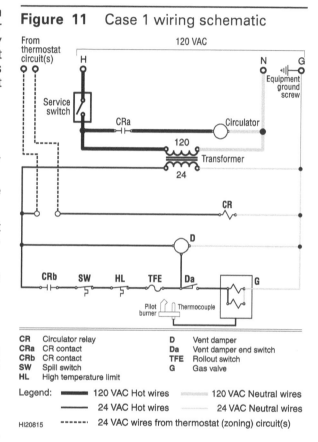

Figure 11 Case 1 wiring schematic

CR	Circulator relay	D	Vent damper
CRa	CR contact	Da	Vent damper end switch
CRb	CR contact	TFE	Rollout switch
SW	Spill switch	G	Gas valve
HL	High temperature limit		

Legend:
— 120 VAC Hot wires ⸻ 120 VAC Neutral wires
— 24 VAC Hot wires ⸻ 24 VAC Neutral wires
HI20815 ------ 24 VAC wires from thermostat (zoning) circuit(s)

opens, providing gas to the main burners.

7. The burners continue to fire until a limit opens or the thermostat opens. The boiler high limit HL may often open during a heat call, particularly on zoned systems, because the boiler heat output is more than needed for most cycles. When the limit opens, the gas valve closes (stopping burner flame) and the vent damper closes. The circulator continues to run because relay **CR** is still powered through the thermostat circuit. When water has flowed through the system enough to cool below the limit setting, the high limit will close again, allowing the burners to fire. The boiler will continue to cycle on limit until the thermostat circuit opens.

8. Call for heat completed — thermostat circuit opens. Relay **CR** drops out and the burners and circulator stop. The vent damper closes.

Case 2:

Steam boiler, atmospheric gas, constant-burning (standing) gas pilot burner (Figure 12)

This is a steam boiler, equipped with a 24 VAC transformer for the low-voltage limit circuit.

The boiler limit circuit includes:

- Low water cutoff (probe type) - connection requires constant 24 VAC to probe circuit. When the probe is covered with water, the circuit will close, provided power is available. If the boiler used a float-type low water cutoff, only contact **LWa** would be used. No additional wiring to float-type low water cutoff is required.

- Spill switch **SW** (mounted on draft diverter, protects against blocked flue conditions).

- High limit **HL** (shuts off burner if pressure rises above setting on switch).

- Thermal fuse element **TFE** (fuse element mounted near burner, protects against excessive heat from the burner area).

- Vent damper (provides contact **Da** to close when damper is open, operated by motor **D**).

Burner

The burner is atmospheric type with a standing pilot gas valve. The thermocouple in the pilot flame must be hot for the valve to open.

Sequence of operation:

1. Pilot must be manually lighted (per boiler lighting instructions in manual).

2. Wiring diagram shows service switch open. Close service switch to allow boiler to fire.

3. Call for heat (thermostat circuit closed through thermostat and/or zone controls).

4. When thermostat circuit closes, 24 VAC power is available to the limit circuit, provided the low water cutoff is satisfied (contact **LWa** closed).

5. If all limits are closed (**LWa, SW, HL,** and **TFE**), vent damper motor receives 24 VAC and damper begins to open. When damper is fully open, damper switch **Da** closes. This provides

Figure 12   Case 2 wiring schematic

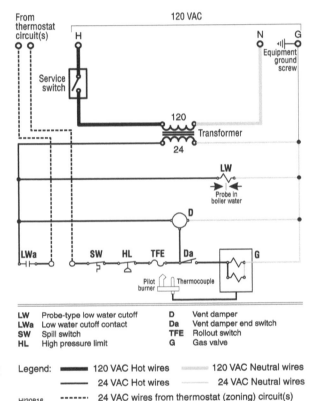

LW	Probe-type low water cutoff	D	Vent damper
LWa	Low water cutoff contact	Da	Vent damper end switch
SW	Spill switch	TFE	Rollout switch
HL	High pressure limit	G	Gas valve

Legend: ▬▬▬ 120 VAC Hot wires   ═══ 120 VAC Neutral wires
────── 24 VAC Hot wires   ──── 24 VAC Neutral wires
HI20816   ------ 24 VAC wires from thermostat (zoning) circuit(s)

24 VAC to the gas valve. NOTE: If any limit is open, the boiler will not fire until the limit is closed.

6. The gas valve receives 24 VAC. If the thermocouple senses pilot flame, the gas valve opens, providing gas to the main burners.

7. The burners continue to fire until a limit or the thermostat opens. When any limit opens, the gas valve closes (stopping burner flame) and the vent damper closes.

8. Call for heat completed — thermostat circuit opens. The gas valve closes. Burners stop firing. The vent damper closes.

Case 3:

Water boiler, atmospheric gas, spark-ignited gas pilot burner (Figure 13)

This boiler is equipped with a circulator and a transformer/relay assembly. The transformer provides 24 VAC to the control circuit. The relay operates the circulator and provides a contact to start the burner. Note that the relay coil CR is powered by 24 VAC when the thermostat circuit closes.

The boiler limit circuit includes:

- Spill switch **SW** (mounted on draft diverter, protects against blocked flue conditions).

- High limit **HL** (shuts off burner if temperature rises above setting on switch).

- Thermal fuse element **TFE** (fuse element mounted near burner, protects against excessive heat from the burner area).

- Vent damper (provides contact **Da** to close when damper is open, operated by motor **D**).

Burner

The burner is atmospheric type with a spark-ignited pilot burner and a main gas valve. The ignition control provided senses pilot flame with a flame rod in the pilot burner flame.

Sequence of operation:

1. Wiring diagram shows service switch open. Close service switch to allow boiler to fire.

2. Call for heat (thermostat circuit closes through thermostat and/or zone controls).

3. When thermostat circuit closes, coil **CR** receives 24 VAC and activates. This closes relay contact **CRa**, starting the circulator. Contact **CRb** also closes, providing power to the limit circuit.

4. If all limits are closed (**SW, HL,** and **TFE**), vent damper motor receives 24 VAC and damper begins to open. When damper is fully open, damper switch **Da** closes. This provides 24 VAC to the ignition control. NOTE: If any limit is open, the boiler will not fire until the limit is closed.

5. The ignition control opens the pilot gas valve **PG** and applies voltage to the spark wire. The spark electrode will create a spark, lighting the pilot gas. If gas is available and the pilot flame starts, the flame rod will be able to conduct current through the pilot flame.

6. The ignition control will sense flame and activate the main gas valve **MG**. NOTE: The flame current and spark rely on a good path to ground. This means the pilot burner must be effectively grounded and the ignition control ground lead must be effectively grounded to the boiler housing. Poor ground connections will often cause nuisance failures. If the ignition control does not sense flame within its time limit (trial for ignition), it may recycle (restart ignition sequence); shutdown, wait and recycle; or lockout (requiring power interruption to restart), depending on control design.

Figure 13 Case 3 wiring schematic

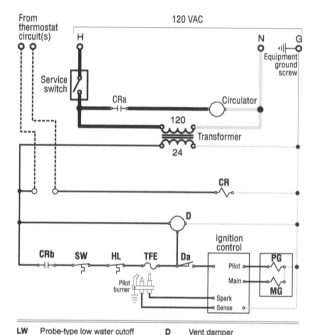

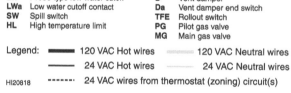

LW	Probe-type low water cutoff	D	Vent damper
LWa	Low water cutoff contact	Da	Vent damper end switch
SW	Spill switch	TFE	Rollout switch
HL	High temperature limit	PG	Pilot gas valve
		MG	Main gas valve

Legend: ▬▬▬ 120 VAC Hot wires ▬▬▬ 120 VAC Neutral wires
▬▬▬ 24 VAC Hot wires ▬▬▬ 24 VAC Neutral wires
HI20818 ------ 24 VAC wires from thermostat (zoning) circuit(s)

7. The burners continue to fire until a limit opens or the thermostat opens. The boiler high limit **HL** may often open during a heat call, particularly on zoned systems, because the boiler heat output is more than needed for most cycles. When the limit opens, the ignition control closes the main and pilot gas valves (stopping burner flame) and the vent damper closes. The circulator continues to run because relay **CR** is still powered through the thermostat circuit. When water has flowed through the system enough to cool below the limit setting, the high limit will close again, allowing the burners to fire. The boiler will continue to cycle on limit until the thermostat circuit opens.

8. Call for heat completed — thermostat circuit opens. Relay **CR** drops out and the circulator stops. Ignition control deactivates main and pilot gas valves. Main and pilot burner flames stop. The vent damper closes.

Case 4:

Steam boiler, atmospheric gas, spark-ignited gas pilot burner (Figure 14)

This is a steam boiler, equipped with a 24 VAC transformer for the low-voltage limit circuit.

The boiler limit circuit includes:

- Low water cutoff (probe type) - connection requires constant 24 VAC to probe circuit. When the probe is covered with water, the circuit will close, provided power is available. If the boiler were equipped with a float-type low water cutoff, only contact **LWa** would be used. No additional wiring to float-type low water cutoff is required.

- Spill switch **SW** (mounted on draft diverter, protects against blocked flue conditions).

- High limit **HL** (shuts off burner if pressure rises above setting on switch).

- Thermal fuse element **TFE** (fuse element mounted near burner, protects against excessive heat from the burner area).

- Vent damper (provides contact **Da** to close when damper is open, operated by motor **D**).

Burner

The burner is atmospheric type with a spark-ignited pilot burner and a main gas valve. The ignition control provided senses pilot flame with a flame rod in the pilot burner flame.

Sequence of operation:

1. Wiring diagram shows service switch open. Close service switch to allow boiler to fire.

2. Call for heat (thermostat circuit closed through thermostat and/or zone controls).

3. When thermostat circuit closes, 24 VAC power is available to the limit circuit, provided the low water cutoff is satisfied (contact **LWa** closed).

4. If all limits are closed (**LWa, SW, HL,** and **TFE**), vent damper motor receives 24 VAC and damper begins to open. When damper is fully open, damper switch Da closes. This provides 24 VAC to the ignition control. NOTE: If any limit is open, the boiler will not fire until the limit is closed.

5. The ignition control opens the pilot gas valve **PG** and applies voltage to the spark wire. The spark electrode will create a spark, lighting the pilot gas. If gas is available and the pilot flame starts, the flame rod will be able to conduct current through the pilot flame.

6. The ignition control will sense flame and activate the main gas valve **MG**. NOTE: The flame current and spark rely on a good path to ground. This means the pilot burner must be effectively

Figure 14   Case 4 wiring schematic

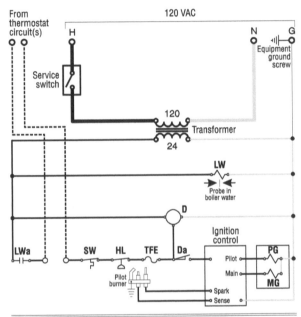

LW	Probe-type low water cutoff	D	Vent damper
LWa	Low water cutoff contact	Da	Vent damper end switch
SW	Spill switch	TFE	Rollout switch
HL	High pressure limit	PG	Pilot gas valve
		MG	Main gas valve

Legend: ▬▬▬ 120 VAC Hot wires   ▭▭▭ 120 VAC Neutral wires

HI20817   ───── 24 VAC Hot wires   ───── 24 VAC Neutral wires

‑‑‑‑‑ 24 VAC wires from thermostat (zoning) circuit(s)

grounded and the ignition control ground lead must be effectively grounded to the boiler housing. Poor ground connections will often cause nuisance failures. If the ignition control does not sense flame within its time limit (trial for ignition), it may recycle (restart ignition sequence); shutdown, wait and recycle; or lockout (requiring power interruption to restart), depending on control design.

7. The burners continue to fire until a limit opens or the thermostat opens. When any limit opens, the ignition control deactivates main and pilot gas valves (stopping burner flame) and the vent damper closes.

8. Call for heat completed — thermostat circuit opens. Ignition control deactivates main and pilot gas valves. Main and pilot burner flames stop. The vent damper closes.

Case 5:

Steam boiler, oil-fired, with tankless coil option (Figure 15)

This is a steam boiler equipped with an oil burner using a typical oil primary control. The boiler can be equipped with a tankless heater if tankless temperature coil **TC** is used. The thermostat circuit connects to the **T-T** terminals on the oil primary control.

The boiler limit circuit includes:

- Low water cutoff (probe type) — connection requires constant voltage to probe circuit. When the probe is covered with water, the circuit will close provided power is available. If the boiler used a float-type low water cutoff, only contact **LWa** would be used. No additional wiring to float-type low water cutoff is required.

- Spill switch **SW** (mounted on draft diverter, protects against blocked flue conditions).

- High limit **HL** (shuts off burner if pressure rises above setting on switch).

Burner

The burner is an oil burner with direct-spark ignition and a typical primary control. The primary control **T-T** terminals are jumpered because the thermostat circuit connects to the combination limit control/relay.

Sequence of operation:

1. Wiring diagram shows service switch open. Close service switch to allow boiler to fire.

2. Call for heat (thermostat circuit closed through thermostat and/or zone controls or tankless heater control **TC** closed).

3. The oil primary control is powered continuously through the limit circuit provided the low water cutoff **LW** and high limit **HL** are closed. If any limit is open, the burner will not fire.

4. When the thermostat circuit closes, relay **FR** is powered.

5. Relay contact **FRb** closes, providing 120 VAC to the burner motor and the ignition transformer or ignitor. The burner motor operates the blower and oil pump. Oil sprays from the nozzle and is ignited by the spark at the electrodes. With this oil primary control, the ignitor or ignition transformer remains on during the entire firing cycle. This is called "intermittent ignition." Other primary controls stop the ignitor or ignition transformer after sensing flame. This ignition system is called "interrupted ignition" because the spark stops during the firing cycle instead of being continuous.

6. When the primary control receives power, current starts to flow through the timer switch heater **SS**. The timer continues to time out unless the cad cell **CC** senses flame. Once the cad cell senses flame, the circuit causes switch **BL** to stop current flowing through the triac **TR**. Timer **SS** stops. If the timer **SS** times out before the cad cell **CC** senses flame, the control locks out, requiring manual reset to restart. The time limit before the switch times out is called the trial for ignition. This is usually 15 or 45 seconds.

7. The burner continues to fire until the limit or thermostat circuit opens.

8. Call for heat completed — thermostat circuit opens. Primary control deactivates burner motor and ignitor or ignition transformer.

Figure 15 Case 5 wiring schematic

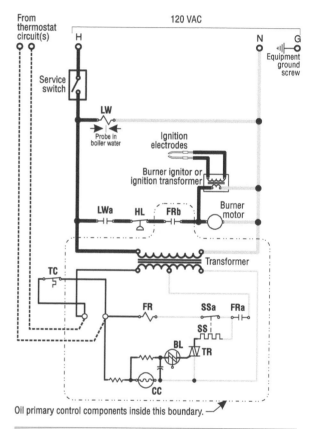

Oil primary control components inside this boundary. —

LW	Probe-type low water cutoff	HL	High temperature limit
LWa	Low water cutoff contact	TC	Low limit control to maintain minimum boiler water temperature for tankless heater (when used)
⏜⏜ Resistor	⋉ Capacitor		

Typical cad cell oil primary

- Relay **FR** activates when thermostat circuit (or **TC**) closes; current flows through relay **FR** coil, safety switch **SS** and triac **TR**, closing relay contacts **FRa** and **FRb**. The burner ignition transformer and motor start (getting voltage through contact **FRb**). The burner motor operates the blower and oil pump. Oil flows to the nozzle, and should be ignited by the spark from the electrodes. The blower pushes air through the burner tube.

- The safety switch begins to heat as current flows through its heater.

- If the cad cell **CC** senses flame, the bilateral switch **BL** opens, stopping current flow through the triac **TR**. This prevents any more current flow through the safety switch. Current continues to flow through relay coil **FR** by flowing through **SSa** and **FRa**. The burner continues to fire until the thermostat is satisfied (or cad cell no longer sees flame).

- If the cad cell **CC** does not sense flame before the safety switch heats up, contact **SSa** opens. This deactivates relay **FR**, opening contacts **FRa** and **FRb**. The burner stops firing. (The safety switch timing is called the trial for ignition.)

Legend:

▬▬ 120 VAC Hot wires	▬▬ 120 VAC Neutral wires
─── 24 VAC Hot wires	▬▬ 24 VAC Neutral wires
------ 24 VAC wires from thermostat (zoning) circuit(s)	

HI20821

Case 6:

Water boiler, oil-fired, combination limit control, no tankless coil (Figure 16)

This boiler is equipped with a circulator and a combination temperature limit control/circulator relay. The relay operates the circulator and provides a contact to start the burner. Note that the relay coil **CR** is powered by 24 VAC when the thermostat circuit closes.

The boiler limit circuit includes:

High limit **HL** (shuts off burner if temperature rises above setting on switch). **HL** is Part of the combination limit control/circulator relay.

Burner

The burner is an oil burner with direct-spark ignition and a typical primary control. The primary control **T-T** terminals are jumpered because the thermostat circuit connects to the combination limit control/relay.

Sequence of operation:

1. Wiring diagram shows service switch open. Close service switch to allow boiler to fire.

2. Call for heat (thermostat circuit closed through thermostat and/or zone controls).

3. When thermostat circuit closes, coil **CR** receives 24 VAC and activates. This closes relay contact **CRb**, starting the circulator. Contact **CRa** also closes, providing 120 VAC power to the limit circuit (high limit **HL**).

4. If the high limit **HL** is closed, power is supplied to the oil primary control (to transformer primary and primary control contact **FRb**). NOTE: If any limit is open, the boiler will not fire until the limit is closed.

5. The primary control relay **FR** activates when the transformer is powered. Relay contact **FRb** closes, providing 120 VAC to the burner motor and the ignition transformer or ignitor.

The burner motor operates the blower and oil pump. Oil sprays from the nozzle and is ignited by the spark at the electrodes. With this oil primary control, the ignitor or ignition transformer remains on during the entire firing cycle. This is called "intermittent ignition." Other primary controls stop the ignitor or ignition transformer after sensing flame. This ignition system is called "interrupted ignition" because the spark stops during the firing cycle instead of being continuous.

6. When the primary control receives power, current starts to flow through the timer switch heater **SS**. The timer continues to time out unless the cad cell **CC** senses flame. Once the cad cell senses flame, the circuit causes switch **BL** to stop current flowing through the triac **TR**. Timer **SS** stops. If the timer **SS** times out before the cad cell **CC** senses flame, the control locks out, requiring manual reset to restart. The time limit before the switch times out is called the trial for ignition. This is usually 15 or 45 seconds.

7. The burner continues to fire until the limit or thermostat circuit opens. The boiler high limit **HL** may often open during a heat call, particularly on zoned systems, because the boiler heat output is more than needed for most cycles. When the limit opens, the primary control and the burner motor lose power, so the motor stops (stopping oil flow). The burner flame is extinguished. The circulator continues to run because relay **CR** is still powered through the thermostat circuit. When water has flowed through the system enough to cool below the limit setting, the high limit will close again, allowing the burner to fire. The boiler will continue to cycle on limit until the thermostat circuit opens.

8. Call for heat completed — thermostat circuit opens. Relay **CR** drops out and the circulator stops. Primary control deactivates burner motor and ignitor or ignition transformer.

Figure 16   Case 6 wiring schematic

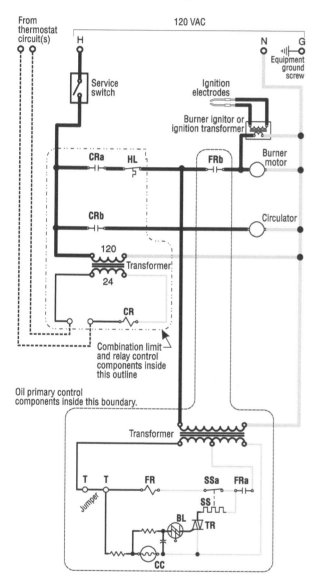

From thermostat circuit(s)

120 VAC

Service switch

Ignition electrodes

Burner ignitor or ignition transformer

CRa   HL   FRb

Burner motor

CRb   Circulator

120

Transformer

24

CR

Combination limit and relay control components inside this outline

Oil primary control components inside this boundary.

Transformer

T   T   FR   SSa   FRa

Jumper

SS

BL   TR

CC

Equipment ground screw

CR   Combination control relay
CRa CR contact
CRb CR contact
HL   High temperature limit

-vvv- Resistor   ≈ Capacitor

Oil primary, typical

- Relay **FR** activates when thermostat calls for heat (powering oil primary transformer through **HL**); current flows through relay **FR** coil, safety switch **SS** and triac **TR**, closing relay contacts **FRa** and **FRb**. The burner ignition transformer and motor start (getting voltage through contact **FRb**). The burner motor operates the blower and oil pump. Oil flows to the nozzle, and should be ignited by the spark from the electrodes. The blower pushes air through the burner tube.

- The safety switch begins to heat as current flows through its heater.

- If the cad cell **CC** senses flame, the bilateral switch **BL** opens, stopping current flow through the triac **TR**. This prevents any more current flow through the safety switch. Current continues to flow through relay coil **FR** by flowing through **SSa** and **FRa**. The burner continues to fire until the thermostat is satisfied (or cad cell no longer sees flame).

- If the cad cell **CC** does not sense flame before the safety switch heats up, contact **SSa** opens. This deactivates relay **FR**, opening contacts **FRa** and **FRb**. The burner stops firing. (The safety switch timing is called the trial for ignition.)

Legend: ▬▬▬ 120 VAC Hot wires   ▬▬▬ 120 VAC Neutral wires
          ——— 24 VAC Hot wires   ——— 24 VAC Neutral wires
HI20819 ------ 24 VAC wires from thermostat (zoning) circuit(s)

Case 7:

Water boiler, oil-fired, tri-function limit control, with tankless coil (Figure 17)

This boiler is equipped with a circulator and a tri-function combination temperature limit control/low limit control/circulator relay. It is also equipped with a tankless domestic water heater. The relay operates the circulator and provides a contact to start the burner. Note that the relay coil **CR** is powered by 24 VAC when the thermostat circuit closes.

The boiler limit circuit includes:

- High limit **HL** (shuts off burner if temperature rises above setting on switch). **HL** is Part of the combination limit control/circulator relay.

- Low limit **LL** (starts burner if temperature of boiler water drops below low limit setting — used for tankless heater boilers to ensure immediate heat available when cold water flows through tankless heater). **LL** is Part of the combination limit control/low limit/circulator relay.

Burner

The burner is an oil burner with direct-spark ignition and a typical primary control. The primary control **T-T** terminals are jumpered because the thermostat circuit connects to the combination limit control/relay.

Sequence of operation:

1. Wiring diagram shows service switch open. Close service switch to allow boiler to fire.

2. Call for heat (thermostat circuit closed through thermostat and/ or zone controls).

3. When thermostat circuit closes, coil **CR** receives 24 VAC and activates. This closes relay contact **CRb**, starting the circulator. Contact **CRa** also closes, providing 120 VAC power to the limit circuit (high limit **HL**).

4. If the high limit **HL** is closed, power is supplied to the oil primary control (to transformer primary and primary control contact **FRb**). NOTE: If any limit is open, the boiler will not fire until the limit is closed.

5. The primary control relay **FR** activates when the transformer is powered. Relay contact **FRb** closes, providing 120 VAC to the burner motor and the ignition transformer or ignitor. The burner motor operates the blower and oil pump. Oil sprays from the nozzle and is ignited by the spark at the electrodes. With this oil primary control, the ignitor or ignition transformer remains on during the entire firing cycle. This is called "intermittent ignition." Other primary controls stop the ignitor or ignition transformer after sensing flame. This ignition system is called

"interrupted ignition" because the spark stops during the firing cycle instead of being continuous.

6. When the primary control receives power, current starts to flow through the timer switch heater **SS**. The timer continues to time out unless the cad cell **CC** senses flame. Once the cad cell senses flame, the circuit causes switch **BL** to stop current flowing through the triac **TR**. Timer **SS** stops. If the timer **SS** times out before the cad cell **CC** senses flame, the control locks out, requiring manual reset to restart. The time limit before the switch times out is called the trial for ignition. This is usually 15 or 45 seconds.

7. The burner continues to fire until the limit or thermostat circuit opens. The boiler high limit **HL** may often open during a heat call, particularly on zoned systems, because the boiler heat output is more than needed for most cycles. When the limit opens, the primary control and the burner motor lose power, so the motor stops (stopping oil flow). The burner flame is extinguished. The circulator continues to run because relay **CR** is still powered through the thermostat circuit. When water has flowed through the system enough to cool below the limit setting, the high limit will close again, allowing the burner to fire. The boiler will continue to cycle on limit until the thermostat circuit opens.

8. Call for heat completed — thermostat circuit opens. Relay **CR** drops out and the circulator stops. Primary control deactivates burner motor and ignitor or ignition transformer.

9. The low limit **LL** can also cause a call for heat. When **LL** closes, though, the burner will fire without the thermostat circuit being closed. When the water temperature in the boiler drops below the low limit setting (usually 180°F), **LL** closes. This directly powers the limit circuit at **HL**. If **HL** is closed, the primary control receives 120 VAC to its transformer and starts a burner firing cycle as above. The circulator does not start because the circulator relay **CR** is only powered when the thermostat circuit is closed. When the low limit **LL** opens, the firing cycle stops.

Figure 17 Case 7 wiring schematic

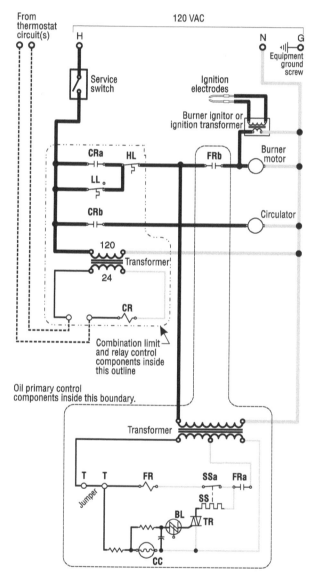

CR Combination control relay
CRa CR contact
CRb CR contact
HL High temperature limit
LL Low limit (starts burner if water temperature drops below setting)

-w- Resistor ≍ Capacitor

Oil primary, typical

- Relay **FR** activates when thermostat calls for heat (powering oil primary transformer through **HL**); current flows through relay **FR** coil, safety switch **SS** and triac **TR**, closing relay contacts **FRa** and **FRb**. The burner ignition transformer and motor start (getting voltage through contact **FRb**). The burner motor operates the blower and oil pump. Oil flows to the nozzle, and should be ignited by the spark from the electrodes. The blower pushes air through the burner tube.
- The safety switch begins to heat as current flows through its heater.
- If the cad cell **CC** senses flame, the bilateral switch **BL** opens, stopping current flow through the triac **TR**. This prevents any more current flow through the safety switch. Current continues to flow through relay coil **FR** by flowing through **SSa** and **FRa**. The burner continues to fire until the thermostat is satisfied (or cad cell no longer sees flame).
- If the cad cell **CC** does not sense flame before the safety switch heats up, contact **SSa** opens. This deactivates relay **FR**, opening contacts **FRa** and **FRb**. The burner stops firing. (The safety switch timing is called the trial for ignition.)

Legend: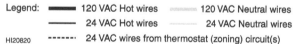

▬▬	120 VAC Hot wires	▭▭	120 VAC Neutral wires
──	24 VAC Hot wires	──	24 VAC Neutral wires
------	24 VAC wires from thermostat (zoning) circuit(s)		

HI20820

Wiring zone controls

In this section, we will discuss components and wiring used for zoning hydronic systems.

Zoning controls

The cases shown in this section use zone valves (4-wire or 3-wire) wired directly to thermostats or circulators operated by circulator relays. The same functions can be applied more easily using zoning controllers, discussed below.

Zoning controllers

Zoning controllers are available to zone with zone valves or circulators. Typical controllers provide control of multiple zones from a single panel or group of panels, greatly simplifying zone applications. You will find this alternative very effective for circulator zoning since it eliminates installing multiple circulator relays.

When applying a zoning controller for zone valve applications, check the zoning controller transformer rating to be sure it can handle all connected zone valves.

The cases in this section use individual controls for zoning. A zone controller could be substituted in any of the cases given.

An added advantage of zone controllers is that most are available with a domestic water heating priority that can be enabled simply by flipping a switch.

Zoning tips:

☐ Pay close attention when sizing control transformers (24 VAC) for use in zoning circuits. The transformer has to be large enough to handle all of the connected loads. Add up the combined va load of all connected devices to determine the minimum size of the transformer. If you undersize the transformer, it will burn out.

The required current for zone valves varies a great deal - typically from 0.2 amp to 0.9 amp per zone valve. At 0.2 amp, each zone valve requires 0.2 x 24 = 4.8 va. But at 0.9 amp, each zone valve requires 0.9 x 24 = 21.6 va. You could only operate a single zone valve from a 40-va transformer using 0.9-amp zone valves.

☐ Pay close attention to voltages. If the relays you are using must handle 120 VAC, be sure the contacts are rated for 120 VAC.

☐ Check current-carrying capacity of relays. Make sure the relay contact rated is greater than the amp load it must carry.

☐ Set the thermostat anticipator to the current it will carry. For zone valves, set the anticipator to the amp rating of the zone valve. For circulator relays, check the relay literature for the suggested anticipator setting.

☐ Comply with wiring codes. Line voltage (120 VAC) wiring must meet the National Electrical Code requirements for Class 1 wiring.

☐ Zone valves are available as 4-wire (having an end switch isolated from any power source) and 3-wire (one side of switch common to the zone valve 24 VAC power source). You must be very careful when wiring 3-wire zone valves not to cross the wiring, causing extra voltage to be applied to the boiler thermostat circuit. This can apply 48 VAC to the boiler control circuit, causing gas valves and relays to burn out.

To be sure the wiring is correct, disconnect the two wires at the boiler thermostat circuit connection. Connect a voltmeter (50 VAC range) across the leads. Activate all thermostats. Under no conditions should you read any voltage across the leads. If you do, check the wiring and correct until you see no reading.

☐ As insurance against possible wiring errors when using 3-wire zone valves, use an isolation relay as shown in some of the wiring examples that follow.

☐ Check zone valve head loss to be sure it won't reduce the flow rate in your circuit below the required minimum. You will find some valve head loss ratings in equivalent feet and others in Cv. To convert Cv to equivalent feet, see Part 5 (Table 11, page 5-40). Note that a low Cv means a high head loss. A high Cv means a low head loss.

Case 8:

Zone valve zoning with 4-wire zone valves (Figure 18)

Figure 18 shows a system with three 4-wire zone valves. You can apply this wiring to any number of zones.

Transformer

Always size the transformer to handle the total va load of all connected devices. If necessary, use multiple transformers, each connected to the allowable number of zone valves. Do not try to connect transformers in parallel to increase capacity. You will encounter problems.

Connections

Each zone valve motor circuit is connected to the 24 VAC side of the control transformer.

You can jumper from end switch to end switch on the zone valves to simplify wiring. Then run a single set of wires from the last valve end switch to the boiler. The thermostat connection wires shown would connect to the thermostat connections at the boiler, as shown in the previous discussion on boiler operation.

Thermostat anticipator

Set each thermostat anticipator to the amp rating of its zone valve.

Operation

When a zone thermostat calls for heat (closes), the corresponding zone valve motor is powered through the thermostat contact. When the zone valve reaches open position, its end switch closes. This closes the thermostat circuit, starting the boiler.

Figure 18 Case 8 wiring schematic

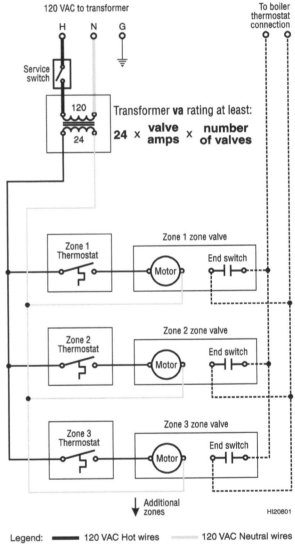

Transformer **va** rating at least:

$$24 \times \frac{\textbf{valve}}{\textbf{amps}} \times \frac{\textbf{number}}{\textbf{of valves}}$$

Legend:
— 120 VAC Hot wires   — 120 VAC Neutral wires
— 24 VAC Hot wires   — 24 VAC Common wires
------ 24 VAC wires to boiler thermostat connection

Case 9:

Zone valve zoning with 4-wire zone valves, domestic water heater, and domestic priority (Figure 19)

"Domestic priority" means when the domestic water heater calls for heat all other heating zones are disabled. The DHW priority relay in Figure 19 provides this feature. Connect the domestic water heater temperature control and zone valve as shown.

Figure 19 shows a system with three 4-wire zone valves. You can apply this wiring to any number of zones.

Transformer

Always size the transformer to handle the total va load of all connected devices (including the DHW priority relay coil). If necessary, use multiple transformers, each connected to the allowable number of zone valves. Do not try to connect transformers in parallel to increase capacity. You will encounter problems.

Connections

Each zone valve motor circuit is connected to the 24 VAC side of the control transformer through the zone thermostat (and the DHW priority relay).

You can jumper from end switch to end switch on the zone valves to simplify wiring. Then run a single set of wires from the last valve end switch to the boiler. The thermostat connection wires shown would connect to the thermostat connections at the boiler, as shown in the previous discussion on boiler operation.

Thermostat anticipator

Set each thermostat anticipator to the amp rating of its zone valve.

Operation

When a zone thermostat calls for heat (closes), the corresponding zone valve motor is powered through the thermostat contact. When the zone valve reaches open position, its end switch closes. This closes the thermostat circuit, starting the boiler.

When the domestic water heater temperature control closes, the DHW zone valve operates, causing the end switch to start the boiler. The DHW priority relay is also powered. When the relay activates, the normally-closed contact opens. This interrupts 24 VAC power to all other zones, preventing other zone valves from opening. All available flow and heat goes only to the domestic water heater.

Figure 19   Case 9 wiring schematic

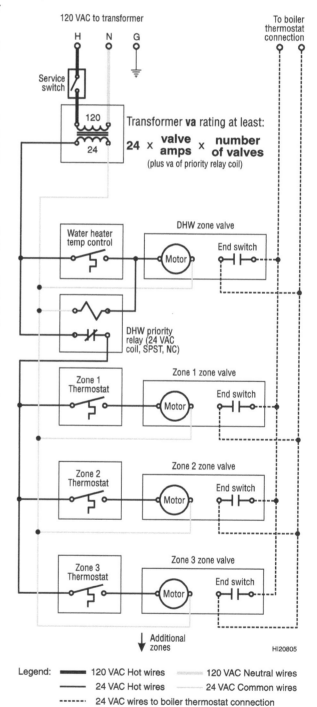

Case 10:

Zone valve zoning with 3-wire zone valves (Figure 20)

Figure 20 shows a system with three 3-wire zone valves. You can apply this wiring to any number of zones.

Transformer

Always size the transformer to handle the total va load of all connected devices. If necessary, use multiple transformers, each connected to the allowable number of zone valves. Do not try to connect transformers in parallel to increase capacity. You will encounter problems.

Connections

Each zone valve motor circuit is connected to the 24 VAC side of the control transformer through the zone thermostat.

You can jumper from end switch to end switch on the zone valves to simplify wiring. Then run a single set of wires from the last valve end switch to the boiler. The thermostat connection wires shown would connect to the thermostat connections at the boiler, as shown in the previous discussion on boiler operation.

Thoroughly test the circuit with a voltmeter to make sure the thermostat circuit never reads a voltage when disconnected at the boiler. For added insurance, you should consider using an isolation relay as shown in Case 11, Page 10–29.

Thermostat anticipator

Set each thermostat anticipator to the amp rating of its zone valve.

Operation

When a zone thermostat calls for heat (closes), the corresponding zone valve motor is powered through the thermostat contact. When the zone valve reaches open position, its end switch closes. This closes the thermostat circuit, starting the boiler.

Figure 20   Case 10 wiring schematic

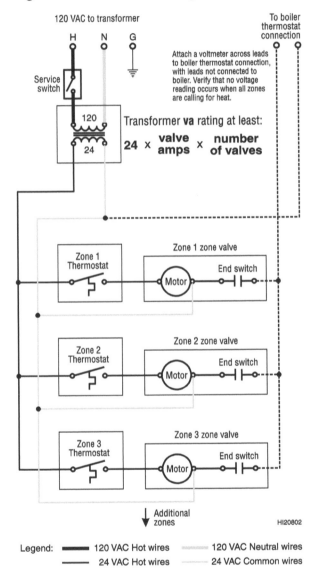

Transformer **va** rating at least:

$$24 \times \frac{\textbf{valve}}{\textbf{amps}} \times \frac{\textbf{number}}{\textbf{of valves}}$$

Legend:

▬▬ 120 VAC Hot wires	⋯⋯ 120 VAC Neutral wires
—— 24 VAC Hot wires	—— 24 VAC Common wires
------ 24 VAC wires to boiler thermostat connection	

Case 11:

Zone valve zoning with 3-wire zone valves, with isolation relay (Figure 21)

Figure 21 shows a system with three 3-wire zone valves and an isolation relay to eliminate possible problems of stray voltage due to 3-wire connections. You can apply this wiring to any number of zones.

Transformer

Always size the transformer to handle the total va load of all connected devices (including the isolation relay coil). If necessary, use multiple transformers, each connected to the allowable number of zone valves. Do not try to connect transformers in parallel to increase capacity. You will encounter problems.

Connections

Each zone valve motor circuit is connected to the 24 VAC side of the control transformer through the zone thermostat (and DHW priority relay).

You can jumper from end switch to end switch on the zone valves to simplify wiring. Then run a single wire from the last valve end switch to the isolation relay coil. The other side of the isolation relay coil goes to the transformer 24 VAC hot side. The thermostat connection wires shown from the isolation relay contact would connect to the thermostat connections at the boiler, as shown in the previous discussion on boiler operation.

Thermostat anticipator

Set each thermostat anticipator to the amp rating of its zone valve.

Operation

When a zone thermostat calls for heat (closes), the corresponding zone valve motor is powered through the thermostat contact. When the zone valve reaches open position, its end switch closes. This activates the isolation relay, closing the thermostat circuit through the relay contact, starting the boiler.

Figure 21   Case 11 wiring schematic

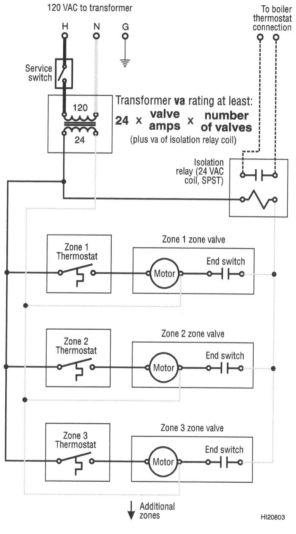

Transformer **va** rating at least:

$$24 \times \frac{\textbf{valve}}{\textbf{amps}} \times \frac{\textbf{number}}{\textbf{of valves}}$$

(plus va of isolation relay coil)

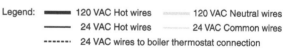

Legend: ▬▬ 120 VAC Hot wires   120 VAC Neutral wires
          ─── 24 VAC Hot wires   24 VAC Common wires
          ------ 24 VAC wires to boiler thermostat connection

Case 12:

Zone valve zoning with 3-wire zone valves, isolation relay, domestic water heater and domestic priority (Figure 22)

Domestic priority means that, when the domestic water heater calls for heat, all other heating zones are disabled. The DHW priority relay in Figure 22 provides this feature. Connect the domestic water heater temperature control and zone valve as shown.

Figure 22 shows a system with three 3-wire zone valves and an isolation relay to eliminate possible problems of stray voltage due to 3-wire connections. You can apply this wiring to any number of zones.

Transformer

Always size the transformer to handle the total va load of all connected devices (including the isolation relay and DHW priority relay coils). If necessary, use multiple transformers, each connected to the allowable number of zone valves. Do not try to connect transformers in parallel to increase capacity. You will encounter problems.

Connections

Each zone valve motor circuit is connected to the 24 VAC side of the control transformer through the zone thermostat and the DHW priority relay.

You can jumper from end switch to end switch on the zone valves to simplify wiring. Then run a single wire from the last valve end switch to the isolation relay coil. The other side of the isolation relay coil goes to the transformer 24 VAC hot side. The thermostat connection wires shown from the isolation relay contact would connect to the thermostat connections at the boiler, as shown in the previous discussion on boiler operation.

Thermostat anticipator

Set each thermostat anticipator to the amp rating of its zone valve.

Operation

When a zone thermostat calls for heat (closes), the corresponding zone valve motor is powered through the thermostat contact. When the zone valve reaches open position, its end switch closes. This activates the isolation relay, closing the thermostat circuit through the relay contact, starting the boiler.

When the domestic water heater temperature control closes, the DHW zone valve operates, causing the end switch to start the boiler. The DHW priority relay is also powered. When the relay activates, the normally-closed contact opens. This interrupts 24 VAC power to all other zones, preventing other zone valves from opening. All available flow and heat goes only to the domestic water heater.

Figure 22   Case 12 wiring schematic

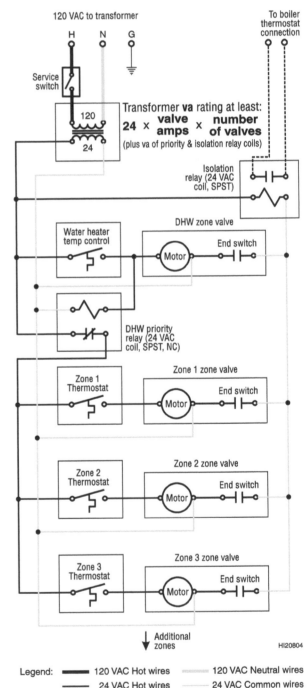

Case 13:

Zoning with circulators (Figure 23)

Figure 23 shows a system with three zones, zoned with circulators. The circulators are 120 VAC, and must be controlled relays, as shown. You can apply this wiring to any number of zones.

Contacts

Pay careful attention to the wiring instructions for the circulator relay. Some internal contacts are jumpered to the 120 VAC connections. If you wired these contacts to the boiler thermostat circuit wiring, you would cause serious damage to boiler components. Use only the isolated contacts in the relays.

Connections

Each circulator relay is connected to the 120 VAC source. When a zone thermostat closes, the corresponding relay coil is powered, closing the relay contacts.

One contact provides 120 VAC to the circulator (other side of circulator wired directly to 120 VAC neutral). The other contact (isolated contacts only) wires to the boiler thermostat circuit as shown.

You can jumper from contact to contact on the circulator relays to simplify wiring the boiler thermostat circuit wiring. Then run a single set of wires from the last relay contact to the thermostat connections at the boiler, as shown in the previous discussion on boiler operation.

Thermostat anticipator setting

Set the thermostat anticipator to the amp rating of the relay.

Operation

When a zone thermostat calls for heat (closes), the corresponding circulator relay coil activates. One set of contacts powers the zone circulator. The isolated contact closes, starting the boiler.

Figure 23   Case 13 wiring schematic

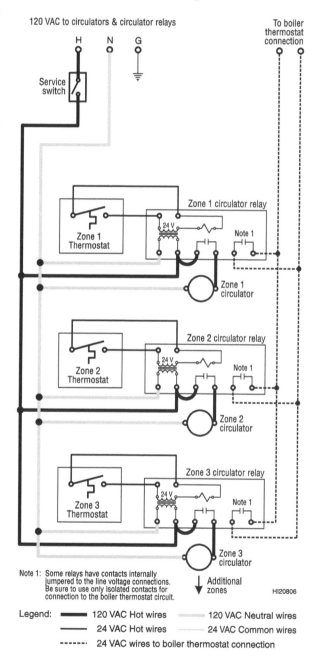

Note 1: Some relays have contacts internally jumpered to the line voltage connections. Be sure to use only isolated contacts for connection to the boiler thermostat circuit.

Legend:
— 120 VAC Hot wires   — 120 VAC Neutral wires
— 24 VAC Hot wires   — 24 VAC Common wires
------ 24 VAC wires to boiler thermostat connection

Case 14:

Zoning with circulators, domestic water heater, and domestic priority (Figure 24)

Domestic priority means that, when the domestic water heater calls for heat, all other heating zones are disabled. The DHW priority relay in Figure 24 provides this feature. Connect the domestic water heater temperature control and circulator relay as shown.

Figure 24 shows a system with three zones, zoned with circulators. The circulators are 120 VAC, and must be controlled relays, as shown. You can apply this wiring to any number of zones.

Contacts

Pay careful attention to the wiring instructions for the circulator relay. Some internal contacts are jumpered to the 120 VAC connections. If you wired these contacts to the boiler thermostat circuit wiring, you would cause serious damage to boiler components. Use only the isolated contacts in the relays.

Connections

Each circulator relay is connected to the 120 VAC source. When a zone thermostat closes, the corresponding relay coil is powered, closing the relay contacts.

One contact provides 120 VAC to the circulator (other side of circulator wired directly to 120 VAC neutral). The other contact (isolated contacts only) wires to the boiler thermostat circuit as shown.

You can jumper from contact to contact on the circulator relays to simplify wiring the boiler thermostat circuit wiring. Then run a single set of wires from the last relay contact to the thermostat connections at the boiler, as shown in the previous discussion on boiler operation.

Thermostat anticipator setting

Set the thermostat anticipator to the amp rating of the relay.

Operation

When a zone thermostat calls for heat (closes), the corresponding circulator relay coil activates. One set of contacts powers the zone circulator. The isolated contact closes, starting the boiler.

When the domestic water heater temperature control closes, the DHW circulator relay activates, starting the DHW zone circulator and the boiler. The DHW priority relay is also powered. When the relay activates, the normally-closed contact opens. This interrupts 120 VAC power to all other zone circulator relays, preventing other zone circulators from operating. All available heat goes only to the domestic water heater.

Figure 24   Case 14 wiring schematic

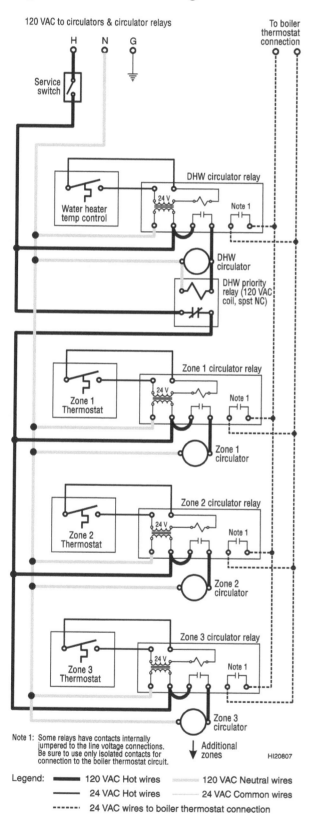

120 VAC to circulators & circulator relays

To boiler thermostat connection

Note 1: Some relays have contacts internally jumpered to the line voltage connections. Be sure to use only isolated contacts for connection to the boiler thermostat circuit.

HI20807

Legend: ▬▬▬ 120 VAC Hot wires   120 VAC Neutral wires

────── 24 VAC Hot wires   ────── 24 VAC Common wires

------ 24 VAC wires to boiler thermostat connection

Outdoor reset controls

Overview

The ideal heating system would operate continuously, supplying only an amount of heat to the space equal to the heat lost to outdoors. The room temperature would never change. It is unlikely any system could operate this way. Heating systems actually supply "bursts" of heat, with the interior temperature moving up and down in response to heat added or taken out. (See Part 7 for more discussion, including the advantages of modulating boilers.)

The effect of these bursts of heat is most noticeable during lighter heat demands (spring and fall, for example). If the boiler water temperature supplied to the terminal units is always the same (200 °F, for example), then the heat output from the terminal units will always be the same — the amount needed for the coldest outdoor conditions. This can cause relatively wide swings in room temperature.

An alternative to constant supply temperature is called outdoor reset. An outdoor reset control senses outdoor temperature and changes the maximum supply water temperature accordingly. The colder the temperature outdoors, the hotter the allowable supply water temperature, and vice versa. This requires an outdoor temperature sensor and a supply water temperature sensor.

Control options

The best method of outdoor reset is to control both the supply water temperature and the boiler return water temperature (when required) with mixing valves. Some controls provide outdoor reset operation without operating a valve, but could cause increased cycling of the boiler because the control turns the boiler on and off depending on the supply water temperature (unless the boiler includes an integral reset control). A mixing valve is likely to provide smoother water temperature control as well as smoother boiler operation. (Another option for controlling both supply and return water temperatures is to use an injection mixing control and injection mixing pump assembly that provides boiler return water temperature control when needed for non-condensing boilers. Refer to the control supplier's instructions for discussion and application information.)

Refer to Part 4 and Part 7 for further discussion and recommended piping.

Advantages of outdoor reset

Because the supply water temperature drops as outdoor temperature increases, the heat output from the terminal units is closer to the actual heat loss. This provides the advantages:

- Less fluctuation of indoor temperature because heat output from terminal units more closely matches room heat loss.

- Less expansion noise from piping because there are fewer heating cycles and cycles last longer.

• Conventional (non-condensing, on/off fired) boiler fuel usage may be reduced as much as 15% due to reduction in heat losses from piping (as a result of lower supply temperature).

There is a catch with some boilers

Conventional (non-condensing) boilers — Outdoor reset is an effective control method, but there is a catch: You must protect the boiler from low return water temperature unless the boiler manufacturer states there is no minimum return temperature. Most conventional boilers have a minimum return temperature, so you have to eliminate potential problems by properly piping and controlling the boiler.

• Some piping recommendations will show fixed bypass piping for low temperature systems. This is not an acceptable method when using outdoor reset because the return temperature is not always low. During low outdoor temperature conditions, the supply water temperature might be 200°F, with 180°F return temperature. But, during spring and fall, return water temperature could be well below 100°F. You must use a piping/control system that accounts for this variation in return temperature.

• A disadvantage of outdoor reset, when using setback operation, is that space temperature pick-up after setback takes longer with the lower supply water temperature of an outdoor reset system.

Condensing boilers — See Part 7 for discussion of outdoor reset with condensing boilers.

Reset ratio

In general, you want the supply water temperature to get closer and closer to room temperature as the outside temperature rises. With supply water temperature equal to room temperature, no heat transfer occurs, so the heat output would be zero when the heat loss was zero.

The rate at which the supply water temperature changes as outside temperature changes is called the "reset ratio." Find the reset ratio for your application by:

• Determine maximum supply temperature, Tmax, needed (for design heating conditions).

• At 70°F outside, you want zero heat input, so supply water temperature should be 70°F at 70°F outside temperature.

• The supply water temperature change is Tmax – 70.

• The outside temperature difference is 70°F minus the outdoor design temperature, ODT.

• Divide supply temperature difference by outside temperature difference to find reset ratio.

Reset ratio = (Tmax – 70) ÷ (70 – ODT)　　　(Eq 8-1)

Figure 25 shows lines representing different reset ratios. See Part 7 for other reset curve examples.

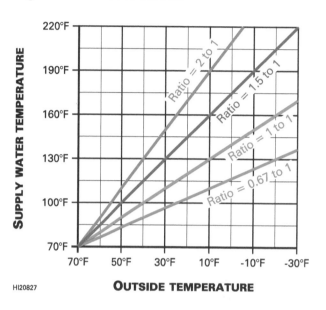

Figure 25　Reset curves

HI20827

Example reset ratio calculation:

Maximum supply water temperature, Tmax is 200°F and outdoor design temperature, ODT, is -10°F.

Reset ratio = (Tmax – 70) ÷ (70–ODT) = (200 – 70)/(70 –(-10)) = 130/80 = 1.6.

Control types

The control can be a mechanical type (using liquid-filled bulbs and capillaries, with a long capillary for the outside sensor) or electronic (using electronic temperature sensors).

Mechanical controls are usually limited to a choice of reset ratios, typically .67 to 1, 1 to 1, or 1.5 to 1.

Electronic outdoor reset controls provide adjustable reset ratio. They may also allow selection of a non-linear reset curve, shaped to match the temperature response of the heating unit used (such as finned-tube baseboard, fan coil, etc.).

Hydronics Institute Section of AHRI

35 Russo Place

Berkeley Heights, NJ 07922-0218

part **11**

I=B=R
Guide
RHH

Components
of hydronic
steam
heating systems

Residential
Hydronic
Heating . . .

Installation
& Design

 AHRI

Hydronics Institute
Section of **AHRI**

I=B=R Guide RHH
Residential Hydronic Heating

Hydronics Institute Section of **AHRI**
35 Russo Place
Berkeley Heights, NJ 07922-0218

Contents – Part 11

Contents *continued*

Illustrations

Contents *continued*

Tables

Background

Overview

In Part 11, we will talk about the components used in hydronic steam heating systems — their applications, sizing and special requirements. Part 13 will deal with steam system design, sizing and operation. You may find it helpful when reading Part 11 to scan ahead in Part 13 to see how the components are incorporated in the systems.

Steam heating has developed a bad reputation at times because of water hammer, banging radiators, and heat distribution problems. But steam systems don't have to behave badly. Steam systems, like anything else, need some maintenance and care over time. Pipes sag or develop leaks, traps and vents fail, and occasionally someone does something to a steam system he shouldn't. When installing a replacement boiler or servicing a steam system, take the time to troubleshoot the whole system and let the owner know what can be done to make it perform better. You will find suggestions throughout Parts 9 and 10 indicating problems that can occur with steam systems that may have been modified over the years, may have been designed for use with a coal-fired boiler or an old boiler with large water content, or may be suffering from leaking or sagging pipes. You will need to at least recommend the work to the owner. The owner may decline having the work done, but you've let him know where the problems are and why. Some extra money spent now could save a lot of money in the future.

Parts 9 and 10 are an introduction to steam heating systems, with emphasis on critical design and installation factors. For more information on steam systems, try Dan Holohan's "The Lost Art of Steam Heating," "The Lost Art of Steam Heating Companion," and "A Pocketful of Steam Problems;" ITT Fluid Handling's "The Steam Book: A Primer for the Non-engineer Installer," and "Hoffman Steam Heating Systems: Design Manual and Engineering Data (Bulletin No. TES-181)." These publications will provide more detail. Dan Holohan's books, in particular, give insight into the original design intent of steam systems, including a wide variety of vapor systems. You may find "A Pocketful of Steam Problems" really helpful when troubleshooting steam systems.

Steam, water and air

Steam heating systems don't just handle steam. Steam condenses. And the system returns this condensate to the boiler. Steam systems also handle air — an unappreciated and vital contributor to steam system operation. Air stops the heating.

So steam system components control the flow of steam, water, and air.

Steam
Steam is produced in the boiler and delivered to heating units through the piping. What makes the steam move? Steam moves because it tries to expand. It moves toward the heating units because the pressure is lower there than at the boiler. When the boiler begins to make steam, pressure begins to build in the boiler header. The piping and radiators are at zero pressure, so steam moves toward them. As in hydronic water systems, nothing moves unless there is a pressure difference to make it move.

Water

Water in steam systems comes from condensate and make-up water. Water moves in steam systems mostly because the installer sloped the pipes in the right direction (and hopefully they are still sloped that way). The main reason water moves in steam systems is gravity (though vacuum pumps speed this up in vacuum systems).

Air

Most hydronic steam systems breathe. They exhale (expel air) as steam moves through the piping and heating units at heating cycle start. They inhale when steam flow stops and the remaining steam condenses, causing a vacuum. Air has to return to the piping and heating units to fill the void left when the steam condenses. (Exceptions: Vacuum and vapor systems are designed to hold a vacuum instead of allowing air to enter.)

Steam systems today may breathe far more often than originally intended. Many existing systems were designed and installed for use with coal-fired boilers, in which the fire was seldom completely off. There were no automatic ignition systems, so the fire had to be maintained almost constantly throughout the heating season. There was a nearly continuous source of steam with a coal-fired boiler. And system start-up was always done manually, not using automatic pressure and burner controls.

Because coal-fired boilers kept some steam available nearly all of the time, the system seldom pulled air in after the initial heating season start-up. Air handling devices (air vents and traps) seldom had to expel air from the system.

Contrast this with today's gas and oil-fired boilers. They can cycle (turn on and off) several times each hour. Every time the boiler turns off, the system steam collapses, pulling in air. Every time the boiler fires, the system air-handling devices have to expel the air again. Consider this when you look at a steam system. Older systems usually had an air vent (or trap) at the end of the steam main, but seldom at the ends of risers. Make sure there is a working main air vent near the end of the steam main AND every riser on one-pipe systems and two-pipe gravity-return systems. Make sure there is a working drip trap at the end of each steam main or downfeed riser on two-pipe pumped-return systems.

Definitions

Boiler feed system

> A condensate storage tank (receiver) with boiler feed pump, on which the feed turns on and off on demand from the boiler. System make-up water is introduced at the feed system, using an automatic water feeder.

Condensate return system

> A condensate storage tank (receiver) with boiler feed pump and a float mechanism that starts the pump on water rise and turns pump off when water level has dropped. System make-up water must be introduced at the boiler, using an automatic water feeder.

Drain (condensate)

> "Drain" condensate means condensate coming from a heating unit. A "drain" trap is any trap (usually thermostatic or float and thermostatic type) used to drain condensate from a heating unit (radiator, convector, heat exchanger, etc.).

Drip (condensate)

> "Drip" condensate is the condensate formed when steam condenses in distribution piping due to heat loss from the piping. A "drip" trap is any trap (thermostatic, float and thermostatic, bucket, etc.) used to handle this condensate. Drip traps handle condensate from steam mains and risers. A "drip line" is a pipe used to connect to the wet return of a gravity-return system. Drip lines handle condensate from heating units as well as piping heat loss condensate.

Gravity return system

> A system in which the condensate returns to the boiler due only to the pressure difference caused by the height of the water accumulated in the return riser(s).

Main

> A horizontal pipe serving two or more branches (or risers) for steam supply (steam main) or condensate return (return main).

Definitions (continued)

One-pipe steam system

> A steam system on which only one pipe is connected to each radiator. This pipe handles both steam supply in and condensate return out. (The one-pipe Paul system also includes pipes from the Paul air vents to the air vacuum pump.)

Pumped-return system

> A steam system in which condensate is returned to the boiler using a boiler feed pump (accompanied by a condensate storage tank).

Riser

> A vertical pipe that supplies steam to, or returns condensate from, heating units.

Runout

> The piping connecting a steam or condensate main to a branch line or riser.

Trap

> A device that prevents steam from entering the condensate return lines. Traps allow air or condensate to flow through, but prevent steam flow.

Two-pipe steam system

> A steam system in which two pipes are connected to each radiator. One pipe handles steam supply; the other handles condensate return.

Vent

> A device used to allow air to flow out of (and back into) steam system piping (main vent) or radiators (radiator air vent).

Water seal

> Also known as a loop seal — Any piping arrangement that provides a column of water separating the return lines from the steam lines. A water seal behaves similarly to a U-tube manometer. Steam cannot overcome the water column if the depth is at least 28 inches for each 1 psig of steam pressure.

Rules of thumb

EDR:

EDR stands for "Equivalent Direct Radiation," meaning the effective heating surface of a heating unit. Each square foot EDR is equivalent to 240 Btuh (with steam at 1 psig).

- To convert from EDR to Btuh, multiply EDR times 240.
- To convert from Btuh to EDR, divide Btuh by 240.
- To convert from Btuh to pph (pounds per hour) steam or pph condensate, divide Btuh by 970.

Condensate load:

- Each square foot EDR will generate about ¼ pound condensate per hour (pph)
- Each 1000 Btuh is equivalent to about 1 pph condensate.
- To convert from pph steam or pph condensate to Btuh, multiply pph times 970.

Steam flow:

- Each square foot EDR requires about ¼ pph steam per hour.

Steam properties:

- See Table 1 for properties of steam under pressure and under vacuum. (Notice that the temperature of steam drops as the pressure is reduced.)
- Steam temperature:

  Steam temperature increases as pressure increases. At 2 psig, steam temperature is 218°F. At 15 psig, steam temperature is 240°F.
- Steam volume:

  Steam volume decreases as pressure increases. At 2 psig, steam occupies about 23 cubic feet per pound. At 15 psig, it occupies only 13.9 cubic feet.
- Latent heat (of vaporization):

  Latent heat means heat required for a change from water to steam or from steam to water. Steam gives up about **970 Btu per pound** when it condenses at low pressure.

Table 1   Properties of steam

Vacuum (inches mercury)	Temperature (°F)	Latent heat (Btu/lb)
1	210	971
2	209	973
3	207	974
4	205	975
5	203	976
6	201	977
7	199	979
8	197	980
9	194	981
10	192	983
11	190	984
12	187	986
13	185	987
14	182	989
15	179	991

Pressure (PSIG)	Temperature (°F)	Latent heat (Btu/lb)
0	212	970
1	215	967
2	218	965
3	222	963
4	224	961
5	227	959
6	230	958
7	232	956
8	235	955
9	237	953
10	239	952
11	242	950
12	244	949
13	246	947
14	248	946
15	250	945

Heating units

Residential hydronic steam heating systems use radiators most often. Some systems use baseboard heaters, convectors, unit heaters, or heat exchangers. See Figure 1.

See Figure 2 to Figure 5, Page 11–10 and Table 2, Page 11–10 to Table 5, Page 11–10 to estimate heating capacity (square feet EDR) of typical heating units you will encounter. For heating capacity of convectors, see Figure 6, Page 11–11.

You will need to know square feet EDR when sizing steam piping and components. Even if you aren't installing a new steam system, you will want to verify the sizing of existing piping when troubleshooting problems. If you add heating units to a system, or revise any piping, you will also need to know how much radiation the piping must support.

Figure 1   Typical steam heating units

Tubular radiator

Columnar radiator

HI20906

Convector

Cast iron baseboard

Recessed radiator

Estimating radiator square feet EDR

Use Table 1 through Table 5 to determine the square feet EDR of radiators or convectors when manufacturer's data is not available.

Figure 2   Cast iron baseboard

To determine EDR of radiator:

1 Measure length of baseboard
2 Read baseboard square feet (EDR) from Table 2

HI20916

Table 2   Square feet EDR

Cast Iron Baseboard		
Depth (inches)	**Height** (inches)	**Square feet per linear foot**
2 1/2	10	3.40

Figure 3   Radiant convectors

To determine EDR of radiator:

1 Measure depth and height of radiator
2 Read square feet (EDR) from Table 3
3 Multiply table value times the number of radiator sections

HI20915

Table 3   Square feet EDR

Radiant Convector		
Depth (inches)	**Height** (inches)	**Square feet per section**
5	20	2.25
7½	20	3.40

Figure 4   Tubular radiators

To determine EDR of radiator:

1 Count the number of tubes (illustration is 5-tube radiator)
2 Measure height of radiator, floor to top
3 Read square feet (EDR) from Table 4
4 Multiply table value times the number of radiator sections

HI20913

Table 4   Square feet EDR

Tubular Radiation					
Height (inches)	3 Tube	4 Tube	5 Tube	6 Tube	7 Tube
14					2.67
17					3.25
20	1.75	2.25	2.67	3.00	3.67
23	2.00	2.50	3.00	3.50	
26	2.33	2.75	3.50	4.00	4.75
32	3.00	3.50	4.33	5.00	5.50
38	3.50	4.25	5.00	6.00	6.75

Figure 5   Columnar radiators

To determine EDR of radiator:

1 Count the number of columns (illustration is a 3-column radiator)
2 Measure height of radiator, floor to top
3 Read square feet (EDR) from Table 5
4 Multiply table value times the number of radiator sections

HI20914

Table 5   Square feet EDR

Columnar Radiation					
Height (inches)	1 column	2 column	3 column	4 column	5 column
14					4.00
17					4.00
18			2.25	3.00	5.00
20	1.50	2.00			5.00
22			3.00	4.00	6.00
23	1.67	2.33			
26	2.00	2.67	3.75	5.00	
32	2.50	3.33	4.50	6.50	
38	3.00	4.00	5.00	8.00	
44				10.00	
45		5.00	6.00		

Figure 6 Convectors, copper or cast iron — estimated square feet EDR for typical units

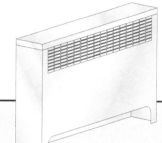

Copper convectors
Cabinet dimensions (inches)
(Typical data — EDR varies slightly with manufacturer)

Approx. Cabinet Depth (inches)	Approx. Cabinet Length (inches)	Cabinet height (inches)					
		18	20	24	26	32	38
4	20	10.4	11.3	13.1	13.3	14.0	14.6
	24	12.8	13.9	16.1	16.4	17.2	17.9
	28	15.2	16.5	19.1	19.4	20.4	21.3
	32	17.6	19.1	22.1	22.5	23.6	24.6
	36	20.0	21.7	25.2	25.5	26.8	28.0
	40	22.4	24.3	28.2	28.6	30.0	31.3
	44	24.8	26.9	31.2	31.6	33.2	34.7
	48	27.2	29.5	34.2	34.7	36.4	38.0
	56	32.0	34.7	40.2	40.8	42.8	44.7
	64	36.8	39.9	46.3	46.9	49.2	51.4
6	20	15.3	16.3	18.4	18.8	19.7	20.6
	24	189.8	20.1	22.7	23.1	24.2	25.4
	28	22.3	23.8	26.9	27.4	28.7	30.1
	32	25.8	27.6	31.1	31.7	33.3	34.8
	36	29.3	31.3	35.4	36.0	37.8	39.6
	40	32.8	35.1	39.6	40.3	42.3	44.3
	44	36.3	38.8	43.8	44.6	46.8	49.0
	48	39.8	42.6	48.1	48.9	51.3	53.8
	56	46.8	50.1	56.5	57.5	60.4	63.2
	64	53.8	57.6	65.0	66.1	69.4	72.7
8	20	18.7	20.0	22.5	23.0	24.5	25.9
	24	22.9	24.5	27.6	28.2	30.1	31.8
	28	27.2	29.1	32.8	33.5	35.6	37.7
	32	31.4	33.6	37.9	38.7	41.2	43.6
	36	35.6	38.1	43.0	43.9	46.8	49.5
	40	39.9	42.7	48.2	49.2	52.3	55.4
	44	44.1	47.2	53.3	54.4	57.9	61.3
	48	48.4	51.8	58.4	59.6	63.5	67.2
	56	56.9	60.9	68.7	70.1	74.6	79.0
	64	65.4	70.0	78.9	80.6	85.8	90.8
10	20	20.4	22.0	25.2	25.7	27.5	29.3
	24	25.1	27.1	31.0	31.7	33.9	36.1
	28	29.8	32.2	36.8	37.7	40.3	42.9
	32	34.6	37.3	42.7	43.7	46.7	49.7
	36	39.3	42.4	48.6	49.6	53.1	56.5
	40	44.0	47.6	54.4	55.6	59.5	63.3
	44	48.8	52.7	60.2	61.6	65.9	70.2
	48	53.5	57.8	66.1	67.6	72.3	77.0
	56	63.0	68.0	77.8	79.6	85.1	90.6
	64	72.5	78.2	89.5	91.5	97.9	104.2

Cast iron convectors
Cabinet dimensions (inches)
(Typical data — EDR varies slightly with manufacturer)

Approx. Cabinet Depth (inches)	Approx. Convector Length (inches)	Cabinet height (inches)					
		18	20	24	26	32	38
4 (No. 3)	18	8.4	9.1	10.5	11.0	11.8	12.3
	23	10.9	11.8	13.5	14.2	15.2	15.9
	28	13.3	14.4	16.5	17.4	18.6	19.4
	33	15.8	17.1	19.7	20.6	22.1	23.0
	38	18.2	19.7	22.7	23.8	25.5	26.5
	43	20.6	22.3	25.7	26.9	28.9	30.1
	48	23.1	25.0	28.7	30.1	32.3	33.6
	53	25.5	27.6	31.8	33.3	35.7	37.2
	58	28.0	30.3	34.8	36.5	39.1	40.7
	63	30.5	33.0	37.9	39.7	42.5	44.3
6 (No. 5)	18	12.3	13.5	15.4	16.2	17.5	18.2
	23	15.9	17.4	19.9	20.9	22.6	23.5
	28	19.5	21.3	24.4	25.6	27.7	28.8
	33	23.1	25.2	28.9	30.4	32.9	34.1
	38	26.7	29.2	33.4	35.1	38.0	39.4
	43	30.3	33.1	37.9	39.8	43.1	44.7
	48	33.9	37.0	42.4	44.5	48.1	50.0
	53	37.5	40.9	46.8	49.2	53.3	55.3
	58	41.1	44.8	51.3	53.9	58.4	60.6
	63	44.7	48.7	55.8	58.7	63.5	65.9
8 (No. 7)	18	16.0	17.1	19.4	20.4	22.5	23.7
	23	20.8	22.2	25.0	26.4	29.1	30.6
	28	25.4	27.2	30.7	32.4	35.7	37.5
	33	30.0	32.2	36.4	38.4	42.3	44.5
	38	34.7	37.2	42.1	44.3	48.9	51.4
	43	39.6	42.3	47.8	50.3	55.5	58.4
	48	44.2	47.3	53.5	56.3	62.0	65.3
	53	48.8	52.3	59.2	62.3	68.6	72.3
	58	53.5	57.3	64.9	68.3	75.2	79.2
	63	58.1	62.3	70.6	74.3	81.8	86.1
10 (No. 9)	18	19.2	20.6	23.4	24.6	27.3	28.8
	23	24.9	26.7	30.3	31.8	35.3	37.2
	28	30.8	32.8	37.2	39.1	43.3	45.7
	33	36.2	38.9	44.2	46.3	51.4	54.2
	38	41.8	45.0	51.1	53.6	59.5	62.7
	43	47.8	51.1	58.0	60.8	67.5	71.2
	48	53.3	57.2	64.9	68.1	75.6	79.6
	53	58.0	63.3	71.8	75.4	83.6	88.1
	58	64.7	69.4	78.7	82.6	91.6	96.6
	63	70.4	75.5	85.6	89.8	99.7	105.1

Components by system

Please see Part 13 for system piping and sizing. Part 11 provides basic information and guidelines for selection of steam system components. See Part 2 for boiler sizing recommendations. You will find some discussion of vapor and vacuum systems in Part 13. Because of the special components required for these systems, they are not covered in Part 11.

One-pipe steam, gravity-return

See Figure 7, Page 11–13.

Air handling

Other than the boiler, the only moving parts in a one-pipe gravity return system are its air vents. The system will have one or more main vents (on steam mains and risers) and an air vent on each radiator or heating unit.

Radiator supply valves

Radiator supply valves should be packless type. Valves must be fully open at all times of operation. Throttling radiator valves will result in water hammer and potential damage.

Zoning

You should only zone most one-pipe gravity-return systems using thermostatic radiator valves. Use only valves designed for one-pipe steam. They install between the radiator and the air vent. These valves regulate radiator heating by preventing the escape of air. They also provide a means to allow air to flow into the radiator when a vacuum develops.

Most one-pipe gravity-return systems do not have enough vertical distance between the boiler water line and the horizontal steam pipes to use zone valves. When a zone valve closes, the full pressure of the boiler pushes on the water in the returns. This causes the water to rise about 30 inches for every 1 psig at the boiler. Even if the boiler only operates at 2 psig maximum, the water will rise 5 feet (60 inches) above the boiler water line. Some installers use a check valve in the returns, intended to prevent the water from rising when using zone valves. The problem is, check valves don't check, at least not forever. Sooner or later, a check valve ends up with sediment on its seat, allowing water to flow back through it. The only reliable way to use zone valves on most one-pipe systems is to convert the system to pumped return.

Make-up water control (if used)

Preferably, use an electronic water feeder connected to the boiler and controlled by the boiler low water cut-off. If using a float-type feeder or a float-controlled electric feeder, you may encounter problems of nuisance flooding in some applications because the feeder can respond prematurely. When cold water enters the boiler, the water level actually falls. Ordinary feeders will add more water, trying to raise the level. Electronic feeders provide waiting times to allow the water feed to stabilize and to wait for normal condensate return.

Figure 7   One-pipe gravity-return

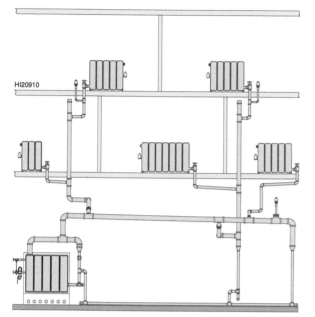

HI20910

Figure 8   One-pipe pumped-return

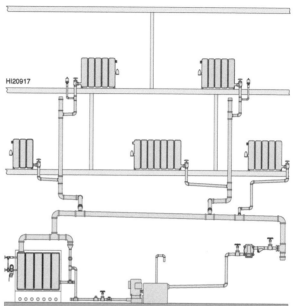

HI20917

One-pipe steam, pumped-return

See Figure 8.

Air handling

One-pipe pumped-return systems use air vents to control air movement. The system will have one or more main vents (on steam main and risers) and an air vent on each radiator.

Radiator supply valves

Radiator supply valves should be packless type. Valves must be fully open at all times of operation. Throttling radiator valves will usually result in water hammer and potential damage.

Condensate handling

Pumped-return one-pipe systems incorporate a boiler feed pump (with condensate receiver) and at least one float and thermostatic drip trap (on steam mains and risers). Some systems may use thermostatic traps in place of the F/T traps, provided the trap is preceded by at least 5 feet of horizontal pipe to cool the condensate.

Zoning

You can zone pumped-return one-pipe systems using either zone valves on the steam supply lines or one-pipe thermostatic radiator valves.

NOTICE: Install a vacuum breaker at the boiler when using zone valves on one-pipe steam lines. When all valves are closed, a vacuum will form when the boiler shuts down unless means are provided to allow air in to break the vacuum. A vacuum at the boiler would pull water from the condensate receiver, flooding the boiler and piping.

NOTICE: Using zone valves — If condensate can accumulate on either side of a steam zone valve when closed, install a drip trap to remove the condensate.

Make-up water control (if used)

Condensate return system — Preferably, use an electronic water feeder connected to the boiler and controlled by the boiler low water cut-off. If using a float-type feeder or a float-controlled electric feeder, you may encounter problems of nuisance flooding in some applications because the feeder can respond prematurely. When cold water enters the boiler, the water level actually falls. Ordinary feeders will add more water, trying to raise the level. Electronic feeders provide waiting times to allow the water feed to stabilize and to wait for normal condensate return.

Boiler feed system — Use an automatic water feeder supplied with, or specified by, the feed system manufacturer. Connect make-up water ONLY to the condensate receiver, NOT to the boiler.

Two-pipe steam, gravity-return (wet return sytstem)

See Figure 9.

Air handling

Other than the boiler, the only moving parts in a two-pipe gravity return system are its air vents. The system will have one or more main vents (on steam main and risers) and an air vent on each radiator.

Radiator supply valves

Radiator supply valves should be packless type. Valves can be used to throttle steam flow to radiators. You must close both radiator valves when isolating a radiator on this system.

Zoning

You should only zone most gravity-return systems using thermostatic radiator valves. Use only valves designed for one-pipe steam. They install between the radiator and the air vent. These valves regulate radiator heating by preventing the escape of air. They also provide a means to allow air to flow into the radiator when a vacuum develops.

Unless the piping provides at least 5 feet (60 inches) between the boiler water line and the lowest horizontal steam-carrying pipe or radiator connection, do not use zone valves on the steam lines or thermostatic radiator valves in the radiator supply piping. (Reason: When the valve closes, the full pressure of the boiler pushes water up the returns. For each psig at the boiler, water will rise about 30 inches.)

Most gravity-return systems do not have enough vertical distance between the boiler water line and the horizontal steam pipes to use zone valves. When a zone valve closes, the full pressure of the boiler pushes on the water in the returns. This causes the water to rise about 30 inches for every 1 psig at the boiler. Even if the boiler only operates at 2 psig maximum, the water will rise 5 feet (60 inches) above the boiler water line. Some installers use a check valve in the returns, intended to prevent the water from rising when using zone valves. The problem is, check valves don't check, at least not forever. Sooner or later, a check valve ends up with sediment on its seat, allowing water to flow back through it. The only reliable way to use zone valves on most one-pipe systems is to convert the system to pumped return. What's more, condensate can't begin to return to the boiler when the zone valve is closed until the water in the returns rises to 5 feet higher than the boiler water level (for 2 psig at the boiler).

Make-up water control (if used)

Preferably, use an electronic water feeder connected to the boiler and controlled by the boiler low water cut-off. If using a float-type feeder or a float-controlled electric feeder, you may encounter problems of nuisance flooding in some applications because the feeder can respond prematurely. When cold water enters the boiler, the water level actually falls. Ordinary feeders will add more water, trying to raise the level. Electronic feeders provide waiting times to allow the water feed to stabilize and to wait for normal condensate return.

NOTICE: Some two-pipe gravity-return systems (including vapor systems) use traps on the radiator condensate outlets. The return lines are piped as for pumped-return systems in Figure 10, Page 11–15. Vapor systems use radiator traps also. Any gravity-return system using traps must be operated at very low pressure (usually ½ psig) to prevent condensate from rising too high in the return lines.

Figure 9   Two-pipe gravity-return

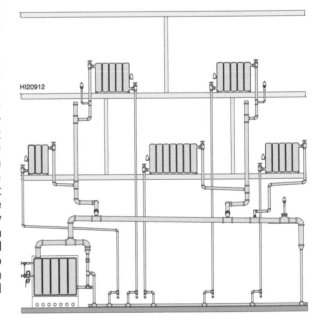

Two-pipe steam, pumped-return

See Figure 10.

Air/condensate handling

Two-pipe pumped-return systems use thermostatic traps on the radiators and float/thermostatic (or thermostatic) traps on mains and risers. These traps handle both condensate and air. The condensate returns must slope continuously down toward the condensate receiver, with no sags or water traps to prevent air movement out of and into the system.

Radiator supply valves

Radiator supply valves should be packless type. These valves can be used to throttle steam to the radiators.

Condensate handling

Pumped-return two-pipe systems incorporate a boiler feed pump (with condensate receiver) and at least one float and thermostatic drip trap (for steam mains and risers). Some systems may use thermostatic traps in place of the F/T traps, provided the trap is preceded by at least 5 feet of horizontal pipe to cool the condensate

Zoning

You can zone pumped-return two-pipe systems using either zone valves on the steam supply lines or thermostatic radiator valves on the radiator supply lines.

NOTICE: Install a vacuum breaker at the boiler when using zone valves or thermostatic radiator valves. When all valves are closed, a vacuum will form when the boiler shuts down unless means are provided to allow air in to break the vacuum. A vacuum at the boiler would pull water from the condensate receiver, flooding the boiler and piping.

NOTICE: Using zone valves — If condensate can accumulate on either side of a steam zone valve when closed, install a drip trap to remove the condensate.

Make-up water control (if used)

Condensate return system: Preferably, use an electronic water feeder connected to the boiler and controlled by the boiler low water cut-off. If using a float-type feeder or a float-controlled electric feeder, you may encounter problems of nuisance flooding in some applications because the feeder can respond prematurely. When cold water enters the boiler, the water level actually falls. Ordinary feeders will add more water, trying to raise the level. Electronic feeders provide waiting times to allow the water feed to stabilize and to wait for normal condensate return.

Boiler feed system — Use an automatic water feeder supplied with, or specified by, the feed system manufacturer. Connect make-up water ONLY to the condensate receiver, NOT to the boiler.

Figure 10   Two-pipe pumped-return

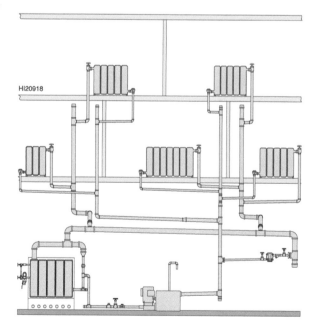

HI20918

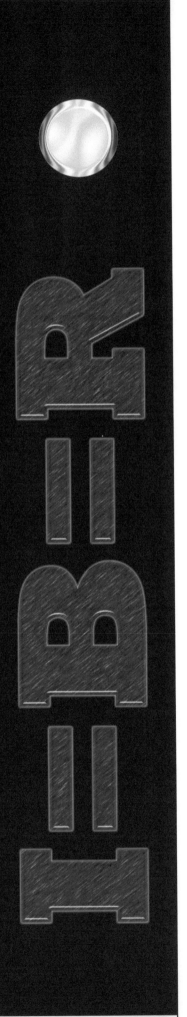

Components

Air vents

Application:
One-pipe steam systems

Two-pipe gravity-return steam systems

Installation
On radiators and ends of steam mains and risers

Operation
Air vents consist of a cylindrical housing containing a lightweight metal float with flexible bottom. The float is partially filled with a mixture of water and alcohol. The float is connected to a valve mechanism.

The air vent valve will close off if water enters the cylinder because the water will lift the float. The valve will also close if the float is heated. Heat causes the alcohol to boil and evaporate. This pressurizes the float, expanding the bottom, causing the float to lift and close the valve.

Figure 11   Radiator air vent, typical

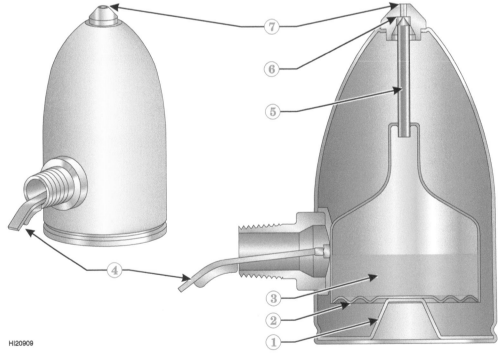

HI20909

(1) Float support   (2) Float diaphragm bottom   (3) Float containing alcohol/water mixture

(4) Tongue (separates air and water flow)   (5) Float needle   (6) Valve seat   (7) Outlet port

Vent ratings

Maximum pressure is the pressure limit for the vent housing. Exceeding this pressure could result in injury should the vent housing split open.

Maximum operating pressure is the highest pressure in the system at which the vent float can drop away, allowing the vent to open. You must set up the boiler limit/operating pressure control(s) so the pressure drops below the vent maximum operating pressure when the boiler cycles off on limit. For one-pipe and two-pipe gravity heating systems, you won't need a boiler steam pressure more than 2 psig. Set the limit to open at 2 psig and to come back on at 1 psig or less. If the boiler pressure were always kept at 2 psig or greater, for example, many air vents would never drop away. They would remain closed indefinitely, preventing air movement from the radiator.

Capacity (selecting vents) — Vents are not rated by capacity except as recommended for "small systems" or "medium to large systems." "Small" means any system under about 70,000 Btuh (290 square feet EDR total radiation). The question really only applies for main vents. Bigger is better for main air vents. You can use "large system" main vents on all steam mains and risers if you want to be sure of best performance for all cases.

NOTE: Use the same make and model vent on all radiators in the system. This ensures uniform performance. For best results, consider using adjustable air vents. They can be adjusted to the needs of each radiator. Use two vents on large radiators (one mounted below the normal location). See following for discussion.

Types:

Radiator steam vents

Radiator vents (Figure 11, Page 11–16) are side-connected for ease of installation. They may incorporate a tongue that separates water flow from air flow to reduce float bounce and vent sputtering.

Radiator steam vents are available in fixed and adjustable types. For most applications, use fixed type vents. You may find the adjustable vents useful for larger radiators. In general, set the venting speed of adjustable vents based on radiator size — smaller radiators can use slower vent speeds while larger radiators use faster vent speeds.

Matching vent rate to radiator size helps even rate of heat distribution in the building since vent rate controls how rapidly the radiator receives steam at start-up.

Some installers use adjustable vents to improve building heat distribution — using slower vent rates on radiators near to the boiler and faster vent rates for those further away. The best way to control heat distribution is to install main vents at the ends of all risers and near the end of the steam main. Main vents vent quickly and ensure that all radiators receive steam at about the same time. A typical main vent can move air up to 16 times faster than a radiator vent. The risk of using proportional venting for heat distribution is, if the vent rate is too fast for the radiator, the steam velocity entering can be too high, causing excessive condensate to be trapped in the radiator. This can cause water hammer in the radiator. Matching the vent rate to the radiator size ensures a smooth flow of condensate and avoids water hammer.

For large radiators, you may find a tendency to encounter water hammer or spitting from the air vent with a fast venting rate using a single adjustable air vent. You can obtain a fast venting rate without water hammer by using two air vents — one in the conventional location on the side of the radiator and the other a few inches below it. You will need to drill and tap the radiator to provide the mounting hole. The double-vent method will cause quick venting at start-up. But, when steam heats the upper vent, the vent closes. The remaining air is moved out at half the original speed this way, reducing the vent rate enough to prevent water hammer and sputtering. The lower-mounted second air vent will also eliminate more air from the radiator, increasing its heat output.

For very large radiators (over 25 square feet EDR), you may find continual problems with heating or tendency to water hammer. You can eliminate this recurring problem by replacing the single large radiator with multiple smaller ones — all no larger than 25 square feet. (For more information, see Chapter 8 of "The Lost Art of Steam Heating," Dan Holohan.)

Other heating unit air vents

In addition to radiator vents you will find vents designed specifically for use on convectors (convector vents) and unit heaters (steam unit heater vents). These are all bottom connected.

Main vents and quick vents

For venting distribution lines (steam mains and risers), use main steam (air) vents or "quick vents." These vents have a capacity up to 16 times higher than a radiator vent. See Figure 12 for typical construction of main vents.

Pay close attention to the mounting of main vents (see Figure 13). They must be protected from water hammer. If a vent is exposed to water hammer, it will fail, almost always failing closed. If you install the vent at the end of a main on top of a tee, you may as well save some money and install a pipe plug instead. After a little system operating time, the vent will be destroyed by water hammer and will no longer allow air to escape.

If a one-pipe or two-pipe gravity system does not have a main vent on every riser, add one. If the riser is piped in an inaccessible chase, then add a tee to a radiator connection near the end of the riser and install a main vent off of the tee.

Figure 12 Main air vent, typical

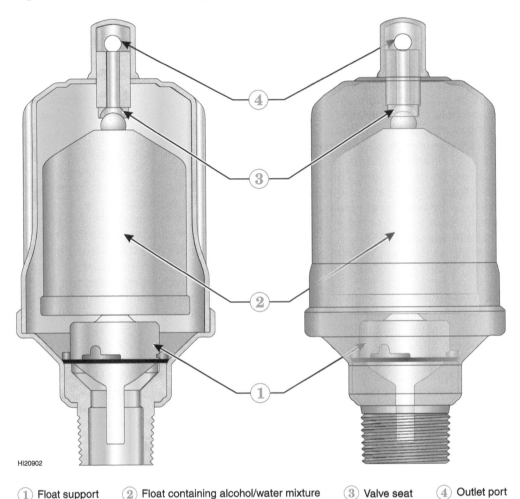

HI20902

① Float support   ② Float containing alcohol/water mixture   ③ Valve seat   ④ Outlet port

Figure 13 Correctly installing main vents

RIGHT WAY

Main vent

6" to 10"

End of steam main or dry return

15" minimum

Main vent is isolated from end of main, where hammer can occur. Using elbow at end of main smooths flow and reduces potential for water hammer. The 6 to 10-inch height above main provides a "snubber" to reduce shock from any water hammer that occurs.

WRONG WAY

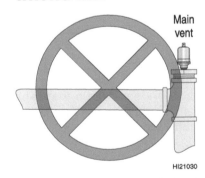

Main vent

HI21030

This is a common installation method, and will result in heating problems. Tee at end of main increases likelihood of water hammer because condensate hits tee and goes both directions. Main vent **WILL BE DAMAGED** by water hammer, rendering vent inoperative.

Traps

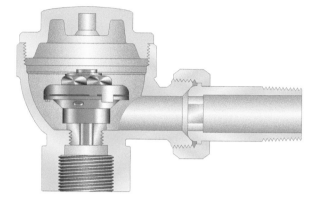

Traps allow condensate and air to pass through, but stop steam (provided they haven't failed open). Note that drip traps on steam mains and risers allow air removal from distribution piping just as main vents on gravity-return systems. Air moves through the drip traps, into the return lines, and out the vent of the condensate receiver on pumped return systems (or out the vent on vapor systems).

The most common traps used in hydronic steam heating systems are the thermostatic trap (upper right and Figure 14, Page 11–20) and float and thermostatic trap (lower right and Figure 15, Page 11–21). You will occasionally find bucket traps (Figure 16, Page 11–22) installed in system trouble spots (areas prone to water hammer) since they are far more resistant to damage. On the other hand, bucket traps can be troublesome for other reasons. They can easily lose their prime (blow out the water that seals around the bucket) and, unless specially equipped with a thermostatic air element, are poor at allowing air to move through.

Trap sizing

NEVER size a trap to match the pipe size (called "line sizing"). It will probably not work for long. The trap valve seat will be damaged from wire drawing because it never opens very far.

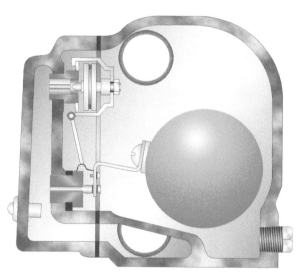

Size the trap for the load it has to handle and the pressure available to push the condensate through. Trap rating data shows capacity in square feet EDR based on available pressure drop across the trap.

What is the pressure drop?
Radiator thermostatic traps

Use a pressure drop of 1 psig. This is all the pressure needed in the radiator to provide 240 Btuh per square foot EDR.

Drip traps on mains and risers

Don't assume any steam pressure available. Size the trap only on the pressure due to the water column height at the trap entrance. This way you know the trap can dump the condensate that occurs when the piping is heated on start-up, before the steam pressure has had a chance to build up. (See Figure 18, Page 11–24.) Always try to install the trap at least 15 inches below the main or header as shown in Figure 18. Divide the available height in inches by 28 to find the pressure in psig. For 15 inches the pressure is 15/28, or about ½ psig.

What's a safety factor and how much should it be?

The safety factor allows for peak loads, trap wear, line blockage, etc., to be sure the trap will handle all conditions.

Heating unit traps — Size for a capacity 2 times the connected radiation.

NOTICE: Thermostatic trap ratings usually include the safety factor already because they are SHEMA ratings. Read the fine print, asterisks and notes. If the ratings are SHEMA (Steam Heating Equipment Manufacturers Association) ratings, they already have a 2-times safety factor built in.

Drip trap sizing — See Figure 18, Page 11–24. The recommended safety factor is 2 times if the start-up load is larger, or 3 times if the running load is larger. If the piping is insulated, the start-up load will probably always be the larger.

NOTICE: All steam piping should be insulated. Uninsulated piping can waste a lot of heat and can cause water hammer problems due to quick vacuum formation on shutdown or excess condensate on gravity-return systems. The only steam piping you shouldn't insulate is the 5 feet (or more) of steam pipe ahead of a thermostatic trap used as a drip trap (See Part 12, Figures 40 and 41, page 12-59). This length of pipe cools the condensate before it reaches the trap, allowing condensate to drain.

Thermostatic traps

See Figure 14, a typical thermostatic trap.

Thermostatic traps are the least expensive, but "smartest" traps used in the system. They are "smart" because they control the temperature of the condensate leaving them.

Thermostatic traps use a mechanism that responds to the temperature of the condensate or gas around it. The mechanism is constructed so it won't open unless the temperature is at least 10°F cooler than the steam temperature. (Condensate passes through float traps at almost the temperature of the steam. The trap simply opens when the condensate lifts the float.)

So how does the trap know what the steam temperature is? Because the pressure on the outside of the trap element affects how much it moves. Increase the pressure and it's harder for the element to expand, so the temperature has to get hotter to make enough movement happen. Decrease the pressure and the element can move enough to close without such a hot temperature.

Why make the condensate cool down? If you drop condensate into a return pipe at a temperature at or above 212°F (sea level), some of the condensate will flash to steam. The condensate falling can create a slight vacuum behind it, making it possible for condensate there to flash to vapor at even lower temperatures. But use a smart trap, like the thermostatic trap, to make the condensate cool off before it gets in the pipe, and you reduce chances of noise due to flashing.

Use thermostatic traps on all of the radiators. You can also use a thermostatic trap to drip a main or riser if you install at least 5 feet of uninsulated pipe ahead of the trap (see Part 12, Figures 40 and 41, page 10-59). The long pipe cools the condensate enough to satisfy the trap. Use this method to replace a float and thermostatic drip trap that has water hammer problems due to flashing in the return line at its exit.

Figure 14   Thermostatic trap, typical

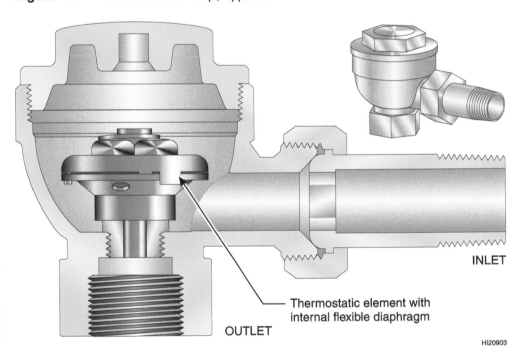

INLET

Thermostatic element with
internal flexible diaphragm

OUTLET

HI20903

Float and thermostatic traps

See Figure 15.

Float and thermostatic traps include a float-operated valve to handle condensate plus a thermostatic element to handle air. They move air quickly, making them ideal for installation on distribution piping (steam mains and risers).

Because the float movement depends on the depth of condensate in the trap, these traps can handle variable loads well. Be careful of oversizing, though. The trap will always be just barely open, causing the valve seat to wiredraw.

The float valve seat is replaceable. Make sure the trap seat is the correct diameter for the application. Too large a diameter and it will wiredraw; too small and it won't handle the load.

Use float and thermostatic traps for drip trapping of mains and risers. Also use them for all unit heater and heat exchanger applications.

Figure 15   Float and thermostatic trap, typical

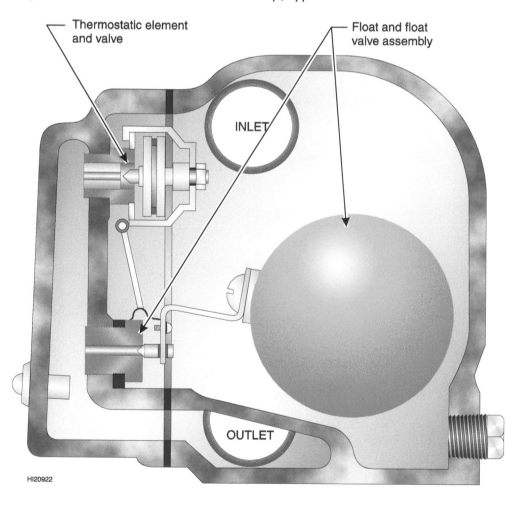

Inverted bucket traps

See Figure 16.

Inverted bucket traps use an inverted "can" that, in service, is immersed in water. As long as there is only water in the trap, the can is heavy, resting on the bottom. Condensate enters the chamber and passes through the valve. (Unless specially equipped with a thermostatic air element, these traps require air to bubble through the condensate, making air movement sluggish.) If steam enters the trap, it rushes to the top of the can, displacing water and making the can buoyant. The can lifts, closing the valve.

You must manually fill these traps (prime them) on initial start-up. They can lose their prime in operation, particularly under low loads. That is, the trap water can blow through, drying out the trap and allowing steam to pass through.

Figure 16   Inverted bucket trap, typical

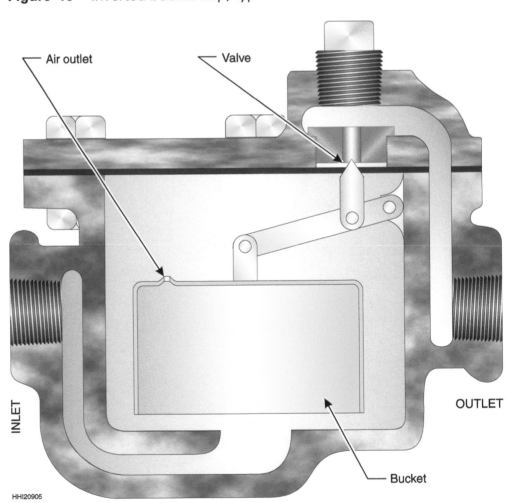

HHI20905

Trap failure modes

Thermostatic traps most often fail open (caused by water hammer damage to the element). This allows steam to flow through, causing steam pressure to build in the return lines of two-pipe systems. See Figure 17.

When troubleshooting heat distribution problems in systems with traps, you'll most likely find the defective traps where the rooms have heat. The radiators with "good" traps can't get heat because steam passes through the "bad" traps, pressurizing the returns and preventing other radiators from moving out the air.

Float and thermostatic traps usually fail closed on the water valve (due to damage to the float when exposed to water hammer). This prevents condensate from passing through. The water valve may also fail due to wire drawing of the seat if not properly sized. This would allow steam to pass through. If the thermostatic element fails closed, the trap will prevent air from exiting system piping, causing heat distribution problems.

Bucket traps most often fail open due to loss of the fill water that provides the seal. Bucket traps must be manually filled with water. (Water loss is most likely to occur due to light loads or over-sizing of the trap.)

For a problem application where a thermostatic or float and thermostatic trap consistently fails due to water hammer damage, consider installing a bucket trap because of its resistance to damage. (If the water hammer is due to flashing in the return line at a float/thermostatic drip trap, first try installing a thermostatic trap with a 5-foot cooling leg ahead of it (see Part 12, Figures 40 and 41, page 12-59). Also, lower the boiler pressure setting to no more than 2 psig if not already set there, unless the system contains a unit heater or heat exchanger that requires higher pressure.

Replacing trap elements

Replace the elements only when the system is down if possible. Replacing traps during heating operation can result in damage to the new elements before you have a chance to replace all of the bad ones. See Figure 17. When a good element opens to a return containing steam, chances of severe water hammer are high.

Set up a trap testing station in the basement or at your workshop. You can test the elements there and replace only the defective ones. Do this during the summer when the heating system is not operating.

Figure 17   Effects of trap failure

Trap failures — Look where the problem isn't!

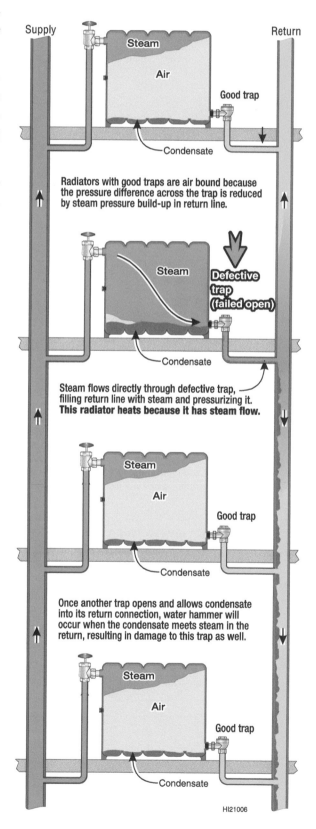

Figure 18 Drip traps handle condensate formed due to heat loss and heat-up of steam distribution pipes. Install drip traps and size as shown below. The illustration shows a float and thermostatic trap. You can also use a thermostatic trap provided you install at least 5 feet of uninsulated horizontal piping ahead of the trap to cool the condensate.

Drip trap installation

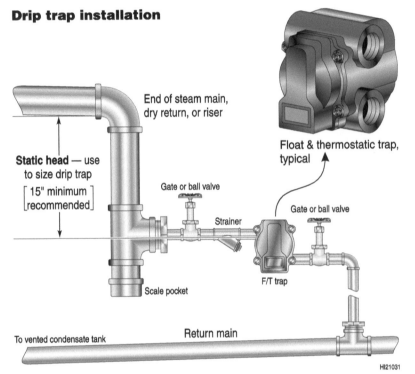

End of steam main, dry return, or riser

Static head — use to size drip trap
[15" minimum recommended]

Gate or ball valve

Strainer

Gate or ball valve

Float & thermostatic trap, typical

F/T trap

Scale pocket

To vented condensate tank

Return main

HI21031

Install drip traps off of stand pipe at end of main (or end of riser or dry return). Make the stand pipe as long as possible, preferably no less than needed to provide a static head of 14" (½ psig). **DO NOT LINE SIZE ANY TRAP.** Size drip traps for the condensate flow caused by piping start-up load or heat loss load, whichever is larger. Size assuming no steam pressure. Use a pressure drop across the trap only equal to the static head provided by the stand pipe to ensure the trap can handle the start-up load, during which there is no available pressure. To find pressure in psig, multiply static head in inches by 0.036; multiply static head in feet by 0.433.

Trap sizing

Determine required trap capacity by calculating both the start-up load and running load, with safety factors applied as recommended in tables. Use the larger of the two results.

If pipes are uninsulated, the running load will usually be the determining load. With insulated pipes, the load will be based on start-up because the running load is significantly lower.

Sizing drip traps for running load
(Multiply calculated load times **3** for safety factor)

Pounds of condensate per linear foot of pipe
(Multiply number below times total length of piping served by trap)

Pipe size (inches)	Bare steel pipe	
	2 psig steam	10 psig steam
¾	0.13	0.15
1	0.15	0.18
1½	0.21	0.24
2	0.25	0.29
2½	0.31	0.36
3	0.39	0.44
4	0.47	0.53
5	0.56	0.64
6	0.66	0.76
8	0.85	0.98
10	1.06	1.21
12	1.23	1.41

Sizing drip traps for start-up load
(Multiply calculated load times **2** for safety factor)

Pounds of condensate per linear foot of pipe
(Multiply number below times total length of piping served by trap)

Pipe Size (Inches)	15-Minute pick-up	30-Minute pick-up
3/4	0.091	0.046
1	0.136	0.068
1 1/4	0.184	0.092
1 1/2	0.220	0.110
2	0.295	0.148
2 1/2	0.469	0.234
3	0.614	0.307
4	0.873	0.437
6	1.536	0.768
8	2.311	1.156
10	3.277	1.638
12	4.333	2.167

Based on 70 ûF room temp, 10 psig steam (239 ûF), schedule 40 steel pipe

Pipe size (inches)	Insulated steel pipe *	
	2 psig steam	10 psig steam
¾	0.02	0.03
1	0.03	0.03
1½	0.03	0.04
2	0.04	0.04
2½	0.04	0.05
3	0.05	0.05
4	0.06	0.07
5	0.07	0.09
6	0.09	0.11
8	0.11	0.13
10	0.14	0.15
12	0.15	0.18

* Insulation equivalent to 2 inches of 85% magnesia insulation

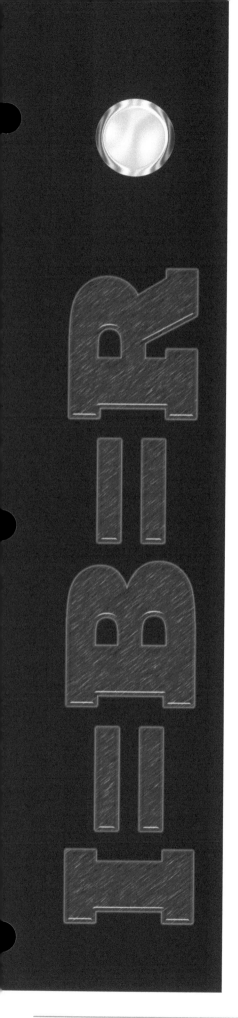

Feed systems & level controls

Water level controls

With pumped return, always provide a square-head cock on the pump line or feed line to throttle the feed rate. When a steam boiler operates, it is no longer filled with water. It contains a mixture of water and steam bubbles. (When water is fed or pumped to the boiler, the cooler entering water collapses some of the steam bubbles, causing the surface water level to actually DROP. Feeding water to the boiler too rapidly will cause overfeeding and result in system or condensate receiver flooding.

The steam/water mixture in a steam boiler causes a water level rise on start-up and a water level drop on shutdown. The amount of this drop will vary between boilers. It will also be affected by impurities in the boiler water. Always thoroughly clean the boiler before use and periodically as directed by the boiler manufacturer's instructions.

Always follow the boiler manufacturer's instructions for acceptable types and installation of water level controls. See Figure 19 and Figure 20, Page 11–26. Notice in Figure 20 that piping a level control lower equalizer connection lower in the boiler than specified by the manufacturer will result in a false water level reading. This can lead to nuisance shutdowns and to flooding (particularly if a float-type feeder is located incorrectly like the left-hand device in Figure 20).

Figure 19   Float-type water level controls

Float-type low water cutoff connections

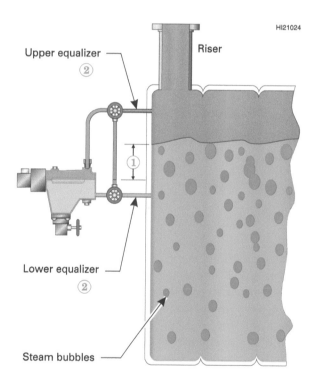

① When the boiler is firing, steam bubbles form and displace water. The mixture of water and steam in the boiler is lighter than water only. But there are not steam bubbles in the gauge glass or low water cutoff water. Consequently, the height of the water in the gauge glass is actually lower than the surface of the water inside the boiler when it is firing. The difference depends on boiler design and water chemistry.

② The pipes that connect the gauge glass and/or low water cutoff to the boiler are called equalizers. Pay careful attention to the boiler manufacturer's and control manufacturer's instructions for mounting these devices. They will not operate correctly if not installed correctly and in the right location.

Figure 20 Always pipe the equalizer connection to level controllers per boiler manufacturer's instructions

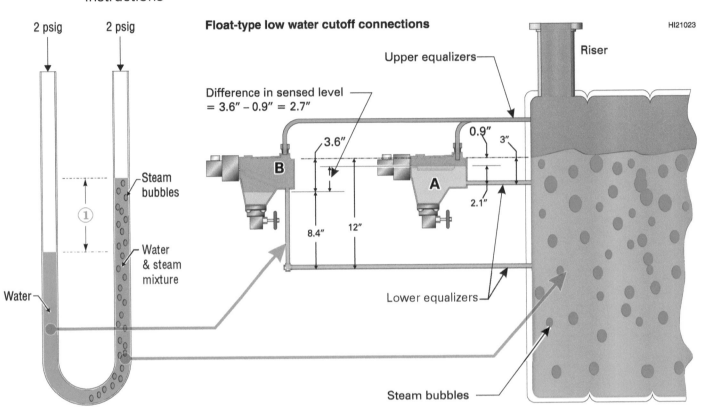

Float-type low water cutoff connections HI21023

2 psig   2 psig

Upper equalizers

Riser

Difference in sensed level
= 3.6" – 0.9" = 2.7"

0.9"   3"

3.6"

B

A

Steam bubbles

2.1"

Water & steam mixture

8.4"   12"

Water

Lower equalizers

Steam bubbles

In this example, the mixture of steam and water in the boiler is 30% steam by volume. The density of this mixture is about 70% of the density of water alone. So it only takes a column of water alone 70% as high as the steam/water mixture to balance the pressure. This means the height of the water in the low water cutoff piping is 70% of the height of the water surface above the lower equalizer, as shown above.

① Notice the comparison to a U-tube manometer at the left. With water on one side and a lighter water/steam mixture on the other, the water/steam column must be taller to balance the pressure of the heavier water-only on the other side. The difference in height increases as the lower equalizer position is placed lower in the boiler.

The water in low water cutoff A is 0.9" below the boiler water surface. The water in low water cutoff B piping is 3.6" below the boiler water surface, a difference of 2.7". If installed as shown, low water cutoff B would shut off the boiler, thinking the water level was too low. If the lower equalizer connection for B was at the same verti-

cal position as that for A, B would sense the same level and would control correctly. You could mount B lower to operate as a back-up device, but its lower equalizer pipe should still be piped at the same height in the boiler as for low water cutoff A.

The water level sensed by a level control drops lower as the lower equalizer is lowered in the boiler. All water level controls and sensing devices should have their lower equalizer connections piped into the boiler only, and at the same vertical height to ensure they sense the same conditions. Only connect level controls to a boiler in accordance with the boiler manufacturer's instructions. Never connect a level control to the boiler steam piping equalizer. This would be the worst possible case of difference in sensed level because the equalizer connects to the boiler return — the lowest possible connection to the boiler.

Height difference will vary based on percentage of steam in water. The more steam in the water, the greater the difference in height.

System time lag

The time lag is the time required for condensate to begin returning from the radiation after the system starts. Problems can arise if this time lag exceeds 10 or 15 minutes, though the time limit may depend on boiler design. If the time lag is longer than 15 minutes, the boiler may run low on water before condensate begins returning from the system. This causes the water feeder to add fresh water. Eventually the boiler will be over-filled, causing serious water carryover to the system. The result will be damage to traps and flooding at the receiver, not to mention significant water hammer in operation. The boiler will be damaged from oxygen corrosion and lime deposits if this continues. (Systems that flood frequently may do so because they have a long time lag.)

To determine time lag, clock the time between initial steaming at the boiler (when steam pipes become hot) and when the condensate return line becomes hot (due to returning condensate). If this time is longer than 15 minutes, you may need to provide additional operating water volume for the boiler. You can do this with an accumulator receiver mounted in the boiler return piping (if recommended by the boiler manufacturer) or by installing a boiler feed system plus a low water cutoff/ pump control on the boiler. A condensate return system (see following) will not solve a time lag problem because it does not feed water to the boiler on demand. The feed pump of a condensate return unit responds only to the level in the condensate receiver.

NOTICE

See "Troubleshooting" in Part 12, pages 12-65 to 12-69, for diagnosing water level problems. When determining the need for a boiler feed system, make sure the problem is actually due to too long a time lag and not to other causes.

Feed pump units

When is a feed pump required?

1. On steam systems that don't have enough height difference between the boiler water line and the lowest steam-carrying pipe (14 inches for systems under 100,000 Btuh or 28 inches for larger systems) to operate with gravity return.

2. On steam systems with zone valves or traps that don't have at least 30 inches per psig operating pressure at boiler between the boiler water line and the lowest steam-carrying pipe.

Pumped-return systems require a boiler feed pump and a receiver for holding condensate. Two types of units are available:

Condensate return systems

The feed pump is controlled by a float-operated switch mounted on the condensate receiver. When the water level rises to a preset height, the switch turns on the pump, feeding condensate to the boiler. When the water level drops to the preset height, the switch stops the pump. Condensate return units are not controlled by the boiler and are unaware of the water level needs of the boiler. Make-up water must be added at the boiler.

Boiler feed systems

The feed pump is controlled by a water level control mounted on the boiler. When the boiler water level drops to a preset level, the boiler level control starts the pump. The boiler level control turns off the pump when the water level rises to a preset height. Make-up water is added not to the boiler but to the condensate receiver.

Condensate return systems

See Figure 21.

A condensate return system includes a condensate receiver, a float mechanism that operates a pump switch, and a feed pump. You adjust the switch to start the pump at the desired height of water in the receiver and to stop the pump when the level drops to the desired point. (Adjust the switch so the volume fed to the boiler doesn't cause overfilling.)

Sizing condensate return systems

You will find most condensate return systems rated by square feet EDR. The condensate return system manufacturer includes a safety factor in this rating to allow for cyclic operation and pump wear, usually sizing the feed pump for 2 to 3 times the boiler steaming rate. (To find gpm (gallons per minute) of water evaporated to steam by a boiler, divide boiler gross output in Btuh by 500,000. Example, a boiler with a gross output rating of 200,000 Btuh will evaporate about $200,000 \div 500,000 = 0.4$ gpm of water into steam.)

Be sure to obtain a condensate return unit with a pump set for 20 psig discharge pressure or less. You will also need to install a square-head cock in the pump discharge piping to adjust the flow so the pressure going to the boiler is no more than about 7 psig (for a boiler operating at 2 psig maximum pressure), or 5 psig above boiler pressure limit setting.

Because the water level actually drops a little when you first inject cooler water into the boiler, set the square-head cock so the boiler water level doesn't drop too severely when the pump comes on.

Most condensate return units are quite large compared to the size of residential steam boilers. You will find the lowest rating typically at around 1000 square feet EDR (equivalent to a boiler with an input of about 400,000 Btuh and gross output of about 320,000 Btuh.) If you are unable to adjust the pump rate low enough, install a by-pass pipe including another square-head cock from the pump discharge back to the condensate receiver return connection. By opening the square-head cock on the by-pass, you divert some of the pump flow and gain more control over feed rate going to the boiler.

Be careful not to over-size condensate return units. The receiver volume increases with rated capacity. When the feed pump operates, it dumps water until the float switch is satisfied. You can adjust the operating range at the switch, but only to a limited degree. The feed system feeds water until the switch is satisfied. It doesn't know and doesn't care what the water level or volume of the boiler is. You can easily flood a residential steam boiler if the feed volume is too high. The other consideration in the size of the receiver is, the larger the receiver, the longer it takes to fill. The boiler could easily run out of water before the receiver water rises high enough to start the pump if the receiver volume is too large. Ask your supplier whether the float switch can be adjusted enough to limit the amount of water pumped per cycle to no more than about five minutes of steaming time for the boiler. To determine this amount in gallons, divide the boiler gross output in MBH by 100. For example, for a boiler with 150 MBH output (150,000 Btuh), the condensate system should pump no more than about $150 \div 100 = 1.5$ gallons of condensate per pump on cycle.

Figure 21 Condensate return system

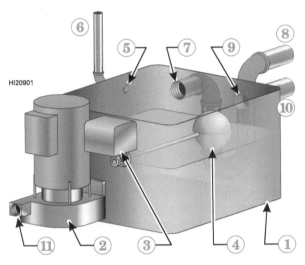

HI20901

① Condensate receiver
② Feed pump and motor (**operated by float switch**)
③ Float-operated switch (starts pump when water rises to preset maximum level; continues to pump until water level drops to preset level.
④ Float
⑤ Vent connection (vent to atmosphere as shown)
⑥ Vent piping
⑦ Overflow connection
⑧ Overflow piping — install loop piping as shown, then to drain
⑨ Condensate return connection (from system)
⑩ Condensate return piping — must slope down continuously from system return; sags or drops would cause water to accumulate, preventing air from escaping system through return line and then out vent piping; result would be heat distribution problems
⑪ Pump discharge connection — pipe to check valve, isolation valve, square-head cock, then to boiler

Note: Fresh water make-up for systems using condensate return units must be piped directly to the boiler, controlled by a make-up water feeder.

DANGER — Condensate receiver tanks are not designed to be pressurized. The vent connection must always be open to atmosphere. Otherwise, an explosion could occur should the tank become pressurized due to leakage from the boiler or system.

Boiler feed systems

See Figure 22.

A boiler feed system includes a condensate receiver and a boiler feed pump. It should also include a make-up water feeder or float-operated switch to operate an electric valve on a make-up water line.

The feed pump operates on demand from the boiler. Make sure the boiler is equipped with a low water cutoff/pump control, installed per the instructions in the boiler manual or installation guide. Use only a water level control recommended by the boiler manufacturer.

A boiler feed system is much "smarter" than a condensate return system. The boiler feed system responds to actual demand from the boiler for the amount of water the boiler needs, and when the boiler needs it. What's more, you size the feed system condensate receiver to provide all of the feed water the boiler needs during the time it takes for the condensate to begin returning. That is, you size the receiver to deal with the system time lag.

Sizing boiler feed systems

You will find most boiler feed systems rated by square feet EDR. The boiler feed system manufacturer includes a safety factor in this rating to allow for cyclic operation and pump wear, usually sizing the feed pump for about 2 times the boiler steaming rate. (To find gpm (gallons per minute) of water evaporated to steam by a boiler, divide boiler gross output in Btuh by 500,000. Example, a boiler with a gross output rating of 200,000 Btuh will evaporate about $200,000 \div 500,000 = 0.4$ gpm of water into steam.)

Be sure to obtain a boiler feed unit with a pump set for 20 psig discharge pressure or less. You will also need to install a square-head cock in the pump discharge piping to adjust the flow so the pressure going to the boiler is no more than about 7 psig (for boiler operating at 2 psig maximum pressure), or 5 psig above boiler pressure limit setting.

Because the water level actually drops a little when you first inject cooler water into the boiler, set the square-head cock so the boiler water level doesn't drop too severely when the pump comes on. The pump should stay on several minutes (preferably 5 minutes or longer if possible) each time the boiler calls for water. Shorter feed times indicate the feed rate may be too fast and is likely to cause water level disturbance in the boiler. Higher feed rates may also cause water hammer in the boiler header piping if the piping includes a Hartford loop as shown in the piping drawings of Part 13.

Most boiler feed units are quite large compared to the size of residential steam boilers. You will find the lowest rating typically at around 1000 square feet EDR (equivalent to a boiler with an input of about 400,000 Btuh and gross output of about 320,000 Btuh.) If you are unable to adjust the pump rate low enough, install a by-pass pipe including another square-head cock from the pump discharge back to the condensate receiver return connection. By opening the square-head cock on the by-pass, you divert some of the pump flow and gain more control over feed rate going to the boiler.

You don't have to worry about over-sizing the condensate receiver of a boiler feed system, except that the larger the receiver, the more expensive the feed system will be. Larger receivers also lose more heat to the boiler room.

Figure 22   Boiler feed system

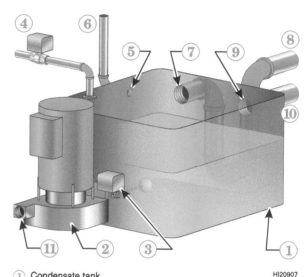

HI20907

1. Condensate tank
2. Feed pump and motor (**operated by boiler level controller**)
3. Float-operated switch (activates solenoid valve to add fresh water make-up if level drops below switch position)
4. Solenoid valve — pipe to fresh water make-up source
5. Vent connection (vent to atmosphere as shown)
6. Vent piping
7. Overflow connection
8. Overflow piping — install loop piping as shown, then to drain
9. Condensate return connection (from system)
10. Condensate return piping — must slope down continuously from system return; sags or drops would cause water to accumulate, preventing air from escaping system through return line and then out vent piping; result would be heat distribution problems
11. Pump discharge connection — pipe to check valve, isolation valve, square-head cock, then to boiler

DANGER — Boiler feed tanks are not designed to be pressurized. The vent connection must always be open to atmosphere. Otherwise, an explosion could occur should the tank become pressurized due to leakage from the boiler or system.

Make-up water

Water feeders

Should you install or use an automatic water feeder?

Many residential steam systems are equipped with automatic water feeders. This makes the system self-reliant, but it can also make it self-destructive.

If a steam system is working properly, with no leaks, the only water that should be lost is through slight water vapor loss at air vents or condensate receiver vent line. Only occasional make-up water is needed to handle this. If system piping were leaking or if water were being lost for some other reason, the problem would show up quickly without a water feeder. The house would get cold because the boiler would be low on water and shut down by its low water cutoff.

In the same situation, with an automatic feeder installed, make-up water would hide the leakage problem. The boiler wouldn't shut down on low water because the automatic feeder would always provide fresh water supply. The result would be serious damage to the boiler due either to the damaging effects of oxygen in the make-up water (oxygen corrosion) or lime accumulation, or both.

If you do install an automatic water feeder or work on a system that already has one, make sure there is a water meter installed on the make-up line and that the owner regularly checks water usage.

Where should water feeders be installed?

Gravity-return systems —

> The feeder must be installed on the boiler, controlled by a boiler water level control.

Pumped-return using a condensate return system —

> The feeder must be installed on the boiler, controlled by a boiler water level control. (Only the float-operated switch on the condensate receiver controls the feed pump of a condensate return unit.)

Pumped-return using a boiler feed system —

> The feeder must be installed on the condensate receiver, using the feeder type and installation supplied by or specified by the feed system manufacturer. Do not install a water feeder on the boiler. It will most always cause more problems than it solves because it will likely "jump to conclusions" about the need for make-up water. Make sure the boiler feed system receiver is sized to provide a steaming time longer than the system time lag (time required for condensate to begin returning to the receiver after the boiler starts steaming). The receiver should usually be sized for no less than 30 minutes steaming time if possible.

Float-type feeders

Float-type feeders use a float-operated valve. When the water level drops below the setpoint, the valve opens, adding make-up water to the boiler. Always include a square-head cock in the make-up line to allow throttling the flow rate to the boiler. Feeding too fast will collapse the water level, causing overfilling of the boiler/system.

NOTICE: Always install only the water feeder(s) recommended by the boiler manufacturer and mounted strictly by the boiler instruction manual. If you connect a feeder to the wrong tapping on the boiler, you can cause flooding due to false water level sensing.

Electric feeders

Electric feeders are generally solenoid-operated valves that activate (open) when the boiler level control/low water cut-off sees a drop in water level.

When the boiler is firing, it no longer contains water. It contains a mixture of water and steam bubbles. When cold water enters the boiler, it collapses some of the bubbles, and the water level actually drops. The faster the water enters, the more the drop. You must install a square-head cock in the feedwater line to allow throttling the flow rate.

Some electric feeders use electronic circuitry to control amount of water fed to the boiler. The advantage of these devices is that they know the boiler water level can react to fresh water introduction. So they usually include a time delay method to allow the water line to stabilize in between shots of fresh water.

An electronic water feeder is usually your best choice to ensure reliable operation.

Install a water meter

Every wet return on every hydronic steam system in the world will have to be replaced or repaired someday due to leakage or blockage. Steam system condensate is acidic and corrodes the return lines constantly exposed to it. When these lines begin to leak, the problem often isn't noticed, particularly with buried lines.

You have to stop systems from leaking. If you don't, the boiler life will be shortened severely due to damage from oxygen corrosion or liming. Oxygen corrosion is the attack of iron and steel components because of the air dissolved in make-up water. Liming occurs because water contains magnesium and calcium salts that drop out of solution when heated. You could supply softened make-up water, but soft water tends to foam more easily and could create other problems for you.

Always install a water meter on the make-up water line. Tell the owner to check the meter reading periodically to make sure the system isn't adding much make-up water. If it is, find the source. If you can't see the source, suspect the buried return lines. Take them out of service by replacing them with new piping.

Hydronics Institute Section of AHRI

35 Russo Place

Berkeley Heights, NJ 07922-0218

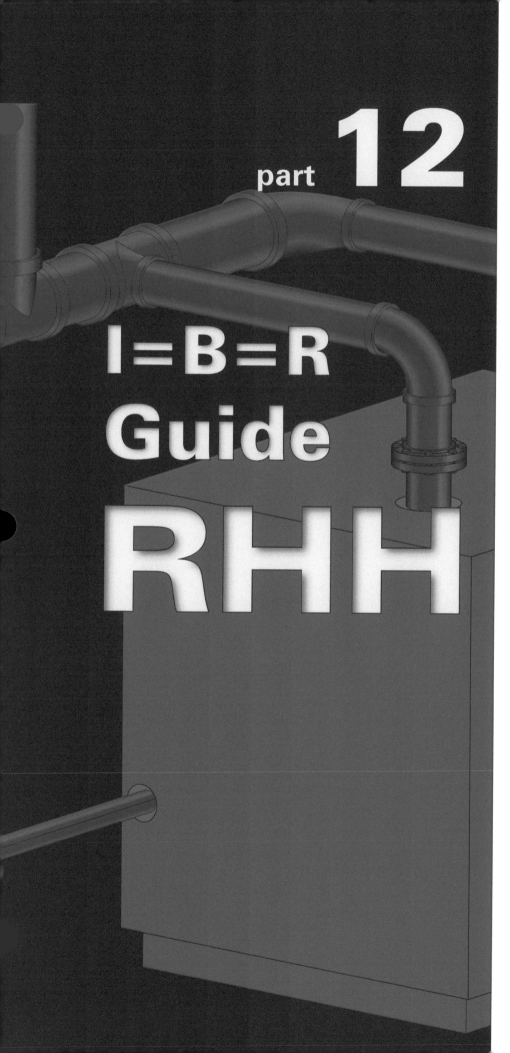

part **12**

I=B=R
Guide
RHH

Piping
hydronic steam heating systems

Residential Hydronic Heating . . .

Installation & Design

Hydronics Institute
Section of **AHRI**

I=B=R Guide RHH
Residential Hydronic Heating

Hydronics Institute Section of **AHRI**
35 Russo Place
Berkeley Heights, NJ 07922-0218

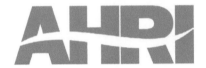

Contents – Part 12

Contents – *continued*

Contents – *continued*

Illustrations

Contents *– continued*

Illustrations (continued)

Contents *– continued*

Illustrations *(continued)*

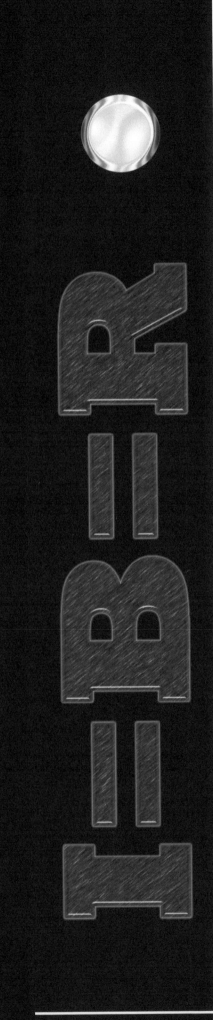

Introduction

Overview

In Part 12, we will discuss steam system design, operation and sizing. For detailed information on components and component selection, see Part 11.

You are unlikely to see many steam systems as simple as the ones shown in the illustrations in Part 12. What you need to do is use the information in Part 12 to capture how each system type is supposed to behave and how the pipes and components should be connected. You might find this will give you a little x-ray vision as you look at the pipes in a real system.

Remember that all of the systems discussed here use gravity to move condensate back to the boiler or condensate receiver. Sizing, installing and sloping the pipe the right way makes this happen. Remember also that you are installing a boiler very different from the boiler that may have been installed originally on the steam system. Apply some of the guidelines in Guide RHH to adapt the system to the equipment you are using today.

Smart piping

More than any other heating system, a steam system performs the way it does because the piping is "smart." Correctly installed, the pipes make the steam and water go when and where they are supposed to go. It may be that, as a hydronic installer, you have more control over performance of a steam system than any other you work with because it's mostly in the piping.

Some of the best engineering in our business was done in the development of steam heating systems. The installers and designers of these systems didn't have the automatic controls we have today. And the boiler heat source was hand-fired coal. They accomplished what they did only by installing "smart" piping.

Probably the single most important improvement in "smart" piping was the contribution by the Hartford Steam Boiler Insurance and Inspection Company in 1919 — called the Hartford loop. Boiler explosions were common at the turn of the century. One of the primary causes was water backing out of the boiler as in Figure 1, page 12–9. Without enough water to cool the boiler properly, boiler surfaces could become red hot. When water returned (or a feeder opened), the water flashed to steam instantaneously when it hit the hot surface. When this happened, it had the power of dynamite. The expanding steam tried to find a place to go, but couldn't. The pressure skyrocketed, tearing the boiler apart. In "The Lost Art of Steam Heating," Dan Holohan relates a quote from a book by Rolla Carpenter in 1895, that a cylinder boiler pressurized to 30 psig could leave the basement at 290 miles per hour and rise to a height of 3,431 feet!

The Hartford loop prevents water from being pushed out of the boiler because the steam pressure pushes equally (through the equalizer pipe) on the return and supply sides of the boiler. Water can only be pushed as low as the Hartford loop connection, normally located from 2 to 4 inches below the boiler water line (as recommended in the boiler instruction manual). This is the only "check valve" that really checks. Check valves can accumulate sediment on the seat and leak back, but not the Hartford loop. Water can still be pushed up in the return, but it can't be pushed out of the boiler.

You see, the Hartford loop is one of the "smart" piping techniques you control as an installer. Pay close attention to the boiler manufacturer's piping instructions to be sure you're using all of the techniques you need for a successful installation.

Figure 1   The Hartford loop

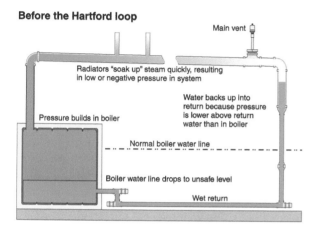

Before the Hartford loop

HI21028

With the Hartford loop — a safer system

Figure 2   Flow symbols used in Part 12

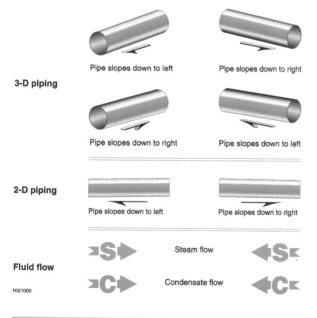

3-D piping

Pipe slopes down to left   Pipe slopes down to right

Pipe slopes down to right   Pipe slopes down to left

2-D piping

Pipe slopes down to left   Pipe slopes down to right

Fluid flow

HI21000

▶S▶ Steam flow   ◀S◀

▶C▶ Condensate flow   ◀C◀

Limit control settings

Cut-in pressure

The pressure at which the limit control will close, allowing the boiler to fire.

Set the cut-in pressure at about ½ psig for gravity-return systems; about 1 psig for pumped-return systems (unless the system piping was designed for a pressure drop higher than ½ psig). The rule-of-thumb is to set the cut-in pressure no less than two times the system pressure drop.

Cut-out pressure

The pressure at which the limit control will open, shutting the boiler down.

Set the cut-out pressure at 1 psig for gravity-return or 2 psig for pumped-return systems using radiators or convectors. If the system contains a heat exchanger or unit heater requiring more than 1 psig steam inside, add the required pressure at the heating unit to the cut-in pressure to find cut-out pressure.

Differential

The difference between the cut-in and cut-out settings of a limit control.

Keep the pressure low

Always set the boiler pressure limit control as low as possible. Never turn up the pressure setting to solve a heat distribution problem. Pressure probably isn't the reason for the heating problem and higher pressure will probably make the problem worse. The piping in most residential steam heating systems is designed for a pressure drop of less than ½ psig, and the radiators will deliver rated capacity of 240 Btuh per square foot EDR with only 1 psig steam inside. If you set the pressure any higher than 2 psig, you can create problems with operating components, particularly on gravity-return systems.

Flow symbols

You will see the symbols in Figure 2 used throughout Part 12. The arrows show the direction of slope, a critical factor in steam system operation. Piping drawings will also show the flow direction of steam and condensate.

Near-boiler piping

For a successful steam system installation, you must install the piping near the boiler exactly as shown by the boiler manufacturer. This piping is as important to boiler and system performance as the boiler burner and controls.

You can't just install the piping "the way the old boiler was installed" when doing a boiler replacement. Too many things are different today. Boilers are smaller for higher efficiency and they behave very differently from the boilers originally installed with the old steam systems. And the needs of a boiler vary with boiler design.

If the boiler manual shows multiple risers, you have to install multiple risers as shown. And the steam supply piping cannot be taken from between the risers — It has to be taken between the equalizer and the last riser.

See Figure 3 for an explanation.

Even when properly cleaned and skimmed, the boiler has to deal with dirty, usually contaminated, water. Steam leaving the boiler carries water droplets with it because of the turbulence at the water surface. As a comparison for what happens inside a boiler, contrast how a pan of water boiling on the stove behaves. It could steam indefinitely until the pot runs dry unless you drop in some macaroni. The starch in the macaroni dissolves in the water. Steam bubbles try to break away at the surface, but surface tension causes liquid to form around the bubbles instead. The surface begins to foam higher and higher with these starchy bubbles until the pan boils over. This behavior happens regularly inside a steam boiler because of sediment and impurities carried over from the system piping. In a steam boiler, impurities or chemical additives cause carryover and foaming.

Liquid separation

Boiler manufacturers provide Part of the solution to this problem of liquid carryover in the steam by showing you how to install the piping to remove this liquid. Locate the steam system supply pipe between the boiler equalizer and the last riser as shown in Figure 3. The piping becomes a mechanical separator. The relatively heavy water drops can't turn as easily as the steam, so they keep moving ahead when the steam turns up into the supply piping, dropping into the equalizer.

If you ignore the boiler manual and pipe the supply from the center, you cause what is shown in the upper right of Figure 3. Water accumulates at the steam supply take-off until it begins to carry over into the system. Result: Water hammer, trap and vent damage, and system flooding because water leaves the boiler too fast, making the system make-up feeder think additional water is needed.

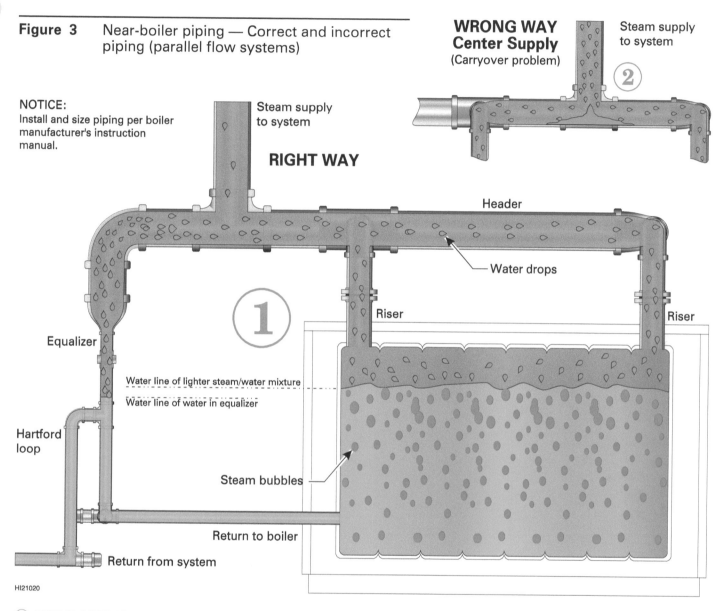

Figure 3 Near-boiler piping — Correct and incorrect piping (parallel flow systems)

NOTICE:
Install and size piping per boiler manufacturer's instruction manual.

Steam supply to system

RIGHT WAY

WRONG WAY Center Supply
(Carryover problem)

Steam supply to system

②

Header

Water drops

Riser

Riser

①

Equalizer

Water line of lighter steam/water mixture

Water line of water in equalizer

Hartford loop

Steam bubbles

Return to boiler

Return from system

HI21020

① **THIS IS CORRECT PIPING**: When the boiler fires, steam bubbles form in the water. These bubbles rise to the surface and exit through the steam supply piping. The mixture of water and steam in the boiler is lighter than water only (since the steam weighs virtually nothing compared to water). So the water line inside the boiler is higher than the water line in the equalizer. For this reason, always locate water level controls only in the tappings specified by the boiler manufacturer.

② **DO NOT CONNECT STEAM SUPPLY BETWEEN BOILER RISERS**: Turbulence and impurities in the water cause water droplets to carry out of the boiler with the steam flow. If the steam supply is piped correctly between the last riser and the equalizer (as in 1), most of the water droplets continue straight ahead to the equalizer as the steam turns up and out. The piping works as a steam separator. Old, large boilers were often installed with the steam supply between the risers. This may have worked for those boilers, but it will only cause problems if done with a modern boiler. Always connect the steam supply between the last riser and the equalizer.

③ **DO NOT REDUCE THE NUMBER OF RISERS SPECIFIED BY THE BOILER MANUFACTURER**: In order for any fluid to flow, the pressure must be higher where it starts than where it goes. If you only install one riser when the boiler instructions call for two or more, you will probably cause the back section of a cast iron boiler to dry out, causing failure from overheating. The water line slopes because the pressure in the back has to be higher than in front for the steam to flow to the front. The higher pressure pushes down on the water more in the back than the front, so the water level slopes. With the water level sloped, there isn't enough water in the back of the boiler.

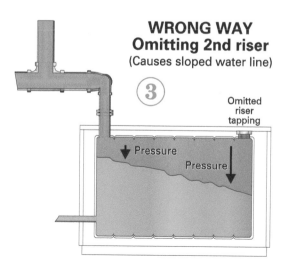

WRONG WAY Omitting 2nd riser
(Causes sloped water line)

③

Omitted riser tapping

Pressure

Pressure

Multiple risers

If you install fewer risers than the boiler manual requires, you can cause what is shown in Figure 3 lower right. For the steam to move toward the outlet pipe inside the boiler, the pressure where the steam starts has to be higher than where it is going. It wouldn't move otherwise. If the boiler needs two risers and you install one, the steam flow inside the boiler is twice what it is supposed to be. If you double the flow, you quadruple the pressure drop. The water inside the boiler is just like a manometer. Apply more pressure on one side than the other and the water will slope as shown. This will cause the right end of the boiler to overheat and would probably cause a cast iron section to crack.

Though the boiler piping may sometimes seem complicated, it isn't optional. The boiler and the system may not work properly if you don't follow the boiler manufacturer's guidelines.

Figure 4, page 12–13 through Figure 6, page 12–15 show near-boiler piping applicable to most cast iron steam boilers. Carefully read the boiler manual when installing a steam boiler and make sure to install the near-boiler piping exactly as shown.

Counterflow systems

Notice the special case of a counterflow system shown in Figure 6. DO NOT connect the system main into the top of the boiler header. You must connect the boiler supply piping into the top of the steam main. Note also that no Hartford loop is required with this piping because there is no dry return. The return piping is all above the water line. Provide at least 14 inches between the boiler water line and the main, as shown, to ensure enough height for condensate to accumulate.

Figure 4   Near-boiler piping — parallel flow steam system; one boiler riser

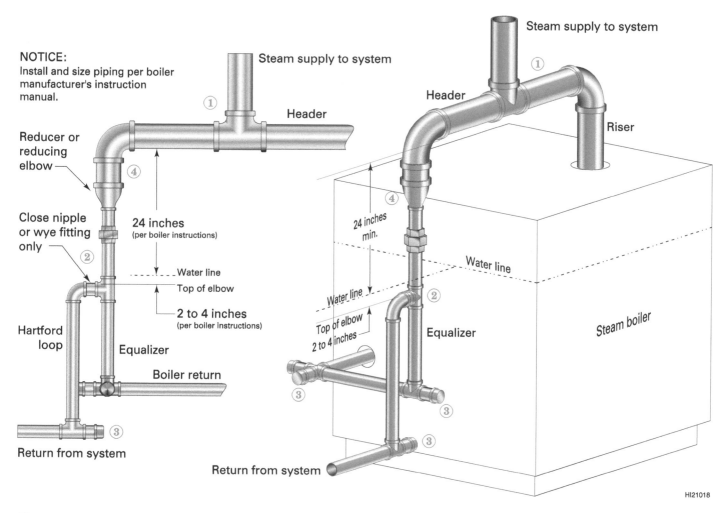

NOTICE:
Install and size piping per boiler manufacturer's instruction manual.

HI21018

① Always connect steam supply between boiler riser and equalizer. This ensures the greatest possible separation of moisture from the steam. Provide at least 24 inches height between the bottom of the header and the boiler water line.

② Connect the Hartford loop using nothing longer than a close nipple or wye fitting. Water level may often drop below top of fitting. A long horizontal connection would cause severe water hammer.

③ Install tee or cross with nipple and cap at all direction changes in return lines. Steam systems typically contain large quantities of sediment. You should periodically inspect and clean return lines to ensure reliable operation.

④ Reduce pipe size only after turning downward. Use a reducing elbow or reducing fitting in the vertical (equalizer) line.

Figure 5 Near-boiler piping — parallel flow steam system; two boiler risers

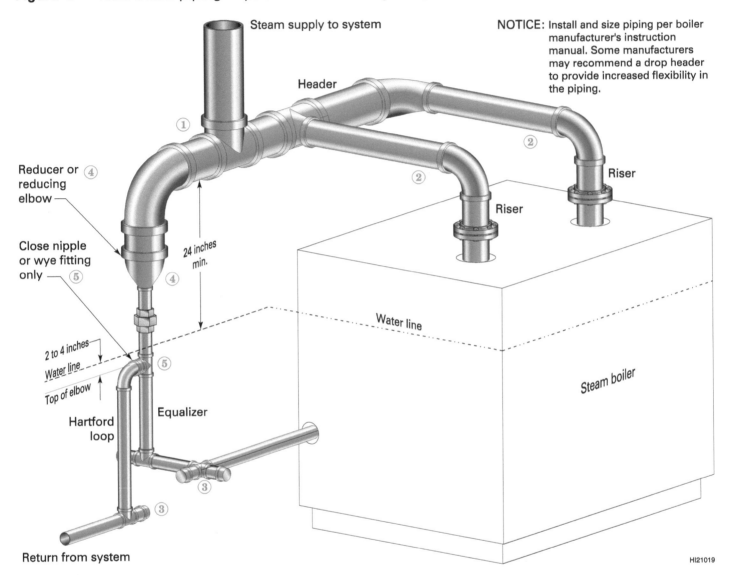

NOTICE: Install and size piping per boiler manufacturer's instruction manual. Some manufacturers may recommend a drop header to provide increased flexibility in the piping.

HI21019

① Always connect steam supply between last boiler riser and equalizer. This ensures the greatest possible separation of moisture from the steam. Connecting the steam supply between the boiler risers will cause water carryover to the system. The large flow of water from the boiler will cause the system to add make-up water unnecessarily, resulting in eventual flooding. Provide at least 24 inches height between the bottom of the header and the boiler water line.

② You must pipe the risers into the header with offset joints as shown. The horizontal piping and elbows provide swing joints to compensate for the expansion and contraction of the header piping. If you don't pipe this way, cast iron boiler sections can crack, or steel boiler welds can fail.

③ Install tee or cross with nipple and cap at all direction changes in return lines. Steam systems typically contain large quantities of sediment. You should periodically inspect and clean return lines to ensure reliable operation.

④ Reduce pipe size only after turning downward. Use a reducing elbow or reducing fitting in the vertical (equalizer) line.

⑤ Connect the Hartford loop using nothing longer than a close nipple or wye fitting. Water level may often drop below top of fitting. A long horizontal connection would cause severe water hammer.

Figure 6 Near-boiler piping — counterflow steam system; one boiler riser

NOTICE:
Install and size piping per boiler manufacturer's instruction manual.

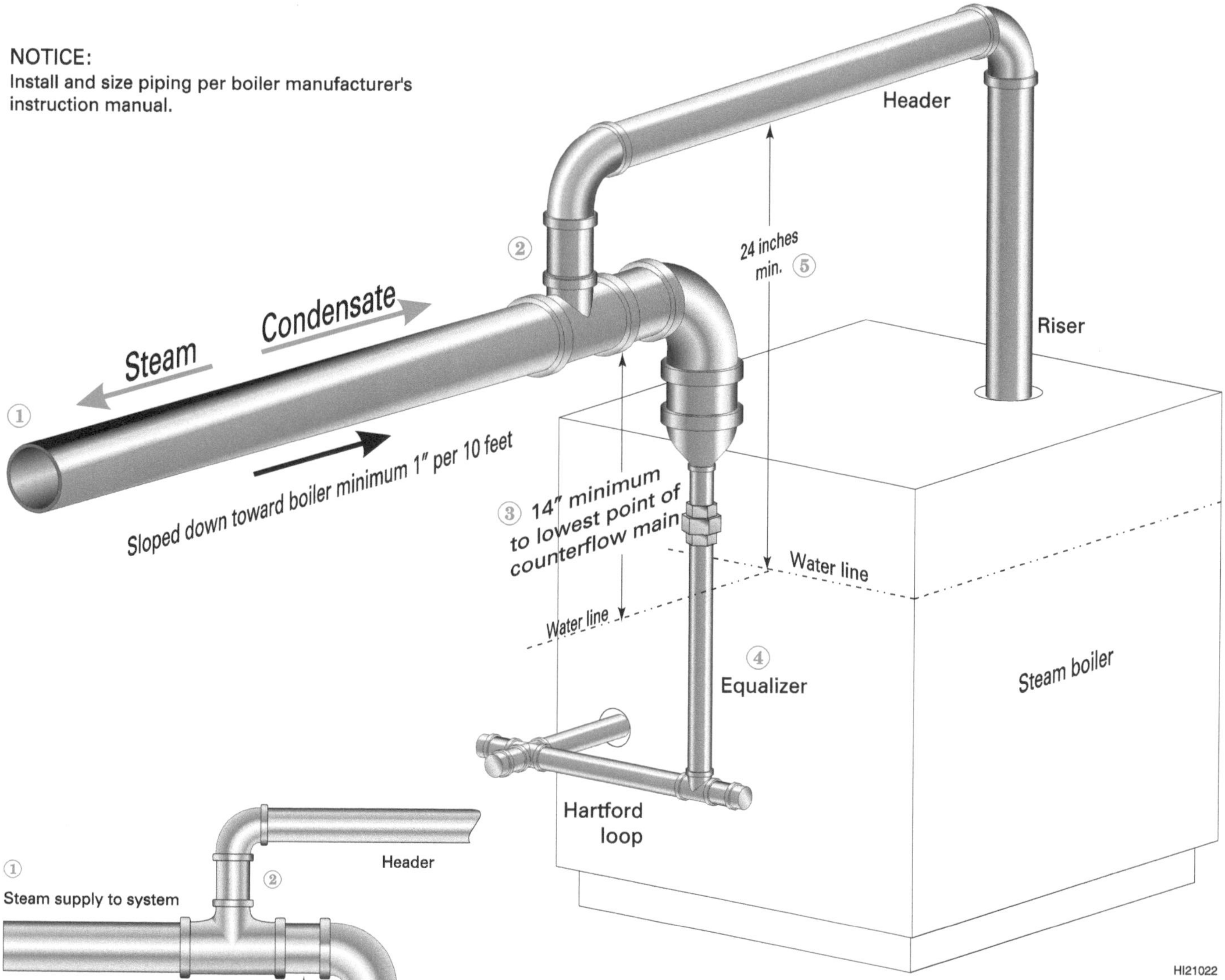

Header

24 inches min. ⑤

Riser

Steam

Condensate

① Sloped down toward boiler minimum 1" per 10 feet

③ 14" minimum to lowest point of counterflow main

②

Water line

Water line

④ Equalizer

Steam boiler

Hartford loop

HI21022

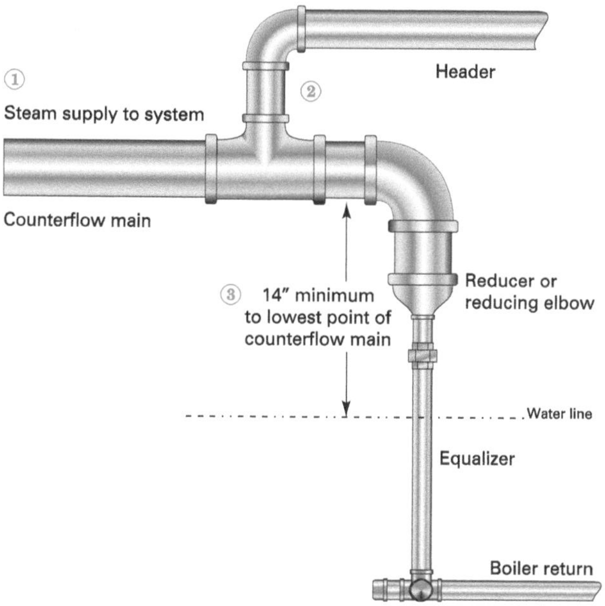

① Steam supply to system

② Header

Counterflow main

③ 14" minimum to lowest point of counterflow main

Reducer or reducing elbow

Water line

Equalizer

Boiler return

① Counterflow mains do double duty. They supply steam to the system going out the pipe, but must also carry the condensate flowing back from the system. Counterflow systems do not have a separate return line.

② The boiler header must connect to the top of the counterflow main. This prevents system condensate from entering the boiler header.

③ The lowest point of the counterflow main should be at least 14 inches above the boiler normal water line. This will prevent the possibility of condensate rising into the counterflow main, causing water hammer.

④ The equalizer connection is all that is required on this system. Since there is no wet return line below the boiler water line, no Hartford loop is needed.

⑤ Provide at least 24 inches height between the bottom of the header and the boiler water line.

Radiators

Figure 7, page 12–17 shows the difference in operation between a one-pipe radiator and a two-pipe radiator. As noted in Part 11, a one-pipe system provides one pipe to the radiator that must handle both steam in and condensate out. Two-pipe systems provide two pipes to the radiator — one for steam, the other for condensate.

Start-up

Figure 7, top illustrations

As steam pushes up the branch piping to the radiator, it pushes air ahead of it. On one-pipe systems, air pushes through the radiator and out the air vent to the room. Air moves through a two-pipe radiator, through the trap and into the return line. From there it finds its way to the condensate receiver and out the tank vent line.

Heating start

When steam reaches the radiator on a one-pipe system, it rises to the top like a hot air balloon. The steam is much lighter than the air. Air continues to flow out the air vent. Notice from the steam flow pattern why the air vent is mounted low on the side of the radiator. If the air vent were mounted on the top of the radiator (as on a hot water system), the air vent would quickly close off from the heat of entering steam.

Two-pipe radiator steam supply comes

in near the top of the radiator. This is possible because the supply piping doesn't have to handle the returning condensate as on one-pipe systems. As steam enters, it stays near the top, being much lighter than the air.

Condensate begins to accumulate in the bottom of the radiator.

Steady heating

When steam pushes out enough air on a one-pipe radiator, it reaches the air vent, expanding the vent float and closing its valve. Condensate flows back down the supply/return line. Note that you cannot throttle a one-pipe radiator valve because there has to be room for the condensate to flow backwards against the steam flow. See illustration below.

The two-pipe radiator thermostatic trap remains open for condensate and air flow until the steam temperature expands the trap element and it closes. Condensate then builds up until the condensate at the trap is from 10 to 30°F lower than the steam temperature. The trap element then opens, allowing condensate to drain until the element is heated up again.

You can throttle the radiator supply valve of the two-pipe radiator if desired to slow down the heating rate. You can do this as fine-tuning through the house to help even out the heat distribution.

With valve fully open, steam can move over condensate without restricting its flow.

If valve is partially closed, flow path is too small. Steam velocity is high, preventing condensate from flowing freely. Condensate backs up into radiator, causing water hammer.

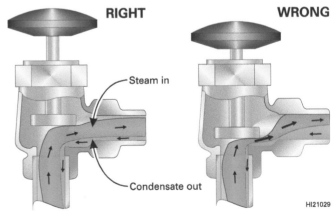

HI21029

Figure 7 Radiator operation on one-pipe and two-pipe systems

One-pipe steam

Two-pipe steam

① **Cycle start**
steam pushes air through radiator
vent allows air to escape

① **Cycle start**
steam pushes air through radiator
trap allows air to flow into return line

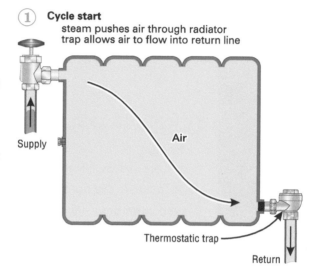

② **Heating begins**
steam pushes into radiator and rises to top
vent continues to allow air to escape

② **Heating begins**
steam pushes into radiator and rises to top
trap allows air and condensate into return line

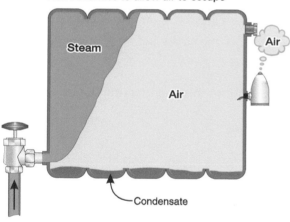
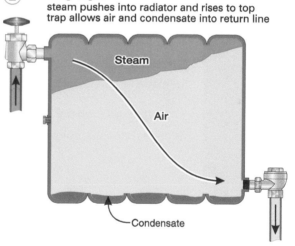

③ **Heating cycle underway**
steam fills radiator until it reaches vent
vent float heats and expands, closing vent valve

③ **Heating cycle underway**
steam fills radiator until it reaches vent
trap closes if heated by steam, opens when
condensate rises and cools

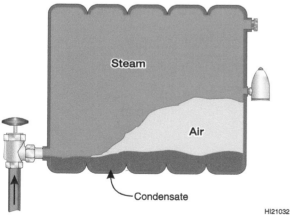
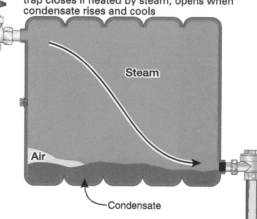

HI21032

Gravity-return systems

U-tube manometers

Gravity-return systems use the same principle as a U-tube manometer to return condensate to the boiler. Water rises high enough in the returns until it overcomes the pressure difference between the boiler and the end of the steam lines. Then water begins to flow back to the boiler.

Figure 8, page 12–19 shows how a U-tube manometer behaves. The height of the column depends on the density of the fluid and the pressure on top of the column.

Most gravity-return steam systems are like Figure 8, example 3, with a pressure difference between the beginning and end of the steam piping of about ½ psig (14 inches water column).

The pressure exerted by a fluid at the bottom of a fluid column equals the fluid density times the column height:

$$P_{fluid} = D \times H/144, \qquad (10\text{-}1)$$

with P in psig, D in pounds per cubic foot, and H in feet.

> Example: What is the pressure exerted by a 3-foot high column of fluid with a density of 62.3 pounds per cubic foot (water at 60°F, for example)?
>
> Use Equation 10-1: $P_{fluid} = D \times H/144 = 62.3 \times 3/144 = 1.30$ psig

The total pressure at the bottom of the fluid column is the pressure due to the fluid (from Equation 10-1) plus the pressure on top of the column.

$$P_{total} = P_{fluid} + P_{top'} \qquad (10\text{-}2)$$

The cross-sectional area of the column makes no difference in the pressure.

The surfaces of the two sides of a manometer adjust until the pressure at the bottom is the same from both sides.

Figure 8   How U-tube manometers behave

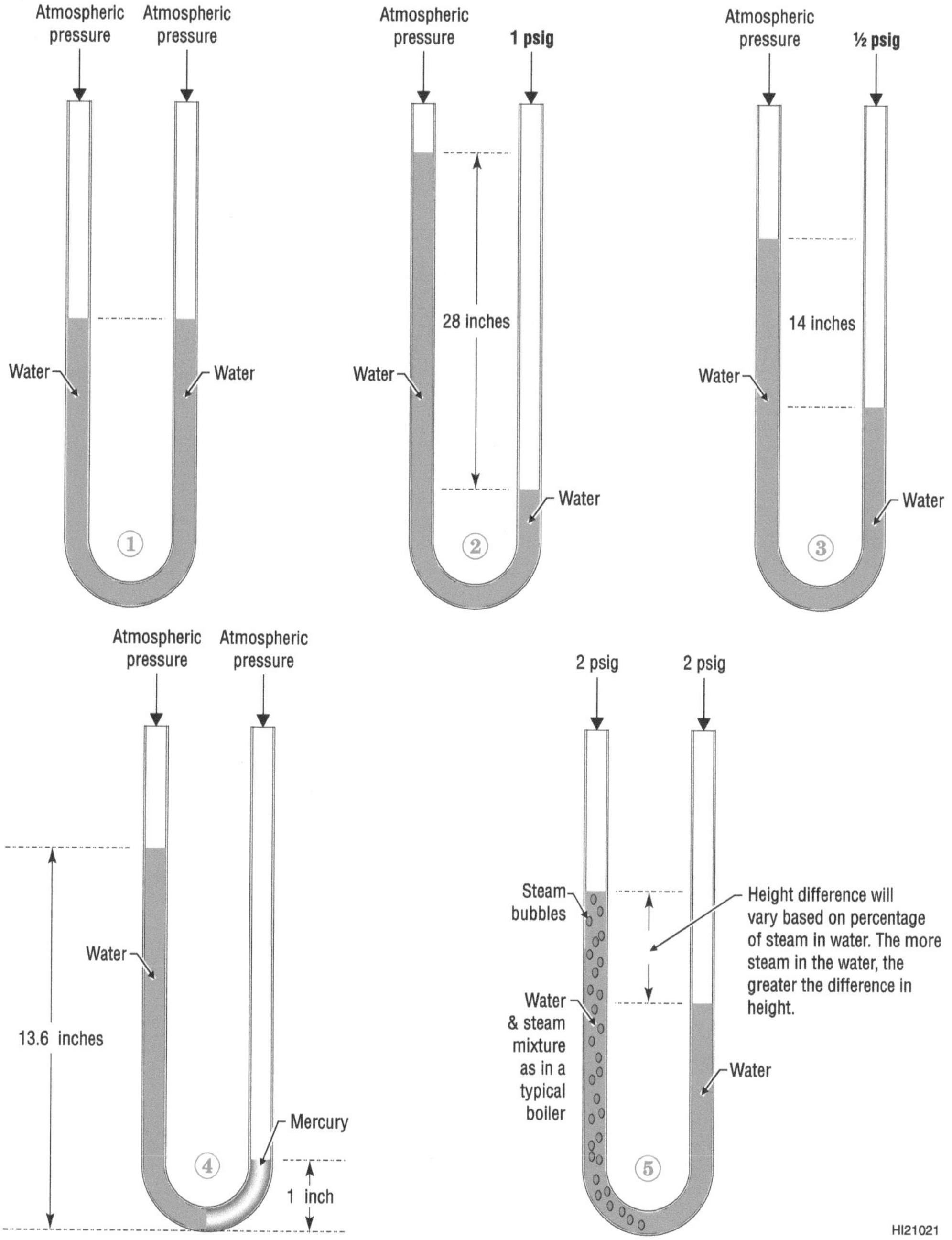

① Water seeks its own level — provided you have water on both sides and the pressure on both sides is the same. A U-tube manometer with water on both sides and equal pressure results in both surfaces at the same height.

② Change the pressure on one side of example 1 and look what happens. The higher pressure side drops. When the surfaces settle, the difference in height will equal the difference in pressure (1 psig is about 28 inches of water).

③ Example 3 is similar to example 2, but with the pressure difference half as much (½ psig is about 14 inches of water).

④ Here the pressure is the same, but we have mercury on one side, water on the other. Mercury is much heavier than water. It takes just a 1-inch-high column of mercury to balance a 13.6-inch column of water.

⑤ Example 5 shows what happens in a steam boiler. When the boiler fires, it contains a mixture of water and steam. A gauge glass or control mounted outside the boiler contains only water. The outside level will actually be lower than the surface level inside the boiler because the steam/water mixture is lighter than water only.

Dimension A

Figure 9 shows three variations of a one-pipe gravity-return system. Parallel flow systems, the top two examples, flow steam and condensate in the same direction in the main. This is why the main slopes downhill toward the return riser (vertical pipe on the right). Counterflow systems, the bottom example, flow steam and condensate in opposite directions in the main. The main slopes back toward the boiler, causing condensate to flow back downhill against the uphill steam flow. See later discussions in Part 12 for detailed piping and operation of gravity-return systems.

Parallel flow systems need enough height in the return riser to allow the condensate to back up in this pipe until the column height overcomes the pressure differences in the system. The distance between the boiler water line and the lowest steam-carrying pipe (the end of the main in Figure 9) is called **Dimension A**.

How big does dimension A have to be?

Pressure difference in main —

> One-pipe steam systems were usually designed so the steam flow in the piping caused about a ½ psig pressure drop. That means the pressure on top of the water in the return riser will be ½ psig lower than the pressure on the water column in the boiler equalizer pipe. As in Figure 8, page 12–19, example 3, this pressure difference will push the water up in the return riser 14 inches (½ psig) higher than in the equalizer.

Pressure loss in return piping—

> When water flows through a pipe, it loses energy because of friction. This causes a pressure drop. Steam systems were usually designed for a return-piping drop of 2 to 4 inches water column. You need to account for the condition of the return piping now, however. The pipes are probably partially blocked by sediment. So the pressure drop will be higher than originally intended. Allow 6 inches water column for this drop. This pressure drop adds to the steam line pressure drop, raising the water line another 6 inches in the return riser.

Start-up condensate rush —

> When the system is running steady, the return riser water column height doesn't have to be much more than the pressure difference in the piping (14 inches plus 6 inches, or 20 inches total). When the system first starts up, though, the piping is cool. This causes an extra steam load at start-up, condensing some of the steam to heat up the pipes. This happens quickly, causing an extra slug of condensate to occur soon after the system begins to steam. You need to allow about 8 inches extra height in the return piping to make room for this start-up condensate load.

Figure 9 Dimension A

Parallel flow system, wet return

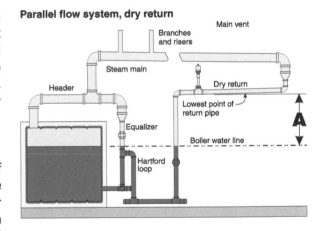

Parallel flow system, dry return

Counterflow system

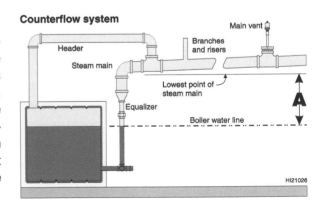

Total required height of Dimension A: **28 inches**.

Systems over 100,000 Btuh should always provide at least 28 inches height above the boiler water line in the return piping. Smaller systems (less than 100,000 Btuh) only need about **14 inches** because the pressure drops are generally smaller than large systems. (Most residential steam boilers are larger than 100,000 Btuh.)

When installing a new boiler or servicing a steam boiler, check out the distance between the boiler water line and the horizontal steam-carrying pipes. Apply the 28-inch rule (14 inches on small systems) to all horizontal steam pipes, including all runouts to radiators and risers. Check all heights with a tape measure.

Be really careful when installing a replacement boiler on a gravity-return system. For example, the old boiler may have been installed in a pit to keep the water line low enough to provide the needed dimension A. If you install the new boiler at surface level, you could cause serious problems. If the water doesn't have enough vertical height in the return risers, it backs into the horizontal piping, causing serious water hammer.

If you cannot install the boiler with a dimension A of at least 28 inches (14 inches for boilers under 100,000 Btuh), you will have to convert the system to pumped return, installing a float and thermostatic trap on the end of the main and any drip line in the system plus a boiler feed system or condensate return system. With pumped return, you no longer need as much vertical height to return the water to the boiler.

If you have to convert a system to pumped return, make sure you include it in your boiler replacement quotation instead of installing the boiler and finding out later that the system has problems, then having to convert the system for free.

Dimension B

Dimension A in a gravity system assumes there is some pressure pushing on top of the return risers. Be careful if you decide to install **zone valves on a gravity-return system**. When the zone valve closes, what happens? The air vents open because the remaining steam condenses, trying to pull a vacuum. The piping downstream of the zone valve fills with air until it reaches room pressure (0 gauge pressure).

See Figure 10. With no pressure on top of the return water columns, all of the pressure at the boiler now pushes back on the water in the return risers. When there is no pressure on top of the return column, the dimension between the water line and the lowest horizontal steam-carrying pipe is called **Dimension B**. The water will try to rise about 30 inches for every psig at the boiler. Even if the boiler limit control opens at only 2 psig, the water backs up in the returns 60 inches (5 feet!). It's unlikely the risers allow that much height. What is more likely is the water will fill up the steam main and horizontal steam pipes. The next time the valve opens, live steam meets really deep water. The piping could be destroyed by the severe water hammer that occurs. The good new is, the valve will probably never work again . The water hammer will likely destroy it also.

Figure 10   Dimension B

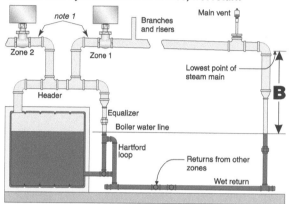

Parallel flow system with zone valves, wet return

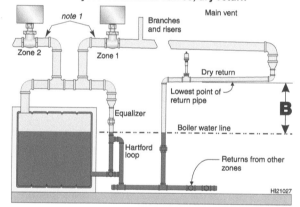

Parallel flow system with zone valves, dry return

HI21027

Note 1: Install a drip line connected to the wet return off the bottom of the steam pipe immediately ahead of each zone valve. These lines drain any condensate that settles ahead of the valves.

> **NOTICE**
>
> Unless you have an exceptionally high system, you can't use zone valves (or traps) on a one-pipe system without converting to pumped return. Exception: Vapor systems can operate with traps because the boiler never operates above ½ psig (requiring a specially pressure control, called a "vaporstat") and the piping is carefully installed to allow the needed height. When using zone valves, you may want to install a vacuum breaker near the main vent to handle the quick vacuum that can form when the zone valve closes.

One-pipe gravity-return system operation

Off cycle

Figure 11, page 12–23.

The air vents on one-pipe gravity systems open at the end of the heating cycle, allowing air to enter the system and fill the void caused by steam condensation. If the system remains off long enough, the entire system fills with air, throughout all radiators and all piping. With no flow in the system piping, the pressure is 0 psig throughout. All return risers, the boiler equalizer and the Hartford loop connection all level out at the same level as the boiler. All air vents and main vents are open.

Heating cycle starts

Figure 12, page 12–24.

When the boiler begins to fire on a call for heat, pressure begins to build at the boiler end of the system. Air flows through the piping, flowing much more quickly out the main vents than out the radiator vents.

The boiler no longer has water inside. It has a mixture of steam bubbles and water, because the steam bubbles form on the boiler heating surfaces below the water line and have to find their way to the top. So the boiler is filled with a mixture that is lighter than water only. The steam bubbles move water up and out of the way, so the surface level rises in the boiler. The boiler gauge glass shows Part of this increase, but not all. The gauge glass is filled with water only — heavier than the mixture inside the boiler (as in Figure 8, page 12–19, example 5). The boiler equalizer doesn't see any change. The total weight of the water column inside the boiler hasn't changed.

Because air is flowing in the distribution piping, the pressure is higher at the boiler than at the end of the steam main. This pushes the water up in the return riser and drip leg by an amount equal to the pressure difference.

Steam fills distribution piping

Figure 13, page 12–25.

The main vents have done their job here. All of the air is out of the distribution piping, providing steam to all radiators at about the same time. The main vents close as steam heats and expands their floats. If you service a one-pipe upfeed system that doesn't have main vents near the tops of the risers, recommend the owner have you install them. They will make a big difference in heat distribution in the house. Without them, the lower floors will heat much faster than the upper floors, causing some areas to be too warm, others too cold.

If the main vent on the steam main were not operating, the first riser (on left) would receive steam much sooner than the one on the right, causing overheating of the rooms connected to the first riser. Make sure all main vents are installed correctly and working.

Steam condenses in the distribution piping as it heats the pipe up to steam temperature. This causes a rush of condensate to the return risers, increasing the water level height.

Radiators begin to heat

Figure 14, page 12–26

The radiators begin to make condensate, which runs down the risers and the steam main into the returns. This is the critical stage for Dimension A. The water level in the risers is at the maximum.

The main vents are closed, but radiator vents remain open until steam reaches them.

Heating cycle steady

Figure 15, page 12–27

All of the air vents have closed. The boiler continues to feed steam to the system and condensate returns back to the drip leg and return riser. The height of the water in the risers should drop slightly as the heating cycle steadies out. (If the wet return happens to be badly clogged, though, the water in the returns can continue to rise, causing it to back into the horizontal steam pipes at this stage of the heating cycle. This will show up as water hammer in the main or runouts late in the heating cycle.)

The boiler will continue to provide steam until the thermostat ends the call for heat or the boiler limit control opens.

At the end of the heating cycle, steam in the system will condense and air will return to the radiators and piping.

NOTICE: See page 12-40 for discussion on how to submerge a wet return that ends up above the boiler water line when a new boiler is installed.

Figure 11 One-pipe gravity-return system operation — Off cycle

Off cycle (no call for heat)

① **Water levels:** When the boiler isn't firing, the fluid inside is only water (no steam). The boiler, gauge glass, equalizer and return lines contain only water. With no pressure differences, water seeks its own level. The column height is equal to the boiler surface level at all locations.

② **Air:** At the end of a firing cycle, the boiler stops making steam. The radiators continue to condense remaining steam. This causes a vacuum to develop. The air vents drop open, allowing air to enter the system, breaking the vacuum. This continues during an off cycle until all radiators and pipes are filled with air (unless another heat call occurs first).

③ **Air vents:** If the pressure at an air vent drops below its drop-out pressure, the float will drop away, opening the valve. This allows air to enter the system, breaking the vacuum that occurs as remaining steam condenses.

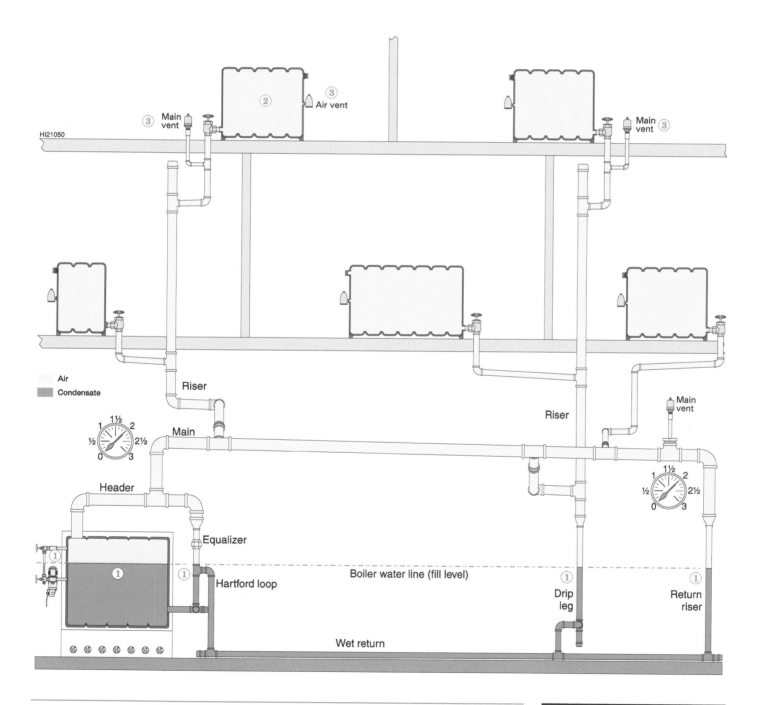

Figure 12   One-pipe gravity-return system operation — Heating cycle start

Heating cycle starts (thermostat calls for heat)

① **Water levels:** When the boiler fires, *it no longer contains water*. It contains a mixture of water and steam bubbles, a frothy mixture that can weigh considerably less than water only. At the beginning of the cycle, the boiler internal surface level rises because the steam bubbles push water out of the way, raising the level. The gauge glass sees part, but not all, of this change because the gauge glass contains only water. So its column height is slightly lower than the boiler internal level. The Hartford loop and return risers don't sense any change in water level height. This is because the total weight of water in the boiler hasn't changed yet. But there is a difference in pressure from the boiler to the end of the steam main. This pressure difference is due to the pressure lost by the air and steam as they pass through the piping. The pressure at the boiler side of the system is higher, so the return piping water levels rise enough to balance the pressure difference.

② **Air:** Air moves out of the piping and radiators through the air vents as pressure begins to rise in the system.

③ **Air vents:** If the air vents and main air vents are operative, they begin to allow air to escape as pressure builds in the system. Main vents can expel air up to 16 times faster than typical radiator vents. If the main vents don't work, it will take a long time to clear air from the risers and main, and the radiators closest to the boiler will receive steam first, causing heat distribution problems in the building.

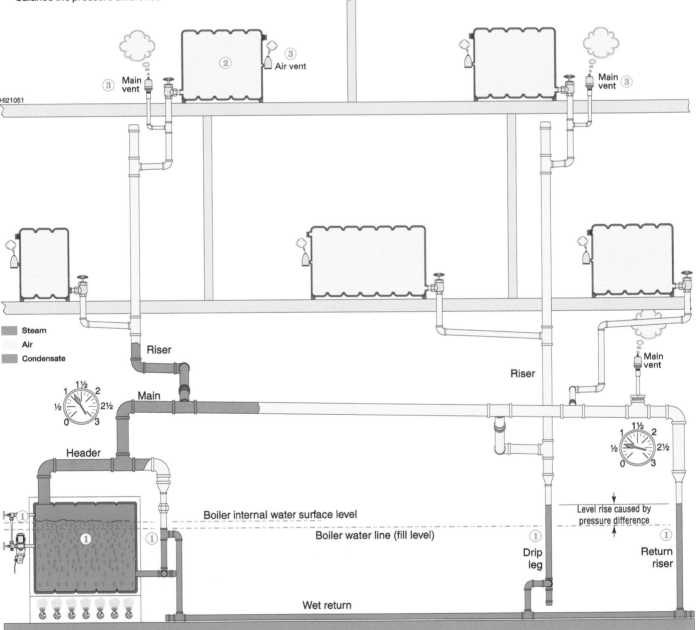

Figure 13 One-pipe gravity-return system operation — Steam fills distribution piping

Steam fills distribution piping (the role of main vents — even heat distribution)

① **Water levels:** The boiler water level has dropped slightly because of the water used to make steam to fill the piping and begin to fill the radiators. The pressure difference between the boiler and the end of the system may fluctuate at this stage of the heating cycle because the piping and radiators can use steam faster than the boiler makes it during the time the piping and radiators are being heated up. This means the flow rate can at times be higher than design flow rate. Condensate has begun to form as the piping and radiators are heated. This rush of condensate causes an additional rise in the water line of the returns.

② **Air:** Air has moved out of the distribution piping (steam main and risers), allowing steam to reach all of the radiators. Air continues to move from the radiators as steam pushes in.

③ **Air vents:** With main vents installed as shown, air is quickly eliminated from the steam main and risers. All radiators receive steam at about the same time, ensuring even heat distribution. [If the main vents were inoperative (failed closed), air would move slowly from the mains. Radiators closest to the boiler would receive steam well before the outermost radiators. The building would suffer from heat distribution problems.] The main vents have closed in the system below because steam has reached them, expanding the floats to close the valves.

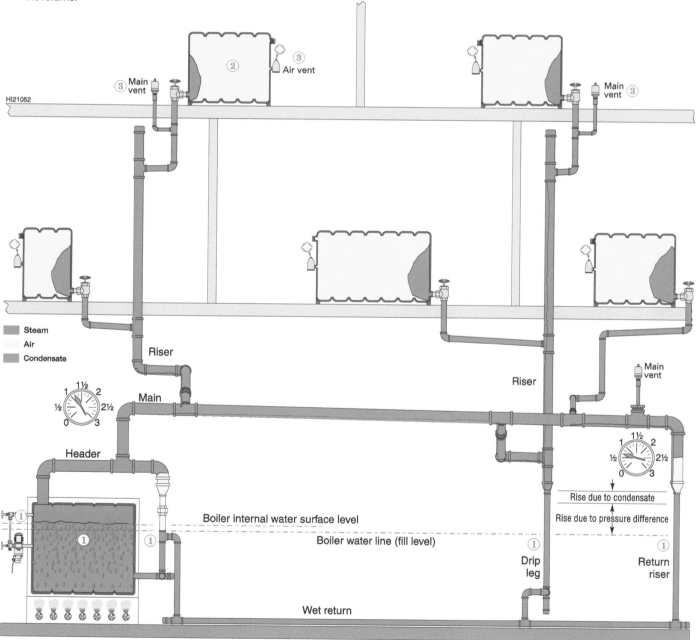

Figure 14 One-pipe gravity-return system operation — Radiators begin to heat

Radiators begin to heat (steam flow inside radiators)

① **Water levels:** The boiler water level continues to drop as steam fills the radiators. The pressure drop through the steam lines will continue to fluctuate until the radiators reach temperature, but will approach the design pressure drop of ½ psig, typical for one-pipe steam systems. The height of condensate in the returns reaches its maximum at this time because of the rush of condensate from heating the pipes added to the beginning of the condensate from radiator heating. This is the time when Dimension **A** is critical. The most likely location to encounter water hammer on the system below would be the runout to the dripped riser (right hand riser) because it is closer to the water line than any other horizontal steam pipe.

② **Air:** Air is pushed toward the air vents by the incoming steam. Air will continue to flow from the radiator air vents until they close due to heating from steam.

③ **Air vents:** The radiator air vents remain open, allowing air to escape. They will remain open until exposed to steam temperature.

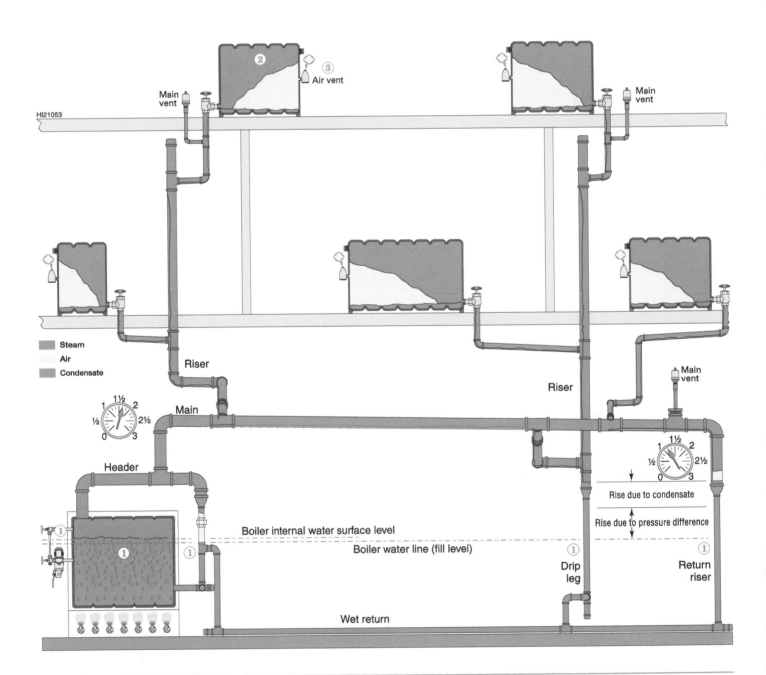

Figure 15 One-pipe gravity-return system operation — Heating cycle steady

Heating cycle steady (air vents closed)

① **Water levels:** Once condensate begins to return steadily, the boiler water level should reach a normal running height. The riser water line will be higher than the boiler by the pressure difference across the steam distribution piping (usually about ½ psig, or 14 inches water column) plus the pressure drop caused by flow of condensate through the return lines (ranging typically from 2 to 6 inches water column, depending on the condition of the wet return piping). So the water level height in the risers should range from about 16 inches to 20 inches higher than the water in the equalizer.

② **Air:** Most of the air has left the system piping and radiators. Air will not return until the boiler shuts down, allowing steam to condense in the radiators and piping, creating a vacuum.

③ **Air vents:** The radiator vents close when exposed to steam temperature. The vents will open again only when pressure is removed when the boiler shuts down.

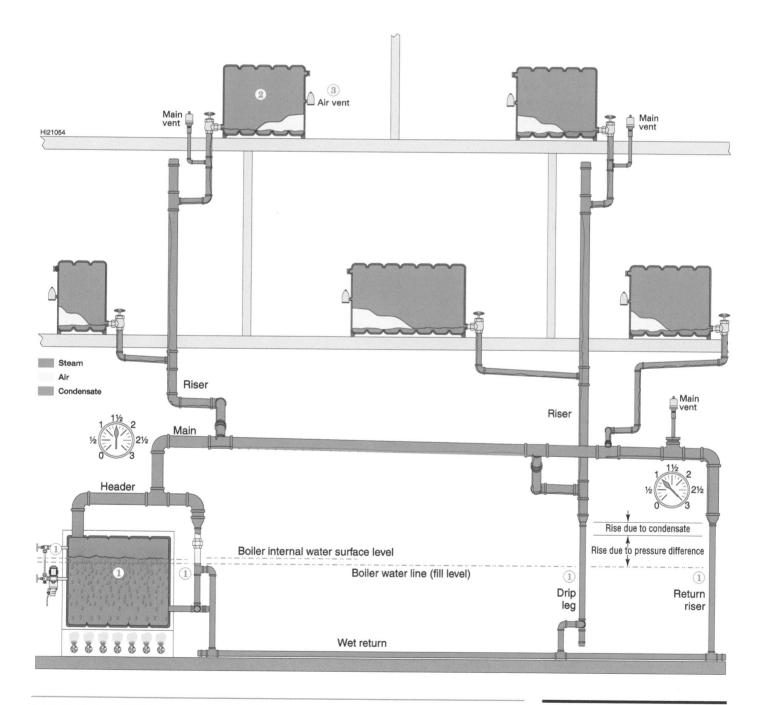

One-pipe parallel upfeed gravity-return system (with wet return)

See Figure 16, page 12–29. See tables on this page to size (or check size of) piping.

1 Make sure the steam main slopes continuously down toward the return riser. The minimum slope should be 1" per 20 feet, as shown. This is hard to detect, particularly if the boiler room floor isn't even. Use a spirit level to check the full length of the main. Any sag in the main will cause water hammer because condensate will either accumulate there or be too deep because it doesn't flow quickly enough.

2 Install the main vent near, but no closer than 15 inches from, the end of the steam main as in Figure 23, page 12–40. If the main vent is installed at the end of the main on a tee, it almost certainly has been damaged by water hammer. Install a new main vent, located where it needs to be to protect it from water hammer. Without the main vent, the system will have very poor heat distribution.

3 If not already installed, find a way to install a main vent at or near the top of every upfeed riser. If you can't get to the pipe, add a tee to an upper-story radiator supply line and install the main vent there as shown in Figure 16. These main vents will help upper-story radiators receive steam at nearly the same time as the lower-story radiators.

4 If the boiler is over 100,000 Btuh, the risers must provide at least 28 inches between the boiler water line and the lowest steam-carrying pipe. The lowest steam pipe in Figure 16 is the runout to the right-hand riser. If you cannot install the boiler low enough to provide the 28 inches, you need to revise the affected piping or convert the system to pumped return (see Figure 29, page 12–45.)

5 Make sure each radiator is equipped with a working air vent. If using adjustable vents, adjust them to the size of the radiator. You might consider fine-tuning the system a bit by closing down more on the near-boiler radiator vents. But heat distribution should not need this if piping is properly vented.

NOTICE

Don't use 1" piping on runouts to undripped upfeed risers. Use 1¼" minimum piping to avoid noise and water hammer problems. Likewise, you should use at least 1¼" radiator supply valves and vertical piping to allow room for condensate to flow freely against the steam flow.

Steam main			
Pipe Size	Square Ft EDR	Btu/Hr Net	Pounds/Hr
2"	432	103,680	108
2½"	696	167,040	174
3"	1,272	305,280	318
4"	2,560	614,400	640
5"	4,800	1,152,000	1,200

Upfeed riser			
Pipe Size	Square Ft EDR	Btu/Hr Net	Pounds/Hr
1¼"	80	19,200	20
1½"	152	36,480	38
2"	288	69,120	72
2½"	464	111,360	116
3"	800	192,000	200
3½"	1144	274,560	286
4"	1520	364,800	380

Runout to riser or radiator (horizontal pipe), not dripped (Increase one pipe size for runouts longer than 8 feet unless pitch is at least ½ inch per foot)			
Pipe Size	Square Ft EDR	Btu/Hr Net	Pounds/Hr
1"	28	6,720	7
1¼"	64	15,360	16
1½"	64	15,360	16
2"	92	22,080	23
2½"	168	40,320	42
3"	260	62,400	65
4"	744	178,560	186
5"	1112	266,880	278

Runout to riser or radiator (horizontal pipe), dripped (Increase one pipe size for runouts longer than 8 feet unless pitch is at least ½ inch per foot)			
Pipe Size	Square Ft EDR	Btu/Hr Net	Pounds/Hr
1"	68	16,320	17
1¼"	144	34,560	36
1½"	224	53,760	56
2"	432	103,680	108
2½"	696	167,040	174
3"	1,272	305,280	318
3½"	1,848	443,520	462
4"	2,560	614,400	640

Wet return piping			
Pipe Size	Square Ft EDR	Btu/Hr Net	Pounds/Hr
1"	700	168,000	175
1¼"	1,200	288,000	300
1½"	1,900	456,000	475
2"	4,000	960,000	1,000

Radiator vertical piping and supply valves			
Pipe Size	Square Ft EDR	Btu/Hr Net	Pounds/Hr
1"	28	6,720	7
1¼"	64	15,360	16
1½"	92	22,080	23
2"	168	40,320	42

Figure 16 One-pipe parallel upfeed gravity-return system (with wet return)

One-pipe parallel upfeed gravity-return system (with wet return)

See opposite page for comments and pipe sizing.

F"xx" below is the Figure number for piping detail.
Example: F39 means Figure 39.

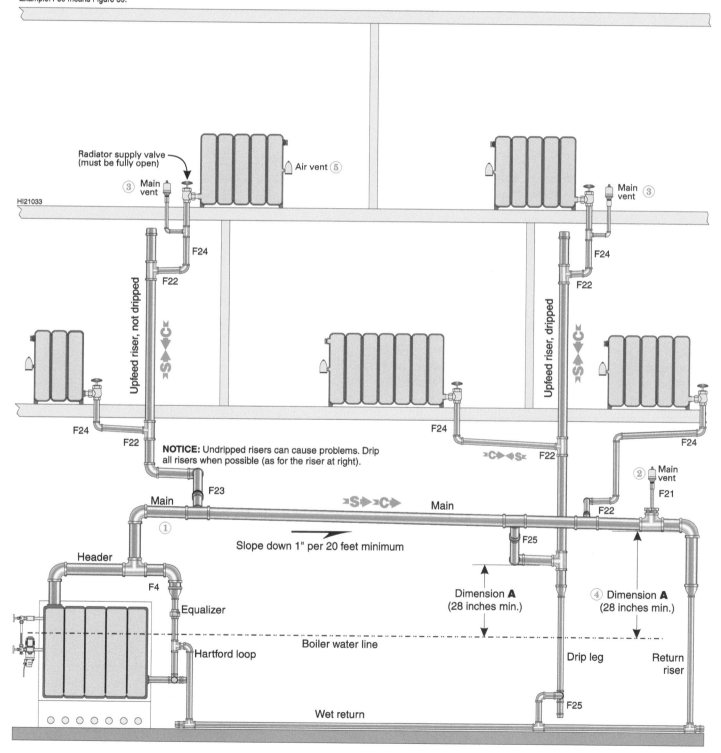

One-pipe parallel upfeed gravity-return system (with dry return)

See Figure 17, page 12–31. See tables on this page to size (or check size of) piping.

This system is similar to Figure 16, page 12–29, except the return line is brought back above the water line. This is why it is called a dry return. You still must have some length of wet return to provide a loop seal, keeping steam from entering the dry return directly from the boiler end of the system.

1 Make sure the steam main slopes continuously down toward the return riser. The minimum slope should be 1" per 20 feet, as shown. This is hard to detect, particularly if the boiler room floor isn't even. Use a spirit level to check the full length of the main. Any sag in the main will cause water hammer because condensate will either accumulate there or be too deep because it doesn't flow quickly enough.

2 Install the main vent near, but no closer than 15 inches from, the end of the dry return as in Figure 23, page 12–40. If the main vent is installed at the end of the dry return on a tee, it almost certainly has been damaged by water hammer. Install a new main vent, located where it needs to be to protect it from water hammer. Without the main vent, the system will have very poor heat distribution.

3 If not already installed, find a way to install a main vent at or near the top of every upfeed riser. If you can't get to the pipe, add a tee to an upper-story radiator supply line and install the main vent there as shown in Figure 17. These main vents will help upper-story radiators receive steam at nearly the same time as the lower-story radiators.

4 If the boiler is over 100,000 Btuh, the risers must provide at least 28 inches between the boiler water line and the lowest steam-carrying pipe (14 inches for smaller boilers). The lowest steam pipe in Figure 17 is the end of the dry return. If you cannot install the boiler low enough to provide the 28 inches, you need to revise the affected piping or convert the system to pumped return (see Figure 29, page 12–45.)

5 Make sure each radiator is equipped with a working air vent. If using adjustable vents, adjust them to the size of the radiator. You might consider fine-tuning the system a bit by closing down more on the near-boiler radiators. But heat distribution should be done with good main and riser venting, not with the radiator vents.

Steam main

Pipe Size	Square Ft EDR	Btu/Hr Net	Pounds/Hr
2"	432	103,680	108
2½"	696	167,040	174
3"	1,272	305,280	318
4"	2,560	614,400	640
5"	4,800	1,152,000	1,200

Upfeed riser

Pipe Size	Square Ft EDR	Btu/Hr Net	Pounds/Hr
1¼"	80	19,200	20
1½"	152	36,480	38
2"	288	69,120	72
2½"	464	111,360	116
3"	800	192,000	200
3½"	1144	274,560	286
4"	1520	364,800	380

Runout to riser or radiator (horizontal pipe), not dripped
(Increase one pipe size for runouts longer than 8 feet unless pitch is at least ½ inch per foot)

Pipe Size	Square Ft EDR	Btu/Hr Net	Pounds/Hr
1"	28	6,720	7
1¼"	64	15,360	16
1½"	64	15,360	16
2"	92	22,080	23
2½"	168	40,320	42
3"	260	62,400	65
4"	744	178,560	186
5"	1112	266,880	278

Runout to riser or radiator (horizontal pipe), dripped
(Increase one pipe size for runouts longer than 8 feet unless pitch is at least ½ inch per foot)

Pipe Size	Square Ft EDR	Btu/Hr Net	Pounds/Hr
1"	68	16,320	17
1¼"	144	34,560	36
1½"	224	53,760	56
2"	432	103,680	108
2½"	696	167,040	174
3"	1,272	305,280	318
3½"	1,848	443,520	462
4"	2,560	614,400	640

Wet return piping

Pipe Size	Square Ft EDR	Btu/Hr Net	Pounds/Hr
1"	700	168,000	175
1¼"	1,200	288,000	300
1½"	1,900	456,000	475
2"	4,000	960,000	1,000

Dry return piping

Pipe Size	Square Ft EDR	Btu/Hr Net	Pounds/Hr
1"	320	76,800	80
1¼"	672	161,280	168
1½"	1,060	254,400	265
2"	2,300	552,000	575
2½"	3,800	912,000	950

Radiator vertical piping and supply valves

Pipe Size	Square Ft EDR	Btu/Hr Net	Pounds/Hr
1"	28	6,720	7
1¼"	64	15,360	16
1½"	92	22,080	23
2"	168	40,320	42

Figure 17 One-pipe parallel upfeed gravity-return system (with dry return)

One-pipe parallel upfeed gravity-return system (with dry return)

See opposite page for comments and pipe sizing.

F"xx" below is the Figure number for piping detail.
Example: F39 means Figure 39.

NOTICE: Don't use 1" piping on runouts to undripped upfeed risers. Use 1¼" minimum piping to avoid noise and water hammer problems. Likewise, you should use at least 1¼" radiator supply valves and vertical piping to allow room for condensate to flow freely against the steam flow.

NOTE:
Drip line from riser to dry return must drop down as shown to provide a loop seal, preventing steam from entering the riser. Provide at least 28 inches between the bottom pipe and the bottom of the dry return piping to provide enough water column height to overcome pressure differences. The largest pressure difference would occur for a riser piped to the beginning of the steam main.

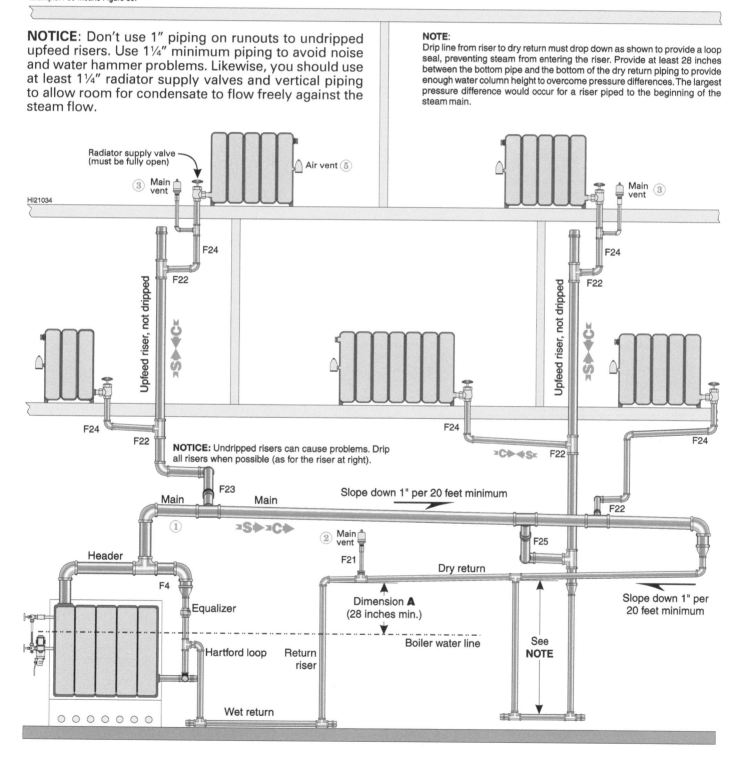

One-pipe parallel downfeed gravity-return system (Mills express riser system)

See Figure 18, page 12–33. See tables on this page to size (or check size of) piping.

Downfeed piping is a clever way to get the steam and condensate flowing in the same direction in the risers as well as the mains. Both the steam and the condensate travel downhill in the risers.

1 A downfeed system uses a steam main located at the top of the building. The express riser supplies steam from the boiler to the main. Downfeed risers supply steam from the main to the heating units below. Each downfeed riser ends in a drip pipe connected to the wet return. Make sure the main slopes continuously downward at no less than 1 inch per 20 feet. Use a spirit level to check the full length of the main. Any sag in the main will cause water hammer because condensate will either accumulate there or be too deep because it doesn't flow quickly enough.

2 Install the main vent near, but no closer than 15 inches from, the end of the steam main as in Figure 23, page 12–40. If the main vent is installed at the end of the steam main on a tee, it almost certainly has been damaged by water hammer. Install a new main vent, located where it needs to be to protect it from water hammer. Without the main vent, the system will have very poor heat distribution.

3 If not already installed, find a way to install a main vent near the end of each downfeed riser (at least 28 inches above the water line, or 14 inches if the boiler is under 100,000 Btuh). These main vents will help lower-story radiators receive steam at nearly the same time as the upper-story radiators. Note that downfeed systems would have just the opposite heat distribution problem of an upfeed system. Here, the top floor would overheat because it would receive steam before the lower floors without good main venting.

4 If the boiler is over 100,000 Btuh, the risers must provide at least 28 inches between the boiler water line and the lowest steam-carrying pipe (14 inches for smaller boilers). The lowest steam pipes in Figure 18 are the main vent connections. Watch out for low radiator branch connections as well. If you cannot install the boiler low enough to provide the 28 inches, you need to revise the affected piping or convert the system to pumped return (see Figure 30, page 12–47.)

5 Make sure each radiator is equipped with a working air vent. If using adjustable vents, adjust them to the size of the radiator. You might consider fine-tuning the system a bit by closing down more on the near-boiler radiators. But heat distribution should be done with good main and riser venting, not with the radiator vents.

NOTICE

Don't use 1" piping on runouts to un-dripped upfeed risers. Use 1¼" minimum piping to avoid noise and water hammer problems. Likewise, you should use at least 1¼" radiator supply valves and vertical piping to allow room for condensate to flow freely against the steam flow.

Steam main			
Pipe Size	Square Ft EDR	Btu/Hr Net	Pounds/Hr
2"	432	103,680	108
2½"	696	167,040	174
3"	1,272	305,280	318
4"	2,560	614,400	640
5"	4,800	1,152,000	1,200

Express riser			
Pipe Size	Square Ft EDR	Btu/Hr Net	Pounds/Hr
2½"	636	152,640	159
3"	1,128	270,720	282
3½"	1,548	371,520	387
4"	2,044	490,560	511
5"	4,200	1,008,000	1,050

Downfeed risers and runouts			
Pipe Size	Square Ft EDR	Btu/Hr Net	Pounds/Hr
1"	68	16,320	17
1¼"	144	34,560	36
1½"	224	53,760	56
2"	432	103,680	108
2½"	696	167,040	174
3"	1,272	305,280	318
3½"	1,848	443,520	462
4"	2,560	614,400	640

Radiator runouts (horizontal piping)			
(Increase one pipe size for runouts longer than 8 feet unless pitch is at least ½ inch per foot)			
Pipe Size	Square Ft EDR	Btu/Hr Net	Pounds/Hr
1"	28	6,720	7
1¼"	64	15,360	16
1½"	64	15,360	16
2"	92	22,080	23
2½"	168	40,320	42
3"	260	62,400	65

Wet return piping			
Pipe Size	Square Ft EDR	Btu/Hr Net	Pounds/Hr
1"	700	168,000	175
1¼"	1,200	288,000	300
1½"	1,900	456,000	475
2"	4,000	960,000	1,000

Radiator vertical piping and supply valves			
Pipe Size	Square Ft EDR	Btu/Hr Net	Pounds/Hr
1"	28	6,720	7
1¼"	64	15,360	16
1½"	92	22,080	23
2"	168	40,320	42

Figure 18 One-pipe parallel downfeed gravity-return system

One-pipe parallel downfeed gravity-return system

② Main vent
F21

① Main

Slope down 1" per 20 feet minimum

F26

F26

⊃S⊃ ⊃C⊃

Runout to riser

Express riser

Downfeed riser

Downfeed riser

See opposite page for comments and pipe sizing.

F"xx" below is the Figure number for piping detail. Example: F39 means Figure 39.

Radiator supply valve (must be fully open)

Air vent ⑤

HI21036

F24

F24

See opposite page for comments and pipe sizing.

⊃S⊃ ⊂C⊂

F22

F22

↓S⊂ ↓C⊂

↓S⊂ ↓C⊂

F24

F22

F24

F22

F24

F22

①

③ Main vent

③ Main vent

③ Main vent

Header

F4

④ Dimension **A** (28 inches min.)

Dimension **A** (28 inches min.)
④

Equalizer

Hartford loop

Boiler water line

Drip leg

Return riser

F26

F26

Wet return

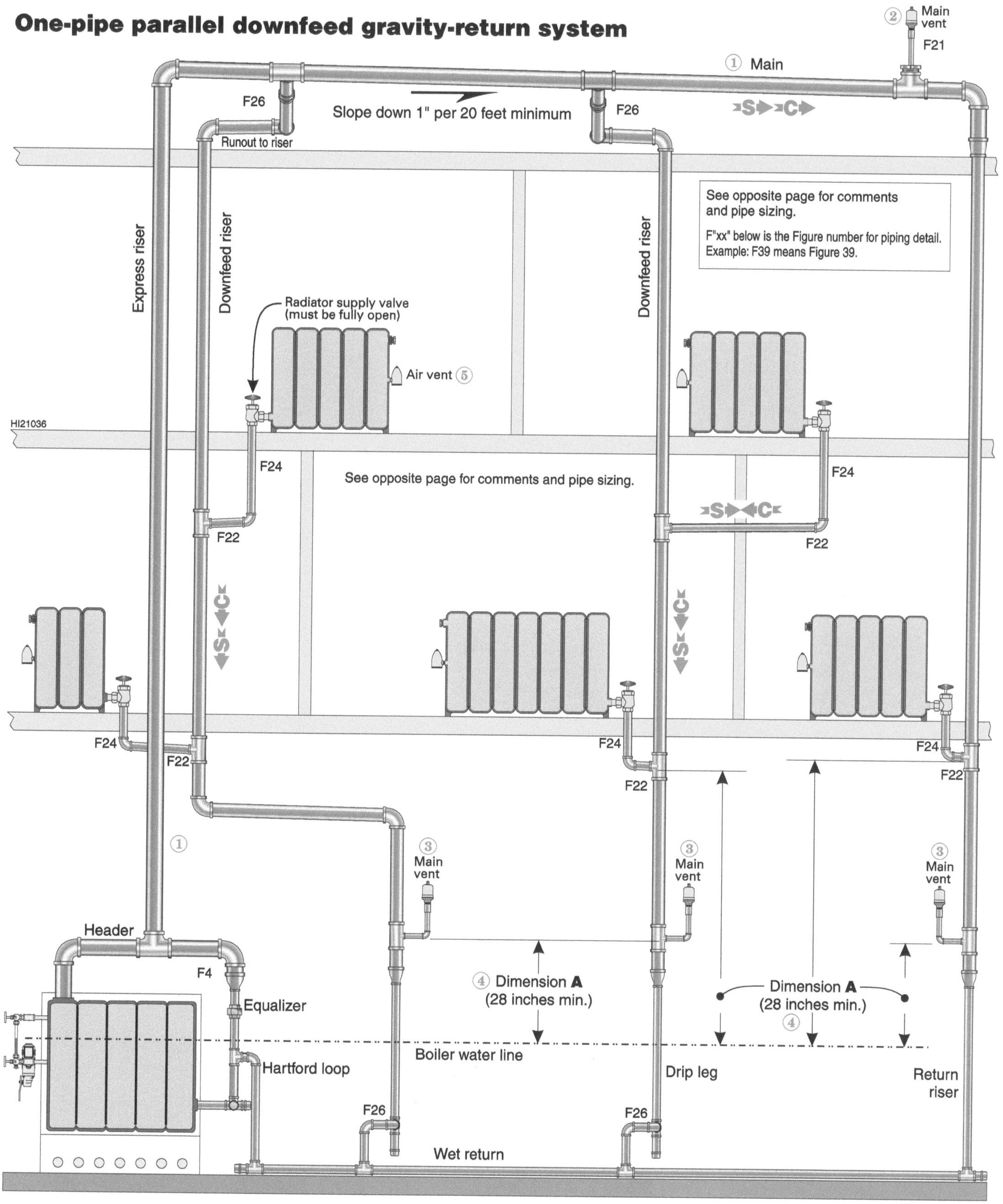

One-pipe counterflow gravity-return system

See Figure 19, page 12–35. See tables on this page to size (or check size of) piping.

Here, the steam main both supplies steam to the system and returns condensate to the system, with the condensate flowing counter to the steam.

1 Notice that the boiler steam supply must enter the top of the steam main. This prevents the condensate from dropping back into the top of the boiler. Since the system has no wet return, it doesn't need a Hartford loop. Connect the end of the steam header to a boiler equalizer pipe as shown. Counterflow headers are sized one pipe size larger than a parallel flow header for the same load (allowed for in the sizing tables). Make sure there are at least 14 inches between the bottom of the steam main and the boiler water line. This allows room for condensate to accumulate at start up without risking condensate back-up into the main that could cause water hammer. Counterflow headers must slope twice as much as parallel flow headers to ensure the condensate won't get too deep. The minimum slope is 1 inch per 10 feet.

2 Install the main vent near, but no closer than 15 inches from, the end of the steam main as in Figure 23, page 12–40. If the main vent is installed at the end of the steam main on a tee, it almost certainly has been damaged by water hammer. Install a new main vent, located where it needs to be to protect it from water hammer. Without the main vent, the system will have very poor heat distribution.

3 If not already installed, find a way to install a main vent at or near the top of every upfeed riser. If you can't get to the pipe, add a tee to an upper-story radiator supply line and install the main vent there as shown in Figure 19. These main vents will help upper-story radiators receive steam at nearly the same time as the lower-story radiators.

4 Make sure each radiator is equipped with a working air vent. If using adjustable vents, adjust them to the size of the radiator. You might consider fine-tuning the system a bit by closing down more on the near-boiler radiators. But heat distribution should be done with good main and riser venting, not with the radiator vents.

NOTICE
Don't use 1" piping on runouts to un-dripped upfeed risers. Use 1¼" minimum piping to avoid noise and water hammer problems. Likewise, you should use at least 1¼" radiator supply valves and vertical piping to allow room for condensate to flow freely against the steam flow.

Steam main

Pipe Size	Square Ft EDR	Btu/Hr Net	Pounds/Hr
2½"	432	103,680	108
3"	696	167,040	174
4"	1,272	305,280	318
5"	2,560	614,400	640
6"	4,800	1,152,000	1,200

Upfeed riser

Pipe Size	Square Ft EDR	Btu/Hr Net	Pounds/Hr
1¼"	80	19,200	20
1½"	152	36,480	38
2"	288	69,120	72
2½"	464	111,360	116
3"	800	192,000	200
3½"	1144	274,560	286
4"	1520	364,800	380

Runout to riser or radiator (horizontal pipe)
(Increase one pipe size for runouts longer than 8 feet unless pitch is at least ½ inch per foot)

Pipe Size	Square Ft EDR	Btu/Hr Net	Pounds/Hr
1"	28	6,720	7
1¼"	64	15,360	16
1½"	64	15,360	16
2"	92	22,080	23
2½"	168	40,320	42
3"	260	62,400	65
4"	744	178,560	186
5"	1112	266,880	278

Radiator runouts (horizontal piping)
(Increase one pipe size for runouts longer than 8 feet unless pitch is at least ½ inch per foot)

Pipe Size	Square Ft EDR	Btu/Hr Net	Pounds/Hr
1"	28	6,720	7
1¼"	64	15,360	16
1½"	64	15,360	16
2"	92	22,080	23
2½"	168	40,320	42
3"	260	62,400	65

Radiator vertical piping and supply valves

Pipe Size	Square Ft EDR	Btu/Hr Net	Pounds/Hr
1"	28	6,720	7
1¼"	64	15,360	16
1½"	92	22,080	23
2"	168	40,320	42

Figure 19 One-pipe counterflow gravity-return system

One-pipe counterflow gravity-return system

See opposite page for comments and pipe sizing.

F"xx" below is the Figure number for piping detail.
Example: F39 means Figure 39.

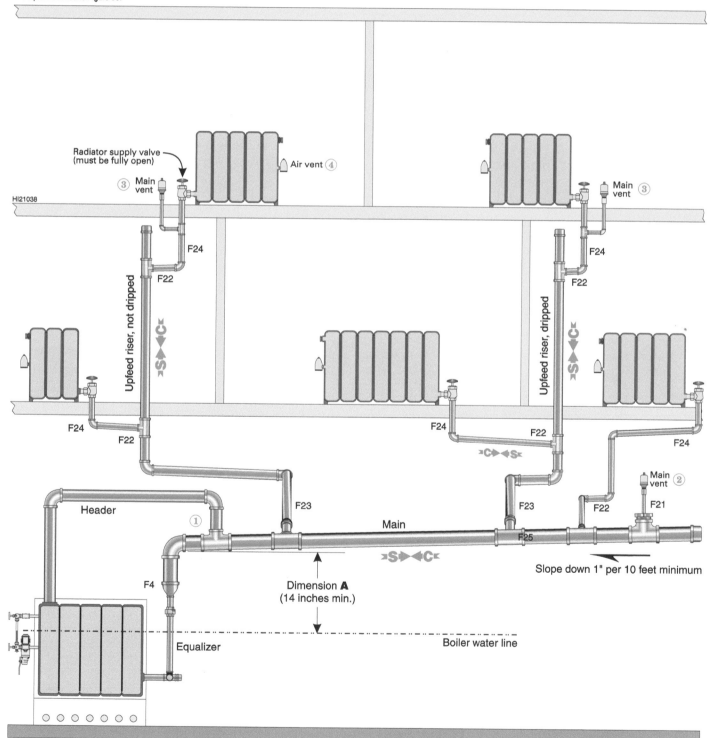

Two-pipe upfeed gravity-return system (with individual drip lines)

See Figure 20, page 12–37. To check sizing, see two-pipe sizing information in Figure 49, page 12–64.

Figure 20 is a two-pipe system that uses air vents. The radiators are equipped with separate steam pipes and condensate pipes. The example shown has only isolation valves on the condensate connections. This system prevents steam from flowing through the radiators to the returns because each return has a loop seal since each return is piped directly to the wet return line. This is an effective, but expensive system to install, because the installer had to pipe from every radiator to the wet return with its own return pipe.

Other variations of the two-pipe gravity system do not individually pipe each condensate outlet to a wet return. They use traps on the condensate connections and/or orifices in the radiator supply valves and do not use air vents. The supply valve orifices, if used, are sized to regulate the steam flow to less than the radiator can handle, preventing any steam from flowing through to the return lines. If you replace these valves, either install the same size orifices or fit each radiator with a thermostatic trap. These systems must be operated at low pressure (less than 2 psig) to ensure water won't be backed up in the returns, because the pressure in the returns is zero. When installing a replacement boiler, make sure the new boiler water line is not higher than the old one. If the new water line will be higher, you must make sure there are no horizontal steam lines closer than 5 feet to the new boiler water line (60 inches, that is, to allow for a rise of 30 inches per psig at the boiler). You will have to convert the system to pumped return if you cannot provide enough height in the returns.

1 The steam main must slope continuously down toward the return riser at no less than 1 inch per 20 feet. Use a spirit level to check the full length of the main. Any sag in the main will cause water hammer because condensate will either accumulate there or be too deep because it doesn't flow quickly enough.

2 Make sure there is a working main vent correctly piped near the end of the steam main as in Figure 23, page 12–40. This vent is responsible for ensuring all risers receive steam at about the same time.

3 If not already installed, find a way to install a main vent near the top of each upfeed riser. These main vents will help upper-story radiators receive steam at nearly the same time as the lower-story radiators.

4 If the boiler is over 100,000 Btuh, the risers must provide at least 28 inches between the boiler water line and the lowest steam-carrying pipe (14 inches for smaller boilers). The lowest steam pipe in Figure 20 is the end of the steam main. Watch out for low radiator branch connections as well. If you cannot install the boiler low enough to provide the 28 inches, you need to revise the affected piping or convert the system to pumped return (see Figure 36, page 12–55.)

5 Make sure each radiator is equipped with a working air vent. If using adjustable vents, adjust them to the size of the radiator. You might consider fine-tuning the system a bit by closing down more on the near-boiler radiators. But heat distribution should be done with good main and riser venting, not with the radiator vents.

6 If replacing a radiator supply valve, inspect the old valve thoroughly. If it is fitted with an orifice, you will need to install the same-sized orifice when you install a new valve or fit the radiator with a trap on the condensate outlet. Also make sure that there are at least 60 inches between the boiler water line and the radiator condensate outlet if the boiler limit control is set for a 2 psig cut-out. Use packless radiator supply valves when replacing.

7 Each radiator condensate return line must terminate below the water line into the wet return.

NOTICE
Another version (not shown) of two-pipe gravity-return piping uses thermostatic traps on the radiators, with return piping from the traps piped as for pumped-return systems (Figure 36, page 12–55, for example). You must operate these systems at very low steam pressure to avoid water being pushed too high in the returns. For most of these systems, do no operate the boiler higher than ½ psig.

Figure 20 Two-pipe upfeed gravity-return system

Two-pipe upfeed gravity-return system

See opposite page for comments and pipe sizing.

F"xx" below is the Figure number for piping detail.
Example: F39 means Figure 39.

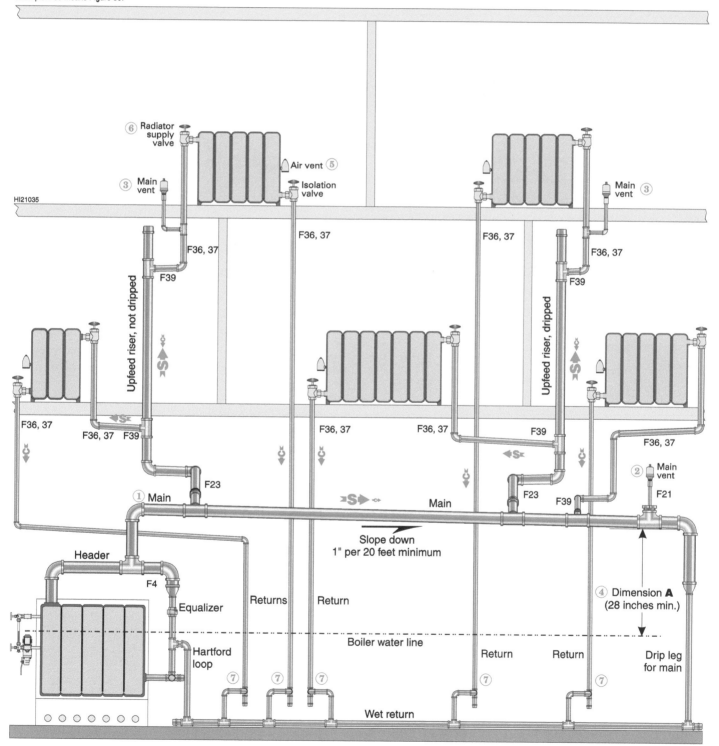

Two-pipe vapor system, typical

See Figure 21, page 12–39. No recommendations on pipe sizing are given here. Maintain existing piping dimensions.

Vapor systems are special cases of two-pipe gravity-return. Each vapor system uses some means to cycle return water flow to the boiler. The radiators are equipped with traps and the return lines are vented to atmosphere. So the entire pressure of the boiler pushes back on the return lines.

To prevent water from entering the return lines, the piping must be installed carefully per the original design, including have the check valves (6 and 7) working and in the proper locations. Vapor system designs varied, but each one used some type of boiler water return mechanism. The one shown (item 2) is called an alternating receiver. This device has a float that operates an internal valve. When the water raises the float high enough, the valve opens, applying steam pressure (from the line connected to the main) on top of the water in the trap. The pressure on this water is no longer full boiler pressure. The water drops out of the trap, through check valve 7, and into the boiler. For a detailed discussion of vapor systems, see "The Lost Art of Steam Heating" and "The Lost Art of Steam Heating Companion" by Dan Holohan. He covers many different systems and explains their operation.

The boiler return mechanisms for vapor systems are no longer available. If you can't get the old one to work, you will have to convert the system to pumped return. That means installing drip traps on return mains and steam mains and adding a boiler feed system or condensate return system.

These systems operate at slight vacuum (vapor) conditions because they force out the air, but don't let any back in. Steam condensing in the radiators causes a slight vacuum condition. It is critical to operate vapor systems at no higher than ½ psig. You will have to install a vaporstat if not already installed to ensure the accuracy needed to operate at such a low maximum pressure.

1   The steam main must slope continuously down toward the return riser at no less than 1 inch per 20 feet. Use a spirit level to check the full length of the main. Any sag in the main will cause water hammer because condensate will either accumulate there or be too deep because it doesn't flow quickly enough.

2   The alternating receiver (or boiler return trap) cycles as the water level rises to apply steam line pressure to the top of the condensate return water column, allowing water to flow back to the boiler. This device must be functioning for the system to operate. If servicing such a system, make careful note of all piping locations and sizes. Restore exactly as original to avoid potential operating problems.

3   This type of vapor system used a special air eliminator, called an air eliminator trap. The air vent on top of the trap is an air check valve that lets air out, but not in.

4   This equalizer connection requires a check valve and isolation valve, as shown.

5   This line allows air to vent from the alternating receiver chamber.

6   Make sure this check valve is in place and working. It allows condensate to flow into the alternating receiver piping column, but prevents it from backing into the returns again.

7   Make sure this check valve is in place and working. If replacing piping or removing and replacing components, return all exactly as originally installed to be sure the system will operate correctly.

8   You will probably find a thermostatic trap between end of the steam main and the end of the return main. This trap allows air to move from the steam main out the air eliminator trap.

9   Some of these systems used special orificed radiator supply valves. The orifice regulates steam flow to the radiator. You will need to make an orifice to replace the original if you can't find a replacement valve. (On some steam systems, the radiator supply orifice was used instead of a trap on the radiator condensate outlet. The orifice was sized small enough that all of the steam going into the radiator was condensed, eliminating the need for a trap since there was no live steam at the outlet.)

10 Trap types vary between systems. Some traps were simply small "P" traps that used a water seal to prevent steam flow. Every system has to provide some means to stop steam from passing into the returns.

Figure 21 Two-pipe vapor system, typical

Two-pipe vapor system, typical

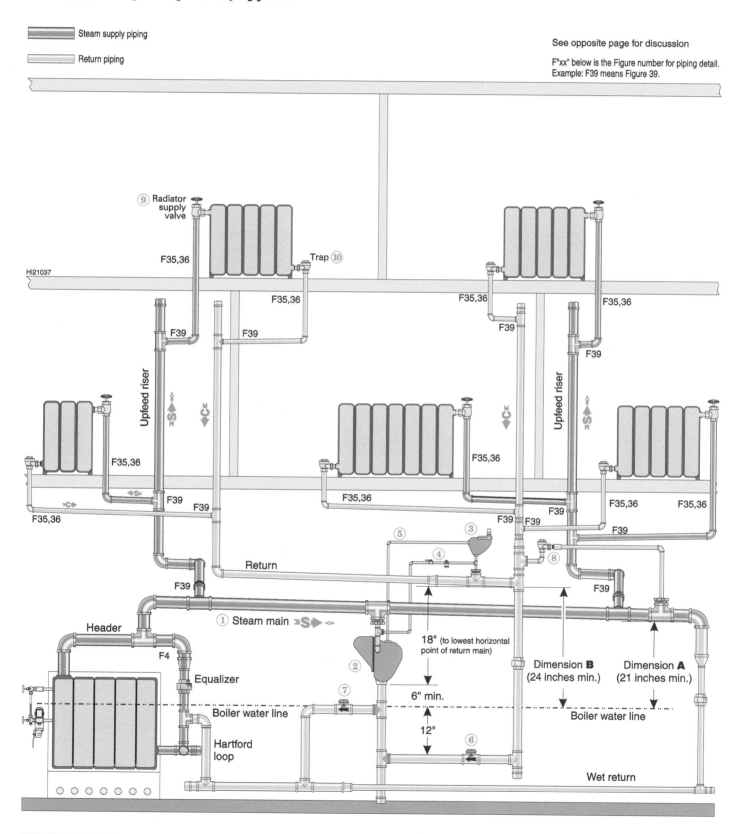

Steam supply piping

Return piping

See opposite page for discussion

F"xx" below is the Figure number for piping detail.
Example: F39 means Figure 39.

⑨ Radiator supply valve

F35,36

Trap ⑩

HI21037

F35,36

F35,36

F39

F35,36

F39

F39

F39

Upfeed riser

Upfeed riser

S

C

C

S

F35,36

F35,36

F39

F39

F35,36

F39

F39

F35,36

F35,36

F35,36

C

S

F39

Return

⑤

③

④

F39 F39

F39

⑧

F39

Header

① Steam main S

F4

②

18" (to lowest horizontal point of return main)

Dimension **B** (24 inches min.)

Dimension **A** (21 inches min.)

Equalizer

⑦

6" min.

Boiler water line

Boiler water line

12"

Hartford loop

⑥

Wet return

Figure 22   Piping a "false water line"

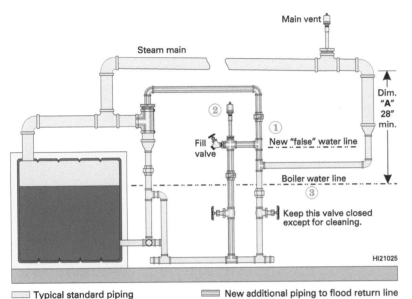

Source: *This concept is found in "The Lost Art of Steam Heating", Holohan, 1994*

① Additional piping causes water to build up in return line to height of tee, as shown. Use the fill valve to manually fill this line to prevent water hammer on initial operation.

② Install an air vent on a nipple at least 12 inches above tee. This will protect the vent from damage due to any water hammer that might occur.

③ Without the addition of the new piping, the return line would be too close to the boiler water line, causing serious water hammer due to periodic back-up of condensate into the horizontal piping.

Keep wet returns wet

An added precaution when installing a replacement on a gravity-return steam system: Make sure that wet returns will remain wet when you install the new boiler. If the new boiler water level is lower than the boiler it replaces, there is a possibility the new water line will be below horizontal return pipes that used to be "wet" (meaning below the water line).

Figure 23, drawn from "The Lost Art of Steam Heating," Dan Holohan, shows a way to modify the piping to cover up a return line intended originally to be wet. If you don't submerge this old return line, it will cause serious water hammer as it alternates between being wet and dry.

Always check the boiler water line for the new boiler before you quote the job. Include any needed revisions in your price.

Figure 23   Branch connections to main

One-pipe steam Runouts to branches (**NOT** risers)

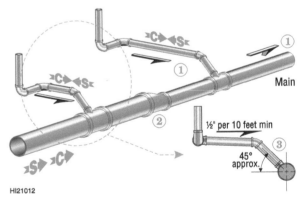

① Always slope horizontal pipes in the direction of condensate flow. If the radiator served is more than one floor above the main or riser connection, this piping is considered a riser, not a branch. You will need to size and install the pipe as for a riser (Figure 23).

② If reducing a steam main, always use a concentric reducer to ensure condensate won't be trapped at the joint. With the eccentric reducer installed as shown, the bottom of the pipe maintains continuous downward slope. If you used a concentric reducer, condensate would build up to the level of the reducer, resulting in severe water hammer.

③ Connect to main at about a 45° angle as shown. Condensate can flow smoothly into main, preventing carryover of liquid in the steam flow.

Figure 24   Correctly installing one-pipe main vents

RIGHT WAY

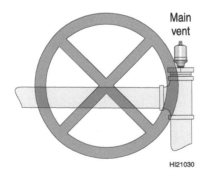

WRONG WAY

Main vent is isolated from end of main, where hammer can occur. Using elbow at end of main smooths flow and reduces potential for water hammer. The 6 to 10-inch height above main provides a "snubber" to reduce shock from any water hammer that occurs.

This is a common installation method, and will result in heating problems. Tee at end of main increases likelihood of water hammer because condensate hits tee and goes both directions. Main vent **WILL BE DAMAGED** by water hammer, rendering vent inoperative.

One-pipe connection details

Figure 23, page 12–40 through Figure 28, page 12–42, show details of critical piping connections on one-pipe systems. These piping designs provide the necessary pitch and flow paths required for smooth flow of condensate and steam, often opposed to one another, in the same pipe. Inspect systems carefully to verify the piping is correctly installed and that the needed slope is there.

Figure 25 Upfeed riser, not dripped

One-pipe steam Runout to upfeed riser when not dripped to wet return

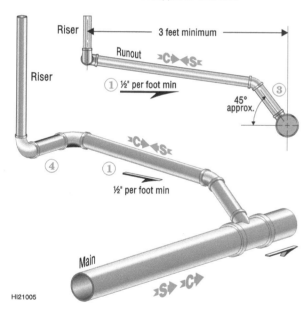

HI21005

① Slope horizontal runout pipe minimum ½" per foot when connected to one-pipe risers with no drip line to wet return. If this slope cannot be maintained, and the length of the horizontal pipe is more than 8 feet, increase the size of the horizontal pipe one pipe size larger than shown in sizing tables.

② The distance from riser center line to main center line should be at least 3 feet. With this distance, the condensate falling down the riser has a chance to stabilize before entering the main. With a shorter distance, the condensate flow would be turbulent as it enters the main, causing water hammer and liquid carryover in the steam.

③ Connect to main at about a 45° angle as shown. Condensate can flow smoothly into main, preventing carryover of liquid in the steam flow.

④ Connect to risers with offset piping to provide for piping expansion and contraction and easier adjustment of slope in runout.

Figure 26 One-pipe radiator piping

One-pipe steam Connecting to radiators

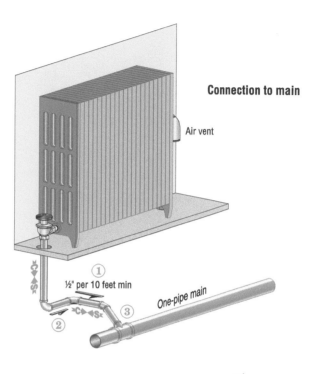

Connection to main

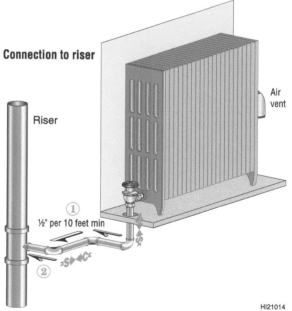

Connection to riser

HI21014

① Slope horizontal runout pipe minimum ½" per 10 feet.

② Connect to risers with offset piping to provide for piping expansion and contraction and easier adjustment of slope in runout.

③ Connect pipe 45° downward from main, ensuring condensate can flow freely and smoothly to the drip line as stem flows up the riser.

Figure 27 Upfeed riser, dripped

One-pipe steam Runout to upfeed riser, dripped to wet return

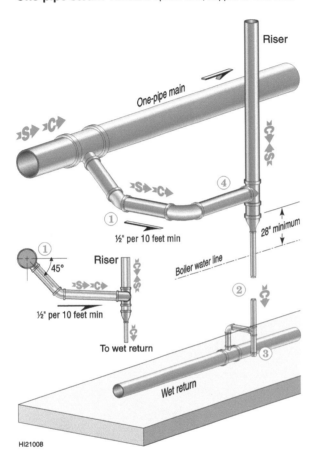

HI21008

Figure 28 Downfeed risers (dripped)

One-pipe steam Runouts to downfeed risers

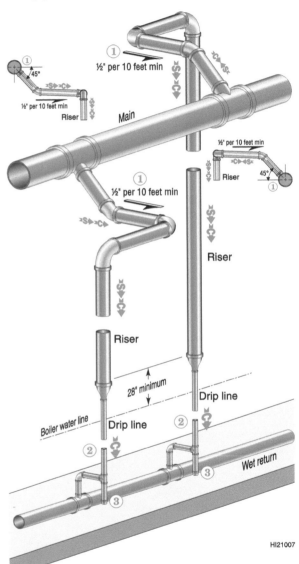

HI21007

1. Slope horizontal runout pipe minimum ½" per 10 feet. Connect pipe 45° downward from main, ensuring condensate can flow freely and smoothly to the drip line as stem flows up the riser.

2. Size drip line to handle the condensate load of all connected radiators. Use ¾" line for up to 192 square feet EDR; 1" for up to 452 square feet EDR; 1¼" for up to 992 square feet EDR; 1½" for up to 1500 square feet EDR.
   If drip line must carry condensate from upstream branches or undripped risers, include the EDR of all of the upstream devices. Install reducer no less than 28" above boiler water line for systems with over 416 square feet EDR (100,000 Btuh); 14" above water line for smaller systems.

3. Terminate drip line with nipple and cap to allow cleanout of sediment leg.

4. Connect to risers with offset piping to provide for piping expansion and contraction and easier adjustment of slope in runout.

1. Slope horizontal runout pipe minimum ½" per 10 feet. Connect pipe 45° into main, ensuring condensate can flow freely and smoothly.

2. Size drip line to handle the condensate load of all connected radiators. Use ¾" line for up to 192 square feet EDR; 1" for up to 452 square feet EDR; 1¼" for up to 992 square feet EDR; 1½" for up to 1500 square feet EDR.
   Install reducer no less than 28" above boiler water line for systems with over 416 square feet EDR (100,000 Btuh); 14" above water line for smaller systems.

3. Terminate drip line with nipple and cap to allow cleanout of sediment leg.

4. Connect to risers with offset piping to provide for piping expansion and contraction and easier adjustment of slope in runout.

Pumped-return systems

General

Pumped-return systems use a boiler feed pump with condensate storage tank, forcing return water into the boiler using the pump pressure. The condensate storage tank is vented to atmosphere (only exception is on vacuum systems). The tank must never be pressurized.

The pump and tank combination will either be a boiler feed system (preferred) or condensate return system. Boiler feed systems are the most reliable because they feed water to the boiler on demand from the boiler low water cutoff/pump control. Condensate return systems simply operate based on the water level in the condensate receiver. See Figure 46, page 12–62, for recommended piping connections.

Most pumped-return systems rely on gravity to return condensate to the tank. The return piping must slope continuously downhill toward the tank, with no drops or areas where water could accumulate in the piping and block the movement of air in and out of the system through the tank vent and return lines.

Vacuum systems are a special case of pumped return, in which the condensate receiver is not vented. The feed pump/tank system pulls a vacuum on the return lines, speeding the movement of condensate and lowering the pressure in the system for lower steam temperature at the radiators. Vacuum systems were seldom installed on residential applications, and are not covered in detail in Guide RHH.

One-pipe parallel upfeed pumped-return system

See Figure 29, page 12–45.
See Figure 49, page 12–64, for sizing recommendations for return piping between traps and condensate receiver.
See page 12–28 for piping sizing of the one-pipe piping.

This is a one-pipe system that has been retrofitted with a boiler feed pump and condensate receiver. To convert to pumped return, notice that you must install a float and thermostatic trap at the end of the main (including a strainer and isolation valves). This controls condensate flow to the atmospheric-vented condensate receiver and prevents steam from blowing through. It also allows air to flow from the main, down the return line and out the tank vent on start-up.

1   The steam main must slope downhill toward the return trap at no less than 1 inch per 20 feet. This is hard to detect, particularly if the boiler room floor isn't even. Use a spirit level to check the full length of the main. Any sag in the main will cause water hammer because condensate will either accumulate there or be too deep because it doesn't flow quickly enough.

2   The upfeed risers supply steam up and condensate down.

3   Size the float and thermostatic trap for both the piping start-up load and the running load of the radiators. Add up the total square feet EDR of all radiators in the system. Divide the total square feet EDR by 2 for the condensate load of the radiators in pph (pounds per hour). This factor allows a two-times safety factor for the trap sizing. Then use Figure 18, page 11-24, in Part 11 to determine the added load the trap must handle to remove the condensate caused by heating the piping at start-up (or running load if piping is not insulated). Add the piping load to the radiator load for total trap capacity in pph. The peak load on the trap probably occurs early after start-up, when there is little pressure at the end of the main. Size the trap using only a ½ psig available pressure drop (if the trap is mounted 15 inches below the header as shown). If the trap is mounted lower than 15 inches, determine available pressure in psig by dividing the height in inches by 28.

4   Install either a boiler feed system (preferred) or a condensate return system. If using a boiler feed system, you will need to mount a low water cutoff/pump control on the boiler. Use only a control recommended by the boiler manufacturer, installed per the boiler manual.

The return piping from the main trap to the condensate receiver must slope continuously downhill. There must be no drops or any location that might allow water to accumulate and block the flow of air in and out of the return line through the condensate receiver vent line. Size the tank and pump as recommended in Part 11. See Figure 46, page 12–62, for recommended piping connections to and from the tank and pump.

5   Install a check valve as shown in Figure 46, page 12–62, to prevent water from being pushed back into the tank when the boiler operates.

6   Install a square-head cock to allow throttling the feed pump flow rate to the boiler. Too high a flow rate will cause water level fluctuations in the boiler and can cause water hammer in the boiler header.

7   The piping illustrations in Part 12 show the use of a Hartford loop even on pumped-return applications. Opinions about this vary. Some argue that this piping method can cause water to splash into the header piping and cause water hammer. If the flow rate is high enough to cause water hammer in the header, then the flow rate is too high. Reduce the flow rate with the square-head cock on the feed piping. The advantage of the Hartford loop is that it is the only "check valve" that always checks. For the best assurance of a successful installation, always carefully follow the boiler manufacturer's recommendations in the boiler manual.

8   If not already installed, find a way to install a main vent at or near the top of every upfeed riser. If you can't get to the pipe, add a tee to an upper-story radiator supply line and install the main vent there as shown in Figure 29. These main vents will help upper-story radiators receive steam at nearly the same time as the lower-story radiators.

9   Make sure each radiator is equipped with a working air vent. If using adjustable vents, adjust them to the size of the radiator. You might consider fine-tuning the system a bit by closing down more on the near-boiler radiators. But heat distribution should be done with good main and riser venting, not with the radiator vents.

Figure 29 One-pipe parallel upfeed pumped-return system

One-pipe parallel upfeed pumped-return system

See opposite page for comments and pipe sizing.

F"xx" below is the Figure number for piping detail.
Example: F39 means Figure 39.

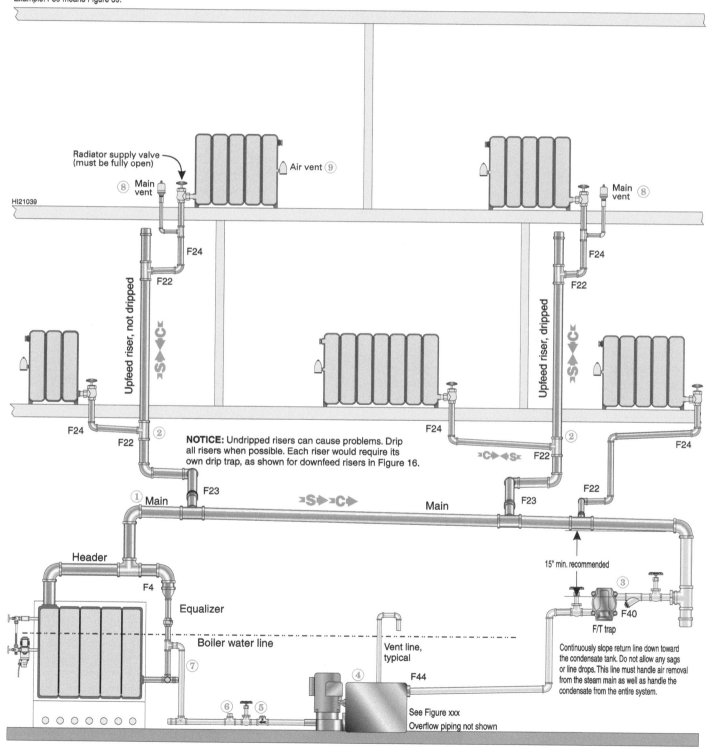

Radiator supply valve
(must be fully open)

Air vent ⑨

⑧ Main vent

Main ⑧ vent

HI21039

F24

F22

Upfeed riser, not dripped

►S► ◄C◄

F24

Upfeed riser, dripped

►S► ◄C◄

F24

F22

②

F24

NOTICE: Undripped risers can cause problems. Drip all risers when possible. Each riser would require its own drip trap, as shown for downfeed risers in Figure 16.

F22

②

►C► ◄S◄

F22

F24

F23

① Main

►S► ◄C►

Main

F23

F22

Header

F4

15" min. recommended

③

Equalizer

F40

F/T trap

Boiler water line

⑦

Vent line, typical

④

F44

Continuously slope return line down toward the condensate tank. Do not allow any sags or line drops. This line must handle air removal from the steam main as well as handle the condensate from the entire system.

⑥ ⑤

See Figure xxx

Overflow piping not shown

One-pipe parallel downfeed pumped-return system

See Figure 30, page 12–47. See Figure 49, page 12–64 for sizing recommendations for return piping between traps and condensate receiver. See page 12–32 for piping sizing of the one-pipe piping.

This is a one-pipe downfeed system that has been retrofitted with a boiler feed pump and condensate receiver. To convert to pumped return, notice that you must install a float and thermostatic trap at the end of the main (end of main return riser) and each downfeed riser (including a strainer and isolation valves). This controls condensate flow to the atmospheric-vented condensate receiver and prevents steam from blowing through. It also allows air to flow from the main, down the return line and out the tank vent on start-up. You cannot use a master trap to handle all risers and the main because they would be connected together above the water line, with risk of reverse steam flow and water hammer.

This example shows only one drip trap — at the end of the steam main. You must also install a drip trap at the lowest point of any steam supply piping that could accumulate condensate (such as a riser runout or radiator runout that runs below the steam main, for example).

1   A downfeed system uses a steam main located at the top of the building. The express riser supplies steam from the boiler to the main. Downfeed risers supply steam from the main to the heating units below. Each downfeed riser ends in a drip pipe connected to the wet return. Make sure the main slopes continuously downward at no less than 1 inch per 20 feet. Use a spirit level to check the full length of the main. Any sag in the main will cause water hammer because condensate will either accumulate there or be too deep because it doesn't flow quickly enough. You can remove the main vent at the end of the steam main or leave it in place. The float and thermostatic trap at the end of the main return riser allows air to move out of the piping.

2   Terminate each downfeed riser with a float and thermostatic trap. Locate the traps so they are at least 15 inches below the lowest steam-carrying piping connected to them. This allows room for condensate to accumulate and push through the trap at start-up.

3   Size each trap for the combined load of the radiators and the heat loss load or start-up load of the piping. Add up the total square feet EDR of all radiators served by the trap. Divide the total square feet EDR by 2 for the condensate load of the radiators in pph (pounds per hour). This factor allows a two-times safety factor for the trap sizing. Then use Figure 18, page 11–24, in Part 11 to determine the added load the trap must handle to remove the condensate caused by heating the piping at start-up (or running load if piping is not insulated). Add the piping load to the radiator load for

total trap capacity in pph. The peak load on the trap probably occurs early after start-up, when there is little pressure at the ends of the main and risers. Size the trap using only a ½ psig available pressure drop (if the trap is mounted 15 inches below the lowest steam-carrying pipe as shown). If the trap is mounted more than 15 inches below, determine available pressure in psig by dividing the height in inches by 28.

4   Install either a boiler feed system (preferred) or a condensate return system. If using a boiler feed system, you will need to mount a low water cutoff/pump control on the boiler. Use only a control recommended by the boiler manufacturer, installed per the boiler manual. The return piping from the traps to the condensate receiver must slope continuously downhill. There must be no drops or any location that might allow water to accumulate and block the flow of air in and out of the return line through the condensate receiver vent line. Size the tank and pump as recommended in Part 11. See Figure 46, page 12–62, for recommended piping connections to and from the tank and pump.

5   Install a check valve as shown in Figure 46, to prevent water from being pushed back into the tank when the boiler operates.

6   Install a square-head cock to allow throttling the feed pump flow rate to the boiler. Too high a flow rate will cause water level fluctuations in the boiler and can cause water hammer in the boiler header.

7   The piping illustrations in Part 12 show the use of a Hartford loop even on pumped-return applications. Opinions about this vary. Some argue that this piping method can cause water to splash into the header piping and cause water hammer. If the flow rate is high enough to cause water hammer in the header, then the flow rate is too high. Reduce the flow rate with the square-head cock on the feed piping. The advantage of the Hartford loop is that it is the only "check valve" that always checks. For the best assurance of a successful installation, always carefully follow the boiler manufacturer's recommendations in the boiler manual.

8   Make sure each radiator is equipped with a working air vent. If using adjustable vents, adjust them to the size of the radiator. You might consider fine-tuning the system a bit by closing down more on the near-boiler radiators. But heat distribution should be done with good main and riser venting, not with the radiator vents.

Figure 30   One-pipe parallel downfeed pumped-return system

One-pipe parallel downfeed pumped-return system

Main vent
F21

① Main

F26   Slope down 1" per 20 feet minimum   F26   ⊳S▸◂C▸

Express riser

Downfeed riser

Downfeed riser

See opposite page for comments and pipe sizing.

F"xx" below is the Figure number for piping detail. Example: F39 means Figure 39.

Radiator supply valve (must be fully open)

Air vent ⑧

F24   See opposite page for comments and pipe sizing.   F24

F22   ⊳S▸◂C◂   F22

HI21041

②

◂S◂▾C◂   ◂S◂▾C◂

F24   F22   F24   F22   F24   F22

②

①   15" min. recommended   15" min. recommended   15" min. recommended

②

③   ③   ③

F41   F41   F40
F/T trap   F/T trap   F/T trap

Header   F4

Equalizer

Boiler water line   Return

⑦   Vent line, typical   F44

Continuously slope return line down toward the condensate tank. Do not allow any sags or line drops. This line must handle air removal from the steam main as well as handle the condensate from the entire system.

⑥   ⑤   ④

Overflow piping not shown

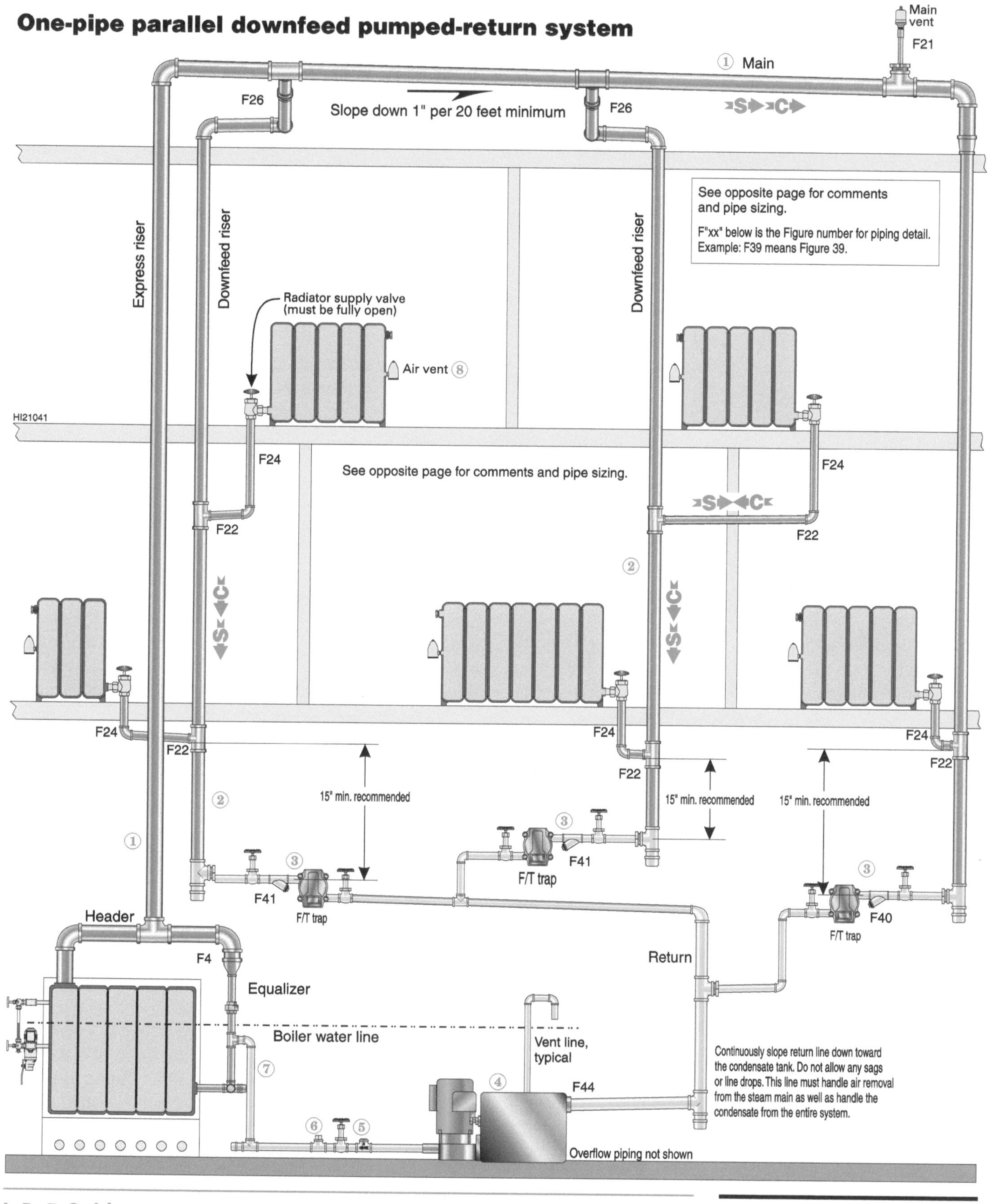

Two-pipe pumped-return system operation

Off cycle

Figure 31, page 12–49.

At the end of a heating cycle, the boiler stops making steam. Steam remaining in the system condenses, causing a vacuum to form. This pulls air into the system — flowing into the vent line on the condensate receiver, then up through the return piping, through the drip traps on mains and risers and into the radiators through the thermostatic traps and supply valves. All condensate remaining in the system will drain back to the condensate receiver through the traps and return lines. All thermostatic traps will be open. The float valve of the float/thermostatic steam main drip trap will close. The entire system will fill with air.

Heating cycle starts

Figure 32, page 12–50.

When the boiler begins to make steam, pressure builds at the boiler end of the system. This pushes air through the piping, through the radiator thermostatic traps and the steam main drip trap, into the return piping and out of the system through the condensate receiver vent line. The thermostatic element in the steam main drip trap is high capacity, and will move air faster than the radiator traps. You might find, though, that you need to slow the flow rate through the lower-floor radiators (by throttling the radiator valves) to help steam reach all radiators at about the same time.

Steam fills distribution piping

Figure 33, page 12–51.

Steam fills the distribution piping as air continues to flow from the system. Adjusting the radiator supply valves on the lower-floor radiators will help to even the heat distribution in the house by slowing air movement from the lower-floor radiators long enough to clear air from the upfeed risers. All radiators would then receive steam at about the same time. When steam heats the thermostatic element of the steam main drip trap, the element closes, preventing any steam from passing through the trap. The radiator traps remain open, allowing radiator air to be pushed out of the system. Steam in the distribution begins to condense as it heats the piping. Condensate flows down the steam piping to the steam main drip trap leg.

Radiators begin to heat

Figure 34, page 12–52.

Steam now enters the radiators, staying near the top because it is much lighter than the air inside. Condensate begins to form in the radiators as the steam condenses. The radiator thermostatic traps remain open as long as the element isn't heated near steam temperature. The float in the steam main drip trap will begin to operate as condensate accumulates in the trap. The float will cycle throughout the operating cycle, allowing condensate to flow through to the condensate receiver. The water level in the condensate receiver rises as condensate returns from the system, while the boiler water level drops because of the water that has been converted to steam.

The example shown in Figure 31, page 12–49 to Figure 35, page 12–53 assumes a condensate return system, not a boiler feed system. The condensate return system will not feed water to the boiler until the condensate receiver water level rises enough to cause the float switch to start the feed pump. If the system time lag isn't too long and the condensate receiver is sized correctly, the condensate return system should soon begin to feed water to the boiler. If the condensate return system fails to feed water soon enough, the water feed (if used) on the boiler will feel fresh water to the boiler. This can eventually cause flooding because the total amount of water in the system will be too large. Use a boiler feed system when possible. It will always provide surer operation than a condensate return system because a level control on the boiler turns the pump on and off.

Heating cycle steady

Figure 35, page 12–53.

As the heating cycle continues, most of the air is pushed out of the radiators. The radiator thermostatic traps will cycle closed and open as they are heated by the steam and hot condensate. These traps will only open when the condensate in them cools to from 10 to 30°F cooler than the steam temperature.

The condensate return system will pump water to the boiler when the condensate receiver level rises enough to turn on the float-operated switch. The pump will continue feeding water to the boiler until the level in the tank drops to the switch-open setpoint. The heating cycle will continue until the thermostat shuts down the boiler or the pressure in the boiler reaches the pressure limit switch setting (usually 2 psig for steam heating systems).

Figure 31 Two-pipe pumped-return system operation — Off cycle

Off cycle (no call for heat)

① **Water levels:** When the boiler isn't firing, the fluid inside is only water (no steam). The boiler, gauge glass, and equalizer contain water only. The column height is equal to the boiler surface level at all locations. The condensate tank level is shown at mid-position. This will depend on size of tank and initial fill conditions.

② **Air:** At the end of a firing cycle, the boiler stops making steam. The radiators continue to condense remaining steam. This causes a vacuum to develop. Air enters the system through the condensate tank vent to break this vacuum. The thermostatic radiator traps and steam main float and thermostatic trap allow air to flow to the piping and radiators.

③ **Traps:** The thermostatic trap elements are cool, so these traps are open. There is no condensate in the float and thermostatic steam main drip trap and the trap is cool. So the F/T trap thermostatic valve is open. The float valve is closed.

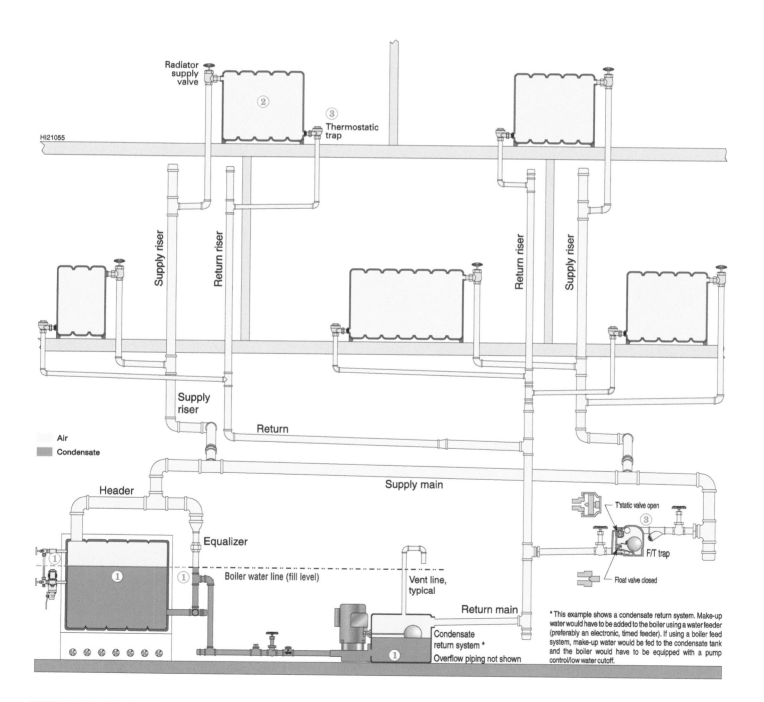

* This example shows a condensate return system. Make-up water would have to be added to the boiler using a water feeder (preferably an electronic, timed feeder). If using a boiler feed system, make-up water would be fed to the condensate tank and the boiler would have to be equipped with a pump control/low water cutoff.

Figure 32 Two-pipe pumped-return system operation — Heating cycle start

Heating cycle starts (thermostat calls for heat)

① **Water levels:** When the boiler fires, *it no longer contains water*. It contains a mixture of water and steam bubbles, a frothy mixture that can weigh considerably less than water only. At the beginning of the cycle, the boiler internal surface level rises because the steam bubbles push water out of the way, raising the level. The gauge glass sees part, but not all, of this change because the gauge glass contains only water. So its column height is slightly lower than the boiler internal level. The Hartford loop and return risers don't sense any change in water level height. This is because the total weight of water in the boiler hasn't changed yet. The condensate tank level remains at mid-position.

② **Air:** Air moves out of the piping and radiators through the thermostatic radiator traps and the thermostatic element of the float and thermostatic steam main drip trap. This air must flow through the return piping, the condensate tank and out the condensate tank vent piping. There must be no low spots in the return piping that could collect water and block the flow of air. Air must be able to move freely into and out of the system through this flow path.

③ **Traps:** The thermostatic elements of the traps remain open as long as they are cool (not exposed to steam temperature).

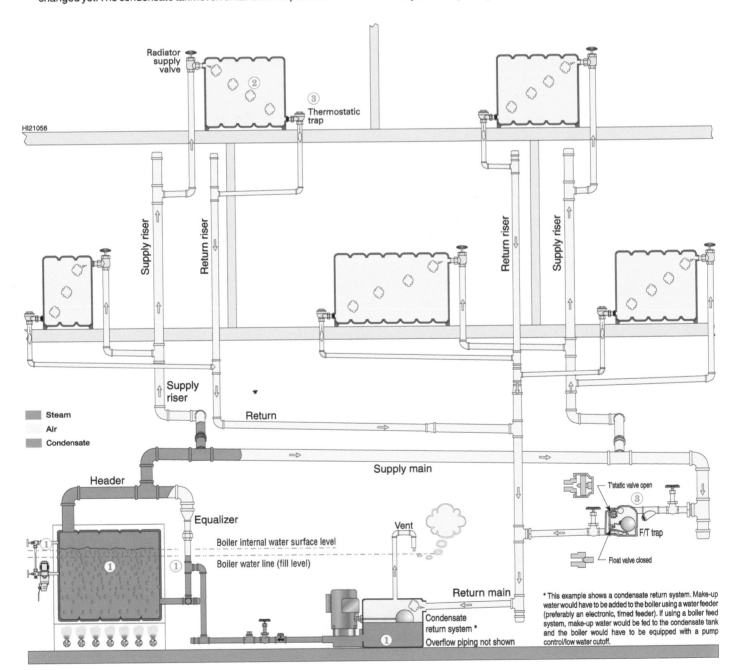

Figure 33 Two-pipe pumped-return system operation — Steam fills distribution piping

Steam fills distribution piping (one role of drip traps — even heat distribution)

① **Water levels**: The boiler water level has dropped slightly because of the water used to make steam to fill the piping. Condensate has begun to form as the piping is heated. This condensate accumulates in the steam main drip leg. The condensate tank water level has not changed since condensate has not begun to return.

② **Air**: Air has moved out of the distribution piping (steam main and risers), allowing steam to reach all of the radiators. Air continues to move from the radiators as steam pushes in. The larger size of the thermostatic element in the F/T drip trap allows air to move more quickly from the distribution piping than from the radiators, helping to even heat distribution.

③ **Traps**: The thermostatic elements of the radiator traps remain open as long as they are cool (not exposed to steam temperature). The thermostatic element of the steam main F/T drip trap is now closed because it has been heated by the steam.

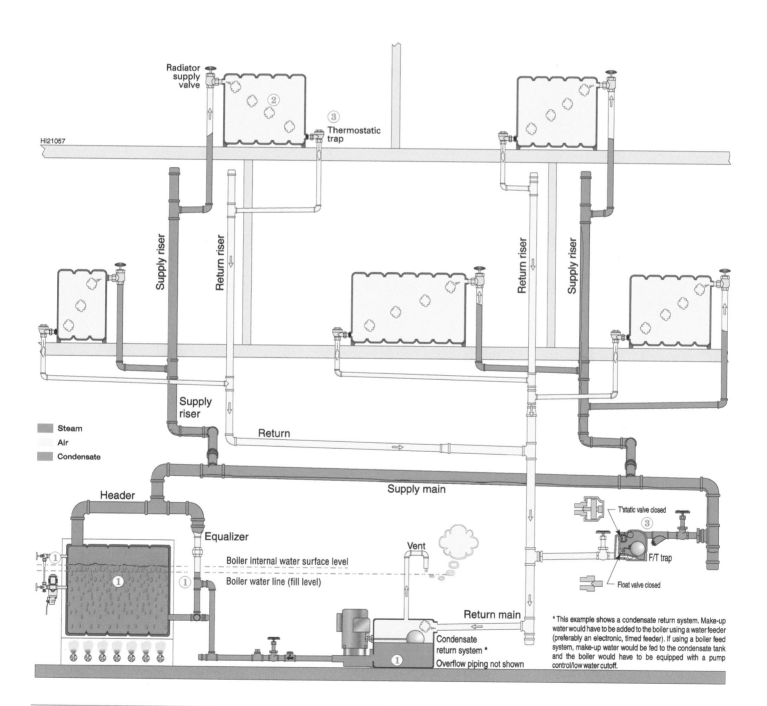

Steam
Air
Condensate

* This example shows a condensate return system. Make-up water would have to be added to the boiler using a water feeder (preferably an electronic, timed feeder). If using a boiler feed system, make-up water would be fed to the condensate tank and the boiler would have to be equipped with a pump control/low water cutoff.

Figure 34 Two-pipe pumped-return system operation — Radiators begin to heat

Radiators begin to heat (steam flow inside radiators)

① **Water levels:** The boiler water level continues to drop as steam fills the radiators. The height of condensate in the steam main drip leg reaches its maximum at this time because of the rush of condensate from heating the pipes. This is the time when the distance between the bottom of the main and the trap entrance is critical. There is very little steam pressure available to push the condensate through the trap. The height of the water column in the drip leg is all that makes the condensate move. Always provide at least 15 inches height when possible. Condensate begins to form in the radiators and flows through the thermostatic traps into the return lines. The water level in the condensate tank begins to rise as condensate returns.

② **Air:** Air continues to move from the radiators through the radiator thermostatic traps until steam reaches the traps and heats the elements.

③ **Traps:** The thermostatic elements of the radiator traps remain open as long as they are cool (not exposed to steam temperature). The thermostatic element of the steam main F/T drip trap is now closed because it has been heated by the steam. The float valve of the steam main drip trap will cycle as condensate raises the float. When the condensate has moved through, the float will drop, closing the float valve.

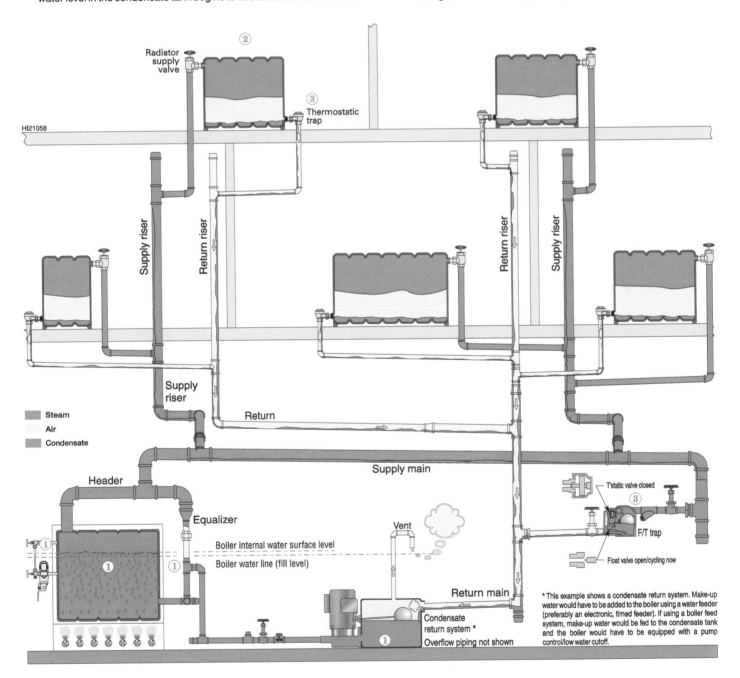

Figure 35 Two-pipe pumped-return system operation — Heating cycle steady

Heating cycle steady (traps and condensate pump begin cycling)

① **Water levels:** The boiler water level has dropped more, having supplied steam to the system to fill all piping and radiators. The condensate return system should now be ready to operate the pump and feed condensate to the boiler. If the feed pump doesn't operate soon, the water feeder on the boiler (not shown) will introduce fresh make-up water. This can result in flooding of the system because water levels will be too high when the boiler shuts down and all condensate returns. Adjust the tank float switch settings so the water is supplied to the boiler before the boiler water feeder introduces fresh water. Make sure to set the operating range so the pump doesn't introduce too much water to the boiler.

② **Air:** Most of the air has been ejected from the system. Only small amounts of air will now exit from the condensate tank vent line.

③ **Traps:** As steam nears the bottoms of the radiators, the thermostatic traps will close. They will open only when the condensate has covered the element and has cooled at least 10°F below the steam temperature. This will allow condensate to exit the radiator. Steam will then heat the element again, closing it off. The traps will cycle this way throughout the heating cycle.

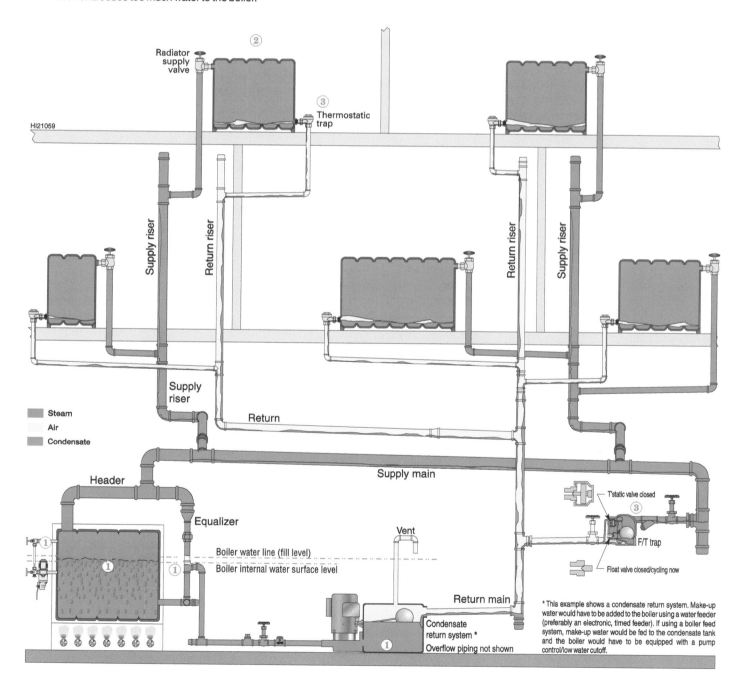

HI21059

Two-pipe upfeed pumped-return system

See Figure 36, page 12–55. See page 12–64 for pipe-sizing recommendations.

This is a typical two-pipe pumped-return system. For systems using radiators and convectors, set the boiler limit control to cut out at 2 psig and cut in at ½ psig. This is all the pressure you should need to obtain rated output from the heating units. Only raise the pressure higher if the system contains a heat exchanger or unit heater that requires higher pressure. The higher the pressure of the steam, the more likely flashing will occur in the return piping.

This example shows only one drip trap — at the end of the steam main. You must also install a drip trap at the lowest point of any steam supply piping that could accumulate condensate (such as a riser runout or radiator runout that runs below the steam main, for example).

1 The steam main must slope downhill toward the return trap at no less than 1 inch per 20 feet. This is hard to detect, particularly if the boiler room floor isn't even. Use a spirit level to check the full length of the main. Any sag in the main will cause water hammer because condensate will either accumulate there or be too deep because it doesn't flow quickly enough.

2 The upfeed risers supply steam to the radiators. A small amount of condensate flows down the risers as steam condenses in heating the riser and supply pipes.

3 Two-pipe systems provide return risers to handle the condensate. The return risers must slope continuously downhill toward the condensate receiver.

4 Radiator supply valves should be packless type. You can use the radiator supply valves to "tune" the system heat. Close off on the radiators in spaces that tend to overheat. This will give time for other radiators to heat up uniformly.

5 Install a thermostatic trap on the condensate connection of each radiator. Thermostatic traps force the condensate to cool at least 10°F below the steam temperature, reducing likelihood of steam flashing in the returns. Size each trap for the square feet EDR of the radiator times two for the SHEMA rating that includes a two-times safety factor, or EDR divided by 2 for the pph (pounds per hour) rating with a safety factor of two. Use an available pressure of 1 psig.

6 Install either a boiler feed system (preferred) or a condensate return system. If using a boiler feed system, you will need to mount a low water cutoff/pump control on the boiler. Use only a control recommended by the boiler manufacturer, installed per the boiler manual. The return piping from the main trap to the condensate receiver must slope continuously downhill. There must be no drops or any location that might allow water to accumulate and block the flow of air in and out of the return line through the condensate receiver vent line. Size the tank and pump as recommended in Part 11. See Figure 46, page 12–62, for recommended piping connections to and from the tank and pump.

7 Install a check valve as shown in Figure 46, to prevent water from being pushed back into the tank when the boiler operates.

8 Install a square-head cock to allow throttling the feed pump flow rate to the boiler. Too high a flow rate will cause water level fluctuations in the boiler and can cause water hammer in the boiler header.

9 The piping illustrations in Part 12 show the use of a Hartford loop even on pumped-return applications. Opinions about this vary. Some argue that this piping method can cause water to splash into the header piping and cause water hammer. If the flow rate is high enough to cause water hammer in the header, then the flow rate is too high. Reduce the flow rate with the square-head cock on the feed piping. The advantage of the Hartford loop is that it is the only "check valve" that always checks. For the best assurance of a successful installation, always carefully follow the boiler manufacturer's recommendations in the boiler manual.

10 Size the float and thermostatic drip trap using the guidelines of Figure 18, page 11-24, Part 11. Remember to install the trap at least 15 inches below the main to provide static head of water to push the condensate through the trap even with no steam pressure in the system. Size the trap assuming only ½ psig available pressure (the equivalent of about 15 inches of water).

Figure 36 Two-pipe upfeed pumped-return system

Two-pipe upfeed pumped return

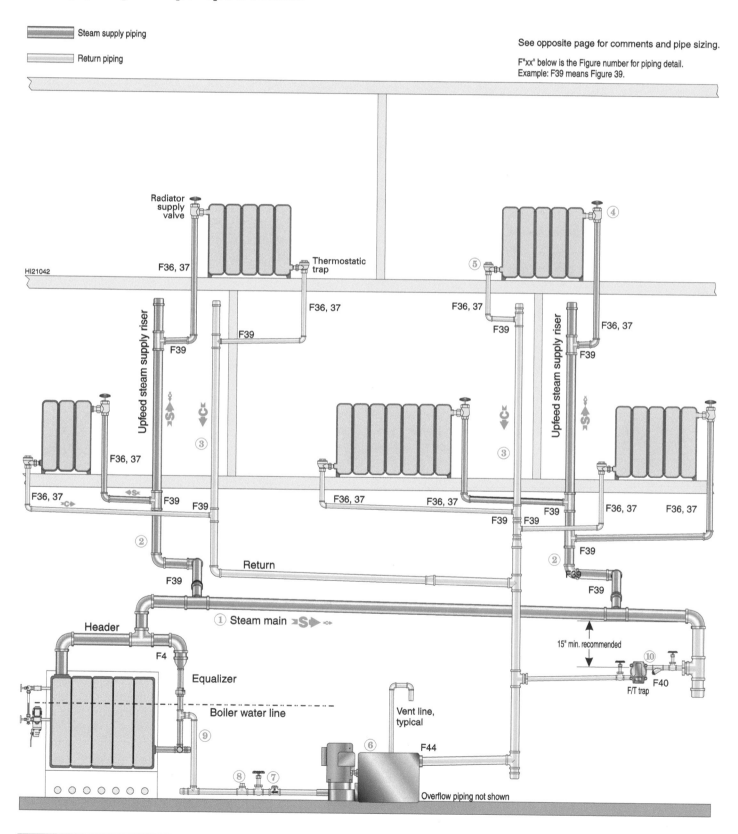

Steam supply piping
Return piping

See opposite page for comments and pipe sizing.

F"xx" below is the Figure number for piping detail.
Example: F39 means Figure 39.

Two-pipe downfeed pumped-return system

See Figure 37, page 12–57. See Figure 49, page 12–64 for pipe-sizing recommendations.

This is a typical two-pipe pumped-return downfeed system. For systems using radiators and convectors, set the boiler limit control to cut out at 2 psig and cut in at ½ psig. This is all the pressure you should need to obtain rated output from the heating units. Only raise the pressure higher if the system contains a heat exchanger or unit heater that requires higher pressure. The higher the pressure of the steam, the more likely flashing will occur in the return piping.

Notice that each downfeed riser must have a drip trap as well as the steam main. In general, any steam supply line that can accumulate condensate requires a drip trap at its low point.

1 The steam main must slope downhill toward the return riser at no less than 1 inch per 20 feet. This is hard to detect, particularly if the boiler room floor isn't even. Use a spirit level to check the full length of the main. Any sag in the main will cause water hammer because condensate will either accumulate there or be too deep because it doesn't flow quickly enough.

2 The downfeed risers supply steam to the radiators. A small amount of condensate flows down the risers as steam condenses in heating the riser and supply pipes. Each riser must have a drip trap to handle this condensate.

3 Two-pipe systems provide return risers to handle the condensate. The return risers must slope continuously downhill toward the condensate receiver.

4 Radiator supply valves should be packless type. You can use the radiator supply valves to "tune" the system heat. Close off on the radiators in spaces that tend to overheat. This will give time for other radiators to heat up uniformly.

5 Install a thermostatic trap on the condensate connection of each radiator. Thermostatic traps force the condensate to cool at least 10°F below the steam temperature, reducing likelihood of steam flashing in the returns. Size each trap for the square feet EDR of the radiator times two for the SHEMA rating that includes a two-times safety factor, or EDR divided by 2 for the pph (pounds per hour) rating with a safety factor of two. Use an available pressure of 1 psig.

6 Install either a boiler feed system (preferred) or a condensate return system. If using a boiler feed system, you will need to mount a low water cutoff/pump control on the boiler. Use only a control recommended by the boiler manufacturer, installed per the boiler manual. The return piping from the main trap to the condensate receiver must slope continuously downhill. There must be no drops or any location that might allow water to accumulate and block the flow of air in and out of the return line through the condensate receiver vent line. Size the tank and pump as recommended in Part 11. See Figure 46, page 12–62, for recommended piping connections to and from the tank and pump.

7 Install a check valve as shown in Figure 46, to prevent water from being pushed back into the tank when the boiler operates.

8 Install a square-head cock to allow throttling the feed pump flow rate to the boiler. Too high a flow rate will cause water level fluctuations in the boiler and can cause water hammer in the boiler header.

9 The piping illustrations in Part 12 show the use of a Hartford loop even on pumped-return applications. Opinions about this vary. Some argue that this piping method can cause water to splash into the header piping and cause water hammer. If the flow rate is high enough to cause water hammer in the header, then the flow rate is too high. Reduce the flow rate with the square-head cock on the feed piping. The advantage of the Hartford loop is that it is the only "check valve" that always checks. For the best assurance of a successful installation, always carefully follow the boiler manufacturer's recommendations in the boiler manual.

10 Size the float and thermostatic drip traps using the guidelines of Figure 18, page 11-24, Part 11, based on the piping connected to each riser or steam main. Remember to install the trap at least 15 inches below the main or lowest branch connection to provide static head of water to push the condensate through the trap even with no steam pressure in the system. Size the trap assuming only ½ psig available pressure (the equivalent of about 15 inches of water).

Figure 37 Two-pipe downfeed pumped-return system

Two-pipe downfeed pumped return

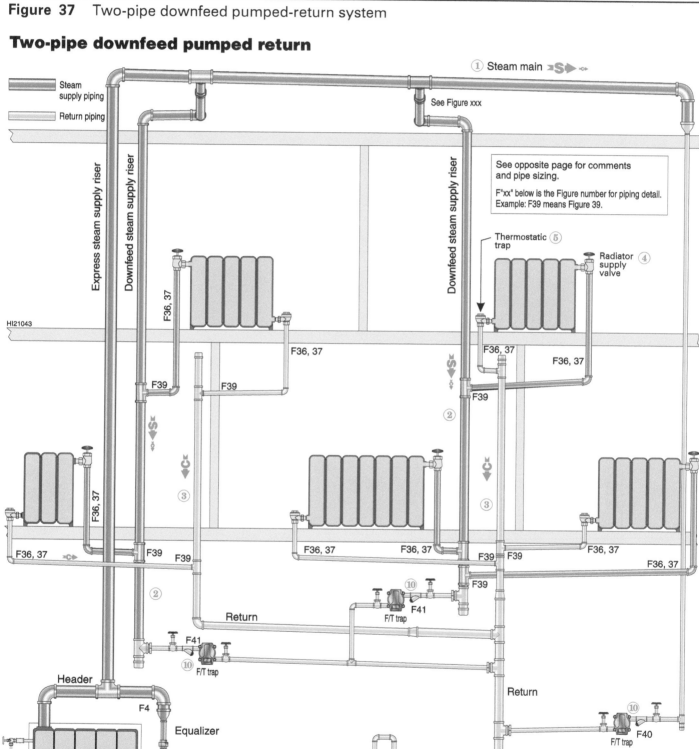

① Steam main ⇒**S**▶ ⊶

Steam supply piping

Return piping

See Figure xxx

See opposite page for comments and pipe sizing.

F"xx" below is the Figure number for piping detail. Example: F39 means Figure 39.

Express steam supply riser

Downfeed steam supply riser

Downfeed steam supply riser

Thermostatic ⑤ trap

Radiator ④ supply valve

HI21043

F36, 37

F36, 37

F36, 37

F36, 37

F39

F39

F39

②

F36, 37

③

F36, 37

F36, 37

③

F36, 37

F36, 37

F39

F39

F39

F36, 37

F36, 37

②

F39

⑩

F41

F/T trap

F41

⑩

F/T trap

Return

F39

F36, 37

Header

F4

Return

⑩

F40

F/T trap

Equalizer

Boiler water line

Vent line, typical

⑨

F44

⑥

⑧ ⑦

Overflow piping not shown

Two-pipe connection details

The illustrations on pages 12–58 to 12–63 show details of critical piping connections on two-pipe systems. These piping designs provide the necessary pitch and flow paths required for smooth flow of condensate and steam. Inspect systems carefully to verify the piping is correctly installed and that the needed slope is there.

Figure 38 Radiator connection to main

Two-pipe steam Piping radiators to mains (horizontal lines)

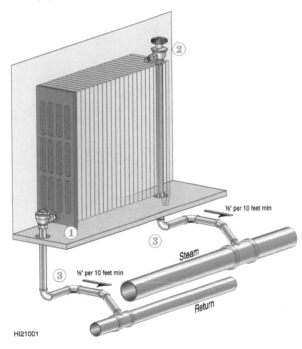

HI21001

① Thermostatic trap.

② Radiator steam valve. Valve may be used to throttle steam flow on two-pipe systems.

③ Slope return line continuously downward toward the riser connection. There must be no sags or loops in the return piping because this pipe is the means by which air flows into and out of the radiator. If the line is blocked by water seals, air cannot leave. The radiator could fail to heat properly. Slope steam supply horizontal piping downward toward steam main.

Figure 39 Radiator connection to riser

Two-pipe steam Piping radiators to risers (vertical lines)

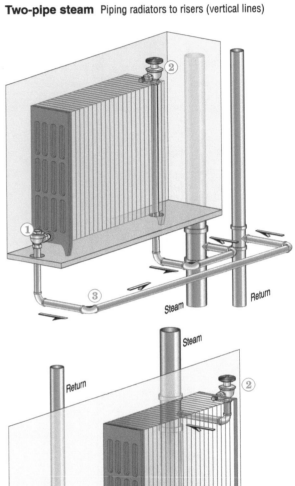

HI21002

① Thermostatic trap. Use swivel trap when connecting to piping behind walls (lower illustration).

② Radiator steam valve. Valve may be used to throttle steam flow on two-pipe systems.

③ Slope return line continuously downward toward the riser connection (minimum ½" per 10 foot pitch). There must be no sags or loops in the return piping because this pipe is the means by which air flows into and out of the radiator. If the line is blocked by water seals, air cannot leave. The radiator could fail to heat properly.

Figure 40 Unit heater connection

Two-pipe steam Unit heater connections

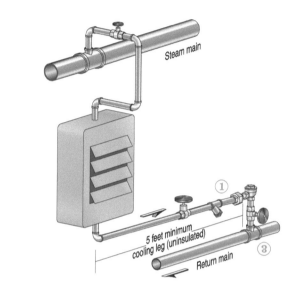

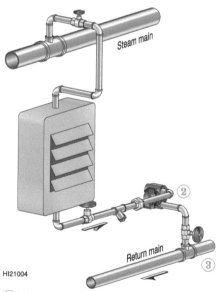

HI21004

① You can use a thermostatic trap as a drain trap for a unit heater provided you install a minimum uninsulated 5-foot length of pipe ahead of the trap as shown. This pipe serves to cool the condensate enough to allow the thermostatic element to open.

② When using a float and thermostatic trap to drain a unit heater, install a strainer ahead of the trap to prevent potential sediment from clogging the water valve orifice. Make sure the strainer is piped to allow easy access for removal of the screen for cleaning.

③ Slope return line continuously downward toward the atmospheric-vented condensate tank. There must be no sags or loops in the return piping because this pipe is the means by which air flows into and out of the system. If the line is blocked by water seals, air cannot leave the system. Heat distribution problems will result.

Figure 41 Branch or undripped riser

Two-pipe steam Runouts to risers or branches, from either steam or condensate mains

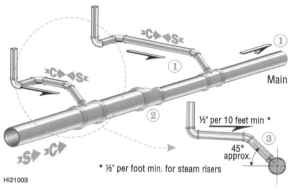

HI21003

① Always slope horizontal pipes in the direction of condensate flow — even two-pipe steam system steam lines. Steam lines always carry some condensate, even on two-pipe systems, because of condensation due to piping heat loss.

② If reducing a steam main, always use a concentric reducer to ensure condensate won't be trapped at the joint. With the eccentric reducer installed as shown, the bottom of the pipe maintains continuous downward slope. If you used a concentric reducer, condensate would build up to the level of the reducer, resulting in severe water hammer.

③ Connect to main at about a 45° angle as shown. Condensate can flow smoothly into main, preventing carryover of liquid in the steam flow.

Two-pipe steam Runouts to branch

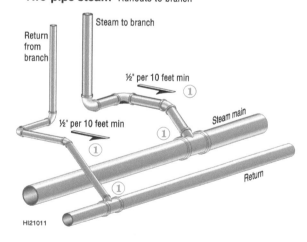

HI21011

① Slope horizontal runout pipes minimum ½" per 10 feet. Connect pipe 45° into main, ensuring drip condensate can flow freely and smoothly.

② Slope return line continuously downward toward the atmospheric-vented condensate tank. There must be no sags or loops in the return piping because this pipe is the means by which air flows into and out of the system. If the line is blocked by water seals, air cannot leave the system. Heat distribution problems will result.

Figure 42   Drip trap piping to main

Two-pipe steam Drip trap connection to steam main

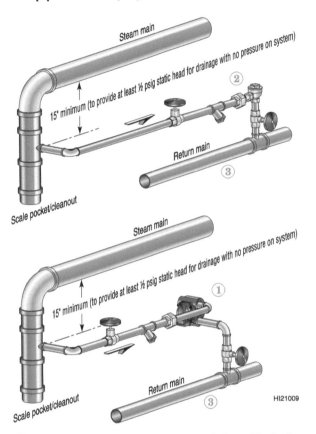

HI21009

① When using a float and thermostatic trap to drain a unit heater, install a strainer ahead of the trap to prevent potential sediment from clogging the water valve orifice. Make sure the strainer is piped to allow easy access for removal of the screen for cleaning.

② You can use a thermostatic trap as a drain trap for a unit heater provided you install a minimum uninsulated 5-foot length of pipe ahead of the trap as shown. This pipe serves to cool the condensate enough to allow the thermostatic element to open.

③ Slope return line continuously downward toward the atmospheric-vented condensate tank. There must be no sags or loops in the return piping because this pipe is the means by which air flows into and out of the system. If the line is blocked by water seals, air cannot leave the system. Heat distribution problems will result.

Figure 43   Drip trap piping to riser

Two-pipe steam Drip trap connection to riser

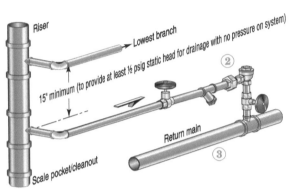

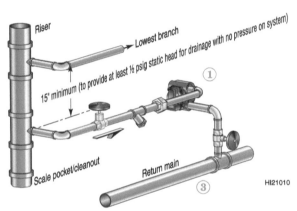

HI21010

① When using a float and thermostatic trap to drain a unit heater, install a strainer ahead of the trap to prevent potential sediment from clogging the water valve orifice. Make sure the strainer is piped to allow easy access for removal of the screen for cleaning.

② You can use a thermostatic trap as a drain trap for a unit heater provided you install a minimum uninsulated 5-foot length of pipe ahead of the trap as shown. This pipe serves to cool the condensate enough to allow the thermostatic element to open.

③ Slope return line continuously downward toward the atmospheric-vented condensate tank. There must be no sags or loops in the return piping because this pipe is the means by which air flows into and out of the system. If the line is blocked by water seals, air cannot leave the system. Heat distribution problems will result.

Figure 44 Downfeed risers (dripped)

Two-pipe steam Runouts to downfeed risers

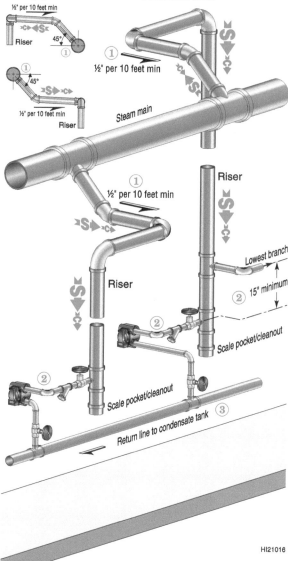

Figure 45 Upfeed risers

Two-pipe steam Runouts to upfeed risers

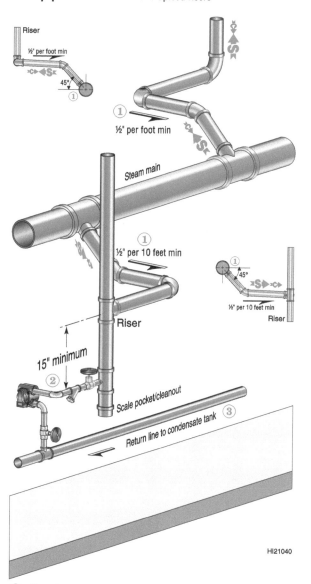

1. Slope horizontal runout pipes minimum ½" per 10 feet. Connect pipe 45° into main, ensuring drip condensate can flow freely and smoothly.

2. Install float and thermostatic trap, strainer and valve on each riser at least 15 inches below lowest branch connection. This ensures at least ½ psig available static pressure to drain riser through trap before pressure builds in system. Route piping to provide clear access to the strainer cleanout.

3. Slope return line continuously downward toward the atmospheric-vented condensate tank. There must be no sags or loops in the return piping because this pipe is the means by which air flows into and out of the system. If the line is blocked by water seals, air cannot leave the system. Heat distribution problems will result.

1. Slope horizontal runout pipes minimum ½" per 10 feet. Connect pipe 45° into main, ensuring drip condensate can flow freely and smoothly.

2. Install float and thermostatic trap, strainer and valve on each riser at least 15 inches below lowest branch connection. This ensures at least ½ psig available static pressure to drain riser through trap before pressure builds in system. Route piping to provide clear access to the strainer cleanout.

3. Slope return line continuously downward toward the atmospheric-vented condensate tank. There must be no sags or loops in the return piping because this pipe is the means by which air flows into and out of the system. If the line is blocked by water seals, air cannot leave the system. Heat distribution problems will result.

Figure 46 Piping to condensate receiver

Condensate receiver piping connections

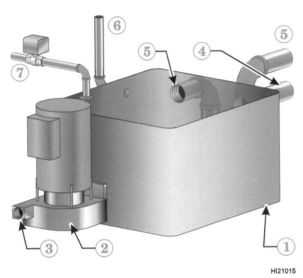

HI21015

① Condensate receiver

② Feed pump and motor

③ Piping connection to boiler — see details at upper right. Use spring-loaded check valves where shown. They are less likely to hammer when closing. Install the square-head cock as shown. Throttle this valve closed to regulate flow of water to boiler. Excess flow can cause hammer and water level collapse. When the boiler feed piping goes up and drops down, add a second spring-loaded check valve. This valve maintains water under pressure in the vertical piping, preventing water hammer due to flashing of hot condensate as it falls in the vertical line.

④ Install strainer in return line to protect feed pump from system sediment. After boiler installation, allow condensate to waste to drain through the by-pass shown in order to eliminate sediment freed by the new boiler installation. When condensate is relatively free of sediment, close by-pass and allow normal operation. Instruct operator/owner to shut down boiler periodically and clean strainer. Should the strainer become clogged, condensate will back up in the return main, causing water hammer and excess make-up water addition to the system.

⑤ Pipe the overflow connection to a floor drain. To prevent escape of vapor or flash steam to the boiler room, install loop piping as shown. Use overflow loop fill valve to prime the overflow loop piping.

⑥ Pipe vent line to outside if desired to prevent escape of vapor or flash steam to the boiler room. Never block or close off the vent line. It must be free and open to atmosphere to prevent any possibility of pressure build-up in the condensate receiver.

⑦ Connect make-up water line to fresh water source, following local plumbing codes regarding backflow prevention if required. The example above shows a boiler feed system condensate receiver, with make-up water fed to the receiver. If the receiver is on a condensate return system, make-up water is fed to the boiler through an automatic water feeder.

③ Piping directly to boiler

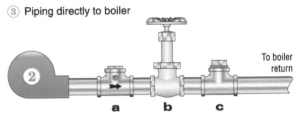

To boiler return

a   b   c

③ Piping to boiler with overhead piping

a Spring-loaded check valves

b Gate valve or ball valve

c Square-head cock

d Strainer

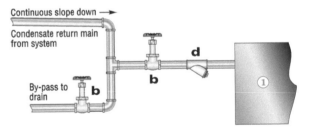

To boiler return

a   b   c   a

④ Condensate return main from system

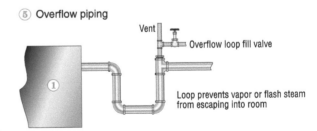

Continuous slope down →

Condensate return main from system

By-pass to drain

b   b   d

⑤ Overflow piping

Vent

Overflow loop fill valve

Loop prevents vapor or flash steam from escaping into room

DANGER — Condensate receivers cannot be pressurized. The vent connection must always be open to atmosphere. Otherwise, an explosion could occur should the receiver become pressurized due to leakage from the boiler or system.

Figure 48 Return piping around obstacles

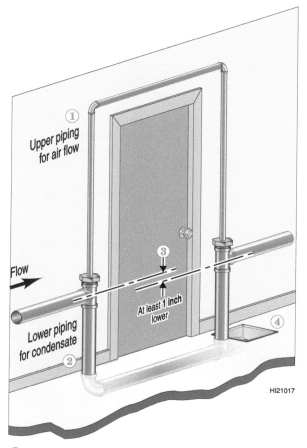

HI21017

Figure 47 Steam piping around obstacles

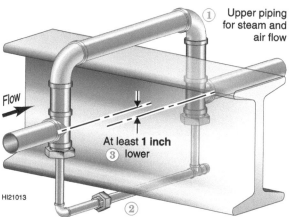

HI21013

Lower piping for condensate — Minimum 1" pipe size size not critical, but large enough to avoid clogging with sediment

① The upper piping, full size of the steam main, carries steam. It also allows air to move through the piping.

② The lower piping allows condensate in the lines to move through. Without this line, condensate would build up in the steam main. Size as for a wet return line, based on full load carried by steam line.

③ Offset the continuation of the steam main at least 1 inch lower than the incoming line. This ensures condensate won't accumulate in the steam main.

① The upper piping allows air to flow through to the vented condensate tank. Size this line ½" or larger for residential applications.

② The lower piping, full size of the condensate return line, allows condensate to flow through under the obstruction.

③ Offset the continuation of the steam main at least 1 inch lower than the incoming line. This ensures condensate won't accumulate in the steam main.

④ Install a nipple and cap on the lower piping, as shown, to allow removal for inspection and cleanout. Make sure the fittings are accessible.

Figure 49 Sizing two-pipe steam system piping

Pipe size (inches)	Steam main/Downfeed riser/Dripped horizontal runout		
	Square Ft EDR	Btu/Hr Net	Pounds/Hr
¾	36	8,640	9
1	68	16,320	17
1¼	144	34,560	36
1½	224	53,760	56
2	432	103,680	108
2½	696	167,040	174
3	1,272	305,280	318
4	2,560	614,400	640
5	4,800	1,152,000	1,200

Upfeed or express riser			
¾	32	7,680	8
1	56	13,440	14
1¼	124	29,760	31
1½	192	46,080	48
2	388	93,120	97
2½	636	152,640	159
3	1,128	270,720	282
4	2,044	490,560	511
5	4,200	1,008,000	1,050

Horizontal riser runout, not dripped			
¾	28	6,720	7
1	56	13,440	14
1¼	108	25,920	27
1½	168	40,320	42
2	362	86,880	91
2½	528	126,720	132
3	800	192,000	200
4	1,700	408,000	425
5	3,152	756,480	788

Pipe size (inches)	Dry return		
	Square Ft EDR	Btu/Hr Net	Pounds/Hr
1	320	76,800	80
1¼	672	161,280	168
1½	1,060	254,400	265
2	2,300	552,000	575
2½	3,800	912,000	950

Wet return (gravity system)			
1	700	168,000	175
1¼	1,200	288,000	300
1½	1,900	456,000	475
2	4,000	960,000	1,000

Radiator vertical piping/Radiator supply valve			
½	25	6,000	6
¾	75	18,000	19
1	150	36,000	38
1¼	200	48,000	50
1½	400	96,000	100

Radiator vertical return piping/Radiator trap size			
½	200	48,000	50
¾	400	96,000	100
1	700	168,000	175

Radiator horizontal return piping			
¾	400	96,000	100
1	700	168,000	175

Notes

1. Sizing above can be applied to low-pressure steam heating systems, allowing a pressure drop of 1 ounce per 100 feet of piping.

2. Steam main piping should be no smaller than 2" pipe size.

3. For additional sizing information, see the ASHRAE "Fundamentals" Handbook or ITT Hoffman "Steam Heating Systems: Design Manual and Engineering Data" (Bulletin No. TES-181).

Troubleshooting

Steam system problems

If a steam system is correctly installed, its pipes sized and sloped correctly, and its components operational, it will work smoothly and quietly. The house will heat uniformly. Like any other mechanical system, problems will arise when the system isn't installed correctly or its components are working incorrectly. The two most common problems are water hammer and heat distribution.

Water hammer

Water hammer occurs when water slams into the piping. When water is too deep in the pipes, it can trap steam pockets as shown in Figure 50. These pockets condense instantaneously with no supply of steam. The vacuum formed causes the water to rush in to fill the void. The resultant collision causes water hammer. Each collision can cause other collisions to occur, and the result can be machine-gun like action. Allowed to become severe enough and water hammer can cause the piping to break apart. At the least, water hammer damages key system components like traps and air vents.

What causes water hammer? Water hammer occurs when steam meets cold water or water that is too deep in horizontal pipes. It can also occur when hot condensate flashes to steam in return lines. See the troubleshooting guide at the end of Part 12 for suggestions on locating the cause of water hammer.

Figure 50   Water hammer

Water hammer

Steam flowing over deep water causes waves

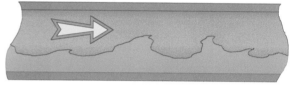

Turbulence causes water to trap steam pockets   HI21060

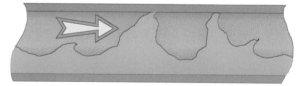

Steam in pockets condenses, causing instantaneous vacuum. Water closes in explosively, causing hammer.

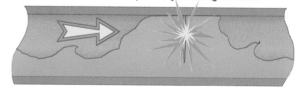

Heat distribution

Heat distribution problems occur when some heating units receive heat much sooner than others. The main culprits are air removal problems or failed traps.

Other problems

The following pages provide suggestions for troubleshooting common steam system problems, including water hammer and heat distribution.

Clean the boiler

Carefully follow the boiler manual guidelines for cleaning and skimming the boiler. Include the time needed for this in your quote.

Troubleshooting

Heat distribution/insufficient heat

System type	Check	Explanation
System type shorthand below means: 1PG = One-pipe gravity-return; 1PP = One-pipe pumped-return; 2PG = Two-pipe gravity-return; 2PP = Two-pipe pumped-return; VP = Vapor system (Two-pipe gravity-return)		
1PG, 1PP, 2PG	Main air vents	Every riser and main should have a working main vent (quick vent) installed correctly to protect it from water hammer.
1PG, 1PP, 2PG	Radiator air vents	Make sure radiator vents are working.
2PG, 2PP	Drip traps	Drip traps on the steam mains and risers must have a working thermostatic element to allow air to move through and out of system. Make sure traps are open for condensate drainage. Make sure strainers in return lines are clean.
VP	End-of-main trap	Check the thermostatic trap between the end of the steam main and the return. This trap allows air to flow from the main into the return, where it leaves through the air vent.
1PP, 2PP	Condensate tank vent	Is condensate tank vent line installed, open, and unobstructed?
1PG, 2PG	Boiler pressure	If pressure is too high, steam can blow through loop seals and prevent air vents from dropping. Limit cut-out pressure must be lower than loop seal maximum (divide loop seal height in inches by 30 to find maximum pressure in psig).Limit cut-in pressure must be below maximum operating pressure of air vents. Don't set cut-in pressure above ½ psig if possible.If boiler limit cut-out pressure is too low, insufficient steam will reach end of system. For any system but vapor or vacuum system, set limit at 2 psig. If system has a heat exchanger or large unit heater, additional pressure may be required. Verify limit is operating correctly by checking cut-in and cut-out pressures against pressure gauge.
All	Steam supply piping	Steam supply piping must be clear, with no obstructions or drops or sags that could accumulate water. Check all steam piping, particularly runouts to risers or branches. Riser runouts must have slope of ½ inch per foot if not dripped. Branch runouts must have slope of ½ inch per 10 feet if not dripped. If piping is sloped (and sized) correctly, check for blockage in runouts.
All	Dry return piping	Dry return piping must be clear, with no obstructions or drops or sags that could accumulate water. Make sure air flows out of condensate tank vent line as system warms up. If piping is correct, open and check for sediment. Make sure all return lines slope downward.
2PG, 2PP	Radiator thermostatic traps	Have any traps failed open? Open traps allow steam to flow through and pressurize returns, slowing air removal from other radiators. Look for defective traps where the room(s) has heat because radiators with open traps get steam flow.
All	Radiator supply valves	Are valves open, in good condition and not blocked?
All	Pipe sizes	Add up total load on each riser and steam main. Are pipes large enough per sizing recommendations in this guide? Undersized piping will cause insufficient heat near the end of the riser or main.
All	Boiler size	Is the boiler sized correctly for the system? Is the boiler firing rate correct per nameplate data?
All	Radiator sizes	Are radiators large enough for room heat losses?

Troubleshooting

Water hammer (Most of the following suggestions for troubleshooting water hammer are from Dan Holohan's book, "The Lost Art of Steam Heating".)

Occurs:	Check	Explanation
At start-up	Near-boiler piping	Is the near-boiler piping installed exactly as shown in boiler manual? If not, water can be carried into the steam piping, causing water hammer.
	Pipe slope	All pipes must slope in the direction of condensate flow. Check for sags or insufficient pitch in supply and return piping. Check slope in runouts to risers and branch piping. Use a 6-inch level along full length of pipes because pitch may be difficult to see.
	Steam main drip lines	Steam and condensate flowing in same direction: Steam mains should not exceed 150 feet between drip lines or drip traps.Steam and condensate flowing opposite directions: Steam mains should not exceed 50 feet between drip lines or drip traps.
	Pipe reducers	Make sure no concentric reducers are used in steam piping. Concentric reducers cause condensate to puddle ahead of the reducer. Only concentric reducers can be used on steam lines.
	Boiler water level	Gravity systems over 100,000 Btuh need at least 28 inches between boiler water line and lowest steam-carrying pipe (or dry return). Smaller systems need 14 inches. If replacement boiler water line is higher than original boiler, could have insufficient height.
	Drip traps	Are drip traps (on steam mains and risers) piped with condensate accumulator piping ahead of trap and high enough to cause gravity drainage? Do traps allow condensate to flow?
	Zone valves	Motorized zone valves must have drip line or trap if condensate can accumulate at the valve.
Mid-cycle	Return piping	Are return lines on gravity system clogged? This will cause condensate to back further and further up in returns because the available pressure of the return column can't push it through the return lines fast enough.Make sure there are no sags or drops in dry returns that could trap water and prevent free flow of air.If a trap has been installed at the condensate tank (attempt to "master trap" the system), remove it. Condensate builds up ahead of the trap. Steam leakage through defective traps in system cause hammer when it contacts condensate in returns.
	All traps	Any trap failed open allows steam to enter returns, causing water hammer in returns.
	Boiler piping	Is near-boiler piping installed per boiler manual? Incorrect piping will cause water carryover to system, making liquid level in pipes too deep.Check the horizontal nipple at the Hartford loop connection. It must be no longer than a close nipple or the connection must be made with a downturned wye fitting. Any longer pipe will cause water hammer as water level bounces and exposes the loop pipe during operation.
	Boiler water	Dirty, foamy boiler water will cause carryover of liquid to system, making liquid level in pipes too deep. Clean boiler if necessary. Make sure pH of water is less than 9.5 (or per boiler manual limit). Higher pH water will foam easily.

Troubleshooting

Water hammer continued (Most of the following suggestions for troubleshooting water hammer are from Dan Holohan's book, "The Lost Art of Steam Heating".)

Occurs:	Check	Explanation
Mid-cycle (continued)	Boiler size	Is boiler over-sized for system or over-fired vs nameplate rating? Boiler net load rating shouldn't be much greater than 10% larger than the total system square feet EDR. If the system has been down-sized, for example, the boiler may be too large, causing excessive flow in the piping. The pressure drop changes by the square of the flow (approximately). Example: If the flow is just 25% higher than intended for the piping, the pressure drop will be 1.25^2, or 1.56 times intended pressure drop. A system designed for ½ psig (14 inches water column) will have a pressure drop of 1.56 times 14 inches = 21.8 inches water column. On a gravity system, this will raise the water line in the returns another 7.8 inches, possibly enough to push water into horizontal steam pipes.
	Radiator size	Water hammer in a one-pipe radiator can occur if the radiator is vented too quickly. This causes steam entering the radiator valve to hold back condensate trying to exit. Water hammer occurs because the water depth in the radiator gets too deep. Correct by installing two smaller air vents, one mounted below the other. The vent rate will slow down when steam reaches the first vent. A better method is to replace the single large radiator with multiple smaller radiators, none of which is larger than 250 square feet EDR.
	Radiator pitch	Make sure radiators pitch toward the condensate outlet. If a radiator pitches the other way, condensate accumulated in the radiator will be too deep.
	Zone valves	If condensate can accumulate on either side of a zone valve, there must be a drip line or trap installed to relieve condensate or hammer can occur.
	Pipe insulation	If insulation has been removed from piping and not replaced, condensate load will be too high because of excessive heat loss from bare piping. Insulate all steam piping (except the 5-foot cooling length of pipe ahead of a thermostatic drip trap).
On shutdown	Boiler piping	The Hartford loop horizontal connection may be too close to the boiler water line. The water level drops slightly when the boiler shuts down. This exposes the loop connection. Check boiler manual for correct location.
	Pipe insulation	If pipe insulation has been removed only in the boiler room: When boiler shuts down, steam in boiler room pipes condenses rapidly. Steam in insulated system piping condenses more slowly, holding some pressure. This causes a momentary vacuum in the boiler room piping. Boiler water will lift up in the header piping, causing hammer.
	Radiator vents	Dirty air vents will not allow air to enter fast enough to break the vacuum at shutdown. This can cause water to be pulled up the return piping, causing water hammer at the Hartford loop.
In boiler	Burner flame	Wet base boilers only: Flame impingement on boiler due to mis-shaped flame will cause steam pockets to form. These pockets collapse, causing water hammer.
	Sediment in boiler	Build-up of sludge in boiler will slow circulation, causing steam pockets to form, resulting in water hammer when pockets collapse.
	Tankless coil	Over-sized tankless coils in steam boilers can collapse the water level when they operate. This yanks the water down quickly in one Part of the boiler. The remaining water rushes into the void, causing hammer.

Troubleshooting

Flooding or overfilling

Check	Explanation
Boiler size	Is replacement boiler smaller than old boiler? New boiler may not have enough steaming time to steam until condensate returns. If so, you will need to add storage by installing an accumulator tank at the boiler water line (if recommended by boiler manufacturer) or install a boiler feed system with a condensate tank large enough to allow steaming during the system time lag.
Water feeder	Throttle the square-head cock on the make-up water line (install one if not already installed) to slow feed rate. Feeding cold water too fast will cause water level collapse and excessive feeding.Is the water feeder installed per boiler manual (or tank instructions if tank-mounted)? If boiler-mounted, make sure it is piped to the correct tappings.If feeder is boiler-mounted float type or electric feeder, using an electronic feeder may provide better control on the amount of make-up water.Make sure feeder is operating correctly and not leaking.
Boiler piping	Is near-boiler piping per boiler manual? Incorrect near-boiler piping will cause liquid carryover and/or water level bounce. This will use excessive water, causing the feeder to act prematurely.
Feed pump rate	Try reducing boiler feed pump rate to boiler. Throttle the square-head cock in feed piping (install one if not already installed) to slow feed rate. Feeding too fast will collapse water level, causing premature feeder operation.
Boiler water	Dirty or foamy water in boiler will cause carryover and water level bounce. Water level bounce will cause float-type feeders or electric feeders to operate prematurely.

Rapid cycling

Check	Explanation
Air vents and drip traps	If vents or traps don't allow quick movement of air from the distribution piping, pressure will build too quickly, causing the boiler limit to shut down.
Boiler pressure control	Boiler limit must be operating correctly. Make sure differential is not too close. Set for at least 1 to 1½ psig difference between cut-out and cut-in pressures if operating at 2 psig or less.
Low water cutoff	Is low water cutoff cycling due to water level bounce? If it is, make sure near-boiler piping is correct. Clean boiler and skim per boiler manual if necessary.
Zoning	If system is zoned, some zones may have longer heat calls than others. When zones close, boiler capacity may be large for remaining load. Try to balance heat flow to zones to slow down heating in the zones that consistently shut down soonest.
Thermostat anticipator	If the thermostat anticipator setting is set too low, the boiler will cycle often. Set the anticipator to its maximum setting for steam systems to avoid this problem.

Water level bounce

Check	Explanation
Note: The boiler water bounces on all steam applications. One-pipe systems, in particular, experience bounce as water drops into the return lines.	
Boiler water	Clean and skim boiler per boiler manual if water is dirty or foamy. Check pH level and verify per boiler manual. A pH higher than 9.5 will cause excessive foaming. Boiler additives will often cause surging.
Boiler piping	Is near-boiler piping per boiler manual? Incorrect piping will causes water level instability.
Feed water rates	Excessive feed rate, either from the boiler feed pump or water make-up line, will cause water level bounce. Throttle feed line square-head cocks for smooth operation.

Part 12
Addenda

I=B=R

Near-boiler piping — using a dropped header

The horizontal piping off of the boiler risers needs to be at least 24 inches above the boiler water line to assist in separation of water from the steam before it enters the system piping.

On some installations, you will find that the steam system piping connection is too low to allow the boiler steam header to be high enough for the 24-inch distance above the water line as shown in Figure 4, page 12–13 and Figure 5, page 12–14. To provide the needed 24 inches from water line to horizontal piping off of the risers and still allow connection to the system piping, use the method shown in Figure 51, page 12-73. This piping applies only to parallel-flow systems, not to counterflow.

In Figure 51, the horizontal piping off of the risers is at least 24 inches above the water line. But the header is installed as a drop header, as shown.

Pipe the header at least 6 inches above the water line to allow for fittings and to prevent possibility of water backing into the header. This method provides very dry steam because in includes more direction changes for the steam flow. Each time the direction changes, more water is slung free from the steam.

Pay close attention to the boiler manufacturer's instructions for pipe sizing and configurations required. Always follow the boiler manufacturer's guidelines.

Figure 51 Using a dropped header to lower connection to system steam supply

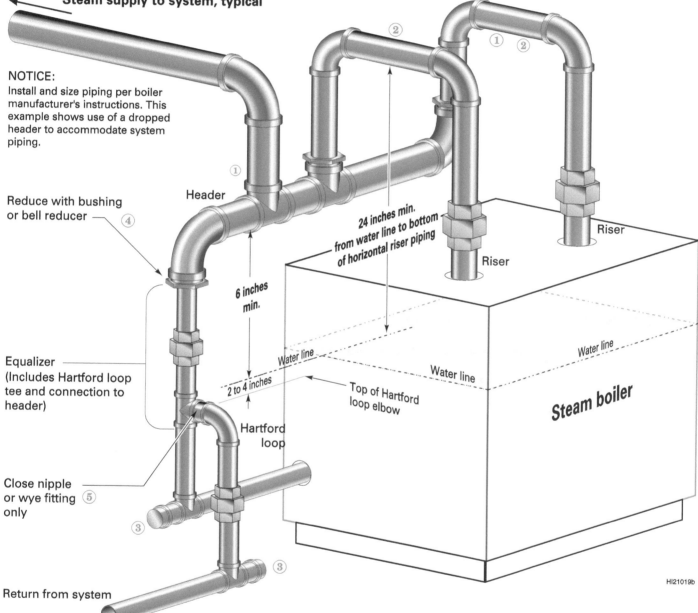

← **Steam supply to system, typical**

NOTICE:
Install and size piping per boiler manufacturer's instructions. This example shows use of a dropped header to accommodate system piping.

Reduce with bushing or bell reducer — ④

Header

Equalizer (Includes Hartford loop tee and connection to header)

Close nipple or wye fitting ⑤ only

Return from system

24 inches min. from water line to bottom of horizontal riser piping

6 inches min.

2 to 4 inches

Water line

Top of Hartford loop elbow

Hartford loop

Riser

Riser

Water line

Water line

Steam boiler

HI21019b

① Always connect steam supply between last boiler riser and equalizer. This ensures the greatest possible separation of moisture from the steam. Connecting the steam supply between the boiler equalizers will cause water carryover to the system. The large flow of water from the boiler will cause the system to add make-up water unnecessarily, resulting in eventual flooding. Provide at least 6 inches height between the bottom of the header and the boiler water line. Install boiler riser piping horizontal lines at least 24 inches above the boiler water line, as shown.

② You must pipe the risers into the header with offset joints as shown. The horizontal piping and elbows provide swing joints to compensate for the expansion and contraction of the header piping. If you don't pipe this way, cast iron boiler sections can crack, or steel boiler welds can fail.

③ Install tee or cross with nipple and cap at all direction changes in return lines. Steam systems typically contain large quantities of sediment. You should periodically inspect and clean return lines to ensure reliable operation.

④ Reduce pipe size only after turning downward. Use a reducing elbow or reducing fitting in the vertical (equalizer) line.

⑤ Connect the Hartford loop using nothing longer than a close nipple or wye fitting. Water level may often drop below top of fitting. A long horizontal connection would cause severe water hammer.

Hydronics Institute Section of AHRI

35 Russo Place

Berkeley Heights, NJ 07922-0218

part 13

System survey

I=B=R Guide

RHH

Residential Hydronic Heating . . .

Installation & Design

Hydronics Institute
Section of **AHRI**

I=B=R Guide RHH
Residential Hydronic Heating

Hydronics Institute Section of **AHRI**
35 Russo Place
Berkeley Heights, NJ 07922-0218

Contents – Part 13

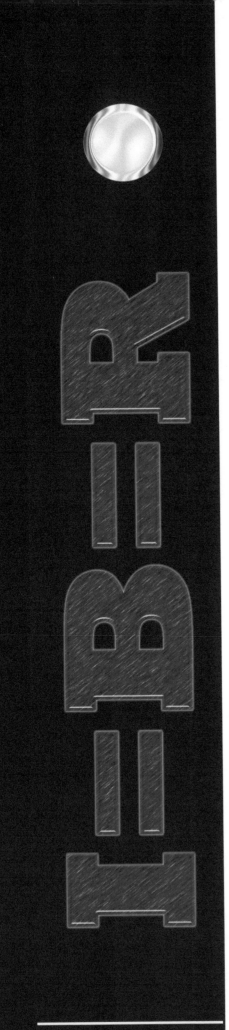

System survey

Overview

Part 13 provides suggested procedures for inspecting existing systems to determine causes of system problems. Also included is a copy of each of Forms 1530-W and 1530-S (water and steam Hydronic System Survey forms). You can obtain these forms from the Hydronics Institute Division of AHRI and use them to provide your customers with a summary of your findings and recommendations for their systems.

Why do a Hydronic System Survey?

The boiler and heating system components can be damaged if the system is allowed to operate with defects (water leaks, incorrect piping, defective components, etc.). Try operating a car without oil, for instance, or leave your portable radio out in the rain. Every mechanical device needs maintenance, attention, and occasional repairs. Nothing lasts forever, including heating systems and components. At some time every heating system will need repairs or component replacements. Allow the system to operate with defects and problems will develop. The boiler may be damaged or other components will fail; flooding can occur, resulting in potential property damage.

Take the opportunity when you quote a boiler replacement to thoroughly inspect the system and interview the customer. Identify problems or emerging problems and recommend the changes you believe are needed to ensure the system will last long and will satisfy the homeowner's needs. This adds value to the boiler replacement job, and should build confidence in your customers by demonstrating your commitment to their satisfaction.

Use Form 1530-W or 1530-S on each of your replacement boiler or system troubleshooting calls. You can give a copy of the survey form to the homeowner along with your quotation, giving them the background work you used for your recommended actions (repairs, replacements, additions).

Don't just replace the boiler — Repair the system.

When replacing a boiler, take time to study the boiler you are replacing. You should find out why the boiler failed before you quote a boiler replacement. A boiler doesn't usually fail because of an inherent defect. It fails because of hostile conditions created by the system (excessive make-up water, sediment, corrosive water, etc.). Use the table on page 13-5 and information available from the boiler supplier to determine the cause of the boiler failure. Use this information, along with the results of your System Survey, to recommend a complete repair program for the homeowner.

Let the homeowner know the risks of not having the system repairs made. Without the repairs, the same problems that caused the old boiler failure can cause the new boiler to fail quickly. The new boiler won't fail faster than the old boiler did. It's just that the hostile conditions weren't there, or as bad, when the old boiler was installed. If the homeowner declines the additional costs, at least you have given a thorough recommendation. If the new boiler does fail due to the existing problems, you did your best to let the homeowner know.

Troubleshoot boiler failure before installing a new boiler

Old boilers don't just die — they get done in. If the old boiler failed, it was probably due to a system problem — **not** a boiler problem — even if the boiler was old. If you don't repair the problem that caused the old boiler to fail, the new boiler may not last long.

Inspection shows	Likely cause	Look for . . .

Boiler cracked or overheated	Metal overheated (but no interior deposits near failure)
	
	Metal discolored due to high temperature exposure
		• low water condition possible:
		  • was system filled?
		  • was auto fill system working?
		  • is boiler higher than radiation? — should have low water cutoff
		• flame impingement also possible
		  • if flame impinges directly on metal surface, will cause local overheating and material fracture
		• copper boilers — probably due to loss of flow or low water condition

	Sludge deposits (carried back to boiler from old system pipes)	• excessive sludge in system piping
		  • drain and chemically clean piping
		  • install strainer
		• glycol used in dirty piping

Holes in the boiler	Lime deposits (caused by excessive makeup water)	• weeping relief valve
		  • waterlogged or undersized expansion tank
		  • wire drawn or blocked valve seat
		  • fill pressure too high
		  • fill valve defective

	Oxygen pitting (caused by excessive makeup water)	• leaking piping or joints (leaking returns on steam)
		• drainage (using system water for other purposes)
		• draining and refilling too often
		• excessive blowdown
		• steam systems — flooded condensate tank or boiler
		  • probably need boiler feed system instead of condensate pump to provide more steaming time
		  • check near-boiler piping — incorrect piping can cause heavy carryover of water to system, causing too much make-up
		  • dirty boiler water or some cleaning chemicals can cause carryover

Boiler corroded in layers from inside	Acid attack	• glycol inhibitor level wrong • acidic water (low pH)
Exchanger eroded inside	Rubbing of surfaces by coarse sediment	• sediment in system being forced through boiler tubes — clean system and install strainer
Boiler corroded from outside	Return water temperature too low	• converted gravity or steam system — install bypass piping • night setback or weekend setback — install mixing valve to regulate return temp to boiler • outdoor reset temperature control on system water — install mixing valve to regulate return temp to boiler • low temp system (radiant or heat pump) — manual bypass may work, mixing valve on boiler return preferred
	Contaminated fuel oil, high acid condensate	• check boiler area — are laundry products, paint or chemicals stored or used in the home? (If these are allowed to contaminate the boiler combustion air supply, they will cause acid formation in the boiler and vent system.) • oil boilers — verify that oil supply is not contaminated — should only be #2 fuel oil, never crank case or waste oils

Boiler flues plugged with soot	Combustion problems	• vent system may be blocked or inoperative due to leaks, or building may be negative pressure due to tight house or use of exhaust fans (or whole-house fan)
		• oil boilers
		  • burner, nozzle, fuel line or fuel problems can cause rapid soot deposition
		  • ensure that problem is not caused by fuel or fuel distribution piping when installing new oil boiler
		• gas boilers
		  • usually due to damaged or dirty burner, burning at the orifice (flashback) or impingement on an object in the chamber (such as insulation on the burner or obstructing flame)
		  • can also result from case of air starvation due to lack of air openings or negative in boiler space
		  • badly oversized gas orifice

Hydronics Institute Section of AHRI

35 Russo Place

Berkeley Heights, NJ 07922-0218

Hydronic System Survey

Form
1530-W

_____
Date

_____
Homeowner

_____
Address

_____
Address

_____
Contractor

_____
Address

_____
Address

Hydronic System Survey — Water system

The purpose of this form is to identify known or potential problems in a residential heating system. Correction of these problems should improve system performance and avoid shortening of life or damage to the boiler and system components.

Inspect	Look for and quote corrective action . . .	Recommendations
Boiler	If replacing a old boiler, follow the troubleshooting suggestions in Part 12 of Guide 2000.	
Vent system	Masonry chimneys • liner must be intact, unobstructed, and mortar must be in good condition; no visible holes • cleanout door must be in place and tightly sealed • vent connector must not protrude into chimney liner • exterior chimneys must be lined for new appliance installations • National Fuel Gas Code may disallow masonry for some applications — refer to code Metal chimneys and vent connectors • verify no corrosion or blockage • verify joints are gas-tight, showing no signs of leakage • metal vents through unheated areas must be rated for condensing use (stainless steel) • metal vents may require condensate traps • many boilers will require new, special vent system (Category II, III or IV appliances)	
Combustion/ ventilation air	Air openings • boiler room and building require openings per National Fuel Gas Code • if house has been tightened (new windows, weatherstripping, etc.), new air openings may be needed • if exhaust fan (or whole-house fan) has been added, air openings may be inadequate • may require installing a direct vent (sealed combustion) boiler to duct air directly to appliance • even if boiler is installed for sealed combustion, air openings may be required for ventilation • if house is particularly tight, consider a makeup air heater	
Circulator	Verify system circulator is functional • listen to pump while operating — should be no high-pitched noises, usually due to cavitation • is pump moving water adequately? • is system piped to avoid dead-heading pump (i.e., with by-pass pressure regulator if pump can run with all zone control valves closed)? • does pump start correctly — not bound up? Verify circulator location • high-head circulators should pump away from the expansion tank connection • fill line must always be connected at the same point as the expansion tank	